Bacterial Endotoxic Lipopolysaccharides

Volume I

Molecular Biochemistry and Cellular Biology

Edited by

David C. Morrison, Ph.D.
Associate Director, Cancer Center
Professor of Microbiology,
Molecular Genetics, and Immunology
University of Kansas Medical Center
Kansas City, Kansas

John L. Ryan, Ph.D., M.D.
Executive Director
Infectious Diseases Department
Merck Research Laboratories
West Point, Pennsylvania

CRC Press
Boca Raton   Ann Arbor   London   Tokyo

Library of Congress Cataloging-in-Publication Data

Bacterial endotoxic lipopolysaccharides / editors, David C. Morrison,
    John Ryan.
        p. cm.
    Includes bibliographical references and indexes.
    Contents: v. 1. Molecular biochemistry and cellular biology — v.
2. Immunopharmacology and pathophysiology.
    ISBN 0-8493-6787-5 (v. 1). — ISBN 0-8493-6788-3 (v. 2)
    1. Endotoxins. I. Morrison, David C., 1941- . II. Ryan, John
Louis.
    [DNLM: 1. Endotoxins. 2. Lipopolysaccharides—physiology. QW
630 B1285]
QP632.E4B23 1992
615.9 5299—dc20
DNLM/DLC
for Library of Congress   92-7865
                                                                                     CIP

© 1992 by CRC Press, Inc.

International Standard Book Number 0-8493-6787-5 (vol. 1)
International Standard Book Number 0-8493-6788-3 (vol. 2)

Library of Congress Card Number 92-7865

Printed in the United States of America   2 3 4 5 6 7 8 9 0

Printed on acid-free paper

PREFACE

It was approximately one hundred years ago that Pfeiffer, working in Robert Koch's laboratory, indentified a novel microbial toxin present in the culture supernatants of *Vibrio* sp. He termed these biologically active microbial constituents "endotoxins". In the ensuing century, investigators have learned much about these fascinating bacterial products, which seem to be endowed with unlimited potential to interact with and perturb mammalian cellular systems. Indeed, the insightful observation of Lewis Thomas some forty years ago in his classic volume *The Lives of a Cell*[1] that "there is nothing intrinsically toxic about endotoxin, but it must look or feel awful when sensed by cells. Cells believe that it signifies the presence of Gram-negative bacteria and cells will do just about anything to avoid this threat," appropriately summarizes our current understanding of the role of endotoxin in the pathogenesis of Gram-negative sepsis.

Although the essential lipopolysaccharide (LPS) component of endotoxin has been clearly recognized for more than fifty years, bacterial endotoxins have traditionally remained as the black sheep in the family of microbial toxins. This view has resulted, in large part, from the fact that endotoxins do not, as pointed out above, manifest their activity through selective toxic effects on host cells, nor do they derive their identity from specific Gram-negatives. Rather these molecules are chemically diverse, sharing only a relatively conserved lipid domain — termed lipid A by Westphal and his colleagues some forty years ago. The fact that lipid A was shown to be responsible for many of the observed endotoxic properties of LPS did little to change this view. The last decade of endotoxin research, however, has witnessed the precise elucidation of the chemical structure of lipid A as well as characterization of the pathways for its biosynthesis. The recent achievement of synthetic lipid A by total organic synthesis and the demonstration that synthetic lipid A manifests biological activities indistinguishable from normal bacterially derived lipid A has moved endotoxins into well-deserved respectability with their microbial protein toxin cousins. Indeed, the capacity to undertake detailed structure-function studies using synthetic lipid A analogs will undoubtedly prove as informative as site-directed mutagenesis studies will be to molecular mechanisms of action of the protein toxins.

Endotoxin researchers have not limited their chemical efforts to lipid A, and much is now being learned about the oligosaccharide regions adjacent to the important lipid A domain. Once perceived as a relatively inert carbohydrate structure which was relatively conserved among Gram-negative genera, these so called "core oligosaccharides" are now prompting considerable interest among investigators. Several core structures have now been prepared by organic synthesis, and it is likely that the forthcoming decade will witness

[1]Thomas, L. (1974) in *The Lives of a Cell*, Viking Press, New York.

the synthesis of a complete lipid A core structure. Coupled with the growing body of information to suggest that mammalian host defense systems may themselves be endowed with the capacity to recognize endotoxins through specific core structures, it is likely that significant new knowledge in our understanding of endotoxin will derive from advances in this area of research.

Paralleling the remarkable growth in information about the chemical structure of endotoxin has been the elucidation of specific host target cells with which endotoxins interact and the biochemical mechanisms by which specific host cell responses are elicited. The recognition of the macrophage/monocyte and the endothelial cell as critical cells responsible for endotoxin responses has prompted very detailed studies designed to define activation pathways, intracellular regulation, and specific responses. While as little as a decade ago information on specific receptors for endotoxin was extremely limited, there have now been identified a variety of candidates. This field is continuing to evolve. Since these volumes went to press, the emergence of CD14 as a critical cell surface constitutent contributing to endotoxin cell responses has provided for new perspectives on cellular activation mechanisms. Given the importance of recognition of endotoxin by host cells, it is likely that multiple pathways will exist by which the host can respond to this bacterial toxin.

The physiologic and pathophysiologic effects of endotoxic lipopolysaccharides are now well recognized to be mediated in large part through the release of cytokines and other mediators. TNF, IL-1, and IL-6, as well as PAF and arachidonic acid metabolites represent the major mediators of endotoxin activity. The central role of TNF has been clarified in recent years, and efforts are now being made to block TNF in order to ameliorate septic shock. Monoclonal antibodies that are widely cross-reactive with diverse endotoxins continue to show promise in clinical trials, but the mechanism by which they protect remains enigmatic. It is possible that combination immunotherapy of these reagents with other monoclonals designed to modify cytokine activity may provide synergistic efficacy in the treatment of endotoxin shock.

It is, in any case, clear that research on endotoxins has been stimulated immensely by the continued recognition of Gram-negative sepsis as a major clinical problem. Recent studies have partially elucidated the mechanisms by which endotoxic lipopolysaccharides bind to cells and trigger responses. Several serum proteins have been found to bind endotoxins, and certain of these facilitate the interaction of endotoxins with cell membrane glycoproteins. CD14 has emerged as a binding site for LPS bound to lipopolysaccharide binding protein and, as pointed out above, this pathway for potentiation of host responses to endotoxin may prove to be very important in understanding the pothogenesis of endotoxin mediated disease. Since recognition of LPS by macrophages results in a complex release of cytokines and other mediators, attention is now appropriately focused on antagonizing these cytokines and mediators in a further effort to abrogate the sepsis syndrome. Clinical trials with anti-TNF and IL-1 receptor antagonist may yield data to complement the promising data from the anti-endotoxin trials.

As with many scientific endeavors, the more we have learned about endotoxins, the more questions arise. The recent progress in the elucidation of endotoxin structure-activity relationships and cellular interactions of endotoxins which have been summarized in these two volumes represent major advances in our understanding of the fundamental chemistry, biology, and pathophysiology of this complex bacterial product. These studies have provided a framework for the more rational design of clinical trials in which endotoxin has been implicated as an important factor in disease. It is, however, readily apparent that, as with any fast-moving scientific area (and endotoxin research would certainly qualify as such), much new has been learned even in the time since these volumes were completed. The next several years will undoubtedly reveal many major new insights into the mechanisms by which endotoxic lipopolysaccharides trigger the synthesis of potentially harmful quantities of cytokines and other mediators. Many more clinical trials will be directed at one or more factors involved in sepsis syndrome. We can be certain, in any event, that the next decade will prove to be as exciting as the last has been for endotoxin research.

THE EDITORS

David C. Morrison, Ph.D. is Associate Director for Basic Research Programs of the Cancer Center and Professor of Microbiology, Molecular Genetics and Immunology at the University of Kansas Medical Center, Kansas City, Kansas.

Dr. Morrison was graduated in 1963 from the University of Massachusetts, Amherst, with a B.S. degree in physics (magna cum laude) and obtained his Ph.D. degree in molecular biophysics and biochemistry in 1969 from Yale University. He did postdoctoral training at both the National Institutes of Health, Bethesda, Maryland and the Research Institute of Scripps Clinic in LaJolla, California. Dr. Morrison has held faculty positions at Scripps Clinic (Department of Immunology), Emory University (Department of Microbiology and Immunology) and the University of Kansas Medical Center (Chairman — Department of Microbiology, Molecular Genetics and Immunology) prior to assuming his current position in 1991.

Dr. Morrison is a member of the American Association for the Advancement of Science, the American Society of Microbiology, the American Association of Immunologists, the American Society of Experimental Pathologists, the American Society of Biochemistry and Molecular Biology, the Society for Leukocyte Biology, the American Academy of Microbiology, the Infectious Disease Society of America, and the New York Academy of Sciences. He is a charter member of the International Endotoxin Society, where he currently holds the office of Councillor.

Among other awards, Dr. Morrison has been the recipient of both a Research Cancer Development Award (RCDA) and Method for Extension of Research in Time (MERIT) Award from the National Institutes of Health. He was invited to give the Enrique E. Ecker Distinguished Lecture at Case Western Reserve and the Thomas A. Mahvi Lecture at the University of Missouri School of Medicine. In 1991, he was presented with the Innovation in Teaching Award by the medical students at the University of Kansas Medical School.

Dr. Morrison is the author of more than 130 publications and has written a number of comprehensive reviews on the subject of bacterial endotoxins. He is currently the recipient of three research awards from the NIH as well as several research grants from biotechnology companies. He serves as co-principal investigator of a multi-institution cancer research and training grant. His major research interests are in the area of structure-function relationships of bacterial endotoxins with particular focus on endotoxin receptors/signal transduction pathways and the pathogenesis of endotoxin shock.

John L. Ryan, Ph.D., M.D. is Executive Director, Clinical Research, Infectious Diseases at Merck Research Laboratories, West Point, Pennsylvania.

Dr. Ryan was graduated in 1964 from Yale University, New Haven, Connecticut, with a B.S. degree in biophysics and obtained his M.S. and Ph.D. degrees in molecular biophysics from Yale University. He attended medical school at the University of California at San Diego and obtained his M.D. in 1972. During medical school, he was a postdoctoral trainee in infectious diseases under the guidance of the late Abraham I. Braude. Following medical school, Dr. Ryan completed his clinical training in Internal Medicine at Yale University and joined the Immunology Branch of the National Cancer Institute as a Clinical Associate and Staff Fellow. He returned to Yale and completed fellowship training in infectious diseases before joining the faculty of the Department of Internal Medicine as an Assistant Professor and Chief of the Infectious Disease Section at the V.A. Medical Center in West Haven, Connecticut. Dr. Ryan remained at Yale as a faculty member in internal medicine until 1989 when he assumed his current position at Merck.

Dr. Ryan is a member of the American Association of Immunologists, the American Society of Experimental Pathologists, the American Academy of Microbiology, and a Fellow of the Infectious Disease Society of America. He is a charter member of the International Endotoxin Society and a member of the Program Committee of the International Congress of Antimicrobial Agents and Chemotherapy.

Dr. Ryan was supported by Merit Review Grants from the Veteran's Administration and U.S. Public Health Service Grants from NIAID during his faculty tenure at Yale University. He has authored over 100 publications, including several reviews on bacterial endotoxins. He currently directs a worldwide clinical research program in infectious diseases, including antibiotics, antivirals, and vaccines.

CONTRIBUTORS

Dolph O. Adams, M.D., Ph.D.
Professor of Pathology, and
Director, Laboratory of Cellular and
  Molecular Biology of Leukocytes
Duke University Medical Center
Durham, North Carolina

Helmut Brade, M.D.
Associate Professor and Head
Division of Biochemical
  Microbiology
Forschungsinstitut Borstel
Borstel
Germany

Lore Brade, Dr.vet.med.
Division of Immunochemistry and
  Biochemical Microbiology
Institute for Experimental Biology
  and Medicine
Forschungsinstitut Borstel
Borstel
Germany

Klaus Brandenburg, Dr.rer.nat.
Division of Biophysics
Forschungsinstitut Borstel
Borstel
Germany

Leora Suprun Brown, M.S.
Medical Program Coordinator
Infectious Diseases/Clinical
  Research
Merck Sharp & Dohme Research
  Laboratories
West Point, Pennsylvania

Jean-Marc Cavaillon, Dr.Sc.
Unité d'Immuno-Allergie
Institut Pasteur
Paris
France

Tai-Ying Chen, Ph.D.
Department of Microbiology,
  Molecular Genetics, and
  Immunology
University of Kansas
Medical Center
Kansas City, Kansas

J. Dijkstra, Ph.D.
Assistant Professor
Department of Physiological
  Chemistry
University of Groningen
Groningen
The Netherlands

Aihao Ding, Ph.D.
Associate Professor
Department of Medicine
Cornell University Medical College
New York, New York

Alice L. Erwin, Ph.D.
Laboratory of Bacterial
  Pathogenesis and Immunology
The Rockefeller University
New York, New York

John M. Harlan, M.D.
Professor of Medicine
Division of Hematology
University of Washington Medical
  Center
Seattle, Washington

Nicole Haeffner-Cavaillon, Dr.Sc.
Director of Research
INSERM
Hôpital Broussais
Paris
France

Otto Holst, Dr.rer.nat.
Forschungsinstitut Borstel
Borstel
Germany

Shozo Kotani, M.D., Ph.D.
Director
Osaka School of Medical
  Technology
Osaka
Japan

Shoichi Kusumoto, Ph.D.
Professor of Organic Chemistry
Department of Chemistry
Faculty of Science
Osaka University
Osaka
Japan

Mei-Guey Lei, Ph.D.
Research Assistant and Professor
Department of Microbiology,
  Molecular Genetics, and
  Immunology
University of Kansas Medical
  Center
Kansas City, Kansas

Buko Lindner, Dr.rer.nat.
Department of Immunochemistry
  and Biochemical Microbiology
Forschungsinstitut Borstel
Borstel
Germany

Robert S. Munford, M.D.
Professor
Departments of Internal Medicine
  and Microbiology
University of Texas —
  Southwestern Medical Center
Dallas, Texas

Masayasu Nakano, M.D., Ph.D.
Professor
Department of Microbiology
Jichi Medical School
Tochigi-ken
Japan

Carl F. Nathan, M.D.
Professor of Medicine
Cornell University Medical College
New York, New York

Naohito Ohno, Ph.D.
Associate Professor
Department of Microbiology
Tokyo College of Pharmacy
Tokyo
Japan

Timothy H. Pohlman, M.D.
Assistant Professor
Department of Surgery
University of Washington
Seattle, Washington

Nilofer Qureshi, Ph.D.
Research Chemist
Mycobacteriology Research
  Laboratory
William S. Middleton Memorial
  Veterans Hospital
Madison, Wisconsin

**Christian R. H. Raetz, M.D.,
  Ph.D.**
Vice President
Biochemistry and Microbiology
Merck Research Laboratories
Rahway, New Jersey

Ernst T. Rietschel, Dr.rer.nat.
Professor and Director
Department of Immunochemistry
and Biochemical Microbiology
Forschungsinstitut Borstel
Borstel
Germany

John L. Ryan, Ph.D., M.D.
Executive Director
Department of Infectious Diseases,
Clinical Research
Merck Research Laboratories
West Point, Pennsylvania

Ulrich Seydel, Dr.rer.nat.
Associate Professor
Division of Biophysics
Department of Immunochemistry
and Biochemical Microbiology
Forschungsinstitut Borstel
Borstel
Germany

Hiroto Shinomiya, M.D., Ph.D.
Assistant Professor
Department of Microbiology
School of Medicine
Ehime University
Ehime
Japan

Peter Stütz, Ph.D.
Head
Department of Scientific
Infrastructure
Sandoz Research Institute
Vienna
Austria

Haruhiko Takada, D.D.S., Ph.D.
Professor
Department of Microbiology and
Oral Microbiology
Kagoshima University Dental
School
Kagoshima
Japan

Kuni Takayama, Ph.D.
Supervisory Research Chemist
Mycobacteriology Research
Laboratory
William S. Middleton Memorial
Veterans Hospital
Madison, Wisconsin

Frank M. Unger, Ph.D.
President
Chembiomed Ltd.
Edmonton, Alberta
Canada

Herbert C. Yohe, Ph.D.
Research Associate
Department of Internal Medicine
Veterans Administration Medical
Center
West Haven, Connecticut

Ulrich Zähringer, Dr.rer.nat.
Head, Division of
Immunochemistry
Institute for Experimental Biology
and Medicine
Forschungsinstitut Borstel
Borstel
Germany

Volume II

IMMUNOPHARMACOLOGY AND PATHOPHYSIOLOGY

Section A: Pharmacology of Endotoxins

Section B: Immunopathophysiology of Endotoxins

TABLE OF CONTENTS

Section A: Biochemistry of Endotoxins

Chapter 1

BIOCHEMISTRY OF LIPOPOLYSACCHARIDES

Ernst Th. Rietschel, Lore Brade, Buko Lindner, and Ulrich Zähringer

TABLE OF CONTENTS

I. LIPOPOLYSACCHARIDES (LPS), THE O-ANTIGENS AND ENDOTOXINS OF GRAM-NEGATIVE BACTERIA

Gram-negative bacteria, which include many human pathogens, express various amphiphilic macromolecules at their surface such as the capsular antigens, lipoproteins, the enterobacterial common antigen, and the *lipopolysaccharides* (LPS). Of these, LPS are of particular microbiological, immunological, and medical significance. Mutants with defects in early steps of LPS biosynthesis are not viable. Therefore, it appears that LPS are essential for bacterial survival, with their vital role being based on the participation in the proper organization and function of the bacterial outer membrane. Further, LPS represent the main surface antigens (O-antigens) of Gram-negative bacteria. In their membrane-associated location they are targets for bacteriophages, they harbor binding sites for antibodies and nonimmunoglobulin serum factors, and, thus, they are involved in the specific recognition and elimination of bacteria by the host organism's defense systems. On the other hand, LPS may function to prevent the activation of complement and uptake of bacteria by phagocytic cells and, by shielding pathogens from cellular host defenses, play an important role in bacterial virulence. In addition, LPS isolated or released from bacteria are endowed with a broad spectrum of biological activities such as pyrogenicity and lethal toxicity. To emphasize these activities, LPS have been termed *endotoxins*. Also, because of their endotoxic properties, LPS contribute to the pathogenic potential of Gram-negative bacteria. Finally, LPS activate B-lymphocytes, granulocytes, and mononuclear cells and, hence, are potent immunostimulators. By virtue of this property they also seem to be involved in certain physiological host-parasite interactions.

In view of their fascinating and diverse spectrum of pathological and physiological activities, as well as their important function in bacteria, LPS have been studied over the last decades in many laboratories, using genetic,

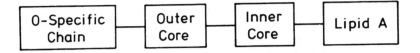

FIGURE 1. Schematic architecture of an enterobacterial lipopolysaccharide.

biological, immunological, chemical, and physical approaches. The goals of these investigations have included

- Analysis of the chemical structure and three-dimensional conformation of LPS
- Definition of biologically active regions of LPS
- Chemical synthesis of these regions
- Establishment of quantitative relationships between the chemical as well as the physical structure and the biological activity of LPS
- Determination of the involvement of LPS in outer membrane architecture and the function of this membrane as a permeation barrier
- Elucidation of the role of LPS in bacterial viability
- Elucidation of the genetic determination and biosynthetic pathways of LPS
- Clarification of the molecular mechanisms involved in endotoxin action *in vivo* and *in vitro*
- Pursuit of the *in vivo* fate and metabolism of the LPS molecule
- Development of strategies to immunologically and pharmacologically control harmful manifestations of endotoxicosis, to separate detrimental from beneficial endotoxin effects, and to make use of the beneficial activities of LPS

Many of these problems have been solved and the progress made during the last decade is summarized in Volumes I and II. Other questions, however, have not yet been answered. In the present volume, our current views on the molecular archiecture and relationships of the chemical and physical structure to biological activity and function are discussed. In the framework of this discussion, this chapter will summarize selected aspects of relationships between structure and activity or function in the field of LPS and indicate possible future directions of LPS research in these areas.

II. STRUCTURE AND ACTIVITY OR FUNCTION OF LPS REGIONS

In the classical model of enterobacterial wild-type (S-form) LPS, four covalently interlinked segments are distinguished, i.e., the O-specific chain, the core oligosaccharide with the inner and outer region, and the lipid A component.[1] Figure 1 shows the schematic structure of an enterobacterial LPS with these segments indicated. A similar architecture is found in LPS of *Pseudomonas aeruginosa, Vibrio cholerae, Rhizobium,* and other bacteria. However, in LPS of a number of other pathogenic Gram-negative bacteria such as *Bordetella pertussis, Bacteroides fragilis, Campylobacter jejuni, Acinetobacter calcoaceticus, Neisseria meningitidis,* and *Haemophilus influenzae,* the O-specific chain is lacking.[2] In their general architecture and migration behavior in polyacrylamide gel electrophoresis, these LPS resemble those of enterobacterial rough-(R-)mutants and, therefore, are termed R-type LPS. It should be noted, however, that R-type LPS, despite their similarity in size, differ significantly from enterobacterial R-mutant (R-form) LPS in the fine structure of their carbohydrate region (see Section VII).

The segments of enterobacterial LPS, i.e., O-specific chain, outer core, inner core, and lipid A, radically differ from each other in their structural makeup and biosynthesis, and different biological activities and functions have been ascribed to each of them. Below, the structure, biosynthesis, and activity or function of these regions will be discussed.

III. O-SPECIFIC CHAIN

A. CHEMICAL STRUCTURE

The architecture of the O-specific chain is unique in that it represents a carbohydrate polymer consisting of up to 50 oligosaccharides, termed repeating units. The chemical structure of the O-specific chain is determined by the repeating unit constituents which are, in general, glycosyl residues. As an example, Figure 2 shows the structure of repeating units present in LPS of *Salmonella abortus equi* (A) and *S. typhi* (B). The nature, ring form, type of linkage, and substitution of the individual monosaccharide residues, as well as their sequence within a repeating unit, is characteristic and unique for a given LPS and the parental bacterial serotype.[1] Today, the structure of more than 100 types of enterobacterial and nonenterobacterial O-chains is known in detail, the elucidation of these structures being greatly facilitated by their regular and blockpolymer-like architecture. With the chemical and physical methodology currently available, in combination with serological techniques, it should be possible to analyze the primary structure of any desired O-specific chain. Because of the diversity of sugar constituents and their possible linkage sites, an enormous number of O-chain structures is conceivable and is also verified in nature. Thus, an immense structural variability is revealed if the O-specific chains of distinct bacterial origin are compared.[3]

The O-specific chains exhibit heterogeneity. Thus, in a given LPS preparation, a family of molecules is present which differ in the presence or absence of O-specific chains and the degree of polymerization of repeating units. Another facet of heterogeneity is due to the fact that certain substituents of repeating units are transferred to the O-specific chain after its biosynthetic completion and that these may be present in nonstoichiometric amounts.[4]

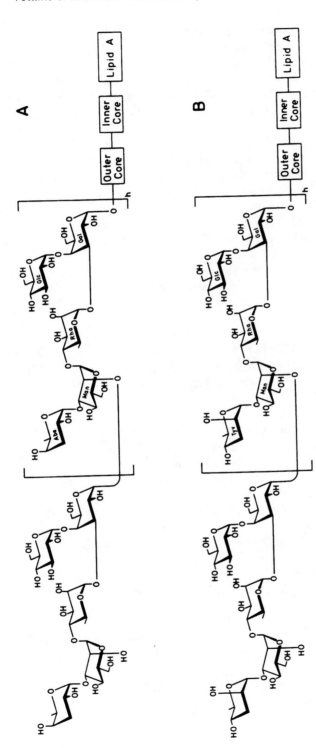

FIGURE 2. Chemical structure of the O-specific chain of (A) *Salmonella abortus equi* and (B) *S. typhi* LPS.

The O-chain biosynthesis in *S. typhimurium* is initiated by the assembly of an oligosaccharide repeating unit, whereby undecaprenyl phosphate serves as antigen carrier lipid (ACP). Polymerization occurs by a continuous transfer of the activated growing chain to a newly formed activated repeating unit, a reaction catalyzed by a polymerase. Thus, the O-specific chain is elongated at the reducing end.[4] The mechanism regulating the polymerase and, thus, determining the length of the O-specific chain is, at present, poorly understood and certainly represents an important area for future research. Also O-specific chains are known which contain repeating units consisting of only one type of sugar (such as D-mannose in *Escherichia coli* O8 and *E. coli* O9), the "repeating units" being characterized as containing identical sequences of specific linkages. Here, as in many other *E. coli* and also *Samonella* serotypes, the biosynthesis of the O-specific chain proceeds by transfer of glycosyl groups to the *non*reducing end of the polysaccharide chain. This mechanism is dependent on the *rfe* gene, the function of which, however, is not yet understood.[4,7] It is not clear and remains to be studied how, in this case, the polymerase is regulated to yield defined repeating units. The completed O-chains are transferred from ACP to the independently synthesized core-lipid A structure yielding mature S-form LPS.

B. BIOLOGICAL ACTIVITY AND FUNCTION

During the last decades it has become clear that the O-specific polysaccharide chain mediates several biological properties of LPS. Thus, the O-specific chain

- Determines the serological O-specificity, i.e., it carries the private epitopes (O-factors) of LPS and the parental bacterial strain[1]
- Functions as a receptor for bacteriophages[5]
- Is involved in binding of nodulating bacteria such as Rhizobiaceae to their leguminous host[6]
- Modulates activation of the alternative complement (C')-pathway, i.e., phagocytosis via C3b[8]
- Inhibits attachment of C5b-9 membrane attack complex to the bacterial outer membrane[8,9]
- Is, therefore, important for bacterial virulence[9]
- Plays a role in the beneficial effects of endotoxin[10]
- Downregulates endotoxic activity of LPS[11,12]

Of these manifold activities, three selected examples of structure-activity relationships are discussed below.

C. STRUCTURE-ACTIVITY RELATIONSHIPS

The O-specific chain determines the *serological specificity* of LPS and the parental bacterial strain (serotype). The O-immunogenic and -antigenic properties of LPS are determined by so-called O-factors which are located within the repeating units. In many cases, the chemical structure of O-factors has been elucidated, and correlations between the structural and serological features of a defined LPS or its parental bacterial strain have been established,[1] i.e., serotypes have been correlated to chemotypes. As an example, O-factor 4 present in the O-specific chain of *S. abortus equi* LPS (Figure 2A) is determined by abequose (Abe, 3,6-dideoxy-D-galactose), whereas O-factor 9 of *S. typhi* LPS (Figure 2B) is represented by tyvelose (Tyv, 3.6-dideoxy-D-mannose). Both *S. abortus equi* and *S. typhi* LPS also contain factor 12 mainly represented by the αDGlc1-4βDGal dissacharide of the repeating unit.

The O-specific chain carries *receptor structures for bacteriophages*. Also, in this case, clear relationships between structure and activity were elucidated.[5] As an example, phage 22 uses the O-specific chain of *Salmonella* serotypes A, B, and D as a receptor, and its associated enzymatic activity hydrolyzes the αLRha1-3αDGal linkage with repeating units (see also Figure 2). However, it was also found that one repeating unit does not suffice as a receptor structure, indicating that the phage most probably recognizes a conformational determinant generated by more than one repeating unit (cited in Reference 5).

The conformation of O-chain segments seems also to be important in *complement activation* and the *virulence of bacteria*.[8,9] It is known that LPS activates the alternative C' pathway through the O-specific chain; however, the efficiency of activation is dependent on the structure of repeating units to different degrees. Also of relevance is the fact that bacteria expressing O-specific chains with factors 4 and 12 are significantly more virulent than bacteria exposing factors 9 and 12. This difference in virulence has been related to the relative inability of factors 4- and 12-expressing bacteria (as compared to factors 9- and 12-expressing bacteria) to activate complement by the alternate pathway and to bind C3b, thereby resisting phagocytic uptake and killing. Abe (factor 4) and Tyv (factor 9) chemically differ only in the orientation of two hydroxyl groups (Figure 2). It will be important to determine what conformational differences exist between O-specific chains harboring either factors 4 and 12 or factors 9 and 12.

These examples demonstrate that antibodies, enzymes, and complement proteins specifically interact with defined regions of the O-specific chain. The primary structure of this region is known, but, in order to obtain deeper insight into host protein recognition, the three-dimensional conformation of O-specific chains must be elucidated. Studies in this direction have been performed on selected examples[13,14] and, in future investigations, synthetic partial structures of O-specific chains will certainly prove to be powerful tools.[14-16] The field of O-chain conformation and its specific interaction with mammalian proteins represent an important aspect of future LPS research.

IV. CORE REGION

The core of enterobacterial LPS is a branched heteropolysaccharide which lacks repeating glycosyl structures. Its biosynthesis starts with lipid A-bound 3-deoxy-D-*manno*-octulosonic acid (Kdo), and the core chain grows at the nonreducing end by sequential addition of activated glycosyl residues.[7] Thus, core and O-chain biosynthesis each proceeds independently and according to fundamentally different pathways. The elucidation of the chemical structure, the genetic determination, and the principles of biosynthesis of the core region were greatly facilitated by the availability and use of rough (R)-mutants which harbor specific defects in LPS biosynthesis.[1,4,7] In the case of an O-chain defect in the synthesis or transfer of a constituent of a repeating unit or the translocating enzyme, an LPS is formed which consists of the lipid A-bound complete core but which lacks the O-specific chain (Ra mutants). Defects in core constituent-related glycosyl transferases, synthetases, epimerases, and also in phosphoryltransferases lead to R-mutants (termed Rb_1 to Re mutants) which express progressively glycosyl-defective LPS structures.[1,4] These R-mutant (R-form) LPS have been and remain of great value for studies aimed at the establishment of the structure of the core oligosaccharide. Such studies have revealed that the core oligosaccharide can be divided into two regions: the *outer core* and the *inner core*.

The structural variability of the core within many species of an enterobacterial genus is limited. Thus, within *Salmonella* species only one core type (Ra core) exists as identified serologically and by phage typing. In *E. coli*, five core types (R1 to R4 and K-12) have been identified to date. In the group of *Proteus* and related genera and in *Citrobacter* five and four core types, respectively, have so far been identified.[17-19] Structural differences between these core types are mainly recognized in the outer core region, while the inner core is structurally rather conserved. Our present knowledge on the chemistry of the LPS core is discussed in depth in Chapter 6.

V. OUTER CORE

A. CHEMICAL STRUCTURE

As an example, the outer core of enterobacterial LPS contains the common pyranosidic hexoses D-glucose (Glc), D-galactose (Gal), *N*-acetyl-D-glucosamine (GlcNAc), and (in *Citrobacter*) *N*-acetyl-D-galactosamine (GalNAc). It is, therefore, also termed the hexose region. In *Salmonella* and *E. coli* these hexoses form a branched pentasaccharide of the general structure:

α/βDHex*p*(V)

$$\overset{1}{\underset{3/4 \qquad\qquad 6}{-----}}$$

αDHex*p*(IV)1→2αDHex*p*(III)1→2αDHex*p*(II)1→3αDHex*p*(I)1→

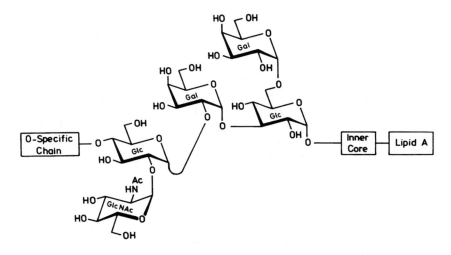

FIGURE 3.   Chemical structure of the outer core region of a *Salmonella* LPS (Ra core).

As a specific example, the *Salmonella* Ra core, is shown in Figure 3.[1] In all enterobacterial core types the outer core is, via αDGlc (1-3)-linked to an L-*glycero*-D-*manno*-heptosyl (L,D-Hep) residue of the inner core (see Figure 4). Very recently, the *E. coli* K-12 outer core structure was shown to contain terminal L,D-Hep. In Chapter 6 the revised structure of the K-12 core is described.[20]

The conformation of the *Salmonella* and *E. coli* outer core types has been studied by minimum energy calculations.[21] It was shown that the linear hexosyl tetrasaccharide (Hex I to Hex IV of the general structure) is relatively rigid in all core types, except in K-12. In the three-dimensional structure of the sterically crowded and rigid outer core region, two sides are recognized, termed the "front" and the "back" sides. Interestingly, the conformational shape of the "front" side is similar for the different cores, whereas the extracatenarian glycosyl residue (Hex V) is located on the back side, rendering this conformational area core-type specific.[21]

In recent years, great progress has been made in the elucidation of outer core structures; however, several questions remain to be answered. Thus, the common assumption that, within the over 200 serotypes of the genus *Salmonella* which have to date been identified and characterized, the concept that all of these area associated with only *one* core type (Ra core) has not been proven unequivocally by chemical and physical methods. In addition, since previous studies were mainly focused on enterobacterial structures, it will certainly prove rewarding to analyze the primary structure and conformation of, notably, nonenterobacterial LPS cores and to compare their chemical, biological, and taxonomic relatedness or difference. This area may receive greater attention in the years to come.

Bacterial Endotoxic Lipopolysaccharides

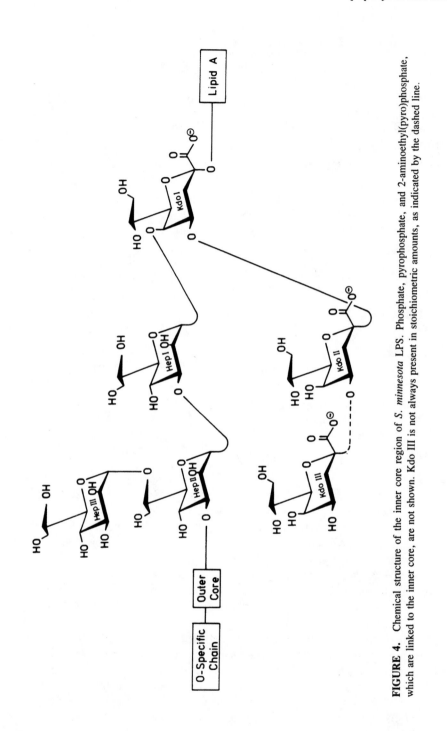

FIGURE 4. Chemical structure of the inner core region of *S. minnesota* LPS. Phosphate, pyrophosphate, and 2-aminoethyl(pyro)phosphate, which are linked to the inner core, are not shown. Kdo III is not always present in stoichiometric amounts, as indicated by the dashed line.

B. BIOLOGICAL ACTIVITY AND FUNCTION

As compared to the other LPS structural components, not much is known about the activity and/or function of the outer core region. Nevertheless, it has been shown that the outer core oligosaccharide

- Determines outer core specificities[1]
- Functions as receptor for outer core phages[22]
- Is involved in binding of certain bacterial mutants to lymphocytes[23,24]
- Recognizes an Ra-specific serum factor[143]

It has been suggested that this interaction between bacteria and lymphoid cells plays a role in the differentiation of T-helper cells or the induction of suppressor cells[23] and the removal of bacteria by lymphatic cells.[24]

C. STRUCTURE-ACTIVITY RELATIONSHIPS

Studies linking the structure of the outer core to bioactivity are rare and are primarily concerned with its interaction with antibodies and phages. Interesting views were revealed by studies concerning the bacteriophage G13.[22] This phage was known to bind to *S. typhimurium*, *E. coli*, and *Shigella flexneri* LPS and thus to exhibit a very broad host range. Surprisingly, the primary structure of various receptor-active outer core-derived oligosaccharides did not show anticipated similarities.[17] However, the conformation of those oligosaccharide molecules interacting with phage G13 exhibited a similarly shaped region ("front site") which is also verified in the heptose region of the inner core.[21] Hence, the receptor site for phage G13 would appear to be a structural conformation which is expressed by various enterobacterial outer core types and also by the heptose region of the inner core.

The concept of three-dimensional carbohydrate determinants being shared by different core types and regions stimulates the question as to whether monoclonal antibodies can be raised against such common determinants. These antibodies would be expected to cross-react with chemically and otherwise serologically unrelated LPS. In the event that these monoclonal antibodies would protect against harmful endotoxin activities or serve as bacterial opsonins, they would be of enormous importance in the prevention and treatment of septicemia and endotoxinemia. Studies along these lines certainly represent a major aspect of future LPS research providing great biochemical and biomedical perspectives.

VI. INNER CORE

A. CHEMICAL STRUCTURE

The inner core region of enterobacterial LPS is characterized by the unusual sugars heptose (Hep), mainly in the L-*glycero*-D-*manno* though occasionally in the D-*glycero*-D-*manno* configuration, and in particular 3-deoxy-

D-*manno*-octulosonic acid (2-keto-3-deoxyoctonic acid, Kdo). LPS of Enterobacteriaceae and the majority of other Gram-negative bacteria studied contain Hep with, for example, the exception of *B. fragilis* and *A. calcoaceticus* (compare Section VII). On the other hand *all* LPS, independent of their bacterial origin, harbor at least one α-bound pyranosidic or furanosidic Kdo residue or a derivative thereof. This Kdo group possesses a free carboxyl group and occupies the lipid A-proximal position of the reducing terminus of the inner core region (Figure 4). Hence, Hep represents a very frequent and Kdo a common and obligatory component of LPS. Structurally, therefore, the inner core region represents a rather conserved region of LPS.

The chemical structure of the glycosyl backbone of the enterobacterial inner core has been established as shown in Figure 4.[1,25] The reducing end of the inner core is represented by a Kdo residue (Kdo I) which is in a α-ketosidic linkage to the primary hydroxyl group (position 6′) of the non-reducing GlcN residue (GlcN II) of the lipid A backbone (compare Figure 6). This Kdo group, being located in the main core oligosaccharide chain, is substituted in position 4 by α-linked Kdo II, which in turn is often non-stoichiometrically substituted at position 4 by a further α-linked Kdo residue (Kdo III). All Kdo groups are present as pyranosides. In its position 5, Kdo I carries α-bound L,D-Hep (Hep I) to which a second L,D-Hep residue (Hep II) is α-linked at position 3. The latter is substituted in position 7 by a third L,D-Hep group (Hep III). In LPS of less-defective mutants and in wild-type bacteria, the saccharide chain extends from the hydroxyl group in position 3 of Hep II (Figure 4), i.e., the outer core is via Glc α(1-3)-linked to Hep II of the inner core. The structure shown in Figure 4 does not take into account the fact that the inner core region is substituted at various sites (such as position 4 of Hep I and Hep II) by phosphate, pyrophosphate, 2-aminoethyl phosphate, and 2-aminoethyl pyrophosphate.[1,26] In addition, *E. coli* LPS possessing the R1 or R3 core type, GlcN may be attached to L,D-Hep III of Figure 4.[27] Kdo, on the other hand, may carry in different bacterial genera a great variety of different substituents, which are discussed in detail in Chapter 6. As examples, the hydroxyl group in position 4 is substituted, in general, by negatively charged groups (Kdo, phosphate, D-galacto-, and D-glucuronic acid), position 5 by pyranosidic neutral sugars (Hep, Man, Rha), position 7 by 2-aminoethyl phosphate and Gal), and position 8 by either negatively or positively charged residues such as Kdo, phosphate, or 4-amino-4-deoxy-L-arabinopyranose (Ara4N). These inner core substituents are often not present in stoichiometric amounts, which further underscores the genuine LPS heterogeneity (for literature compare References 28 and 29). It should be noted that, in part of *A. calcoaceticus* LPS, Kdo is positionally replaced by D-*glycero*-D-*talo*-octulosonic (Ko) acid which is isosteric to Kdo.[30] Due to the presence of the carboxyl groups of Kdo (and Ko) and negatively charged L,D-Hep and Kdo substituents, the overall charge of the inner core and, thus, of LPS is negative. This is true for enterobacterial and nonenteric strains and, therefore, is a general feature of both the inner core and the intact LPS molecule.

Structural studies on the inner core, despite the development of suitable analytical procedures and the availability of modern analytic techniques, are still very difficult and cumbersome. This is mainly due to the problems associated with the analytical chemistry of the polyfunctional deoxy sugar Kdo, both in its free form and more particularly in its substituted (i.e., glycosylated or phosphorylated) form. In most studies to address core structures, the Kdo region of the inner core is, therefore, excluded. In view of the biological significance of Kdo for LPS activity and function (compare Sections VI.B and VI.C), however, a detailed knowledge of the structural organization of the Kdo region is highly desirable. As a consequence, the elucidation of the primary and three-dimensional structure of the Kdo-containing inner core of various LPS represents a most important challenge for future LPS research.

The core structures known today were all derived from analysis of R-mutants, i.e., R-form LPS. It is commonly assumed that identical structures are also present in the corresponding S-form LPS. This hypothesis is certainly reasonable as far as the main core glycosyl chain is concerned. Chemical proof, however, is lacking that this identity also extends to substituents of the chain such as phosphate, ethanolamine, 2-aminoethyl phosphate, Kdo, and other glycosyl residues. This question is of great importance in view of present efforts to generate core specific and protective antibodies which are expected to cross-react with the corresponding S-form LPS or wild-type bacteria. This area is of great biochemical importance and the structural characterization of the inner core in Enterobacteriaceae and notably nonenteric human pathogens will certainly remain attractive in the years to come.

B. BIOLOGICAL ACTIVITY AND FUNCTION

During recent years, it has become evident that the Kdo-containing inner core is of great significance for many of the biological activities and functions of the LPS molecule. Thus, structures in the inner core

- Carry common (shared) LPS epitopes for antibodies and serum factors.[31-34] Some aspects of the immunoreactivity of the inner core are discussed below (Section VI.C).
- Modulate biological activity of lipid A.[12,35] While it is established (compare Section VIII.A.3) that the lipid A component is essential for endotoxin effects *in vivo* (e.g., pyrogenicity, lethality) and *in vitro* (e.g., elaboration of toxic peptide and lipid mediators by mononuclear cells), the degree of lipid A bioactivity may be modulated by the polysaccharide portion,[11] particularly by the Kdo-containing inner core. As examples, the release of interleukin 1[35] or leukotriene C4[12] from adherent human or murine mononuclear cells, respectively, cannot be provoked by

polysaccharide-free lipid A, but is readily induced by Re mutant LPS.
- Are essential for bacterial viability.[36,40]

Several systems of relationships between the chemical constitution of the inner core and its biological activity or function have been studied and this area is discussed in Chapter 8. Below, two examples, i.e., immunoreactivity and significance for bacterial viability, are briefly described.

C. STRUCTURE-ACTIVITY RELATIONSHIPS

The *immunoreactivity* of the inner core has recently received great scientific attention. Being a common structural element of LPS, the Kdo-containing inner core has been considered an attractive target for antibodies which cross-react with LPS of different bacterial origin and which possibly provide cross-protection against pathogenic Gram-negative bacteria and their endotoxins.[37,38] With the aid of chemically synthesized Kdo-containing antigens, the epitope specifically of four monoclonal antibodies raised against enterobacterial Re-mutant bacteria, the LPS of which consits of an α(2-4)-linked Kdo disaccharide being α(2-6)-linked to lipid A, has been established.[31-33] Two of these antibodies (clones 17 and 22) bind to the Kdo disaccharide and parts of the lipid A region. Clone 17 does not require phosphoryl groups of lipid A for the binding; by contrast, clone 22 does require these groups. In polyclonal antisera raised against Re mutant bacteria, the majority of antibodies were of the same specificity as clones 17 or 22. These monoclonal and polyclonal antibodies specifically bind to Re mutant LPS and do not cross-react with other R- or S-form preparations.[32,33] In contrast to clones 17 and 22, which represent anti-Re LPS antibodies, two other clones (20 and 25) were characterized as true anticore antibodies.[31] Thus, clone 25 antibody reacts specifically with an α(2-4)-linked terminal or lateral Kdo disaccharide which is expressed, for example, by *S. minnesota* Re and Rb_2 mutant LPS. Clone 20, on the other hand, recognizes an α-pyranosidically linked Kdo monsaccharide which is present in a terminal (Re mutants) or lateral position in various non-Re LPS. As expected, this antibody cross-reacts with a series of LPS. Recent studies indicate that this antibody binds also to wild-type bacteria and that it exhibits protective activity in certain experimental infection models.[39]

Bacteria with a defect in Kdo biosynthesis do not grow. This observation provides strong support for the concept that Kdo (and LPS in general) is essential for microbial *viability*.[36,40] The LPS of wild-type *Chlamydia trachomatis* contains three Kdo residues,[41] that of enterobacterial Re mutants two,[42] and that of an *H. influenzae* R mutant one (phosphorylated) Kdo residue, demonstrating that only *one* Kdo residue, in addition to lipid A, can suffice for the survival of a Gram-negative bacterium.[43] That Kdo is an obligatory constituent of LPS, that it plays a vital role for bacteria, and that it is probably not present in mammalian cells have led to the idea that pharmacologic reagents interfering with its biosynthesis may provide a new class of antibiotics, active against Gram-negative bacteria in general. So far, Kdo derivatives have been constructed which exhibit good antibacterial activity *in vitro*,[44,45] justifying the hope that this line of research will lead to new chemotherapeutic agents with a novel mode of action.

In none of the systems where Kdo contributes to or is essential for biological activity and function of LPS have the exact structural parameters required for the expression of its bioactivity been clearly defined. Some studies, however, make the carboxyl group likely to be important. Hence, due to the presence of negatively charged carboxyl groups, the Kdo region represents a high-affinity binding site for bivalent cations. A high concentration of Ca^{2+} and Mg^{2+} at the bacterial surface may be essential for the function and structural integrity of the outer membrane and thus for survival of the microbial cell.[40] The carboxyl group also appears to be necessary in the immunoreactivity[46] and the IL-1-inducing properties of LPS.[35]

It can be foreseen that synthetic structures containing modified or derivative Kdo groups[47-50] will further our understanding of the molecular requirements of Kdo bioactivity and the inner core in general. The synthesis of such structures is in progress (Chapter 7) and will provide the necessary tools in order to perform such studies. It will be of equal importance to determine the conformation of the inner core in order to understand its interaction with humoral and cellular host components and its biological significance in general.

VII. POLYSACCHARIDE REGION OF SOME NONENTEROBACTERIAL LPS

A number of nonenterobacterial Gram-negative human pathogens such as *N. meningitidis*,[2] *H. influenzae*,[51,52] *B. pertussis*,[53] *V. parahaemolyticus*,[54,60] *A. calcoaceticus*,[30] and *B. fragilis*[55,56] synthesize LPS which lack O-specific chains and, thus, repeating units. Because of their relatively shorter glycosyl chains these LPS have been termed lipooligosaccharides (LOS). For reasons discussed in detail in an earlier publication[57] this term will not be employed in the present paper where, instead, the more-established terms LPS (or alternatively R-type LPS) will be used. In these O-chain-deficient R-type LPS, as in the enterobacterial core, a hexose region and a Kdo (and often L,D-Hep)-containing portion are recognized. As examples, Figure 5 shows the structures of the L,D-Hep-containing LPS of *N. meningitidis* L5 (Reference 58; Figure 5A) and *V. parahaemolyticus* 012 (References 59 and 60; Figure 5B) and the L,D-Hep-lacking LPS of *B. fragilis* (Refrence 55; Figure 5C) and *A. calcoaceticus* (Reference 30; Figure 5D). On comparison of these LPS with enterobacterial R-mutant preparations, similarities but also important differences are recognized. Thus, in *N. meningitidis* and *V. parahaemolyticus* LPS, the hexose chain contains exclusively β-anomers and is

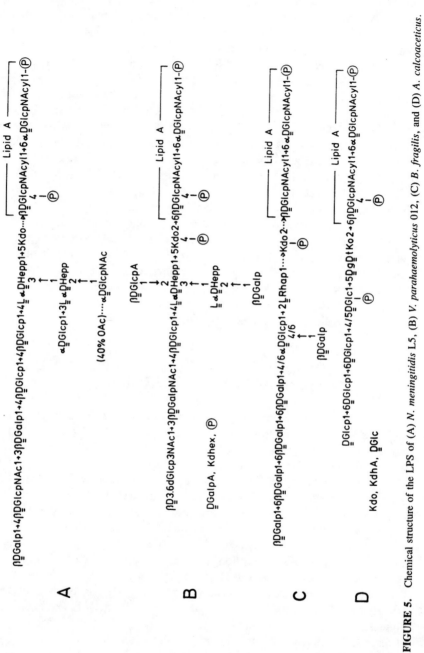

FIGURE 5. Chemical structure of the LPS of (A) *N. meningitidis* L5, (B) *V. parahaemolyticus* 012, (C) *B. fragilis*, and (D) *A. calcoaceticus*.

β(1-4) linked to L,D-Hep. In the enterobacterial outer core, only α-hexopyr-anosides are present and hexose segment is linked α(1-3) to Hep II of the inner core (Figure 4). One could, however, view the structural arrangement in these R-type LPS differently in that (1) their basic architecture corresponds structurally to the LPS of an enterobacterial RcP⁻-mutant (as in *N. menin-gitidis*) and an Rd_1P^--mutant (as in *V. parahaemolyticus*) and that (2) the oligohexose region as well as the various other groups (e.g., Gal, GlcNAc, OAc) represent substituents of this basal structure. Interestingly, in *Neisseria* the specificity of serotypes is *not* determined by the hexose region, but rather by the structure of the inner core. The lack of immunogenicity of the hexosyl region may be related to the fact that its nonreducing terminus (lacto-*N*-neotetraose) is a known constituent of a human and animal tissue antigen and, thus, subject to immunological tolerance.[61] In addition, *N*-acetyl neura-minic acid may be transferred to the terminal Gal-residue of bacteria-associated LPS *in vivo*,[62,63] rendering the oligohexosyl domain immuno-logically nonreactive. In contrast, in *B. fragilis*, the hexose region determines the serospecificity. Here, an immunodominant β(1-6)-interlinked D-Gal oli-gomer is present and it has been speculated that this oligomer represents a rudimentary O-specific side chain.[56] A similar argument could be put forward for *A. calcoaceticus* LPS, which contains a (1-6)-interlinked D-Glc oligomer (Figure 5D).

With few exceptions, the polysaccharide region of the O-chain-lacking LPS has received considerably less scientific attention than enterobacterial LPS. Thus, despite ongoing intensive work, much remains to be learned about the chemical structure, serological properties, and pathogenic significance of LPS of such important human pathogens as *Neisseria*, *Haemophilus*, and *Bordetella*. It is evident that work in these areas will be highly rewarding and provide new insights into the architecture, genetic determination, bio-synthesis, immunoreactivity, and endotoxicity of LPS.

VIII. LIPID A

A. PRIMARY CHEMICAL STRUCTURE

Lipid A represents the covalently bound lipid component of LPS. The term lipid A denotes a family of (phospho)glycolipid molecules which are closely related in structure but which are not identical among various LPS.[64] This becomes evident on anlysis of lipid A of one defined bacterial strain as well as on comparison of lipid A from different bacterial origins. Lipid A obtained from a given LPS preparation by acid hydrolysis contains, in addition to a predominant form (of which often the structural formula is displayed in scientific communications), partial structures of this mature form. The latter are generated either by incomplete biosynthesis or by chemical degradation which takes place during the preparation of free lipid A. Hence, a lipid A preparation exhibits both inherent and chemically induced heterogeneity.

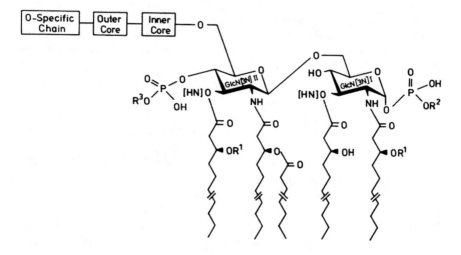

FIGURE 6. Generalized structure of the lipid A component of bioactive LPS. R^1 represents acyl group or hydrogen; R^2 and R^3 denote substituents of phosphate groups. For details see text.

On the other hand, comparison of endotoxic lipid A from different Gram-negative genera and families reveals that they not only manifest many constituents which are the same but in addition are made up according to a very similar architectural principle illustrated in Figure 6. Thus, all endotoxic lipid A studied so far contain *gluco*-configurated and pyranosidic D-hexosamine residues which, in general, are present as a β(1-6)-linked disaccharide. Only in rare cares can lipid A be found as monosaccharides (not shown). As a general rule, the disaccharide carries a glycosidic and a nonglycosidic phosphoryl group and, in ester and amide linkage, medium- to long-chain (R)-3-hydroxy fatty acids (C10 to C28), some of which are acylated at their 3-hydroxyl group. Structurally, therefore, lipid A represents the least variable region of LPS.

The structural conservation of lipid A is reflected by the uniform mechanism of its biosynthesis in various Gram-negative bacteria (see Chapter 3). As elucidated in *E. coli* and *Salmonella*, UDP-*N*-acetyl-D-glucosamine is, after addition of two 14:0(3-OH) residues, condensed with 2,3-bisacyl-D-glucosamine 1-phosphate to a tetraacyl disaccharide 1-phosphate (a reaction catalyzed by the enzyme lipid A synthase). This molecule is then phosphorylated at position 4′ to yield the so-called precursor Ia of lipid A biosynthesis.[65] To this precursor, Kdo is transferred prior to the completion of the lipid A structure which is accomplished by addition of dodecanoic (12:0) and tetradecanoic (14:0) acid. In *Pseudomonas*, however, mature lipid A is formed first, to which Kdo is added in a subsequent step.[66] The lipid A synthase also accepts UDP-activated and acylated 2,3-diamino-2,3-dideoxy-D-glucose (GlcN3N). It, therefore, is understandable that backbone structures containing a β(1-6)-linked GlcN3N disaccharide or hybrid disaccharides such as GlcN3N-GlcN may also be formed *in vitro*[67] and *in vivo* (see Sections 2 and 3).

Although lipid A structures of different bacterial origin are constructed according to the same general architectural principle, they may nevertheless differ in their fine structure. Variation in structure may result from the type of hexosamine present, the nature, chain length, number, and location of acyl residues, the degree of phosphorylation, and the nature of phosphate substituents (residues R^2 and R^3 of Figure 6). In the following, selected, representative examples of types of lipid A structures are described that are classified according to the nature of the constituent hexosamine (either GlcN or GlcN3N) and certain features of the acylation pattern. With few exceptions, the chemical structure of these lipid A have been determined by analyzing R-mutant LPS. The chemical structure of lipid A is discussed in detail in Chapter 2 and here only some general principles are summarized.

1. Lipid A Structures Containing a GlcN-GlcN Disaccharide

The *hexaacyl* lipid A of *E. coli* is a classical representative of enterobacterial lipid A and its chemical structure has been elucidated in great detail during the last decade.[64,68-70] In Figure 7A the primary structure of the predominant component of *E. coli* lipid A, as present in the LPS of the Re mutant strain F515, is shown. *E. coli* lipid A is composed of a β-D-glucosaminyl-(1-6)-α-D-glucosaminyl disaccharide which carries two phosphoryl groups: one in position 4' (of the distal GlcN residue, GlcN II) and one in the glycosidic position 1 (of the reducing GlcN residue, GlcN I). This hydrophilic lipid A backbone contains two nonsubstituted hydroxyl groups in positions 4 and 6'. The latter primary hydroxyl group is only free in polysaccharide-deprived lipid A (''free lipid A''), since in LPS it represents the attachment site of Kdo, i.e., the polysaccharide component (compare Figure 6). Positions 2, 3, 2', and 3' of the backbone carry four (R)-3-hydroxytetradecanoic acid residues [14:0(3-OH)]. Since they are directly linked to the GlcN disaccharide they are termed primary acyl groups.[71] The hydroxyl groups of the two 14:0(3-OH) residues bound to GlcN II are acylated by secondary fatty acids: that at position 2' by 12:0 and that at position 3' by 14:0. In *E. coli* lipid A six fatty acids are present, of which four are associated with GlcN II and two with GlcN I. Thus, the overall acylation pattern is *asymmetric* [4 + 2].

The glycosidic phosphate carries a further phosphoryl group in nonstoichiometric amounts.[68] In addition to the main structure shown in Figure 7A, molecules are present with a smaller number of acyl groups. The same structure as that of *E. coli* lipid A has been identified in *S. typhimurium* (Reference 72; see also Chapter 1); this type of lipid A is also present in other enterobacterial and nonenterobacterial bacteria such as *H. influenzae*[43] and *Providencia rettgeri*.[73] Hexaacyl *E. coli* lipid A has been successfully synthesized (Reference 50; see Chapter 2).

In contrast to hexaacyl *E. coli* lipid A, which manifests as asymmetric [4 + 2] acylation pattern, a *symmetrical* fatty acid distribution was identified in *hexaacyl Chromobacterium violaceum* lipid A (References 74 and 75;

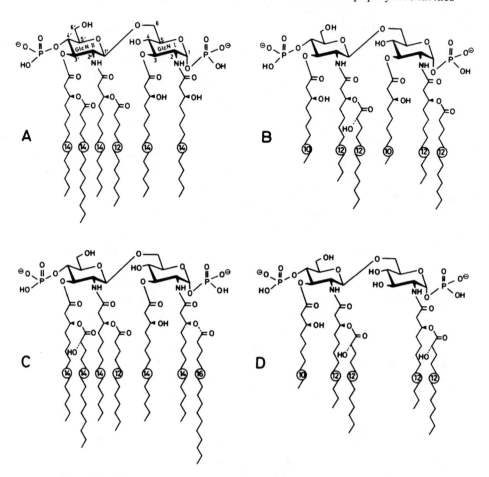

FIGURE 7. Chemical structure of major lipid A species of (A) *E. coli*, (B) *Chromobacterium violaceum*, (C) *S. minnesota*, and (D) *P. aeruginosa*. Substituents of phosphate groups are not shown. The hydroxyl group in position 6' represents the attachment site of Kdo. Dotted lines indicate nonstoichiometric substitution. Numbers in circles refer to the number of carbon atoms in acyl chains. For details see text.

Figure 7B). Here, GlcN I and GlcN II each carry three fatty acids [3 + 3]. Lipid A of *C. violaceum* further differs from that of *E. coli* in the chain length of the fatty acids present. It should be noted that the amide-linked 12:0(3-OH) at GlcN II is a 3-O-acylated by either 12:0 or (*S*)-2-hydroxydodecanoic acid [12:0(2-OH)] yielding two molecular species. In *C. violaceum* lipid A, the glycosidic phosphate carries nonsubstituted GlcN and the nonglycosyl phosphate nonsubstituted Ara4N as substituents in stoichiometric amounts (not shown in Figure 7B).[74] This type of hexaacyl lipid A exhibiting a symmetrical acylation pattern is also present in *N. gonorrhoeae*,[76] *N. meningitidis*,[77] and most likely in *Xanthomonas*[75] strains.

As a representative of *heptaacyl* lipid, Figure 7C shows the structure of *S. minnesota* lipid A.[64,78] This lipid A is of particular historical note since all essential structural features of enterobacterial and other lipid As, including the β(1-6)-linked GlcN disaccharide,[79] the location of phosphate groups,[80] and the ester- and amide-bound 3-acyloxyacyl groups,[81,82] were discovered for the first time in the *S. minnesota* Re-mutant strain R595.[83] Two major species of *S. minnesota* lipid A have been recognized: a heptaacyl species differing from *E. coli* by the presence of an additional hexadecanoic acid (16:0) bound to the hydroxyl group of the 14:0(3-OH) residue amide linked to GlcN I and a hexaacyl species identical to *E. coli* lipid A (Figure 7A). Heptaacyl components have also been encountered in other lipid As such as *Proteus mirabilis*.[84] *S. minnesota* lipid A further differs from that of *E. coli* by the nonstoichiometric substitution of the backbone phosphate groups with Ara4N (bound to nonglycosidic phosphate of GlcN II) and 2-aminoethyl phosphate (bound to glycosidic phosphate of GlcN I) (for literature see Reference 64, but also Reference 85). Also in *S. minnesota* lipid A, acyl-deficient components are present.[86] Heptaacyl *S. minnesota* lipid A has been chemically synthesized.[50,78]

A *pentaacyl* lipid A is present in *P. aeruginosa* (Reference 87; Figure 7D). Here, the main lipid A species contains a total of five fatty acids and a minor hexaacyl species corresponding in structure to lipid A of *C. violaceum* (Figure 7B). The prominent pentaacyl component which constitutes approximately 75% of *P. aeruginosa* lipid A harbors three structural forms, all of which possess an identical β(1-6)-linked GlcN backbone with three primary 3-hydroxy fatty acids attached to positions 2, 2′ and 3′ (Figure 7D). These structural forms differ by the 3-O-acylation of each of the two amide-linked 12:0(3-OH) residues by the secondary acyl groups 12:0 or 12:0(2-OH), as indicated in Figure 7D by the dotted lines. Of the four conceivable structural types, the one harboring two 12:0(2-OH) residues has not been found.

In the major component of *B. fragilis* lipid A (Figure 8A) also, only five, comparatively long-chain and, in part, isobranched fatty acids are present[88] but in a location different from that in *P. aeruginosa* lipid A. The most striking difference to *E. coli* and other lipid A structures shown, however, concerns the findings that in *B. fragilis* lipid A the hydroxyl group at position 4′ of GlcN II is free, i.e., that the nonglycosidic phosphate group is absent.

A pentaacyl lipid A is also found in *Rhodobacter sphaeroides*[89] (previously termed *Rhodopseudomonas sphaeroides*. This lipid A has the same distribution of fatty acids as *B. fragilis* but, as a peculiar feature, it contains the unsaturated fatty acids *cis*-Δ^7-14:1 and 3-oxotetradecanoic acid [14:0(3-Oxo)], the former being present as a 3-acyloxytetradecanoic acid group amide bound to GlcN I and the latter being amide linked to GlcN I. Unsaturated and 3-keto fatty acids are rarely encountered in lipid A and the acylation pattern of *R. sphaeroides*, therefore, has to be considered as unusual. Also,

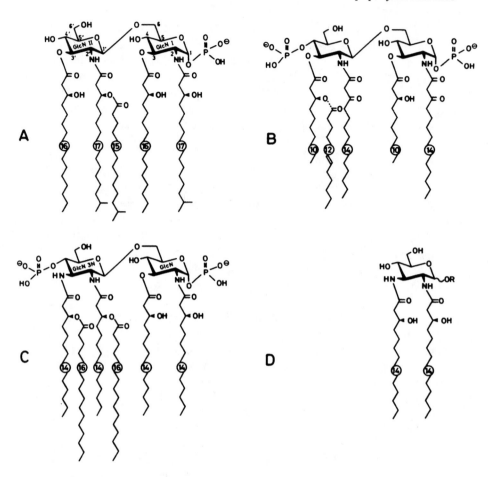

FIGURE 8. Chemical structure of major lipid A species of (A) *B. fragilis*, (B) *Rhodobacter capsulatus*, (C) *Campylobacter jejuni*, and (D) *R. viridis*. For details, compare legend to Figure 7 and text.

in *R. capsulatus*, a pentaacyl lipid A (Figure 8B) represents a main compo-nent.[90] As in *R. sphaeroides*, 14:0(3-Oxo) is present in amide linkage. Two residues of 10:0(3-OH) are ester linked, of which the one bound to GlcN II carries dodecenoic acid (12:1) in nonstoichiometric amounts at its 3-hydroxyl group. The location of *R. capsulatus* fatty acids, however, differs significantly from that of the other pentaacyl lipid A in that the only secondary fatty acid (12:1) is bound to the ester-linked 3-hydroxy fatty acid at position 3′. Another lipid A structure identified in *R. capsulatus* lacks 12:1 and, thus, represents a *tetraacyl* lipid A species.

2. Lipid A Containing a GlcN3N-GlcN Disaccharide

In recent years a number of lipid A have been encountered in which GlcN3N is present in addition to GlcN. As an example, Figure 8C shows the chemical structure of the major component of *C. jejuni* lipid A.[91] In this hexaacyl species, GlcN3N positionally replaces the GlcN II residue of the lipid A structures discussed earlier (compare Section 1). The hexosamine disaccharide is β(1-6)-linked, and the location of phosphoryl and acyl groups is similar to the *E. coli* lipid A structure (Figures 7A and 8C). It is possible that this type of hybrid backbone is also present in *Brucella* and in phototrophic bacteria such as the Chromatiaceae.[73] So far, lipid A structures in which the backbone disaccharide contains GlcN3N at the reducing and GlcN at the nonreducing position have not been identified.

3. Lipid A Containing a GlcN3N-GlcN3N Disaccharide

In *P. diminuta* lipid A, a GlcN3N disaccharide was shown to represent the lipid A backbone.[93] The nature of the glycosidic linkage of the disaccharide, however, has not been established. At position 4' of the nonreducing GlcN3N residue, a phosphoryl group is present, whereas the nature of the substituent at the glycosidic position has not been identified. Besides various 3-hydroxy fatty acids, 5,9-dihydroxytetradecanoic acid represents a major component of the acyl spectrum of *P. vesicularis* and *P. diminuta*.[94]

It is noteworthy that in *C. jejuni* lipid A (see Section 2) a minor fraction contains a β(1-6)-linked GlcN3N disaccharide.[91]

4. Lipid A Containing a GlcN3N Monosaccharide

As a constituent of lipid A, GlcN3N was first detected in *Rhodopseudomonas viridis*.[73,95] In this case, it is present as a nonphosphorylated but bis-N-acylated monosaccharide which carries the polysaccharide at position 6 (Figure 8D). A monosaccharidic GlcN3N residue also forms the lipid A backbone in *R. palustis*,[73,95] *P. fluorescens*,[96] and *Phenylobacter immobile*.[97] Lipid A containing only a GlcN monosaccharide have not been identified.

5. Lipid A Containing Other Backbones

Results of a recent study[98] have suggested that, in *Rhizobium trifolii*, D-glucosaminuronic acid (GlcNA) which carries amide-linked 27-hydroxyoctacosanoic acid [28:0(27-OH)] represents the lipid A backbond, and not a β(1-6)-linked GlcN disaccharide as proposed previously.[92] Unpublished studies in our own laboratory, however, were unable to confirm this. Instead our results indicate that 28:0(27-OH) is ester linked to GlcN. Whether this GlcN residue is part of a GlcN-disaccharide backbone remains to be investigated.

In *summary*, the above sections and the data presented in Chapter 2 show that much has been learned in recent years about the primary structure of lipid A, notably, of human pathogenic bacteria. With a few exceptions, structural studies on lipid A have been performed on R-mutant LPS preparations. Although it is commonly assumed that these lipid A structures are also present in S-form LPS, this assumption has yet to be rigorously documented. For example, in enterobacterial S-form LPS, the β(1-6)-linked D-GlcN disaccharide forms, as in R-form LPS, the lipid A backbone[99] but the latter was shown to greatly differ in its acylation pattern in S- and R-form preparations.[100] Systematic studies on the lipid A component of enterobacterial S-form LPS have not, to date, been performed. In this respect and in view of the enormous variety of Gram-negative bacteria, our knowledge on the lipid A architecture remains limited. It is, therefore, reasonable to predict that further studies on the chemical structure of lipid A of different bacterial origin will not only reveal new structural features but will also contribute to our understanding of the taxonomic position and evolution of bacterial genera and families.[28,73]

B. THREE-DIMENSIONAL PHYSICAL STRUCTURE

From their primary chemical structure comprizing polar (hydrophilic, e.g., Kdo, phosphate) groups as well as apolar (hydrophobic, i.e., acyl residues) regions, it is obvious that LPS and free lipid A are amphiphilic molecules.[101,102] On the one hand, this characteristic is a prerequisite for the incorporation of LPS in the bacterial outer membrane and LPS, thus, influences the physical properties of this membrane as a permeation barrier. In aqueous solution, on the other hand, above a critical concentration, LPS as an amphiphile does not exist in the form of individual molecules, but tends to form aggregate structures. It is well known from studies on other amphiphiles such as phospholipids, which have been extensively investigated for many years, that such three-dimensional supramolecular configurations can form micellar, lamellar, hexagonal, and nonlamellar cubic, as well as inverted hexagonal, structures.[103] Which particular physical structure is formed depends on intrinsic chemical parameters, i.e., the primary chemical structure, and extrinsic factors, i.e., ambient conditions such as temperature, pH, and concentration of divalent cations, in particular Mg^{2+} and Ca^{2+}. Increase of the temperature within one type of structure may cause a phase transition at the phase transition temperature T_c from the gel (β) to the liquid-crystalline (α) state, reflecting an increase in the fluidity of the amphiphile's hydrophobic region, i.e., of the acyl groups. Frequently, this $\beta \leftrightarrow \alpha$ phase transition is associated with a transition between different three-dimensional structures. There are experimental data deriving from small-angle X-ray measurements that LPS and free lipid A as structurally complex amphiphiles also adopt such supramolecular structures.[104-107] However, they exhibit a complex structural polymorphism depending on temperature, water content, and divalent cation concentrations, as was shown recently for free lipid A of *S. minnesota* and *E. coli.*[108] Under physiological conditions, different supramolecular configurations coexist among which cubic and inverted phases play a dominant role. Furthermore, the fluidity of the fatty acid chains at physiological temperature is an important parameter with respect to the expression of biological activities. This follows from the observation that free lipid A and LPS preparations with low-phase transition temperature are, in general, endotoxically more active than compounds exhibiting high T_c values.[12,28] A more detailed review on the three-dimensional physical structure of endotoxins and its correlation to biological activity is presented in Chapter 9.

C. BIOLOGICAL ACTIVITY AND FUNCTION

As postulated since the early 1950s[109] and proven during the last decade,[110] lipid A contains the endotoxic principle of LPS and is responsible for the induction of the various classical endotoxin effects such as fever, local Shwartzman reactivity, and lethal toxicity. Today, lipid A is known to represent the general immunomodulating center of LPS.

In addition, the lipid A component

- Carries common LPS epitopes, although in a cryptic form[29,111]
- Activates the classical but antibody-independent complement cascade[112,113]
- Binds to serum proteins such as high- or low-density lipoproteins[114,116] and LPS binding protein (LBP; References 117 and 118)
- Interacts with recognition proteins (receptors) exposed on monocytes/macrophages and other host cells[119]
- Is essential for bacterial viability, that is, for the function of the outer membrane in growth and survival of Gram-negative bacteria[40]

D. STRUCTURE-ACTIVITY RELATIONSHIPS

The successful elucidation of the chemical and, to a certain extent, physical structure of lipid A, as well as the chemical synthesis of corresponding (partial) structures, has provided both a theoretical and an experimental basis to establish quantitative relationships between structure and activity at the molecular level. In the following paragraphs, various aspects of the relationship of primary and tertiary structure to the biological activity of lipid A will be discussed in the fields of humoral immunoreactivity and endotoxicity.

The *immunoreactive* properties of lipid A have been studied with polyclonal antisera and monoclonal antibodies. Using synthetic lipid A and partial structures as antigens, our laboratory has been able to define five distinct antigenic determinants, all of which are localized to the hydrophilic lipid A backbone region.[29,120,121] One epitope comprises the acylated and 1,4'-bis-phosphorylated β(1-6)-linked-D-glucosamine disaccharide (Figure 7A), a second the 4'-monophosphorylated, and a third the 1-monophosphorylated disaccharide. The other two immunoreactive determinants are located in acylated D-hexosamine 1- and 4-phosphate monosaccharides, respectively. For the latter, the specificity is also expressed by phosphorylated preparations differing in the nature of the acylated D-*gluco*-configurated monosaccharide (Glc, GlcN, Glc3N, GlcN3N). The specificity of the lipid A antibodies is independent of the acylation pattern, that is, fatty acids are not part of the epitopes, but due to their important contribution to the conformational state of lipid A (Section B), acyl groups may modulate the exposure of lipid A determinants.[121] Phosphate groups, on the other hand, are essential in lipid A epitope expression; however, it remains to be established whether these phosphate groups are actually part of the antigenic determinants or whether they generate, perhaps due to their negative charge, a particular immunoreactive conformational state of lipid A.

The characterization of lipid A epitopes is of great significance in view of the present developments in the field of anti-endotoxic antibodies and their application in the treatment of a Gram-negative septicemia.[122-126] A human and a murine monoclonal antibody (HA-1A and E5), both of the IgM class, have recently been shown to be of therapeutic efficacy[127,128] and their immunoreactivity has been described as lipid A specific, that is, these antibodies are described to cross-react with LPS of different bacterial origin. Crossreactivity was determined in the ELISA assay system which, however, unless certain precautions are taken, may lead to false results simulating crossreactivity.[129] The five anti-lipid A antibody specificities described above, however, do not exhibit crossreactivity with LPS, i.e., with lipid A carrying the polysaccharide portion. On the other hand, they do cross-react with a variety of polysaccharide-deprived free lipid A preparations, provided these contain a bisphosphorylated D-hexosamine disaccharide. It could be concluded that, in LPS, the lipid A epitopes are either not expressed (i.e., free lipid A may represent a neoantigen having determinants not present in LPS) or not exposed (i.e., Kdo or other LPS constituents may sterically shield the binding site for lipid A antibodies). In view of these conflicting experimental data on the one hand and the potential biomedical significance of crossreactive antibodies on the other, further studies are highly indicated which should aim at a detailed characterization of the structural and conformational epitopes of lipid A, both isolated from LPS and in the intact LPS molecule.

It is now established that *endotoxicity* of LPS is dependent on, and mediated by, the lipid A component and great progress has been made in defining the structural prerequisites for bioactivity of lipid A. These structural parameters were elucidated by analyzing the biological activity of synthetic lipid A and, notably, partial structures in both *in vivo* (e.g., pyrogenicity) and *in vitro* systems [e.g., tumor necrosis factor α (TNF) or interleukin-1 (IL-1) production in mononuclear cells]. Collectively, the results of various studies (for a detailed discussion, compare Chapter 5 and References 28, 29, 130, and 131) show that full endotoxic activity requires a molecule containing *two* hexosamine residues (i.e., a hexosamine disaccharide), *two* phosphoryl groups, and *six* fatty acids in a defined location as it is present in *E. coli* lipid A (Figure 7A). As far as the hexosamine groups are concerned, only synthetic D-*gluco*-configurated residues were biologically analyzed in the form of either β(1-6)-linked disaccharides or monosaccharides. Monosaccharidic hexosamine derivatives were shown to completely lack endotoxic activity *in vivo*. Of the disaccharide compounds (i.e., synthetic *E. coli* lipid A partial structures) those with only one phosphoryl group were significantly less active (up to 100-fold) than bisphosphorylated compounds. Fatty acids were found to be of great importance in the expression of lipid A bioactivity and it is now established that their number and location, and most likely their chain length and stereochemistry, play significant roles. As an example, preparations with an identical bisphosphorylated backbone as present in *E. coli* lipid A (Figure 7) but with only five (preparations LA-20-PP and LA-21-PP) and four (preparation LA-14-PP) fatty acids are 10-fold[132] and 100-fold,[133] respectively, less pyrogenic than hexaacyl *E. coli* lipid A. A compound with only two amide-bound acyl groups (preparation LA-19-PP, resembling alkali-treated lipid A) does not induce fever. A similar hierarchy of the biological potency of synthetic lipid A and partial structures was identified when assessed by IL-1 and TNF induction in human peripheral mononuclear cells.[134,135]

In the *in vitro* system of IL-1 and TNF production in human mononuclear cells, it was found that some inactive compounds were potent inhibitors (antagonists) of LPS-stimulatory activity.[134,136,137] The potency of different compounds to antagonize LPS-induced mediator production also appears to be governed by strict structural parameters. As an example, in the human system the tetraacyl compound LA-14-PP is a strong inhibitor of IL-1 and TNF induction, whereas the bisacyl compound LA-19-PP is a weak antagonist, indicating that antagonistic activity is influenced by the *number of acyl chains* present. If the human and the murine system are compared, a dependency of activity *on the chain lengths of fatty acids* or their different *nature* is observed. In human cells the tetraacyl compound LA-14-PP, harboring four fatty acids [14:0(3-OH)], lacks IL-1- and TNF-inducing capacity, but exhibits strong antagonistic activity.[134-137] In murine cells, the same compound shows reduced but still significant stimulatory or agonistic activity,[133,138] i.e., poor antagonist properties. However, a major component of lipid A of *Rhodobacter capsulatus* (compare Section A.1) also having four fatty acids of which two possess shorter acyl chains [ester-bound 10:0(3-OH)], are not stimulatory in the murine system, but serve as excellent antagonists *in vitro* and *in vivo*.[138] As discussed (Section A.1), *R. capsulatus* contains two amide-bound 14:0(3-Oxo) residues instead of 14:0(3-OH), and it is also possible that (in addition to or instead of the different acyl chain length) this particular feature is important for the lack of agonistic and the expression of antagonistic activity (compare also Reference 139).

In addition, compounds with three acyl groups [14:0(3-OH)] have been investigated.[67] These preparations are partial structures of synthetic preparation LA-14-PP in that they lack the 4′-phosphoryl and one acyl group. It was found that a compound carrying two acyl groups at GlcN II (positions 2′ and 3′, Figure 7A) and one at GlcN I (position 2, Figure 7A) induced TNF release in mouse macrophages, whereas the preparation with two acyl groups at GlcN I and one at GlcN II was inactive.[67] This latter compound, however, proved to be a potent antagonist in LPS-induced TNF release, suggesting that the bisacylated GlcN I unit of lipid A is important for onistic activity. In fact, the monosaccharidic partial structure lipid X, corresponding to bisacylated GlcN I of lipid A (compare Chapters 2 and 3), is completely devoid of endotoxicity, but has been reported to exhibit significant antagonistic properties *in vitro*[67,136] and *in vivo*.[140,141]

In *summary*, these results show that lipid A manifesting a very unique primary structure having the constituent D-*gluco*-hexosaminyl, phosphoryl, and acyl groups in a defined location (as in *E. coli* lipid A) is necessary for full expression of endotoxic activity. Partial structures lacking only one of these constituents, molecules containing different although chemically related constituents, or analogs with a different arrangement of constituents are either much less endotoxically active or not endotoxic. Such inactive preparations, on the other hand, may strongly antagonize LPS or lipid A effects *in vivo* or *in vitro*. Both agonistic activity and antagonistic properties of lipid A are linked to unique primary structures which, in turn, are likely to determine peculiar conformations and supramolecular arrangements. The presently available evidence favors the view that, for the expression of endotoxic activity, a unique conformation and a particular supramolecular structure allowing, at least, partial melting of acyl chains at the physiological temperature are required.[28,130] Whether a single lipid A molecule[142] being in a defined conformation or a supramolecular lipid A complex determines endotoxicity remains to be shown. Results of our laboratory are compatible with both models and allow the conclusion that neither micellar, lamellar, nor inverted hexagonal or cubic arrangements *per se* are responsible for bioactivity. We have, however, hypothesized that the ability of *E. coli* lipid A to reversibly adopt lamellar and inverted conformations may be related to its potent endotoxic activity.[130]

It is obvious that the analysis of relationships between the chemical and physical structure of lipid A to endotoxic potency or antagonistic activity will represent a field of important and fruitful research in the years to come. Knowledge of the bioactive minimal determinants of the primary structure as well as of the conformation of lipid A is a necessity to understand, at the molecular level, its stimulatory activity and to make use of the antagonistic properties of partial structures and analogs. In such studies, the modulatory role of Kdo (Section VI.B) must also be considered and elucidated. This type of research will bring together biology and synthetic chemistry and may ultimately provide a rational basis for the development of molecular strategies for the prevention and treatment of endotoxinemic and septicemic states.

IX. CONCLUDING REMARKS

For almost a century lipopolysaccharides have fascinated researchers of various scientific disciplines. It was through their efforts that endotoxin was brought out of the dark of an undefined mysterious bioactive material to the light of a molecule well defined in genetic, biological, and, notably, structural terms. The elucidation of the primary structure of LPS was an important event for progress in endotoxin research in general and for the understanding of the molecular mechanisms involved in endotoxic activity in particular. It was during the last 4 decades that the chemical structure of the LPS region O-specific chain, outer core, inner core, and lipid A were elucidated, first for Enterobacteria and later for many other nonenterobacterial genera and families. This type of analytical research is still going on in many laboratories worldwide, since various aspects of the primary LPS structure remained obscure.

With progress made in analyzing the primary LPS structure, synthetic chemistry entered the field, and partial structures of the macromolecule LPS were prepared based on the results of analytical chemistry. Thus, O-specific oligosaccharides, partial structures of the outer and inner core, and lipid A have been synthesized, and many examples exist where both the bacterial and synthetic structure have been tested comparatively and found to exhibit identical biological activity. The highlight of this development will be the first synthesis of a mature LPS molecule, which is to be accomplished in the near future. Synthetic LPS partial structures are undoubtedly important as reference compounds to verify the structural results of analytical chemistry. They proved, however, to be of even greater value for studies aiming at the elucidation of relationships between chemical structure and biological activity. For such studies to be successful, chemical pure and homogenous, i.e., well-defined, structures are indispensable, since laborious purification procedures have to be applied to obtain such structures from bacterial LPS. The use of synthetic partial structures will also allow the identification of LPS antibody epitopes, receptors, complement-binding sites, substrate structures of enzymes, and determinants of endotoxicity, bacterial pathogenicity, and membrane-forming properties. Such studies are presently performed and they will occupy various research groups for many years to come.

Synthetic partial structures as well as natural LPS components have just been recognized as being active in inhibiting harmful endotoxin effects *in vivo* and *in vitro*. The elucidation of the structural prerequisites of LPS antagonistic activity will be important, as will be the development of optimized nontoxic LPS partial structures and new structures, i.e., LPS analogs, which possibly act as even stronger endotoxin antagonists. Quantitative structure-activity relationship analysis may also lead to lipid A-related drugs of low toxicity but high immunostimulatory potency. Lipid A and defined partial structures will also serve as structures to define the physicochemical properties and conformational states of toxic (agonistic) and nontoxic (antagonistic) structures. The final aim of these efforts will be to chemically and physically define the structure and conformation of the endotoxically active LPS domain and the humoral and cellular receptors to which this domain specifically binds. The interaction between these structures probably represents the central event of endotoxicity, and its molecular characterization will be the major challenge in the field of molecular biochemistry of LPS in years to come.

ACKNOWLEDGMENTS

The financial support of the Fonds der Chemie (Ernst Th. Rietschel) is greatly appreciated. We thank M. Lohs, F. Richter, B. Köhler, and I. Stegelmann for preparing drawings and photographs and I. Bendt for excellent secretarial assistance. We also thank E. Coats and O. Lüderitz for critically reading this manuscript and for their helpful suggestions.

REFERENCES

1. **Lüderitz, O., Freudenberg, M. A., Galanos, C., Lehman, V., Rietschel, E. Th., and Shaw, D. H.**, Lipopolysaccharides of Gram-negative bacteria, in *Microbial Membrane Lipids*, Vol. 17, Razin, S. and Rottem, S., Eds., Academic Press, New York, 1982, 79.
2. **Griffis, J. M., Schneider, H., Mandrell, R. E., Yamasaki, R., Jarvis, G. A., Kim, J. J., Gibson, B. W., Hamadeh, R., and Apicella, M. A.**, Lipooligosaccharides: the principal glycolipids of the neisserial outer membrane, *Rev. Infect. Dis.*, 10, 287, 1988.
3. **Rietschel, E. Th., Galanos, C., Lüderitz, O., and Westphal, O.**, Chemical structure, physiological function and biological activity of lipopolysaccharides and their lipid A component, in *Immunopharmacology and the Regulation of Leukocyte Function*, Webb, D., Ed., Marcel Dekker, New York, 1982, 183.
4. **Mäkelä, P. H. and Stocker, B. A. D.**, Genetics of lipopolysaccharide, in *Chemistry of Endotoxin*, Rietschel, E. Th., Ed., (*Handbook of Endotoxin*, Vol. 1, Proctor, R. A., Ed.), Elsevier Science, Amsterdam, 1984, 59.
5. **Lindberg, A. A., Wollin, R., Bruse, G., Ekwall, E., and Svenson, S. B.**, Immunology and immunochemistry of synthetic and semisynthetic *Salmonella* O-antigen specific glycoconjugates, *ACS Symp. Ser.*, 231, 83, 1983.
6. **Wolpert, J. S. and Albersheim, P.**, Host symbiont interactions. I. The lectins of legumes interact with the O-antigen containing lipopolysaccharides of their symbiont rhizobia, *Biochem. Biophys. Res. Commun.*, 70, 729, 1976.
7. **Jann, K. and Jann, B.**, Structure and biosynthesis of O-antigens, in *Chemistry of Endotoxin*, Rietschel, E. Th., Ed., (*Handbook of Endotoxin*, Vol. 1, Proctor, R. A., Ed.), Elsevier Science, Amsterdam, 1984, 138.
8. **Grossman, N. and Leive, L.**, Complement activation via the alternative pathway by purified *Salmonella* lipopolysaccharide is affected by its structure but not its O-antigen length, *J. Immunol.*, 132, 376, 1984.
9. **Mäkelä, P. H., Hovi, M., Saxen, H., Muotiala, A., and Rhen, M.**, Role of LPS in the pathogenesis of salmonellosis, in *Cellular and Molecular Aspects of Endotoxin Reactions*, Nowotny, A., Spitzer, J. Y., and Ziegler, E. J., Eds., (Endotoxin Research Series, Vol. 1, Elsevier Science, Amsterdam, 1990, 537.
10. **Nowotny, A.**, Review of the molecular requirements of endotoxic actions, *Rev. Infect. Dis.*, 9, 503, 1987.
11. **Morrison, D. C., Vukajlovich, S. W., Goodman, S. A., and Wollenweber, H.-W.**, Regulation of lipopolysaccharide biologic activity by polysaccharide, in *The Pathogenesis of Bacterial Infections (Bayer Symp. VIII)*, Jackson, G. G. and Thomas, H., Eds., Springer-Verlag, Berlin, 1985, 68.

12. **Lüderitz, Th., Brandenburg, K., Seydel, U., Roth, A., Galanos, C., and Rietschel, E. Th.**, Structural and physicochemical requirements of endotoxins for the activation of arachidonic acid metabolism in mouse peritoneal macrophages *in vitro*, *Eur. J. Biochem.*, 179, 11, 1989.

13. **Bock, K., Meldal, M., Bundle, D. R., Iversen, T., Pinto, B. M., Garegg, P. J., Kvanström, I., Norberg, T., Lindberg, A. A., and Svenson, S. B.**, The conformation of *Salmonella* O-antigenic oligosaccharides of serogroups A, B, and D_1 inferred from ^{1}H- and ^{13}C-nuclear magnetic resonance spectroscopy, *Carbohydr. Res.*, 130, 35, 1984.

14. **Bundle, D. R. and Josephson, S.**, Artificial carbohydrate antigens: the synthesis of a tetrasaccharide hapten, a *Shigella flexneri* O-antigen repeating unit, *Carbohydr. Res.*, 80, 75, 1980.

15. **Kochetkov, N. K., Betaneli, V. I., Ovchinnikov, M. V., Backinovsky, L. V.**, Synthesis of the O-antigenic polysaccharide of *Salmonella newington* and of its analogue differing in configuration at the only glycosidic centre, *Tetrahedron*, 37, 149, 1981.

16. **Paulsen, H. und Bünsch, H.**, Bausteine von Oligosacchariden. XXXII. Synthese der verzweigten Pentasaccharid-Einheit der O-spezifischen Seitenkette des Lipopolysaccharides von *Shigella dysenteriae*, *Chem. Ber.*, 114, 3126, 1981.

17. **Jansson, P. E., Lindberg, A. A., Lindberg, B., and Wollin, R.**, Structural studies on the hexose region of the core in lipopolysaccharides from *Enterobacteriaceae*, *Eur. J. Biochem.*, 115, 571, 1981.

18. **Radziejewska-Lebrecht, J. and Mayer, H.**, The core region of *Proteus mirabilis* R110/1959 lipopolysaccharide, *Eur. J. Biochem.*, 183, 573, 1989.

19. **Romanowska, E., Gamian, A., Lugowski, C., Romanowska, A., Dabrowski, J., Hauck, M., Opferkuch, H. J., and van der Lieth, C.-W.**, Structure elucidation of the core regions from *Citrobacter* 04 and 036 lipopolysaccharides by chemical and enzymatic methods, gas chromatography/mass spectrometry, and NMR spectroscopy at 500 MHz, *Biochemistry*, 27, 4153, 1988.

20. **Holst, O., Zähringer, U., Brade, H., and Zamojski, A.**, Structural analysis of the heptose/hexose region of the lipopolysaccharide from *Escherichia coli* K-12 strain W3100, *Carbohydr. Res.*, 215, 323, 1991.

21. **Jansson, P.-E., Wollin, R., Bruse, G. W., and Lindberg, A. A.**, The conformation of core oligosaccharides from *Escherichia coli* and *Salmonella typhimurium* lipopolysaccharides as prediced by semi-empirical calculations, *J. Mol. Recogn.*, 2, 25, 1989.

22. **Lindberg, A. A.**, Bacterial surface carbohydrates and bacteriophage adsorption, in *Surface Carbohydrates of the Procaryotic Cell*, Sutherland, I. W., Ed., Academic Press, New York, 1977, 289.

23. **Lehmann, V., Streck, H., Minner, I., Krammer, P. H., and Ruschmann, E.**, Selection of bacterial mutants from *Salmonella* specifically recognizing determinants on the cell surface of activated T-lymphocytes, *Eur. J. Immunol.*, 10, 685, 1980.

24. **Jirillo, E., De Simone, C., Covelli, V., Kiyono, H., McGhee, J. R., and Antonaci, S.**, LPS-mediated triggering of T-lymphocytes in immune response against gram-negative bacteria, *Adv. Exp. Med. Biol.*, 256, 417, 1990.

25. **Tacken, A., Rietschel, E. Th., and Brade, H.**, Methylation of the heptose/3-deoxy-D-manno-2-octulosonic acid region (inner core) of the lipopolysaccharide from *Salmonella minnesota* rough mutants, *Carbohydr. Res.*, 149, 279, 1986.

26. **Holst, O., Röhrscheidt-Andrzejewski, E., Brade, H., and Charon, D.**, Isolation and characterization of 3-deoxy-D-*manno*-2-octulopyranosonate 7-(2-aminoethyl phosphate) from the inner core region of *Escherichia coli* K-12 and *Salmonella minnesota* lipopolysaccharides, *Carbohydr. Res.*, 204, 93, 1990.

27. **Kaca, W., de Jongh-Leuvenink, J., Zähringer, U., Rietschel, E. Th., Brade, H., Verhoef, J., and Sinnwell, V.**, Isolation and chemical analysis of 7-O-(2-amino-2-deoxy-α-D-*gluco*-pyranosyl)-L-*glycero*-D-*manno*-heptose as a constituent of the lipopolysaccharides of the UDP-galactose epimerase-less mutant J-5 of *Escherichia coli* and *Vibrio cholerae*, *Carbohydr. Res.*, 179, 289, 1988.

28. **Rietschel, E. Th., Brade, L., Schade, U. F., Seydel, U., Zähringer, U., and Brade, H.,** Bacterial endotoxins: properties and structure of biologically active domains, in *Surface Structures of Microorganisms and their Interaction with the Mammalian Host,* Schrinner, E., Richmond, M., Seibert, G., and Schwarz, U., Eds., Verlag Chemie, Weinheim, Germany, 1988, 1.

29. **Brade, H., Brade, L., and Rietschel, E. Th.,** Structure-activity relationships of bacterial lipopolysaccharides (endotoxins). Current and future aspects, *Zentralbl. Bakteriol. Hyg.,* A268, 151, 1988.

30. **Kawahara, K., Brade, H., Rietschel, E. Th., and Zähringer, U.,** Studies on the chemical structure of the core-lipid A region of the lipopolysaccharide of *Acinetobacter calcoaceticus* NCTC 10305. Detection of a new 2-octulosonic acid interlinking the core oligosaccharide and lipid A component, *Eur. J. Biochem.,* 163, 489, 1987.

31. **Brade, L., Kosma, P., Appelmelk, B. J., Paulsen, H., and Brade, H.,** Use of synthetic antigens to determine the epitope specificities of monoclonal antibodies against the 3-deoxy-D-*manno*-octulosonate region of bacterial lipopolysaccharide, *Infect. Immun.,* 55, 462, 1987.

32. **Rozalski, A., Brade, L., Kosma, P., Appelmelk, B. J., Krogman, C., and Brade, H.,** Epitope specificities of murine monoclonal and rabbit polyclonal antibodies against enterobacterial lipopolysaccharides of the Re chemotype, *Infect. Immun.,* 57, 2645, 1989.

33. **Rozalski, A., Brade, L., Kuhn, H.-M., Brade, H., Kosma, P., Appelmelk, B. J., Kusumoto, S., and Paulsen, H.,** Determination of the epitope specificity of monoclonal antibodies against the inner core region of bacterial lipopolysaccharides by use of 3-deoxy-D-*manno*-octulosonate-containing synthetic antigens, *Carbohydr. Res.,* 193, 257, 1989.

34. **Brade, L. and Brade, H.,** A 28,000 dalton protein of normal mouse serum binds specifically to the inner core region of bacterial lipopolysaccharide, *Infect. Immun.,* 50, 687, 1985.

35. **Haeffner-Cavaillon, N., Caroff, M., and Cavaillon, J.-M.,** Interleukin-1 induction by lipopolysaccharide: structural requirement of the 3-deoxy-D-*manno*-2-octulosonic acid (Kdo), *Mol. Immunol.,* 26, 485, 1989.

36. **Vaara, M. and Nikaido, H.,** Molecular organization of bacterial outer membrane, in *Chemistry of Endotoxin*, Rietschel, E. Th., Ed., (*Handbook of Endotoxin*, Vol. 1, Proctor, R. A., Ed.), Elsevier Science, Amsterdam, 1984, 1.

37. **McCabe, W. R., Bruins, S. C., Craven, D. E., and Johns, M.,** Cross-reative antigens: their potential for immunization-induced immunity to gram-negative bacteria, *J. Infect. Dis.,* 136, 161, 1977.

38. **Ziegler, E. M., McCutchan, J. A., Fierer, J., Glauser, M. P., Sadoff, J. C., Douglas, H., and Braude, A. I.,** Treatment of gram-negative bacteriemia and shock with human antiserum to a mutant *Escherichia coli, N. Engl. J. Med.,* 307, 1225, 1982.

39. **Silva, A. T., Appelmelk, B. J., Buurman, W. A., Bayston, K. F., and Cohen, J.,** Monoclonal antibody to endotoxin core protects mice from *Escherichia coli* sepsis by a mechanism independent of tumor necrosis factor and interleukin-6, *J. Infect. Dis.,* 162, 454, 1990.

40. **Osborn, M. J.,** Biosynthesis and assembly of lipopolysaccharide of the outer membrane, in *Bacterial Outer Membranes, Biogenesis and Functions*, Inouye, M., Ed., John Wiley & Sons, New York, 1979, 15.

41. **Brade, H., Brade, L., and Nano, F. E.,** Chemical and serological investigations on the genus-specific lipopolysaccharide epitope of *Chlamydia, Proc. Natl. Acad. Sci. U.S.A.,* 84, 2508, 1987.

42. **Brade, H. and Rietschel, E. Th.,** α-2→4-Interlinked 3-deoxy-D-*manno*-octulosonic acid disaccharide. A common constituent of enterobacterial lipopolysaccharides, *Eur. J. Biochem.,* 145, 231, 1984.

43. **Helander, I., Lindner, B., Brade, H., Altmann, K., Lindberg, B. B., Rietschel, E. Th., and Zähringer, U.**, Chemical structure of the lipopolysaccharide of *Haemophilus influenzae* strain I-69 Rd⁻/b⁺. Description of a novel deep rough chemotype, *Eur. J. Biochem.*, 177, 483, 1988.
44. **Hammond, S. M., Claesson, A., Jansson, A. M., Larsson, L.-G., Pring, B. G., Town, C. M., and Ekström, B.**, A new class of synthetic antibacterials acting on lipopolysaccharide biosynthesis, *Nature*, 327, 730, 1987.
45. **Goldman, R. C., Kohlbrenner, W. F., Lartey, P., and Pernet, A. G.**, Antibacterial agents specifically inhibiting lipopolysaccharide synthesis, *Nature*, 329, 162, 1987.
46. **Rietschel, E. Th., Galanos, C., Tanaka, A., Ruschmann, E., Lüderitz, O., and Westphal, O.**, Biological activities of chemically modified endotoxins, *Eur. J. Biochem.*, 22, 218, 1971.
47. **Paulsen, H. and Schüller, M.**, Synthese von Kdo-haltigen Lipoid A Analoga, *Liebigs Ann. Chem.*, p. 249, 1987.
48. **Kosma, P., Schulz, G., and Brade, H.**, Synthesis of a trisaccharide of 3-deoxy-D-*manno*-2-octulopyranosylonic acid (Kdo) residues related to the genus-specific lipopolysaccharide epitope of *Chlamydia*, *Carbohydr. Res.*, 183, 183, 1988.
49. **Unger, F. M.**, The chemistry and biological significance of 3-deoxy-D-*manno*-2-octulosonic acid (Kdo), *Adv. Carbohydr. Chem. Biochem.*, 38, 323, 1981.
50. **Kusumoto, S., Kusunose, N., Imoto, M., Shimamoto, T., Kamikawa, T., Takada, H., Kotani, S., Rietschel, E. Th., and Shiba, T.**, Synthesis and biological function of bacterial endotoxin, *Pure Appl. Chem.*, 61, 461, 1989.
51. **Moxon, E. R.**, Antigen expression influencing tissue invasion of *Haemophilus influenzae* type b, in *Bayer-Symposium VIII. The Pathogenesis of Bacterial Infections*, Jackson, G. G. and Thomas, H., Eds., Springer-Verlag, Berlin, 1985, 17.
52. **Patrick, C. C., Pelzel, S. E., Miller, E. E., Haanes-Fritz, E., Radolf, J. D., Gulig, P. A., McCracken, G. H., Jr., and Hansen, E. J.**, Antigenic evidence for simultaneous expression of two different lipooligosaccharides by some strains of *Haemophilus influenzae* type b, *Infect. Immun.*, 57, 1971, 1989.
53. **Caroff, M., Chaby, R., Karibian, D., Perry, J., Deprun, C., and Szabo, L.**, Variations in the carbohydrate regions of *Bordetella pertussis: electrophoretic, serological, and structural features*, *J. Bacteriol.*, 172, 1121, 1990.
54. **Hisatsune, K., Kondo, S., Iguchi, T., Yamamoto, F., Inaguma, M., Kokubo, S., and Arai, S.**, Lipopolysaccharide of the family *Vibrionaceae*, in *Bacterial Endotoxin: Chemical, Biological and Clinical Aspects*, Homma, J. Y., Kanegasaki, S., Lüderitz, O., Shiba, T., and Westphal, O., Eds., Verlag Chemie, Weinheim, Germany, 1984, 187.
55. **Weintraub, A., Zähringer, U., and Lindberg, A. A.**, Structural studies on the polysaccharide part of the cell wall lipopolysaccharide from *Bacteroides fragilis* NCTC 9343, *Eur. J. Biochem.*, 151, 657, 1985.
56. **Lindberg, A. A., Weintraub, A., Zähringer, U., and Rietschel, E. Th.**, Structure-activity relationships in lipopolysaccharides of *Bacteroides fragilis*, *Rev. Infect. Dis.*, 12, 133, 1990.
57. **Hitchcock, P., Leive, L., Mäkelä, P. H., Rietschel, E. Th., Strittmatter, W., and Morrison, D.**, Lipopolysaccharide nomenclature — past, present and future, *J. Bacteriol.*, 166, 699, 1986.
58. **Michon, F., Beurret, M., Gamian, A., Brisson, J.-R., and Jennings, H. J.**, Structure of the L5 lipopolysaccharide core oligosaccharides of *Neisseria meningitis*, *J. Biol. Chem.*, 265, 7243, 1990.
59. **Kondo, S., Zähringer, U., Rietschel, E. Th., and Hisatsune, K.**, Isolation and identification of 3-deoxy-D-*threo*-hexulosonic acid as a constituent of the lipopolysaccharide of *Vibrio parahaemolyticus* serotypes 07 and 012, *Carbohydr. Res.*, 188, 97, 1989.

60. **Kondo, S., Zähringer, U., Seydel, U., Hisatsune, K., and Rietschel, E. Th.**, Chemical structure of the carbohydrate backbone of *Vibrio parahaemolyticus* serotype 012 lipopolysaccharide, *Eur. J. Biochem.*, 200, 689, 1991.

61. **Mandrell, R. E., Griffiss, J. McL., and Macher, B. A.**, Lipooligosaccharides (LOS) of *Neisseria gonorrhoeae* and *Neisseria meningitidis* have components that are immunochemically similar to precursors of human blood group antigens, *J. Exp. Med.*, 168, 107, 1988.

62. **Mandrell, R. E., Lesse, A. J., Sugai, J. V., Shero, M., Griffis, J. McL., Cole, J. A., Parsons, N. J., Smith, H., Morse, S. A., and Apicella, M. A.**, *In vitro* and *in vivo* modification of *Neisseria gonorrhoeae* lipopolysaccharide epitope structure by sialylation, *J. Exp. Med.*, 171, 1649, 1990.

63. **Smith, H.**, Pathogenicity and the microbe *in vivo*, *J. Gen. Microbiol.*, 136, 377, 1990.

64. **Rietschel, E. Th., Wollenweber, H.-W., Brade, H., Zähringer, U., Lindner, B., Seydel, U., Bradaczek, H., Barnickel, G., Labischinski, H., and Giesbrecht, P.**, Structure and conformation of the lipid A component of lipopolysaccharides, in *Chemistry of Endotoxin*, Rietschel, E. Th., Ed., (*Handbook of Endotoxin*, Vol. 1, Proctor, R., Ed.), Elsevier Science, Amsterdam, 1984, 187.

65. **Raetz, C. R. H.**, Structure and biosynthesis of lipid A, in *Escherichia coli and Salmonella typhimurium. Cellular and Molecular Biology*, Neidhardt, C., Ingraham, J. L., Brooks Low, K., Magasanik, B., Schaechter, M., and Umbarger, H. E., Eds., American Society for Microbiology, Washington, DC, 1987, 498.

66. **Goldman, R. C., Doran, C. C., Kadam, S. K., and Capobianco, J. O.**, Lipid A precusor from *Pseudomonas aeruginosa* is completely acylated prior to addition of 3-deoxy-D-*manno*-octulosonate, *J. Biol. Chem.*, 263, 5217, 1988.

67. **Stütz, P. L., Aschauer, H., Hildebrandt, J., Lam, C., Loibner, H., Macher, I., Scholz, D., Schütze, E., and Vyplel, H.**, Chemical synthesis of endotoxin analogs and some structure activity relationships, in *Cellular and Molecular Aspects of Endotoxin Reactions*, Nowotny, A., Spitzer, J. Y., and Ziegler, E. J., Eds., (Endotoxin Research Series, Vol. 1, Nowotny, A., Ed., Elsevier Science, Amsterdam, 1990, 129.

68. **Rosner, M. R., Khorana, H. G., and Satterthwait, A. C.**, The structure of lipopolysaccharide from a heptose-less mutant of *Escherichia coli* K-12. II. The application of ^{31}P NMR spectroscopy, *J. Biol. Chem.*, 254, 5918, 1979.

69. **Zähringer, U., Lindner, B., Seydel, U., Rietschel, E. Th., Naoki, H., Unger, F. M., Imoto, M., Kusumoto, S., and Shiba, T.**, Structure of de-O-acylated lipopolysaccharide from the *Escherichia coli* Re mutant strain F515, *Tetrahedron Lett.*, 26, 6321, 1985.

70. **Seydel, U., Lindner, B., Wollenweber, H.-W., and Rietschel, E. Th.**, Structural studies on the lipid A component of enterobacterial lipopolysaccharides by laser desorption mass spectrometry. Location of acyl groups at the lipid A backbone, *Eur. J. Biochem.*, 145, 505, 1984.

71. **Erwin, A. L. and Munford, R. S.**, Deacylation of structurally diverse lipopolysaccharides by human acyloxyacyl hydrolase, *J. Biol. Chem.*, 265, 16444, 1990.

72. **Qureshi, N., Takayama, K., and Ribi, E.**, Purification and structural determination of nontoxic lipid A obtained from the lipopolysaccharide of *Salmonella typhimurium*, *J. Biol. Chem.*, 257, 11808, 1982.

73. **Mayer, H., Bhat, R. U., Masoud, H., Radziejewska-Lebrecht, J., Widemann, C., and Krauss, J. H.**, Bacterial lipopolysaccharides, *Pure Appl. Chem.*, 61, 1271, 1989.

74. **Hase, S. and Rietschel, E. Th.**, The chemical structure of the lipid A component of lipopolysaccharides from *Chromobacterium violaceum* NCTC 9694, *Eur. J. Biochem.*, 75, 23, 1977.

75. **Wollenweber, H.-W., Seydel, U., Lindner, B., Lüderitz, O., and Rietschel, E. Th.**, Nature and location of amide-bound (R)-3-acyloxyacyl groups in lipid A of lipopolysaccharides from various gram-negative bacteria, *Eur. J. Biochem.*, 145, 265, 1984.

76. **Takayama, K., Qureshi, N., Hyver, K., Honovich, J., Cotter, R. J., Mascagni, P., and Schneider, H.**, Characterization of a structural series of lipid A obtained from the lipopolysaccharides of *Neisseria gonorrhoeae*, *J. Biol. Chem.*, 261, 10624, 1986.

77. **Kulshin, V. A., Zähringer, U., Lindner, B., Frasch, C., Tsai, C.-R., Dmitriev, B., and Rietschel, E. Th.**, Structural analysis of the lipid A component of *Neisseria meningitidis*, *J. Bacteriol.*, 174, 1793, 1992.

78. **Galanos, C., Lüderitz, O., Freudenberg, M. A., Brade, L., Schade, U., Rietschel, E. Th., Kusumoto, S., and Shiba, T.**, Biological activity of synthetic heptaacyl lipid A representing a component of *Salmonella minnesota* R959 lipid A, *Eur. J. Biochem.*, 160, 55, 1986.

79. **Gmeiner, J., Lüderitz, O., and Westphal, O.**, Biochemical studies on lipopolysaccharides of *Salmonella* R mutants. VI. Investigations on the structure of the lipid A component, *Eur. J. Biochem.*, 7, 370, 1969.

80. **Batley, M., Packer, N. H., and Redmond, J. W.**, Configurations of the glycosidic phosphates of lipopolysaccharide from *Salmonella minnesota* R595, *Biochemistry*, 21, 6580, 1982.

81. **Rietschel, E. Th., Gottert, H., Lüderitz, O., and Westphal, O.**, Nature and linkages of fatty acids present in the lipid A component of *Salmonella* lipopolysaccharides, *Eur. J. Biochem.*, 28, 166, 1972.

82. **Wollenweber, H.-W., Broady, K., Lüderitz, O., and Rietschel, E. Th.**, The chemical structure of lipid A: demonstration of amide-linked 3-acyloxyacyl residues in *Salmonella minnesota* Re lipopolysaccharide, *Eur. J. Biochem.*, 124, 191, 1982.

83. **Rietschel, E. Th., Wollenweber, H.-W., Sidorczyk, Z., Zähringer, U., and Lüderitz, O.**, Analysis of the primary structure of lipid A, in *Bacterial Lipopolysaccharides: Structure, Synthesis, Biological Activities*, Am. Chem. Soc. Symp. Ser., Vol. 231, Anderson, L. and Unger, F., Eds., American Chemical Society, Washington, DC, 1983, 195.

84. **Sidorczyk, Z., Zähringer, U., Rietschel, E. Th.**, Chemical structure of the lipid A component of lipopolysaccharide from a *Proteus mirabilis* Re mutant, *Eur. J. Biochem.*, 137, 15, 1983.

85. **Strain, S. M., Armitage, I. M., Anderson, L., Takayama, K., Qureshi, N., and Raetz, C. R. H.**, Location of polar substituents and fatty acyl chains on lipid A precursors from a 3-deoxy-D-*manno*-octulosonic acid-deficient mutant of *Salmonella typhimurium*, *J. Biol. Chem.*, 260, 16089, 1985.

86. **Qureshi, N., Mascagni, P., Ribi, E., and Takayama, K.**, Monophosphoryl lipid A obtained from the lipopolysaccharides of *Salmonella minnesota* R595. Purification of the dimethyl derivative by high performance liquid chromatography and complete structural determination, *J. Biol. Chem.*, 260, 5271, 1985.

87. **Kulshin, V. A., Zähringer, U., Lindner, B., Jäger, K.-E., Dmitriev, B., and Rietschel, E. Th.**, Structural characterization of the lipid A component of *Pseudomonas aeruginosa* lipopolysaccharides, *Eur. J. Biochem.*, 198, 692, 1991.

88. **Weintraub, A., Zähringer, U., Wollenweber, H.-W., Seydel, U., and Rietschel, E. Th.**, Structural characterization of the lipid A component of *Bacteroides fragilis* strain NCTC 9343 lipopolysaccharide, *Eur. J. Biochem.*, 183, 425, 1989.

89. **Qureshi, N., Honovich, J. P., Hara, H., Cotter, R. J., and Takayama, K.**, Location of fatty acids in lipid A obtained from lipopolysaccharide of *Rhodopseudomonas sphaeroides* ATCC 17023, *J. Biol. Chem.*, 263, 5502, 1988.

90. **Krauss, J. H., Seydel, U., Weckesser, J., and Mayer, H.**, Structural analysis of the nontoxic lipid A of *Rhodobacter capsulatus* 37b4, *Eur. J. Biochem.*, 180, 519, 1989.

91. **Moran, A. P., Zähringer, U., Seydel, U., Scholz, D., Stütz, P., and Rietschel, E. Th.**, Structural analysis of the lipid A component of *Campylobacter jejuni* CCUG 10936 (serotype 0:2) lipopolysaccharide. Description of a lipid A containing a hybrid backbone of 2-amino-2-deoxy-D-glucose and 2,3-diamino-2,3-dideoxy-D-glucose, *Eur. J. Biochem.*, 198, 459, 1991.

92. **Russa, R., Lüderitz, O., and Rietschel, E. Th.**, Structural analyses of lipid A from lipopolysaccharides of nodulating and non-nodulating *Rhizobium trifolii, Arch. Microbiol.*, 141, 284, 1985.

93. **Kasai, N., Arata, S., Mashimo, J., Akiyama, Y., Tanaka, C., Egawa, K., and Tanaka, S.**, *Pseudomonas diminuta* LPS with a new endotoxic lipid A structure, *Biochem. Biophys. Res. Commun.*, 142, 972, 1987.

94. **Kasai, N., Arata, S., Mashimo, J.-I., Ohmori, M., Mizutani, T., and Egawa, K.**, Structure-activity relationships of endotoxic lipid A containing 2,3-diamino-2,3-dideoxy-D-glucose, in *Cellular and Molecular Aspects of Endotoxin Reactions*, Nowotny, A., Spitzer, J. Y., and Ziegler, E. J., Eds., (Endotoxin Research Series, Vol. 1, Nowotny, A., Ed., Elsevier Science, Amsterdam, 1990, 121.

95. **Roppel, J., Mayer, H., and Weckesser, J.**, Identification of a 2,3-diamino-2,3-dideoxyhexose in the lipid A component of lipopolysaccharides of *Rhodopseudomonas viridis* and *Rhodopseudomonas palustris, Carbohydr. Res.*, 40, 31, 1975.

96. **Wilkinson, S. G. and Taylor, D. P.**, Occurrence of 2,3-diamino-2,3-dideoxy-D-glucose in lipid A from lipopolysaccharide of *Pseudomonas diminuta, J. Gen. Microbiol.*, 109, 367, 1978.

97. **Weisshaar, R. and Lingens, F.**, The lipopolysaccharide of a chloridazon-degrading bacterium, *Eur. J. Biochem.*, 137, 155, 1983.

98. **Hollingsworth, R. I. and Lill-Elghanian, D. A.**, Isolation and characterization of the unusual lipopolysaccharide component, 2-amino-2-deoxy-2-*N*-(27-hydroxyoctacosanoyl)-3-O-(3-hydroxytetradecanoyl)-*gluco*-hexuronic acid, and its de-O-acylation product from the free lipid A of *Rhizobium trifolii* ANU 843, *J. Biol. Chem.*, 264, 14039, 1989.

99. **Hase, S. and Rietschel, E. Th.**, Isolation and analysis of the lipid A backbone: lipid A structure of lipopolysaccharides from various bacterial groups, *Eur. J. Biochem.*, 63, 101, 1976.

100. **Jiao, B., Freudenberg, M. A., and Galanos, C.**, Characterization of the lipid A component of genuine smooth-form lipopolysaccharide, *Eur. J. Biochem.*, 180, 515, 1989.

101. **Galanos, C. and Lüderitz, O.**, Lipopolysaccharide: properties of an amphipathic molecule, in *Chemistry of Endotoxin*, Rietschel, E. Th., Ed., (*Handbook of Endotoxin*, Vol. 1, Proctor, R. A., Ed.), Elsevier Science, Amsterdam, 1984, 46.

102. **Shands, J. W., Jr.**, Affinity of endotoxin for membranes, *J. Infect. Dis.*, 128, 189, 1973.

103. **Luzzati, V., Mariani, P., and Gulik-Krzywicki, T.**, The cubic phases of liquid-containing systems and biological implications, in *Physics of Amphiphilic Layers*, Springer Proceedings in Physics 21, Mennier, J., Langevin, D., and Boccara, N., Eds., Springer-Verlag, Berlin, 1987, 131.

104. **Labischinski, H., Barnickel, G., Bradaczek, H., Naumann, D., Rietschel, E. Th., and Giesbrecht, P.**, High state of order of isolated bacterial lipopolysaccharide and its possible contribution to the permeation barrier property of the outer membrane, *J. Bacteriol.*, 162, 9, 1985.

105. **Labischinski, H., Naumann, D., Schultz, C., Kusumoto, S., Shiba, T., Rietschel, E. Th., and Giesbrecht, P.**, Comparative X-ray and Fourier-transform-infrared investigations of conformational properties of bacterial and synthetic lipid A of *Escherichia coli* and *Salmonella minnesota* as well as partial structures and analogues thereof, *Eur. J. Biochem.*, 179, 659, 1989.

106. **Naumann, D., Schultz, C., Born, J., Labischinski, H., Brandenburg, K., von Busse, G., Brade, H., and Seydel, U.**, Investigations on the polymorphism of lipid A from lipopolysaccharides of *Escherichia coli* and *Salmonella minnesota* by Fourier-transform infrared spectroscopy, *Eur. J. Biochem.*, 164, 159, 1987.

107. **Seydel, U., Brandenburg, K., Koch, M. H. J., and Rietschel, E. Th.**, Supramolecular structure of lipopolysaccharide and free lipid A under physiological conditions as determined by synchrotron small-angle X-ray diffraction, *Eur. J. Biochem.*, 186, 325, 1989.

108. **Brandenburg, K. and Seydel, U.**, Investigations into the fluidity of lipopolysaccharide in free lipid A membrane systems by Fourier-transform infrared spectroscopy and differential scanning calorimetry, *Eur. J. Biochem.*, 191, 229, 1990.
109. **Westphal, O. and Lüderitz, O.**, Chemische Erforschung von Lipopolysacchariden Gram-negativer Bakterien, *Angew. Chem.*, 66, 407, 1954.
110. **Galanos, C., Lüderitz, O., Rietschel, E. Th., Westphal, O., Brade, H., Brade, L., Freudenberg, M. A., Schade, U., Imoto, M., Yoshimura, S., Kusumoto, S., and Shiba, T.**, Synthetic and natural *Escherichia coli* free lipid A express identical endotoxic activities, *Eur. J. Biochem.*, 148, 1, 1985.
111. **Galanos, C., Freudenberg, M. A., Jay, F., Nerkar, D., Veleva, K., Brade, H., and Strittmatter, W.**, Immunogenic properties of lipid A, *Rev. Infect. Dis.*, 6, 546, 1984.
112. **Morrison, D. C. and Kline, L. F.**, Activation of the classical and properdin pathways of complement by bacterial lipopolysaccharides, *J. Immunol.*, 118, 362, 1977.
113. **Loos, M., Eutenneuer, B., and Clas, F.**, Interaction of bacterial endotoxin (LPS) with fluid phase and macrophage membrane associated C1q, the Fc-recognizing component of the complement system, *Adv. Exp. Med. Biol.*, 256, 301, 1990.
114. **Ulevitch, R. J., Johnston, A. R., and Weinstein, D. B.**, New function for high density lipoproteins, *J. Clin. Invest.*, 64, 1516, 1979.
115. **Freudenberg, M. A., Bog-Hansen, T. C., Back, U., and Galanos, C.**, Interaction of lipopolysaccharides with plasma high density lipoproteins in rats, *Infect. Immun.*, 28, 373, 1980.
116. **Flegel, W. A., Wölpl, A., Männel, D. N., and Northoff, H.**, Inhibition of endotoxin-induced activation of human monocytes by human lipoproteins, *Infect. Immun.*, 57, 2237, 1989.
117. **Schumann, R. R., Leong, S. R., Flaggs, G. W., Gray, P. W., Wright, S. D., Mathison, J. C., Tobias, P. S., and Ulevitch, R. J.**, Structure and function of lipo-polysaccharide binding protein, *Science*, 249, 1429, 1990.
118. **Wright, S. D., Ramos, R. A., Tobias, P. S., Ulevitch, R. J., and Mathison, J. C.**, CD14, a receptor for complexes of lipopolysaccharide (LPS) and LPS binding protein, *Science*, 429, 1431, 1990.
119. **Morrison, D. C.**, The case for specific lipopolysaccharide receptors expressed on mammalian cells, *Microb. Pathogen.*, 7, 389, 1989.
120. **Brade, L., Rietschel, E. Th., Kusumoto, S., Shiba, T., and Brade, H.**, Immuno-genicity and antigenicity of synthetic *Escherichia coli* lipid A, *Infect. Immun.*, 51, 110, 1986.
121. **Brade, L., Brandenburg, K., Kuhn, H.-M., Kusumoto, S., Macher, I., Rietschel, E. Th., and Brade, H.**, The immunogenicity and antigenicity of lipid A are influenced by its physiochemical state and environment, *Infect. Immun.*, 55, 2636, 1987.
122. **Bogard, W. C., Jr., Dunn, D. L., Abernethy, K., Kilgarriff, C., and Kung, P. C.**, Isolation and characterization of murine monoclonal antibodies specific for gram-negative bacterial lipopolysaccharide: association of cross-genus reactivity with lipid A specificity, *Infect. Immun.*, 55, 899, 1987.
123. **Teng, N. N. H., Kaplan, H. S., Hebert, J. M., Moore, C., Douglas, H., Wunderlich, A., and Braude, A. I.**, Protection against gram-negative bacteremia and endotoxemia with human monoclonal IgM antibodies, *Proc. Natl. Acad. Sci. U.S.A.*, 82, 1790, 1985.
124. **Baumgartner, J. D. and Glausner, M. P.**, Controversies in the use of passive immu-notherapy for bacterial infections in the critically ill patient, *Rev. Infect. Dis.*, 9, 194, 1987.
125. **Calandra, T., Glauser, M. P., Schellekens, J., Verhoef, J., and the Swiss-Dutch J5 Immunoglobulin Study Group**, Treatment of gram-negative septic shock with human IgG antibody to *Escherichia coli* J5: a prospective double-blind, randomized trial, *J. Infect. Dis.*, 158, 312, 1988.
126. **Ziegler, E. J.**, Protective antibody to endotoxin core: the emperor's new clothes?, *J. Infect. Dis.*, 158, 286, 1988.

127. Ziegler, E. J., Fisher, C. J., Sprung, C. L., Straube, R. C., Sadoff, J. C., Foulke, G. E., Wortel, C. H., Fink, M. P., Dellinger, R. P., Teng, N. N. H., Allen, I. E., Berger, H. J., Knatterud, G. L., LoBuglio, A. F., Smith, C. R., and the HA-1A Sepsis Study Group, Treatment of gram-negative bacteremia and septic shock with HA-1A human monoclonal antibody against endotoxin, a randomized, double-blind, placebo-controlled trial, *N. Engl. J. Med.*, 324, 1991.

128. Gorelick, K., Scannon, P. J., Hannigan, J., Wedel, N., Ackerman, S. K., Randomized placebo-controlled study of E5 monoclonal antiendotoxin antibody, in *Therapeutic Monoclonal Antibodies*, Borrebaeck, C. A. K. and Larrick, J. W., Eds., Stockton Press, New York, 1990, 16.

129. Freudenberg, M. A., Fomsgaard, A., Mitov, I., and Galanos, C., ELISA for antibodies to lipid A, lipopolysaccharides and other hydrophobic antigens, *Infection*, 17, 322, 1989.

130. Rietschel, E. Th., Brade, L., Schade, U., Seydel, U., Zähringer, U., Brandenburg, K., Helander, I., Holst, O., Kondo, S., Kuhn, H.-M., Lindner, B., Röhrscheidt, E., Russa, R., Labischinski, H., Naumann, D., and Brade, H., Bacterial lipopolysaccharides: relationship of structure and conformation to endotoxic activity, serological specificity and biological function, *Adv. Exp. Med. Biol.*, 256, 81, 1990.

131. Rietschel, E. Th., Brade, H., Brade, L., Brandenburg, K., Schade, U., Seydel, U., Zähringer, U., Galanos, C., Lüderitz, O., Westphal, O., Labischinski, H., Kusumoto, S., and Shiba, T., Lipid A, the endotoxic center of bacterial lipopolysaccharides: relation of chemical structure to biological activity, *Prog. Clin. Biol. Res.*, 231, 25, 1987.

132. Rietschel, E. Th., Brade, L., Schade, U., Galanos, C., Freudenberg, M. A., Lüderitz, O., Kusumoto, S., and Shiba, T., Endotoxic properties of synthetic pentaacyl lipid A precursor Ib and a structural isomer, *Eur. J. Biochem.*, 169, 27, 1987.

133. Galanos, C., Lehmann, V., Lüderitz, O., Rietschel, E. Th., Westphal, O., Brade, H., Brade, L., Freudenberg, M. A., Hansen-Hagge, T., Lüderitz, T., McKenzie, G., Schade, U., Strittmatter, W., Tanamoto, K., Zähringer, U., Imoto, M., Yoshimura, H., Yamamoto, M., Shimamoto, T., Kusumoto, S., and Shiba, T., Endotoxic properties of chemically synthesized lipid A part structures: comparison of synthetic lipid A precursor and synthetic analogues with biosynthetic lipid A precursor and free lipid A, *Eur. J. Biochem.*, 140, 221, 1984.

134. Loppnow, H., Brade, H., Dürrbaum, I., Dinarello, C. A., Kusumoto, S., Rietschel, E. Th., and Flad, H.-D., Interleukin 1 induction-capacity of defined lipopolysaccharide partial structures, *J. Immunol.*, 142, 3229, 1989.

135. Feist, W., Ulmer, A. J., Musehold, J., Brade, H., Kusumoto, S., and Flad, H.-D., Induction of tumor necrosis factor alpha-release by lipopolysaccharide and defined lipopolysaccharide partial structures, *Immunobiology*, 179, 293, 1989.

136. Wang, M.-H., Feist, W., Herzbeck, H., Brade, H., Kusumoto, S., Rietschel, E. Th., Flad, H.-D., and Ulmer, A. J., Suppressive effect of lipid A partial structures on lipopolysaccharide or lipid A-induced release of interleukin 1 by human monocytes, *FEMS Microb. Immunol.*, 64, 179, 1990.

137. Kovach, N. L., Yee, E., Munford, R. S., Raetz, C. R. H., and Harlan, J. M., Lipid IV$_A$ inhibits synthesis and release of tumor necrosis factor induced by lipopolysaccharide in human whole blood *ex vivo*, *J. Exp. Med.*, 172, 77, 1990.

138. Loppnow, H., Libby, P., Freudenberg, M. A., Krauss, J. H., Weckesser, J., and Mayer, H., Cytokine induction by lipopolysaccharide (LPS) corresponds to lethal toxicity and is inhibited by nontoxic *Rhodobacter capsulatus* LPS, *Infect. Immun.*, 58, 3743, 1990.

139. Takayama, K., Qureshi, N., Beutler, B., and Kirkland, T. N., Diphosphoryl lipid A from *Rhodopseudomonas sphaeroides* ATCC 17023 blocks induction of cachectin in macrophages by lipopolysaccharide, *Infect. Immun.*, 57, 1336, 1989.

140. **Proctor, R. A., Will, J. A., Burhop, K. E., and Raetz, C. R. H.,** Protection of mice against lethal endotoxemia by a lipid A precursor, *Infect. Immun.,* 52, 905, 1986.
141. **Aschauer, I., Grob, A., Hildebrandt, J., Schütze, E., and Stütz, P.,** Highly purified lipid X is devoid of immunostimulatory activity. Isolation and characterization of immunostimulating contaminants in a batch of synthetic lipid X, *J. Biol. Chem.,* 265, 9159, 1990.
142. **Takayama, K., Din, Z. Z., Mukerjee, P., Cooke, P. H., and Kirkland, T. N.,** Physicochemical properties of the lipopolysaccharide unit that activates B-lymphocytes, *J. Biol. Chem.,* 265, 14023, 1990.
143. **Ihara, I., Harada, Y., Ihara, S., and Kawakami, M.,** A new complement-dependent bactericidal factor found in nonimmune mouse sera: specific binding to polysaccharide of Ra chemotype *Salmonella, J. Immunol.,* 128, 1256, 1982.

Chapter 2

CHEMICAL STRUCTURE OF LIPID A

Kuni Takayama and Nilofer Qureshi

TABLE OF CONTENTS

I. INTRODUCTION

Lipopolysaccharide (LPS, also called endotoxin) is an amphipathic gly-colipid associated with the outer surface of the outer membrane of Gram-negative bacteria by hydrophobic interaction.[1,2] LPS from the *Salmonella* strains, *Escherichia coli*, and related enteric bacteria has three distinct structural regions: the hydrophobic lipid A, the core oligosaccharide, and the O-specific antigen (polysaccharide).[3] Isolated LPS have a wide range of pathophysiological and immunological activities,[4] most of which are directly attributed to the lipid A region.[5] When LPS is hydrolyzed under mild acid conditions, several forms of "free" lipid A are released.[6] Because of the clear biological significance of this region, much work has been done in determining the structure and related biological activities of both isolated bacterial and synthetic lipid A. In this review, we shall describe how the correct structure of lipid A was finally determined in 1983, explain methods used in the analysis of lipid A, and cover both full and partial structures of all lipid A preparations of bacterial orgin reported in the literature as of January 1991. These include the toxic, nontoxic, and 2,3-diamino-2,3-dideoxyglucose (DAG)-containing lipid A structures.

II. HISTORY OF STRUCTURAL ELUCIDATION

The early studies of Westphal, Lüderitz, and associates since 1954 laid the foundation which eventually led to the final elucidation of the structure of lipid A moiety of LPS in 1983. The following facts were established concerning the structure of lipid A as of 1981 (see Figure 1): (1) the backbone sugar is a glucosamine disaccharide with $\beta(1\rightarrow6)$linkage;[7] (2) phosphate groups are attached to the sugar at both the 1- and 4'-positions;[8] (3) 3-hydroxytetradecanoic, dodecanoic, tetradecanoic, and hexadecanoic acids are amide and ester linked to the glucosamine disaccharide;[9] and (4) the 3-hydroxytetradecanoic acid has a D-configuration.[9] The proposed LPS structure at the time included a fatty acid ester linked to the 6'-position and 2-keto-3-deoxyoctonate (KDO) attached to the 3'-position of the glucosamine disaccharide.[10] Then, beginning in 1982, a number of important new findings were reported concerning this proposed structure. Wollenweber et al.[11] showed the presence of dodecanoyloxytetradecanoate and tetradecanoyloxytetradecanoate as fatty acyl groups in lipid A, Batley et al.[12] showed that the anomeric configuration of the reducing-end sugar was α, and Strain et al.[13] presented ^{13}C-nuclear magnetic resonance (NMR) evidence that the 3-deoxy-D-*manno*-octulosonic acid (KDO) was not linked to the 3'-position of the sugar. A new fast atom bombardment mass spectrometry (FAB-MS) method was used by Qureshi et al.[14] to determine the precise molecular weight of highly purified hexaacyl monophosphoryl lipid A (MPLA) from *S. typhimurium* LPS, which was established to be 1717 Da.

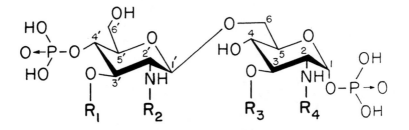

FIGURE 1. General structure of lipid A. R_1 to R_4 represent fatty acyl groups.

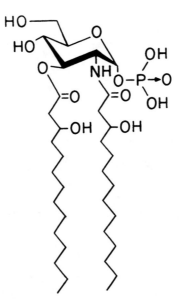

FIGURE 2. Structure of lipid X. It is the monosaccharide precursor to both the reducing and the distal subunits of lipid A.

Important new information of lipid A partial structures also contributed to the final elucidation of the structure of lipid A. In 1981, Nishijima and Raetz[15] isolated and partially characterized a glucosamine-derived phospholipid, termed lipid X, from *E. coli* MN7 (*pgsA, pgsB* mutant), which they thought might be a disaccharide precursor of lipid A. However, when Takayama et al.[16] determined the complete structure of lipid X, they found it to be a monosaccharide that was closely related structurally to lipid A. FAB-MS and ^{1}H-NMR were crucial in establishing its structure, which is shown in Figure 2. Since this structure appeared to represent the monosaccharide precursor (later confirmed by Raetz[17]), it immediately suggested that the fatty acyl groups should be at the 2-, 3-, 2′-, and 3′-positions. This was

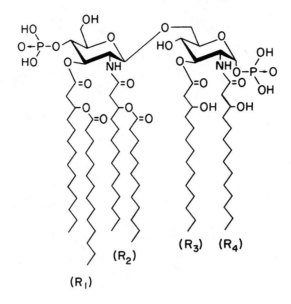

FIGURE 3.   Structure of the hexaacyl DPLA obtained from the LPS of *Salmonella* strains and *E. coli.*

an important clue at the time and was consistent with a location for KDO at the 6′-position.

In 1983, Sidorczyk et al.[18] confirmed that the site of attachment of the oligosaccharide to lipid A is the 6′-position in *Proteus* LPS. Imoto et al.[19] showed by ¹H-NMR that the fatty acids are esterified to the 3- and 3′-positions of *E. coli* lipid A. Qureshi et al.[20] and Takayama et al.[21] obtained the necessary evidence to assign the precise location of all of the fatty acyl groups in lipid A, thus completing the structural determination. In these latter studies, high-performance liquid chromatography (HPLC)-purified lipid A from the LPS of *S. typhimurium* G30/C21 (deep rough mutant) was analyzed using the techniques of FAB-MS and ¹H-NMR. The structure of the hexaacyl diphosphoryl lipid A (DPLA) is shown in Figure 3. Based on this information, lipid A was prepared by total organic synthesis by Imoto et al.[22] When tested for biological activities, the results indicated that this synthetic lipid A was indistinguishable from the corresponding lipid A obtained from natural bacterial sources.[23,24] This is the final proof of structure.

III. PREPARATION AND ANALYSIS OF LIPID A

Isolated bacterial lipid A is generally prepared by performing mild acid hydrolysis of LPS, which selectively cleaves the KDO-glucosamine linkage. Lipid A should be released from the LPS in a gentle manner so that the acid-labile phosphate group at the anomeric position and the fatty acid ester bonds are not disrupted. Thus, DPLA is prepared by treating the LPS with 0.02 *M* sodium acetate buffer, pH 4.5, at 100°C for 1 h.[6,25,26] MPLA, in which the phosphate at the 1-position is selectively hydrolyzed, is prepared by treating the LPS with 0.1 *M* HCl at 100°C for 15 to 30 min.[6,14] Hydrolysis of LPS under acid condition leads to the release of some fatty acyl groups. This was shown by the reverse-phase HPLC analysis of the *E. coli* deep rough chemotype LPS (ReLPS) and the lipid A resulting from the treatment of ReLPS with 0.1 *M* HCl, 100°C, for 30 min, where HPLC separation occurs based on the number of fatty acyl groups.[36] The mass ratio of hexaacyl/pentaacyl/tetraacyl fatty acyl groups for ReLPS was >90:<10:trace, whereas for the lipid A it was 56:25:20.[36] The labile fatty acyl groups are the 3-hydroxytetradecanoate at the 3-position, the tetradecanoyloxytetradecanoate at the 3'-position, or the tetradecanoate in acyloxyacyl linkage at the 3'-position. The ester linkage in dodecanoyloxytetradecanoate at the 2'-position appears to be stable to acid hydrolysis.

To purify the resulting hydrolysis products, an initial fractionation is carried out using a diethylaminoethyl cellulose column,[6] which results in a clear separation of MPLA from DPLA. Each of these two components can then be further fractionated according to the number of fatty acyl groups by using silica gel thin-layer chromatography or column chromatography.[6] Since the purification of MPLA is easier than DPLA, it is usually employed for analytical purposes. Further purification can be accomplished, if necessary, by using reverse-phase HPLC of methylated lipid A (phosphate methyl esters) or underivatized lipid A with paired ion (tetrabutylammonium phosphate) present in the solvent.[6] The purified methylated lipid A is useful only for analytical purposes (NMR and mass spectrometry), whereas lipid A fractionated with paired ion can be recovered and used for biological tests.

The first step in the characterization of purified lipid A subfractions is to perform chemical analyses for total phosphorus,[27] KDO,[28] glucosamine,[29] and fatty acids.[30] These analyses can be performed on silica gel thin-layer chromatography-purified MPLA or DPLA. This is followed by methylation analysis of the reduced disaccharide[31] to establish the 1→6 glycosidic linkage. HPLC-purified dimethyl MPLA or tetramethyl DPLA can be further analyzed by ^{1}H-NMR in benzene d_6/dimethylsulfoxide-d_6 (9:1 v/v).[21,32,33] From such an analysis, it is possible to obtain the following information: (1) position of the ester-linked fatty acids on the sugar (3- and 3'-positions); (2) lack of substitution at the 4- and 6'-positions; (3) the β anomeric configuration of the distal sugar; (4) the α anomeric configuration of the reducing-end sugar; and (5) position of the phosphate group on the distal sugar (4'-position). ^{31}P-NMR and [1]H-decoupled experiments can also be employed to confirm the presence and location of the phosphate groups on the distal and reducing-end sugars.[32]

FAB-MS of HPLC-purified dimethyl MPLA or tetramethyl DPLA should provide the molecular weights and the fatty acid distribution in the distal and

TABLE 1
DPLA Whose Complete Structures Are Established

DPLA source	No. of fatty acids	Fatty acyl group[a]				Ref.
		R_1	R_2	R_3	R_4	
S. typhimurium	6[b]	$C_{14}OC_{14}$	$C_{12}OC_{14}$	OHC_{14}	OHC_{14}	20,21
G30/C21	5	OHC_{14}	$C_{12}OC_{14}$	OHC_{14}	OHC_{14}	
S. minnesota	7	$C_{14}OC_{14}$	$C_{12}OC_{14}$	OHC_{14}	$C_{16}OC_{14}$	33
R595	6[b]	$C_{14}OC_{14}$	$C_{12}OC_{14}$	OHC_{14}	OHC_{14}	
	5	OHC_{14}	$C_{12}OC_{14}$	OHC_{14}	OHC_{14}	
E. coli	6[b,c]	$C_{14}OC_{14}$	$C_{12}OC_{14}$	OHC_{14}	OHC_{14}	37
R. sphaeroides	5[d]	OHC_{10}	$\Delta^7\text{-}C_{14}OC_{14}$	OHC_{10}	$3kC_{14}$	38-40
ATCC 17023	5	OHC_{10}	$C_{14}OC_{14}$	OHC_{10}	$3kC_{14}$	

Note: Refer to Figure 1. In all cases, the backbone sugar is a glucosamine disaccharide with a $\beta(1\rightarrow6)$ linkage where the reducing-end sugar has an α anomeric configuration. Phosphate groups occupy the 1- and the 4'-position. Fatty acyl groups (R) occupy the 2-, 3-, 2'-, and 3'-positions. In the LPS of the *Salmonella* strains (but not *E. coli*), polar substituents of 4-amino-arabinose and ethanolamine are attached to the phosphate groups of the lipid A.

[a] OHC_{10}, 3-hydroxydecanoate; OHC_{14}, 3-hydroxytetradecanoate; $\Delta^7\text{-}C_{14}OC_{14}$, Δ^7-tetradecenoyloxytetradecanoate; $C_{16}OC_{14}$, hexadecanoyloxytetradecanoate; $3kC_{14}$, 3-keto- or 3-oxo-tetradecanoate.

[b] The structure of the hexaacyl DPLA is shown in Figure 3.

[c] Based on the direct analysis of LPS, lipid A containing five fatty acids is also present in *E. coli*.[36]

[d] The structure of the most abundant component of pentaacyl DPLA from *R. sphaeroides* is shown in Figure 4.

reducing-end subunits.[20] This is based on the formation of the oxonium ion (a distal sugar fragment). Laser desorption mass spectrometry (LD-MS) of HPLC-purified samples should reveal the molecular weights as well as cause three types of 2-bond cleavages of the reducing-end sugar.[34,35] These cleavages are especially informative because they allow one to identify the fatty acyl groups at the 2- and 3-positions of the lipid A. Californium plasma desorption mass spectrometry is useful in determining the molecular weights of LPS and lipid A.[36] Both fully and partially O-deacylated lipid A can be analyzed by FAB-MS and LD-MS to determine the distribution of the N-linked fatty acyl groups.[14,20] In FAB-MS, the sample is desorbed from the matrix containing an alkali salt with a neutral beam of xenon atom. In LD-MS and plasma desorption mass spectrometry, the sample is desorbed with an ultraviolet laser beam and a ^{252}Cf fission fragment ionization source, respectively.

IV. LIPID A WHOSE STRUCTURES HAVE BEEN COMPLETELY ESTABLISHED

The complete structures of lipid A from the LPS of only four sources are established at present (Table 1). These include *S. typhimurium* G30/C21,

Salmonella minnesota R595, *E. coli*, and *Rhodobacter sphaeroides* ATCC 17023. The structures of the hexaacyl DPLA (a major lipid A component) of the *Salmonella* strains and *E. coli* are identical (Figure 3). Because this lipid A has the highest endotoxic activities, it is considered to be the model "toxic" lipid A.[41] The minor lipid A component of these bacteria is the pentaacyl DPLA where the tetradecanoate of tetradecanoyloxytetradecanoate at the 3'-position is missing. The LPS of *S. typhimurium* R595 is unique in that in contains a lipid A with seven fatty acids and, because of this, it has a greater degree of heterogeneity than in the lipid A of *E. coli* and *S. typhimurium* LPS. This heptaacyl lipid A is a prominent but not the major component of the LPS of *S. minnesota*.[33]

Two major lipid A components can be obtained from the LPS of *R. sphaeroides*, and these are shown in Table 1. They differ by the presence or absence of unsaturation in the acyl group R_2 and are present in mass ratio of 55:45, respectively. A trace of a third component was also detected, where $R_2 = C_{14}OC_{14}$ and $R_4 = OHC_{14}$.[39] The *R. sphaeroides* DPLA was biologically inactive (activation of B cells, induction of tumor necrosis factor and interleukin-1 by macrophages, and priming of phorbol myristate acetate-stimulated superoxide anion release).[40,42-44] However, it was an effective antagonist in the activation of macrophages and B cells by LPS and lipid A. The major structural differences between the lipid A from this source and that from *E. coli* and *Salmonella* LPS are that the lipid A from *R. sphaeroides* LPS contains (1) only five rather than six fatty acyl groups, (2) 3-hydroxydecanoate rather than 3-hydroxytetradecanoate at the 3- and 3'-positions, (3) 3-oxotetradecanoate and not 3-hydroxytetradecanoate at the 2-position, and (4) an unsaturated fatty acyl group.[42-44] Complete reduction of this DPLA did not restore the biological activity associated with lipid A,[40] thus suggesting the presence of the 3-oxo and unsaturated fatty acids is not responsible for the lack of toxicity and biological activity of this lipid A. This suggested that the proper chain length of the ester-linked fatty acyl groups at R_1 and R_3 (3-hydroxytetradecanoate) and six fatty acyl groups are necessary for endotoxic activity. We consider the pentaacyl DPLA from *R. sphaeroides* to be the model "nontoxic" lipid A (Figure 4).

V. LIPID A WHOSE STRUCTURES HAVE BEEN PARTIALLY ESTABLISHED

Table 2 shows a list of lipid A with varying degrees of characterization, from the almost completely characterized (lipid A from *Haemophilus influenzae*, *Rhizobium meliloti*, *Rhodocyclus gelatinosus*, and *Rhodobacter capsulatus*) to the less characterized (lipid A from *Neisseria gonorrhoeae*). The structures of these lipid A are as yet incomplete, but, in many cases, the fatty acid distributions have been assigned. Although assumed to be α, the anomeric configuration of the reducing-end sugar of these lipid A has not been established.

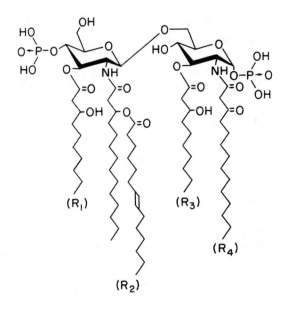

FIGURE 4. Structure of the major component of pentaacyl DPLA obtained from the LPS of
R. sphaeroides.

The lipid A from *Bacteroides fragilis* was reported to be a pentaacyl
MPLA with the phosphate group at the anomer position of the reducing-end
sugar. The fatty acids include the unusual *iso*-pentadecanoic and *iso*-hepta-
decanoic acids. The lipid A from this source has low endotoxic activity,[45]
which might be expected for a lipid A with five fatty acyl groups and one
phosphate (pentaacyl MPLA).

The lipid A from *Rh. meliloti* contains only four fatty acyl groups, including a very unusual 27-hydroxyoctacosanoate and fatty acyl groups whose
chain length is greater than 3-hydroxytetradecanoate. The LPS of *Rh. meliloti*
showed high lethal toxicity in galactosamine-treated mice.[50] Although it is a
tetraacyl DPLA (which is normally much less toxic), the presence of the
unusual fatty acyl group (R_1) and 3-hydroxyoctadecanoyl group (R_2 and R_4)
might be responsible for this high degree of toxicity.

There are two lipid A containing short-chain hydroxy fatty acyl groups
(3-hydroxydecanoate) from the LPS of *Rd. gelatinosus* and *Rb. capsulatus*.
The lipid A from *Rb. capsulatus* contains two molecules of 3-oxotetradeca-
noate, two molecules of 3-hydroxydecanoate, one molecule of Δ^5-dodeca-
noate, and two phosphate groups. The LPS and lipid A from this source are
nontoxic. As in *R. sphaeroides*, this lack of toxicity might be due to the
presence of only five fatty acyl groups and the reduced chain length of the
hydroxy fatty acyl group 3-hydroxydecanoate.[40]

The lipid A from *Rd. gelatinosus* also has 3-hydroxydecanoate at the
3- and 3'-positions of the glucosamine disaccharide backbone. However, it

TABLE 2
Glucosamine Disaccharide-Containing Lipid A Whose Structures Are Partially Established

Chemical structure (glucosamine disaccharide with substituents): R_5-O, positions 4', 5', 6', 1', 2', 3' on the non-reducing sugar bearing R_1 (via O) and R_2 (via NH); positions 6, 4, 5, 3, 2, 1 on the reducing sugar bearing HO, R_3 (via O), R_4 (via NH), and $\sim O-P(=O)(OH)(OH)$; free OH groups as shown.

Lipid A source	N. of PO_4s	N. of fatty acids	\multicolumn{5}{c}{Fatty acyl and phosphate groups[a]}					Ref.
			R_1	R_2	R_3	R_4	R_5	
Bacteroides fragilis NCTC 9343[b]	1[c]	5[d]	OHC_{16}	$iso\text{-}C_{15}O\text{-}iso\text{-}C_{17}$	$OH\text{-}iso\text{-}C_{15}$	OHC_{16}	H	45
Chromobacterium violaceum[b]	2[e]	6[d]	OHC_{10}	$C_{12}OC_{12}$	OHC_{10}	$C_{12}OC_{12}$	PO_4	31, 46
Haemophilus influenzae I-69 Rd⁻/b⁺[b]	2[f]	6	OHC_{10}	$OHC_{12}OC_{12}$	OHC_{10}	$C_{12}OC_{12}$	PO_4	47
			$C_{14}OC_{14}$	$C_{14}OC_{14}$	OHC_{14}	OHC_{14}	PO_4	
Neisseria gonorrhoeae F62[g]	2	6	OHC_{12}	$C_{12}OC_{14}$	OHC_{12}	$C_{12}OC_{14}$	PO_4	34
		5	OHC_{12}	$C_{12}OC_{14}$	OHC_{12}	OHC_{14}	PO_4	
Proteus mirabilis R45[b]	2[f]	7[h]	$C_{14}OC_{14}$	$C_{14}OC_{14}$	OHC_{14}	$C_{16}OC_{14}$	PO_4	18
Providencia rettgeri 6572[b]	2[e]	6	$C_{14}OC_{14}$	$C_{14}OC_{14}$	OHC_{14}	OHC_{14}	PO_4	48
Pseudomonas aeruginosa PA01[b]	2[f]	6[h]	$C_{14}OC_{14}$	$C_{14}OC_{14}$	OHC_{14}	OHC_{14}	PO_4	49
			OHC_{10}	$C_{12}OC_{12}$	OHC_{10}	$C_{12}OC_{12}$	PO_4	
Rhizobium meliloti 10406[b]	2[f]	4[d]	$27\text{-}OHC_{28}$	OHC_{18}	OHC_{14}	OHC_{18}	PO_4	50
Rhodocyclus gelatinosus Dr_2[b]	2[f]	6[d]	OHC_{10}	$C_{12}OC_{10}$	OHC_{10}	$C_{12}OC_{10}$	PO_4	51

TABLE 2 (continued)
Glucosamine Disaccharide-Containing Lipid A Whose Structures Are Partially Established

Lipid A source	N. of PO_4s	N. of fatty acids	Fatty acyl and phosphate groups[a]					Ref.
			R_1	R_2	R_3	R_4	R_5	
Rhodobacter capsulatus 37b4[b]	2[f]	5[d]	Δ^5-$C_{12}OC_{10}$	$3kC_{14}$	OHC_{10}	$3kC_{14}$	PO_4	52
			OHC_{10}	$3kC_{14}$	OHC_{10}	$3kC_{14}$	PO_4	
Shigella sonnei (phase II)[b]	2[f]	6[h]	OHC_{14}	$C_{14}OC_{14}$	OHC_{14}	$C_{12}OC_{14}$	PO_4	53

Note: Although the details of the distribution of fatty acyl groups are known for the lipid A listed in this table, the anomeric configuration of the reducing-end sugar has not been established.

a OHC_{12}, 3-hydroxydodecanoate; OH-*iso*-C_{15}, 3-hydroxy-13-methyltetradecanoate; OHC_{16}, 3-hydroxyhexadecanoate; OHC_{18}, 3-hydroxyoctadecanoate; 27-OHC_{28}, 27-hydroxyoctacosanoate; $C_{12}OC_{10}$, dodecanoyloxydecanoate; Δ^5-$C_{12}OC_{10}$, Δ^5-dodecenoyloxydecanoate; $C_{12}OC_{12}$, dodecanoyloxydodecanoate; $OHC_{12}OC_{12}$, 2-hydroxydodecanoyloxydodecanoate; *iso*-C_{15}O-*iso*-C_{17}, 13-methyltetradecanoyloxy-15-methylhexadecanoate; PO_4, phosphate. Other abbreviations are as in Table 1.

b   The two glucosamine residues are $\beta(1\rightarrow6)$ linked.

c   The phosphate group is linked to the anomeric position of the reducing-end sugar.

d   The fatty acyl distribution was determined by LD-MS and is based on the assumption that these groups occupy the 2-, 3-, 2'-, and 3'-positions.

e   The phosphate group is linked to the anomeric position of the reducing-end sugar and ester linked to the distal sugar.

f   The phosphate group is linked to the anomeric position of the reducing-end sugar and ester linked to the 4'-position.

g   Neither the glycosidic linkage of the two glucosamine residues nor the positions of the two phosphate groups are established. The structure was based on the model lipid A from *Salmonella minnesota*. The anomeric configuration of the distal sugar is not established.

h   The fatty acyl distribution is based on the assumption that these groups occupy the 2-, 3-, 2'-, and 3'-positions.

exhibits high lethal toxicity for mice and high pyrogenicity in rabbits.[54] The basis for the toxicity of this structure appears to be due to the presence of six fatty acyl groups.

From the pattern of fatty acyl distribution shown in Tables 1 and 2, there are two distribution patterns for hexaacyl lipid A. One series, represented by the lipid A of the *Salmonella* strains, *E. coli*, *H. influenzae*, *P. mirabilis*, and *Providencia rettgeri* and having a different fatty acyl distribution in the distal and reducing-end sugars, would be asymmetrical. Another series, represented by the lipid A of *Chromobacterium violaceum*, *N. gonorrhoeae*, *Pseudomonas aeruginosa*, and *Shigella sonnei*, would be symmetrical. Although both series are toxic, the relative biological activities have not been determined.

VI. OTHER LIPID A WITH GLUCOSAMINE BACKBONE

Table 3 shows the fatty acyl composition of lipid A from a wide variety of sources where a glucosamine backbone is involved. In many of these lipid A, the locations of the fatty acyl groups have been partially determined (lipid A from *Acinetobacter calcoaceticus*, *Anabaena variabilis*, *Fusobacterium nucleatum*, *Paracoccus denitrificans*, *Rhodobacter sulfidophilus*, *Rhodopseudomonas blastica*, *Rhodospirillum molischianum*, *Rhodospirillum rubrum*, *Thiobacillus* strains, *Veillonella parvula*, *Vibrio cholerae*, *V. metshnikovii*, *V. parahaemolyticus*, and *Yersinia pestis*), whereas in others only the minimal structural information is available (lipid A from *Bordetella parapertussis*, *Coxiella burnetii*, *Moraxella duplex*, *Pseudomonas paucimobilis*, *Selenomonas ruminantium*, and *Veillonella alcalescens*). The phosphate contents have not been studied in detail. This list includes lipid A that contain other sugars (DAG, galactosamine, mannose, and glucose), presumably in glycosidic linkage to the backbone sugar. The arabinose and arabinosamine are attached to the phosphate groups. An excellent article by Mayer et al. provides a more comprehensive review of unusual lipid A.[77]

VII. LIPID A WITH DIAMINOGLUCOSE (DAG) BACKBONE

Table 4 lists the lipid A containing DAG as the backbone component. Although the fatty acid compositions are known, these lipid A remain essentially uncharacterized with respect to the backbone structure, fatty acid distribution, location of the phosphate groups, and the anomeric configuration of the two sugars. Kasai et al.[101] obtained evidence for the presence of DAG disaccharide in the lipid A of *Pseudomonas diminuta*, but it has not been established whether the backbone sugar is a monosaccharide or a disaccharide. In the case where the glucosamine is also present, as in the lipid A of

TABLE 3
Fatty Acid Composition of Other Lipid A in the Glucosamine Backbone

Lipid A source	Hydroxy or keto fatty acid[a]	Normal fatty acid	Polar substituent[b]	Ref.
Acinetobacter calcoaceticus NCTC 10305	OHC$_{12}$ (A), 2-OHC$_{12}$ (E)	C$_{12}$	—	55
Agmenellum quadruplicatum	OHC$_{14}$, OHC$_{16}$, OHC$_{18}$	C$_{16}$, C$_{22}$	—	56
Agrobacterium tumefaciens	OHC$_{12}$, OHC$_{14}$	—	—	57
Anabaena variabilis	OHC$_{14}$ (E), OHC$_{16}$ (E), OHC$_{18}$ (A)	C$_{16}$	—	58
Bordetella parapertussis	OHC$_{14}$	C$_{14}$, C$_{16}$, iso-C$_{16}$	—	59
B. pertussis	OHC$_{14}$	C$_{14}$, C$_{16}$, iso-C$_{16}$	—	59, 60
Chlorobium vibrioforme[c]	OHC$_{14}$, OHC$_{16}$, OH-iso-C$_{18}$	C$_{12}$, C$_{14}$, C$_{16}$	—	61
Chromatium vinosum[c]	OHC$_{14}$	C$_{12}$, C$_{14:1}$	Man	62
C. tepidum MC[c]	OHC$_{14}$	C$_{12}$, C$_{14}$, C$_{16}$, C$_{18}$	DAG, man	63
Coxiella burnetii	50 different amide-linked fatty acids			64, 65
Edwardsiella ictaluri	OHC$_{14}$	C$_{14}$, C$_{16}$, C$_{16:1}$, cp-C$_{17}$	—	66
Fusobacterium mortiferum	OHC$_{14}$	C$_{14}$, C$_{16}$	—	31
F. nucleatum	OHC$_{14}$ (E), OHC$_{16}$ (A)	C$_{14}$, C$_{16}$	—	67
Legionella pneumophila	OHC$_{13}$-OHC$_{23}$, 2 dihydroxy fatty acids	C$_{14}$-C$_{20}$	—	68
Moraxella duplex	OHC$_{12}$, OHC$_{14}$	C$_{16}$, C$_{18:1}$, cp-C$_{17}$	—	69
Myxococcus fulvus	OH-iso-C$_{17}$, OH-iso-C$_{15}$, OHC$_{14}$, OHC$_{16}$	iso-C$_{15}$, C$_{12}$, C$_{14}$, C$_{16}$	—	70
Neisseria catarrhalis	OHC$_{12}$	C$_{16}$, C$_{18}$, C$_{18:1}$	GalN	71
N. meningitidis	OHC$_{12}$ (E), OHC$_{14}$ (A)	C$_{12}$	—	72
N. perflava	OHC$_{12}$ (E), OHC$_{14}$ (A)	C$_{12}$, C$_{16}$	—	73
Paracoccus denitrificans ATCC 13543	OHC$_{10}$ (E), OHC$_{14}$ (A), 3kC$_{14}$	C$_{12:1}$	—	74
Pseudomonas paucimobilis	2-OHC$_{14}$	C$_{16}$, C$_{18:1}$	—	75
P. pavonacea	OHC$_{12}$, OHC$_{13}$, OHC$_{14}$	C$_{12}$, C$_{12:1}$, C$_{14:1}$	—	76
P. rubescens	OHC$_{11}$, OHC$_{12}$, OHC$_{13}$	iso-C$_{13}$, C$_{13}$	—	76

Organism				Ref.
P. stuzeri	OHC$_{10}$, OHC$_{12}$	C$_{12}$	—	76
P. syncyanea	OHC$_{10}$, OHC$_{12}$, 2-OHC$_{12}$	C$_{12}$, C$_{12:1}$	—	76
Rhizobium leguminosarum	OHC$_{12}$, OHC$_{14}$, 27-OH-C$_{28}$	—	GalA	57
R. trifolii	OHC$_{12}$, OHC$_{14}$, 27-OH-C$_{28}$	—	Ara	77, 78
Rhodobacter sulfidophilus	OHC$_{10}$ (E), OHC$_{14}$ (A), 3kC$_{14}$ (A)	—	Man	79
Rb. veldkampii	OHC$_{10}$, OHC$_{14}$, 3kC$_{14}$	—	—	77
Rhodocyclus purpureus	OHC$_{10}$	—	—	77
Rd. tenuis	OHC$_{10}$	—	—	77
Rhodomicrobium vannielii ATCC 17100c	OHC$_{14}$, OHC$_{16}$	C$_{14}$, C$_{22:1}$	Man	80
Rhodopseudomonas acidophilac	OHC$_{16}$	—	—	79
Rhodopseudomonas blastica	OHC$_{10}$ (E), OHC$_{14}$ (A), 3kC$_{14}$ (A)	—	—	79
Rhodospirillum molischianum ATCC 14031	OHC$_{14}$ (E), OHC$_{16}$ (A)	C$_{12}$, C$_{16}$	—	81
Rhodospirillum rubrum ATCC 11170	OHC$_{14}$ (E), OHC$_{16}$ (A), OHC$_{22}$, OHC$_{24}$	C$_{14}$, C$_{16}$	—	81
Rhodospirillum tenue 2761	OHC$_{10}$	C$_{14}$, C$_{16}$	AraN, Ara, GlcN	82
Selenomonas ruminantium	OHC$_{13}$	—	—	83
Serratia marcescens	OHC$_{12}$ (E), OHC$_{14}$ (A)	C$_{12}$, C$_{14}$, C$_{16}$	—	84
Thiobacillus ferrooxidans IFO 14262	OHC$_{14}$	C$_{12}$	DAG	85-87
Thiobacillus sp. IFO 14569	OHC$_{10}$ (E), OHC$_{14}$ (A), 3kC$_{14}$ (A)	—	—	77, 86
T. thiooxidans	OHC$_{14}$	—	DAG	77, 87
T. versutus ATCC 25364	OHC$_{10}$ (E), OHC$_{14}$ (A), 3kC$_{14}$ (A)	C$_{18:1}$	DAG, man	86
Thiocapsa pfennigii 9111c	OHC$_{14}$	C$_{12}$, C$_{14}$, C$_{16}$, C$_{18}$	Man	63
T. roseopersicinac	OHC$_{10}$, OHC$_{14}$	C$_{12}$, C$_{16}$, C$_{18:1}$	DAG, man, Glc	88
Thiocystis violacea 2711c	OHC$_{14}$	C$_{12}$, C$_{14}$, C$_{16}$, C$_{18}$	—	63
Veillonella alcalescens	OHC$_{13}$ (E, A), OHC$_{15}$ (A)	C$_{13}$	—	89
V. parvula	OHC$_{13}$ (E, A), OHC$_{15}$ (A)	C$_{13}$	—	89
Vibrio cholerae 95R	OHC$_{12}$ (E), OHC$_{14}$ (A)	C$_{12}$, C$_{14}$, C$_{16}$	—	90

TABLE 3 (continued)
Fatty Acid Composition of Other Lipid A in the Glucosamine Backbone

Lipid A source	Hydroxy or keto fatty acid[a]	Normal fatty acid	Polar substituent[b]	Ref.
V. metschnikovii	OHC$_{12}$ (E), OHC$_{14}$ (A)	C$_{12}$, C$_{14}$, C$_{16}$	—	91
V. parahaemolyticus	OHC$_{12}$ (E), OHC$_{14}$ (A)	C$_{12}$, C$_{14}$, C$_{16}$	—	91
Yersinia pestis	OHC$_{14}$	C$_{12}$, C$_{14}$, C$_{16}$, C$_{16:1}$	—	92

Note: Lipid A from these sources are thought to have a glucosamine backbone but its nature is not known. The phosphate groups in these lipid A have not been studied. The positions of the fatty acyl groups on the glucosamine backbone have not been completely established. 2-OHC$_{12}$, 2-hydroxydodecanoate; 2-OHC$_{14}$, 2-hydroxytetradecanoate; OHC$_{11}$ to OHC$_{24}$, 3-hydroxyundecanoate to 3-hydroxytetracosanoate; OH-*iso*-C$_{15}$, 3-hydroxy-13-methyl-tetradecanoate; OH-*iso*-C$_{17}$, 3-hydroxy-15-methyl-hexadecanoate; OH-*iso*-C$_{18}$, 3-hydroxy-16-methyl-heptadecanoate; cp-C$_{17}$, cyclopropane-heptadecanoate; C$_{12}$ to C$_{28}$, normal dodecanoate to octacosanoate.

a   (A) indicates amide linked; (E) indicates ester linked.

b   The polar substituents in some lipid A are nonstoichiometric in relation to glucosamine. GalA is galacturonic acid.

c   Lipid A is free of phosphate.

TABLE 4
Lipid A Containing the DAG Backbone and Their Fatty Acid Composition

Lipid A source	Hydroxy fatty acid	Normal fatty acid	Ref.
Bradyrhizobium japonicum[a]	OHC_{12}, OHC_{14}, 27-OH-C_{28}	C_{12}, C_{14}, C_{16}, C_{18}, $C_{18:1}$	57, 93, 94
B. lupini	OHC_{12}, OHC_{14}	C_{14}, C_{16}, $C_{16:1}$, $C_{18:1}$	57
Brucella melitensis	OHC_{12}, OHC_{13}, OHC_{14}, OHC_{16}	C_{16}, C_{18}	95
Campylobacter jejuni[b]	OHC_{14}	C_{14}, C_{16}	96
Ectothiorhodospira vacuolata[b]	OHC_{10}, OHC_{12}	C_{14}, C_{16}	97
Nitrobacter strain X_{14}[c]	OHC_{14}	—	98
Phenylobacterium immobile	OHC_{12}, OH-Δ^5-C_{12}	C_{12}	99
Pseudomonas carboxydovorans	OHC_{12}, OHC_{14}, OHC_{18}	C_{14}, C_{18}, $C_{18:1}$	57
P. diminuta	OHC_{12}, OHC_{13}, OHC_{14}, OHC_{16}	$C_{12:1}$, C_{14}, $C_{14:1}$	76, 100, 101
P. vesicularis[c]	OHC_{14}	—	77
Rhodopila globiformis	OHC_{14}, OHC_{18}, OHC_{19}	C_{16}, C_{18}	81
Rhodopseudomonas palustris	OHC_{14}, OHC_{16}	C_{16}	102
R. sulfoviridis[c]	OHC_{14}	Δ^2-$C_{14:1}$	103
R. viridis[c]	OHC_{14}	—	104, 105
Thiobacillus novellus[a]	OHC_{14}	C_{16}, $C_{18:1}$	86
Thiobacillus sp.[a]	2-OHC_{12}, OHC_{13}, OHC_{18}	C_{16}, C_{17}, C_{18}, $C_{18:1}$, C_{19}	86

Note: These lipid A all have DAG backbones of unknown structure. The number and location of phosphate(s) and fatty acyl groups are not known. For abbreviations, see previous tables.

[a] Phosphate content was not determined.

[b] Lipid A contains both DAG and GlcN.

[c] Lipid A is devoid of phosphate group.

Campylobacter jejuni, it is not known whether the glucosamine is acylated, present separately as a disaccharide, or associated with DAG. The location and number of phosphate and fatty acyl groups are not known. The lipid A from *Nitrobacter, Pseudomonas vesicularis*, and *Rhodopseudomonas viridis* are devoid of phosphate groups. The LPS containing DAG in the backbone sugar can be endotoxic as was demonstrated with the LPS of *C. jejuni*[96] and the LPS and lipid A of *P. diminuta*.[101] However, the LPS of *R. viridis*[104] and *Rhodopseudomonas palustris*[102] were nontoxic.

VIII. CONCLUSIONS

LPS is an important component that is unique to Gram-negative bacteria. It is now possible to assign functional and structural roles for the partial structures of LPS. Thus, the O-antigen region is involved in the host-to-parasite interactions.[106] The distal core region serves as the linker arm connecting the O-antigen to the rest of the LPS molecule, the proximal (inner) core region (rich in charged groups) appears to maintain the barrier property of the outer membrane,[106] and, finally, the lipid A is the hydrophobic moiety that anchors the LPS to the outer membrane.

Lipid A is a highly conserved structure. When the structure of this glycolipid from many bacterial sources is examined, as in this review, certain common features are revealed. They appear to have the following: (1) $\beta(1\rightarrow6)$-linked disaccharide backbone containing amino sugars, (2) five to seven fatty acyl groups of which the sugar-linked acids are 3-hydroxy fatty acids, (3) acyloxyacyl fatty acids, and (4) one to two phosphate groups at the 1- and 4'-positions.

The structural role of lipid A as a hydrophobic anchor is provided by the abundant fatty acyl groups attached to the sugar. The presence of acyloxyacyl group further enhances the hydrophobicity of the lipid A. It is not known why a disaccharide would be needed instead of a monosaccharide. Perhaps a higher degree of hydrophobicity is required and can be achieved by a disaccharide, due to greater capacity for fatty acyl groups. Amino sugars could provide more stable amide bond with the fatty acids rather than the labile ester bond provided by neutral sugars. This might be an advantage for survival of the bacteria. It is not known why the glycosidic linkage should be 1→6 rather than 1→4. This might be a biosynthetic problem. Because of the need for acylation at the 3- and 2'-positions, the formation of the 1→4 glycosidic bond may not be possible due to steric hindrance. The presence of hydroxy fatty acids is not clear. They might be involved in the intermolecular hydrogen bonding (i.e., with outer membrane proteins) to provide further stability to the outer membrane. The two phosphate groups would be expected to stabilize the outer membrane by cross-linking adjacent lipid A molecules via divalent cation.[106]

It is the pathophysiological effects of lipid A on the mammalian system that have created great interest.[4] Thus, the structure-to-function relationship of lipid A in this context has been studied extensively using both synthetic and bacterial lipid A.[41] These studies have shown that most toxic lipid A can be represented by the *E. coli* and *Salmonella* hexaacyl DPLA (see Figure 3). Reduction in the number of phosphate groups from two to one, reduction in the number fatty acyl groups from six to five or four, or an increase in the number of fatty acyl groups from six to seven all reduced the toxicity and general biological activities of lipid A. Lipid A was also found to be more active than its monosaccharide analogs. The reduction in the chain length of the 3-hydroxy fatty acids at the 3- and 3'-positions from tetradecanoate to decanoate and the reduction in the number of fatty acyl groups from six to five made the DPLA nontoxic and eliminated several biological activities.[40] It thus appears that the entire lipid A molecule is important for the biological activities. If the interaction with a receptor and signal transduction are involved in the activation of B lymphocytes and macrophages,[107-109] one might expect the receptor-to-lipid A interaction to be hydrophobic as well as ionic and involve hydrogen bonding.

It also appears possible to enhance the biological activity of the model toxic lipid A by adding other substituents. We have shown that the ReLPS has a tenfold greater activity in priming of phorbol myristate acetate-stimulated superoxide anion release as compared to the model toxic hexaacyl DPLA.[111] The only difference between the two structures is the presence of the KDO_2 unit in ReLPS.[36] This difference in the biological activity between ReLPS and hexaacyl DPLA may be due to a change in structure and not to greater solubility of ReLPS over DPLA. The effective concentration range tested for activity of ReLPS in this assay was about 100 times lower than its solubility value of $3 \times 10^{-8}\ M$.[110] It is quite possible that the addition of other polar substituents (i.e., 4-amino-arabinose, ethanolamine, or phosphate groups) to the basic structure of hexaacyl DPLA might either enhance or suppress its biological activities. The biochemical basis for the positive modulation of the biological activities of hexaacyl DPLA requires further study.

Finally, a question might be asked as to why lipid A is toxic to mammals. One might speculate that it has a special function. The Gram-negative bacteria, as part of the intestinal flora, might have developed a symbiotic relationship with mammals in which the bacteria provide trace levels of biologically active lipid A to maintain a high state of immunity in the host. The problem of toxicity would then arise only when the production of LPS becomes excessive as in infection.

ACKNOWLEDGMENTS

This review and related work were supported in part by the Medical Research Service of the Department of Veterans Affairs and by National Institutes of Health grants GM-36054 and AI-25856. We thank Carol Steinhart for editorial assistance.

REFERENCES

1. **Mühlradt, P. F. and Golecki, J. R.**, Asymmetrical distribution and artificial reorientation of lipopolysaccharide in the outer membrane bilayer of *Salmonella typhimurium*, *Eur. J. Biochem.*, 51, 343, 1975.
2. **Funahara, Y. and Nikaido, H.**, Asymmetrical localization of lipopolysaccharides on the outer membrane of *Salmonella typhimurium*, *J. Bacteriol.*, 141, 1463, 1980.
3. **Galanos, C., Lüderitz, O., Rietschel, E. T., and Westphal, O.**, Newer aspects of the chemistry and biology of bacterial lipopolysaccharides, with special reference to their lipid A component, *Int. Rev. Biochem.*, 14, 239, 1977.
4. **Nowotny, A.**, *Beneficial Effects of Endotoxin*, Plenum Press, New York, 1983.
5. **Morrison, D. C. and Ryan, J. L.**, Bacterial endotoxins and host immune response, *Adv. Immunol.*, 28, 293, 1979.
6. **Qureshi, N., Cotter, R. J., and Takayama, K.**, Application of fast atom bombardment mass spectrometry and nuclear magnetic resonance on the structural analysis of purified lipid A, *J. Microbiol. Methods*, 5, 65, 1986.
7. **Gmeiner, J., Lüderitz, O., and Westphal, O.**, Biochemical studies on lipopolysaccharides of *Salmonella* R mutants. VI. Investigations on the structure of the lipid A component, *Eur. J. Biochem.*, 7, 370, 1969.
8. **Gmeiner, J., Simon, M., and Lüderitz, O.**, The linkage of phosphate groups and of 2-keto-3-deoxyoctonate to the lipid A component in a *Salmonella minnesota* lipopolysaccharide, *Eur. J. Biochem.*, 21, 355, 1971.
9. **Rietschel, E. Th., Gottert, H., Lüderitz, O., and Westphal, O.**, Nature and linkages of the fatty acids present in the lipid-A component of *Salmonella* lipopolysaccharides, *Eur. J. Biochem.*, 28, 166, 1972.
10. **Rietschel, E. T., Galanos, C., Lüderitz, O., and Westphal, O.**, Chemical structure, physiological function and biological activity of lipopolysaccharides and their lipid A component, in *Immunopharmacology and the Regulation of Leukocyte Function*, Webb, D. R., Ed., Marcel Dekker, New York, 1982, 183.
11. **Wollenweber, H.-W., Broady, K. W., Lüderitz, O., and Rietschel, E. T.**, The chemical structure of lipid A. Demonstration of amide-linked 3-acyloxy-acyl residues in *Salmonella minnesota* Re lipopolysaccharide, *Eur. J. Biochem.*, 124, 191, 1982.
12. **Batley, M., Packer, N. H., and Redmond, J. W.**, Configuration of glycosidic phosphates of lipopolysaccharide from *Salmonella minnesota* R595, *Biochemistry*, 21, 6580, 1982.
13. **Strain, S. M., Fesik, S. W., and Armitage, I. M.**, Characterization of lipopolysaccharide from a heptoseless mutant of *Escherichia coli* by carbon 13 nuclear magnetic resonance, *J. Biol. Chem.*, 258, 2906, 1983.
14. **Qureshi, N., Takayama, K., and Ribi, E.**, Purification and structural determination of nontoxic lipid A obtained from the lipopolysaccharide of *Salmonella typhimurium*, *J. Biol. Chem.*, 257, 11808, 1982.
15. **Nishijima, M. and Raetz, C. R. H.**, Characterization of two membrane-associated glycolipids from an *Escherichia coli* mutant deficient in phosphatidylglycerol, *J. Biol. Chem.*, 256, 10690, 1981.
16. **Takayama, K., Qureshi, N., Mascagni, P., Nashed, M. A., Anderson, L., and Raetz, C. R. H.**, Fatty acyl derivatives of glucosamine 1-phosphate in *Escherichia coli* and their relation in lipid A. Complete structure of a diacyl GlcN 1-P found in a phosphatidylglycerol-deficient mutant, *J. Biol. Chem.*, 258, 7379, 1983.
17. **Raetz, C. R. H.**, Structure and biosynthesis of lipid A in *Escherichia coli*, in *Escherichia coli and Salmonella typhimurium. Cellular and Molecular Biology*, Vol. 1, Neidhardt, F. C., Ed., American Society for Microbiology, Washington, DC, 1987, 498.
18. **Sidorczyk, Z., Zähringer, U., and Rietschel, E. T.**, Chemical structure of the lipid A component of the lipopolysaccharide from *Proteus mirabilis* Re-mutant, *Eur. J. Biochem.*, 137, 15, 1983.

19. **Imoto, M., Kusumoto, S., Shiba, T., Naoki, H., Iwashita, T., Rietschel, E. Th., Wollenweber, H.-W., Galanos, C., and Lüderitz, O.,** Chemical structure of *E. coli* lipid A: linkage site of acyl groups in the disaccharide backbone, *Tetrahedron Lett.,* 24, 4017, 1983.

20. **Qureshi, N., Takayama, K., Heller, D., and Fenselau, C.,** Position of ester groups in the lipid A backbone of lipopolysaccharides obtained from *Salmonella typhimurium, J. Biol. Chem.,* 258, 12947, 1983.

21. **Takayama, K., Qureshi, N., and Mascagni, P.,** Complete structure of lipid A obtained from the lipopolysaccharides of the heptoseless mutant of *Salmonella typhimurium, J. Biol. Chem.,* 258, 12801, 1983.

22. **Imoto, M., Yoshimura, H., Sakaguchi, N., Kusumoto, S., and Shiba, T.,** Total synthesis of *Escherichia coli* lipid A, *Tetrahedron Lett.,* 26, 1545, 1985.

23. **Kotani, S., Takada, H., Tsujimoto, M., Ogawa, T., Harada, K., Mori, Y., Kawasaki, A., Tanaka, A., Nagao, S., Tanaka, S., Shiba, T., Kusumoto, S., Imoto, M., Yoshimura, H., Yamamoto, M., and Shimamoto, T.,** Immunobiologically active lipid A analogs synthesized according to a revised structural model of natural lipid A, *Infect. Immun.,* 45, 293, 1984.

24. **Galanos, C., Lüderitz, O., Rietschel, E. T., Westphal, O., Brade, H., Brade, L., Freudenberg, M., Schade, U., Imoto, M., Yoshimura, H., Kusumoto, S., and Shiba, T.,** Synthetic and natural *Escherichia coli* free lipid A express identical endotoxic activities, *Eur. J. Biochem.,* 148, 1, 1985.

25. **Rosner, M. R., Tang, J.-Y., Barzilay, I., and Khorana, H. G.,** Structure of the lipopolysaccharide from *Escherichia coli* heptose-less mutant. I. Chemical degradation and identification of products, *J. Biol. Chem.,* 254, 5906, 1979.

26. **Takayama, K., Qureshi, N., Ribi, E., and Cantrell, J. L.,** Separation and characterization of toxic and nontoxic forms of lipid A, *Rev. Infect. Dis.,* 6, 439, 1984.

27. **Bartlett, G. R.,** Phosphorus assay in column chromatography, *J. Biol. Chem.,* 234, 466, 1959.

28. **Osborn, M. J.,** Studies on the gram-negative cell wall. I. Evidence for role of 2-keto-3-deoxyoctonate in the lipopolysaccharide of *Salmonella typhimurium, Proc. Natl. Acad. Sci. U.S.A.,* 50, 499, 1963.

29. **Enhofer, E. and Kress, H.,** An evaluation of the Morgan-Elson assay for 2-amino-2-deoxy sugars, *Carbohydr. Res.,* 76, 233, 1979.

30. **Wollenweber, H.-W. and Rietschel, E. T.,** Analysis of lipopolysaccharide (lipid A) fatty acids, *J. Microbiol. Methods,* 11, 195, 1990.

31. **Hase, S. and Rietschel, E. T.,** The chemical structure of the lipid A component of lipopolysaccharides from *Chromobacterium violaceum* NCTC 9694, *Eur. J. Biochem.,* 75, 23, 1977.

32. **Strain, S. M., Armitage, I. M., Anderson, L., Takayama, K., Qureshi, N., and Raetz, C. R. H.,** Location of polar substituents and fatty acyl chains on lipid A precursors from a 3-deoxy-D-*manno*-octulosonic acid-deficient mutant of *Salmonella typhimurium.* Studies by ^{1}H, ^{13}C, and ^{31}P nuclear magnetic resonance, *J. Biol. Chem.,* 260, 16089, 1985.

33. **Qureshi, N., Mascagni, P., Ribi, E., and Takayama, K.,** Monophosphoryl lipid A obtained from lipopolysaccharides of *Salmonella minnesota* R595. Purification of the dimethyl derivative by high performance liquid chromatography and complete structural determination, *J. Biol. Chem.,* 260, 5271, 1985.

34. **Takayama, K., Qureshi, N., Hyver, K., Honovich, J., Cotter, R. J., Mascagni, P., and Schneider, H.,** Characterization of a structural series of lipid A obtained from the lipopolysaccharides of *Neisseria gonorrhoeae.* Combined laser desorption and fast atom bombardment mass spectral analysis of high performance liquid chromatography-purified dimethyl derivatives, *J. Biol. Chem.,* 261, 10624, 1986.

35. **Cotter, R. J., Honovich, J., Qureshi, N., and Takayama, K.,** Structural determination of lipid A from Gram-negative bacteria using laser desorption mass spectrometry, *Biomed. Environ. Mass Spectrom.,* 14, 591, 1987.

36. **Qureshi, N., Takayama, K., Mascagni, P., Honovich, J., Wong, R., and Cotter, R. J.,** Complete structural determination of lipopolysaccharide obtained from deep rough mutant of *Escherichia coli.* Purification by high performance liquid chromatography and direct analysis by plasma desorption mass spectrometry, *J. Biol. Chem.,* 263, 11971, 1988.

37. **Imoto, M., Kusumoto, S., Shiba, T., Rietschel, E. T., Galanos, C., and Lüderitz, O.,** Chemical structure of *Escherichia coli* lipid A, *Tetrahedron Lett.,* 25, 2667, 1985.

38. **Salimath, P. V., Weckesser, J., Strittmatter, W., and Mayer, H.,** Structural studies on the non-toxic lipid A from *Rhodopseudomonas sphaeroides* ATCC 17023, *Eur. J. Biochem.,* 136, 195, 1983.

39. **Qureshi, N., Honovich, J. P., Hara, H., Cotter, R. J., and Takayama, K.,** Location of fatty acids in lipid A obtained from lipopolysaccharide of *Rhodopseudomonas sphaeroides* ATCC 17023, *J. Biol. Chem.,* 263, 5502, 1988.

40. **Qureshi, N., Takayama, K., Meyer, K. C., Kirkland, T. N., Bush, C. A., Chen, L., Wang, R., and Cotter, R. J.,** Chemical reduction of 3-oxo and unsaturated groups in fatty acids of diphosphoryl lipid A from the lipopolysaccharide of *Rhodopseudomonas sphaeroides.* Comparison of biological properties before and after reduction, *J. Biol. Chem.,* 266, 6532, 1991.

41. **Qureshi, N. and Takayama, K.,** Structure and function of lipid A, in *The Bacteria,* Vol. 9, Iglewski, B. J. and Clark, V. L., Eds., Academic Press, San Diego, 1990, 319.

42. **Takayama, K., Qureshi, N., Beutler, B., and Kirkland, T. N.,** Diphosphoryl lipid A from *Rhodopseudomonas sphaeroides* ATCC 17023 blocks of induction of cachectin in macrophages by lipopolysaccharide, *Infect. Immun.,* 57, 1336, 1989.

43. **Kirkland, T. N., Qureshi, N., and Takayama, K.,** Diphosphoryl lipid A derived from lipopolysaccharide (LPS) of *Rhodopseudomonas sphaeroides* inhibits activation of 70Z/3 cells by LPS, *Infect. Immun.,* 59, 131, 1991.

44. **Qureshi, N., Takayama, K., and Kurtz, R.,** Diphosphoryl lipid A obtained from the nontoxic lipopolysaccharide of *Rhodopseudomonas sphaeroides* is an endotoxin antagonist in mice, *Infect. Immun.,* 59, 441, 1991.

45. **Weintraub, A., Zähringer, U., Wollenweber, H.-W., Seydel, U., and Rietschel, E. Th.,** Structural characterization of the lipid A component of *Bacteroides fragilis* strain NCTC 9343 lipopolysaccharide, *Eur. J. Biochem.,* 183, 425, 1989.

46. **Wollenweber, H.-W., Seydel, U., Lindner, B., Lüderitz, O., and Rietschel, E. T.,** Nature and location of amide-bound (R)-3-acyloxyacyl groups in lipid A of lipopolysaccharides from various gram-negative bacteria, *Eur. J. Biochem.,* 145, 265, 1984.

47. **Helander, I. M., Lindner, B., Brade, H., Altmann, K., Lindberg, A. A., Rietschel, E. Th., and Zähringer, U.,** Chemical structure of the lipopolysaccharide of *Haemophilus influenzae* strain 1-69 Rd$^-$/B$^+$ chemotype, *Eur. J. Biochem.,* 177, 483, 1988.

48. **Basu, S., Radziejewska-Lebrecht, J., and Mayer, H.,** Lipopolysaccharide of *Providencia rettgeri.* Chemical studies and taxonomical implications, *Arch. Microbiol.,* 144, 213, 1986.

49. **Bhat, R., Marx, A., Galanos, C., and Conrad, R. S.,** Structural studies of lipid A from *Pseudomonas aeruginosa* PA01: occurrence of 4-amino-4-deoxyarabinose, *J. Bacteriol.,* 172, 6631, 1990.

50. **Urbanik-Sypniewska, T., Seydel, U., Greck, M., Weckesser, J., and Mayer, H.,** Chemical studies on the lipopolysaccharide of *Rhizobium meliloti* 10406 and its lipid A region, *Arch. Microbiol.,* 152, 527, 1989.

51. **Masoud, H., Lindner, B., Weckesser, J., and Mayer, H.,** The structure of lipid A component of *Rhodocyclus gelatinosus* Dr$_2$ lipopolysaccharide, *System. Appl. Microbiol.,* 13, 227, 1990.

52. **Krauss, J. H., Seydel, U., Weckesser, J., and Mayer, H.,** Structural analysis of the nontoxic lipid A of *Rhodobacter capsulatus* 37b4, *Eur. J. Biochem.,* 180, 519, 1989.

53. **Bath, U. R., Kontrohr, T., and Mayer, H.,** Structure of *Shigella sonnei* lipid A, *FEMS Microbiol. Lett.,* 40, 189, 1987.

54. **Galanos, C., Roppel, J., Weckesser, J., Rietschel, E. T., and Mayer, H.,** Biological activities of lipopolysaccharides and lipid A from *Rhodospirillaceae, Infect. Immun.,* 16, 407, 1977.

55. **Brade, H. and Galanos, C.,** Isolation, purification, and chemical analysis of the lipopolysaccharide and lipid A of *Acinetobacter calcoaceticus* NTC 10305, *Eur. J. Biochem.,* 122, 233, 1982.

56. **Buttke, T. M. and Ingram, L. O.,** Comparison of lipopolysaccharides from *Agmenellum quadruplicatum* to *Escherichia coli* and *Salmonella typhimurium* by using thin-layer chromatography, *J. Bacteriol.,* 124, 1566, 1975.

57. **Mayer, H., Krauss, J. H., Urbanik-Sypniewska, T., Puvanesarajah, V., Stacey, G., and Auling, G.,** Lipid A with 2,3-diamino-2,3-dideoxy-glucose in lipopolysaccharides from slow-growing members of Rhizobiaceae and from "Pseudomonas carboxydovorans," *Arch. Microbiol.,* 151, 111, 1989.

58. **Weckesser, J., Katz, A., Drews, G., Mayer, H., and Fromme, I.,** Lipopolysaccharide containing L-acofriose in the filamentous blue-green alga *Anabaena variabilis, J. Bacteriol.,* 120, 672, 1974.

59. **Amano, K.-I., Fukushi, K., and Watanabe, M.,** Biochemical and immunological comparison of lipopolysaccharides from *Bordetella* species, *J. Gen. Microbiol.,* 136, 481, 1990.

60. **Starkloff, A. and Szabo, L.,** The fatty acid content of the *Bordetella pertussis* endotoxin, *J. Gen. Microbiol.,* 132, 97, 1986.

61. **Meissner, J., Fischer, U., and Weckesser, J.,** The lipopolysaccharide of the green sulfur bacterium *Chlorobium vibrioforme* f. thiosulfatophilum, *Arch. Microbiol.,* 149, 125, 1987.

62. **Hurlbert, R. E., Weckesser, J., Mayer, H., and Fromme, I.,** Isolation and characterization of the lipopolysaccharide of *Chromatium vinosum, Eur. J. Biochem.,* 68, 365, 1976.

63. **Meissner, J., Pfennig, N., Krauss, J. H., Mayer, H., and Weckesser, J.,** Lipopolysaccharide of *Thiocystis violacea, Thiocapsa pfennigii,* and *Chromatium tepidum,* species of the family Chromatiaceae, *J. Bacteriol.,* 170, 3217, 1988.

64. **Wollenweber, H.-W., Schramek, S., Moll, H., and Rietschel, E. T.,** Nature and linkage type of fatty acids present in lipopolysaccharides of phase I and phase II *Coxiella burnetii, Arch. Microbiol.,* 142, 6, 1985.

65. **Mayer, H., Radziejewska-Lebrecht, J., and Schramek, S.,** Chemical and immunological studies on lipopolysaccharides of *Coxiella burnetii* phase I and phase II, *Adv. Exp. Med. Biol.,* 228, 577, 1988.

66. **Weete, J. D., Blevins, W. T., Chitrakorn, S., Saeed, M. O., and Plumb, J. A.,** Chemical characterization of lipopolysaccharide from *Edwardsiella ictaluri,* a fish pathogen, *Can. J. Microbiol.,* 34, 1224, 1988.

67. **Hase, S., Hofstad, T., and Rietschel, E. Th.,** Chemical structure of the lipid A component of lipopolysaccharides from *Fusobacterium nucleatum, J. Bacteriol.,* 129, 9, 1977.

68. **Sonesson, A., Jantzen, E., Bryn, K., Larsson, L., and Eng, J.,** Chemical composition of a lipopolysaccharide from *Legionella pneumophila, Arch. Microbiol.,* 153, 72, 1989.

69. **Adams, G. A., Quadling, C., Yaguchi, M., and Tornabene, T. G.,** The chemical composition of cell-wall lipopolysaccharides from *Moraxella duplex* and *Micrococcus calco-aceticus, Can. J. Microbiol.,* 16, 1, 1970.

70. **Rosenfelder, G., Lüderitz, O., and Westphal, O.,** Composition of lipopolysaccharides from *Myxococcus fulvus* and other fruiting and nonfruiting Myxobacteria, *Eur. J. Biochem.,* 44, 411, 1974.

71. **Adams, G. A., Tornabene, T. G., and Yaguchi, M.,** Cell wall lipopolysaccharides from *Neisseria catarrhalis, Can. J. Microbiol.,* 15, 365, 1969.

72. **Jennings, H. J., Hawes, G. B., Adams, G. A., and Kenny, C. P.,** The chemical composition and serological reactions of lipopolysaccharides from serogroups A, B, X, and Y *Neisseria meningitidis, Can. J. Biochem.,* 51, 1347, 1973.

73. **Adams, G. A., Kates, M., Shaw, D. H., and Yaguchi, M.,** Studies on the chemical constitution of cell-wall lipopolysaccharides from *Neisseria perflava, Can. J. Biochem.,* 46, 1175, 1968.

74. **Wilkinson, B. J., Hindahl, M. S., Galbraith, L., and Wilkinson, S. G.,** Lipopolysaccharide of *Paracoccus denitrificans* ATCC 13543, *FEMS Microbiol. Lett.,* 37, 63, 1986.

75. **Kawahara, K., Uchida, K., and Aida, K.,** Isolation of an unusual 'lipid A' type glycolipid from; Pseudomonas paucimobilis, *Biochim. Biophys. Acta,* 712, 571, 1982.

76. **Wilkinson, S. G., Galbraith, L., and Lightfoot, G. A.,** Cell walls, lipids, and lipopolysaccharides of *Pseudomonas* species, *Eur. J. Biochem.,* 33, 158, 1973.

77. **Mayer, H., Krauss, J. H., Yokota, A., and Weckesser, J.,** Natural variant of lipid A, in *Endotoxin, Advances in Experimental Medicine and Biology,* Vol. 256, Friedman, H., Klein, T. W., Nakano, M., and Nowotny, A., Eds., Plenum Press, New York, 1990, 45.

78. **Hollingsworth, R. I., and Carlson, R. W.,** 27-Hydroxyoctacosanoic acid is a major structural fatty acyl component of the lipopolysaccharide of *Rhizobium trifolii* ANU 843, *J. Biol. Chem.,* 264, 9300, 1989.

79. **Tegtmeyer, B., Weckesser, J., Mayer, H., and Imhoff, J. F.,** Chemical composition of the lipopolysaccharides of *Rhodobacter sulfidophilus, Rhodopseudomonas acidophila,* and *Rhodopseudomonas blastica, Arch. Microbiol.,* 143, 32, 1985.

80. **Holst, O., Borowiak, D., Weckesser, J., and Mayer, H.,** Structural studies on the phosphate-free lipid A of *Rhodomicrobium vannielii* ATCC 17100, *Eur. J. Biochem.,* 137, 325, 1983.

81. **Pietsch, K., Weckesser, J., Fischer, U., and Mayer, H.,** The lipopolysaccharides of *Rhodospirillum rubrum, Rhodospirillum molischianum,* and *Rhodopila globiformis, Arch. Microbiol.,* 154, 433, 1990.

82. **Tharanathan, R. N., Weckesser, J., and Mayer, H.,** Structural studies on the D-arabinose-containing lipid A from *Rhodospirillum tenue* 2761, *Eur. J. Biochem.,* 84, 385, 1978.

83. **Kamio, Y., Kim, K. C., and Takahashi, H.,** Chemical structure of lipid A of *Selenomonas ruminantium, J. Biochem.,* 70, 187, 1971.

84. **Adams, G. A. and Singh, P. P.,** The chemical constitution of lipid A from *Serratia marcescens, Can. J. Microbiol.,* 48, 55, 1970.

85. **Hirt, W. E. and Vestal, J. R.,** Physical and chemical studies of *Thiobacillus ferroxidans* lipopolysaccharides, *J. Bacteriol.,* 123, 642, 1975.

86. **Yokota, A., Schlect, S., and Mayer, H.,** Lipopolysaccharides of chemolithotrophic bacteria *Thiobacillus versutus* and a related *Thiobacillus* species, *FEMS Microbiol. Lett.,* 44, 197, 1987.

87. **Yokota, A., Rodriguez, M., Yamada, Y., Imai, K., Borowiak, D., and Mayer, H.,** Lipopolysaccharides of *Thiobacillus* species containing lipid A with 2,3-diamino-2,3-dideoxyglucose, *Arch. Microbiol.,* 149, 106, 1987.

88. **Hurlbert, R. E., Weckesser, J., Tharanathan, R. N., and Mayer, H.,** Isolation and characterization of the lipopolysaccharide of *Thiocapsa roseopersicina, Eur. J. Biochem.,* 90, 241, 1978.

89. **Hewett, M. J. and Knox, K. W.,** Biochemical studies on lipopolysaccharides of *Veillonella, Eur. J. Biochem.,* 19, 169, 1971.

90. **Broady, K. W., Rietschel, E. Th., and Lüderitz, O.,** The chemical structure of the lipid A component of lipopolysaccharides from *Vibrio cholerae, Eur. J. Biochem.,* 115, 463, 1981.

91. **Rietschel, E. Th., Palin, W. J., and Watson, D. W.,** Nature and linkages of the fatty acids present in lipopolysaccharides from *Vibrio metshnikovii* and *Vibrio parahaemolyticus, Eur. J. Biochem.,* 37, 116, 1973.

92. **Dalla Venezia, N., Minka, S., Bruneteau, M., Mayer, H., and Michel, G.,** Lipopolysaccharides from *Yersinia pestis*. Studies on lipid A of lipopolysaccharides I and II, *Eur. J. Biochem.,* 151, 399, 1985.

93. **Puvanesarajah, V., Schell, F. M., Gerhold, D., and Stacy, G.,** Cell surface polysaccharides from *Bradyrhizobium japonicum* and a nonnodulating mutant, *J. Bacteriol.,* 169, 137, 1987.

94. **Carrion, M., Bhat, U. R., Reuhs, B., and Carlson, R. W.,** Isolation and characterization of the lipopolysaccharides from *Bradyrhizobium japonicum, J. Bacteriol.,* 172, 1725, 1990.

95. **Moreno, E., Borowiak, D., and Mayer, H.,** *Brucella* lipopolysaccharides and polysaccharides, *Ann. Inst. Pasteur (Microbiol.),* 138, 102, 1987.

96. **Moran, A. P., Rietschel, E. T., Kusunen, T. U., and Zähringer, U.,** Chemical characterization of *Campylobacter jejuni* lipopolysaccharides containing *N*-acetylneuramic acid and 2,3-diamino-3,4-dideoxy-D-glucose, *J. Bacteriol.,* 173, 618, 1991.

97. **Meissner, J., Borowiak, D., Fischer, U., and Weckesser, J.,** The lipopolysaccharide of the phototrophic bacterium *Ectothiorhodospira vacuolata, Arch. Microbiol.,* 149, 245, 1988.

98. **Mayer, H., Bock, E., and Weckesser, J.,** 2,3-Diamino-2,3-dideoxyglucose containing lipid A in the *Nitrobacter* strain X_{14}, *FEMS Microbiol. Lett.,* 17, 93, 1983.

99. **Bellmann, W. and Lingens, F.,** Structural studies on the core oligosaccharide of *Phenylobacterium immobile* strain K_2 lipopolysaccharide. Chemical synthesis of 3-hydroxy-5c-dodecenoic acid, *Biol. Chem. Hoppe-Seyler,* 366, 567, 1985.

100. **Wilkinson, S. G. and Taylor, D. P.,** Occurrence of 2,3-diamino-2,3-dideoxy-D-glucose in lipid A from lipopolysaccharide of *Pseudomonas diminuta, J. Gen. Microbiol.,* 109, 367, 1978.

101. **Kasai, N., Arata, S., Mashimo, J., Akiyama, Y., Tanaka, C., Egawa, K., and Tanaka, S.,** *Pseudomonas diminuta* LPS with a new endotoxic lipid A structure, *Biochem. Biophys. Res. Commun.,* 142, 972, 1987.

102. **Weckesser, J., Drews, G., Fromme, I., and Mayer, H.,** Isolation and chemical composition of lipopolysaccharides of *Rhodopseudomonas palustris* strains, *Arch. Microbiol.,* 92, 123, 1973.

103. **Ahamed, N. M., Mayer, H., Biebl, H., and Weckesser, J.,** Lipopolysaccharide with 2,3-diamino-2,3-dideoxyglucose containing lipid A in *Rhodopseudomonas sulfoviridis, FEMS Microbiol. Lett.,* 14, 27, 1982.

104. **Roppel, J., Mayer, H., and Weckesser, J.,** Identification of a 2,3-diamino-2,3-dideoxyhexose in the lipid A component of lipopolysaccharides of *Rhodopseudomonas viridis* and *Rhodopseudomonas palustris, Carbohydr. Res.,* 40, 31, 1975.

105. **Weckesser, J., Drews, G., and Mayer, H.,** Lipopolysaccharides of photosynthetic prokaryotes, *Annu. Rev. Microbiol.,* 33, 215, 1979.

106. **Nikaido, H. and Vaare, M.,** Outer membrane, in *Escherichia coli and Salmonella typhimurium. Cellular and Molecular Biology,* Vol. 1, Neidhardt, F. C., Ed., American Society for Microbiology, Washington, DC, 1987, 7.

107. **Lei, M.-G. and Morrison, D. C.,** Specific endotoxic lipopolysaccharide-binding proteins on murine splenocytes. I. Detection of lipopolysaccharide-binding sites on splenocytes and splenocyte subpopulations, *J. Immunol.,* 141, 996, 1988.

108. **Hara-Kuge, S., Amano, F., Nishijima, M., and Akamatsu, Y.,** Isolation of a lipopolysaccharide (LPS)-resistant mutant with defective LPS binding of cultured macrophage-like cells, *J. Biol. Chem.,* 265, 6606, 1990.

109. **Kirkland, T. N., Virca, G. D., Kuus-Reichel, T., Multer, F. K., Kim, S. Y., Ulevitch, R. J., and Tobias, P. S.,** Identification of lipopolysaccharide-binding proteins in 70Z/3 cells by photoaffinity cross-linking, *J. Biol. Chem.,* 265, 9520, 1990.

110. **Takayama, K., Din, Z. Z., Mukerjee, P., Cooke, P. H., and Kirkland, T. N.,** Physicochemical properties of the lipopolysaccharide unit that activates B lymphocytes, *J. Biol. Chem.,* 265, 14023, 1990.

111. **Qureshi, N., Meyer, K. C., and Takayama, K.,** unpublished results.

Chapter 3

BIOSYNTHESIS OF LIPID A

Christian R. H. Raetz

TABLE OF CONTENTS

I. OVERVIEW

The lipid A domain of lipopolysaccharide (LPS) is a unique, glucosamine-based phospholipid that makes up the outer monolayer of the outer membrane of Gram-negative bacteria.[1,2] Following the discovery of acylated monosaccharide precursors of lipid A in 1983, the pathway for the enzymatic synthesis of lipid A was elucidated.[1,3,4] Most studies of lipid A biosynthesis have employed extracts of *Escherichia coli*, but it is now apparent that the major features of the *E. coli* system apply to diverse Gram-negative organisms.[5] The conditional lethality of *E. coli* mutants defective in the first committed step of lipid A biosynthesis confirms the essentiality of lipid A for cell viability,[6] and it suggests that lipid A biosynthesis inhibitors might be novel antibiotics. The cloning and sequencing of certain key enzymes of lipid A biosynthesis[7-9] are providing insights into lipid A/protein recognition and have facilitated the preparation of radiolabeled probes[10] with which to study the interaction of lipid A with animal cells.[10-12]

II. THE MINIMAL LPS REQUIRED FOR GROWTH

As discussed elsewhere in this volume, the minimal LPS structure required for the growth of *E. coli* and most other Gram-negative bacteria is termed Re LPS.[1,2] It consists of lipid A derivatized with two KDO moieties (Figures 1 and 2).[1,2] The LPS molecules of wild-type cells contain additional core and O-antigen sugars (Figure 1).[1,2] Strains possessing the minimal Re LPS are hypersensitive to hydrophobic antibiotics and detergents.[2] These observations indicate that there are functional advantages to a complete LPS, even though Re LPS is sufficient to support outer membrane biogenesis and cell growth.

That Re LPS is relatively conserved between bacterial species may explain why cells of the immune system respond to lipid A. Recognition of lipid A, rather than other more variable regions of LPS, might reduce the complexity of the molecular machinery required by an animal cell to detect the presence of diverse Gram-negative bacteria.[1,13]

III. BIOSYNTHESIS OF LIPID A

A. GENERAL FEATURES

Since 1984, a unique enzymatic pathway has been identified in *E. coli* that accounts for the formation of Re LPS (Figure 2) from the precursors UDP-GlcNAc, *R*-3-hydroxymyristoyl-ACP, ATP, CMP-KDO, lauroyl-ACP, and myristoyl-ACP.[1,10,14-18] All key intermediates (Figure 2) have been isolated, characterized, and chemically synthesized.[1] Similar enzymes have been found in Gram-negative pathogens and in photosynthetic bacteria.[5]

One cell of *E. coli* contains ~2×10^6 lipid A molecules and ~2×10^7 glycerophospholipids (Figure 1).[1,19] Approximately one quarter of the fatty

ESCHERICHIA COLI ENVELOPE

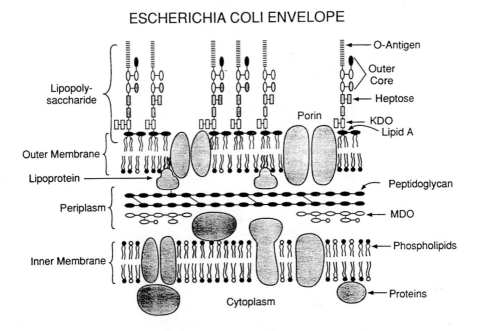

FIGURE 1. Organization of lipids and proteins in the *E. coli* envelope. Ovals and rectangles represent sugar residues. Circles are the polar headgroups of phospholipids. MDO are membrane-derived oligosaccharides[1,19] and KDO is 3-deoxy-D-*manno*-octulosonic acid.[1] KDO and heptose make up the inner core of LPS.[1]

acyl chains of the envelope are associated with LPS. In wild-type cells, the lipid A precursors and the enzymes that interconvert them (Figure 2) are present at 100 to 1000 copies per cell.[1,19,20]

B. THE FATTY ACYLATION OF UDP-GlcNAc

The first step of lipid A biosynthesis in *E. coli* is the acylation of UDP-GlcNAc with *R*-3-hydroxymyristate (Figure 2).[14,21] UDP-GlcNAc acyltransferase is the product of the *lpxA* gene.[1,8] The enzyme displays a remarkable specificity for *R*-3-hydroxymyristate, consistent with the composition of lipid A.[1,5,14] It has an absolute requirement for the *R*-3-OH function. How the enzyme measures fatty acyl chain length with such accuracy is unknown, but it clearly differs in this respect from the relatively nonspecific glycerol-3-phosphate acyltransferases of *E. coli* involved in glycerophospholipid formation.[1,19] How *R*-3-hydroxymyristate is kept out of glycerophospholipids is not clear, but it may involve compartmentalization.[1] The fatty acylation of a sugar nucleotide has not been described in any other biosynthetic pathway besides the one that generates lipid A (Figure 2).

UDP-GlcNAc is situated at a branchpoint in the biogenesis of envelope polymers, since it is also a key precursor of peptidoglycan (Figure 1).[22] *R*-3-Hydroxymyristoyl-ACP is located at yet another metabolic branchpoint

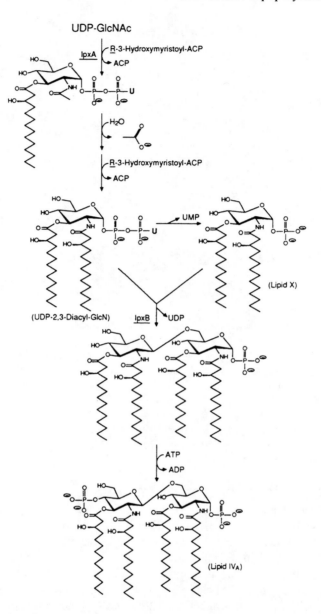

FIGURE 2. Enzymatic synthesis of Re LPS in *E. coli*. UDP-GlcNAc acyltransferase is the first committed enzyme.[1] The *lpxA, lpxB,* and *kdtA* genes code for the enzymes indicated and have been sequenced.[7-9] The structure shown at the bottom (Re endotoxin) is the minimal LPS required for growth. ACP, acyl carrier protein; U, uridine.

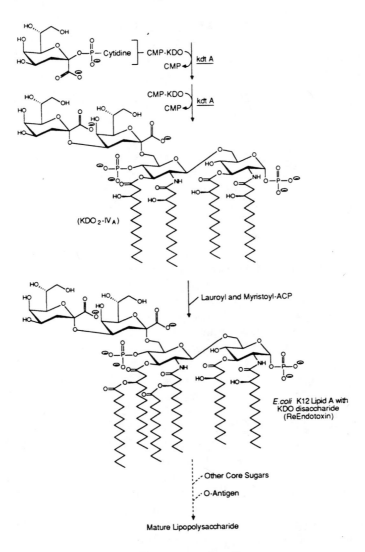

FIGURE 2 (continued).

because it can be elongated to palmitoyl-ACP and incorporated into glycerophospholipids.[1,23] The central roles of these lipid A precursors in envelope biogenesis may account for the fact that two genes of the lipid A pathway (*lpxA* and *lpxB*, Figure 2) are part of an operon near minute 4 on the *E. coli* chromosome.[24,25] This operon contains several internal promoters interspersed among twelve genes, including *dnaE* (the catalytic subunit of DNA polymerase III),[24] *ompH* (a basic outer membrane protein that associates with LPS),[25,26] and *cds* (a key step in the generation of glycerophospholipids).[25,27] The biological significance and possible regulation of this operon require further investigation.

The acylation of UDP-GlcNAc is readily reversible, suggesting that
R-3-hydroxymyristoyl-ACP (a thioester) may be more stable than the product
UDP-3-O-(R-3-hydroxymyristoyl)-GlcNAc (an oxygen ester).[54] This anomaly
might be explained by the relatively low pKa of sugar hydroxyls, steric
constraints, or an especially stable conformation of R-3-hydroxymyristoyl-
ACP.

Next, UDP-3-O-(R-3-hydroxymyristoyl)-GlcNAc is deacetylated and
subsequently N-acylated with another R-3-hydroxymyristate moiety[15] to generate UDP-2,3-diacylglucosamine (Figure 2). Again, coenzyme A thioesters
are not utilized.[14,15] As expected from the composition of lipid A, the
N-acyltransferase displays considerable specificity for R-3-hydroxymyristate,[5,14]
but it is somewhat less selective than the O-acyltransferase.[5,14] Given the
reversibility of the O-acyltransferase, it appears that the deacetylase is actually
the first committed reaction in the biogenesis of lipid A.

C. FORMATION OF THE LIPID A DISACCHARIDE

UDP-2,3-diacylglucosamine (Figure 2) is the precursor of the nonreducing
end unit of lipid A.[11,16] However, some of it is cleaved to generate
2,3-diacylglucosamine-1-phosphate (lipid X) (Figure 2).[16] Two pyrophos-
phatases are known to catalyze this reaction. One of these enzymes also
cleaves CDP-diacylglycerol (a molecule with some formal resemblance to
UDP-2,3-diacylglucosamine).[20]

Lipid X (Figure 2) is the direct precursor of the reducing end unit of lipid
A. Disaccharide formation involves the condensation of UDP-2,3-diacylglu-
cosamine with lipid X, catalyzed by the lipid A disaccharide synthase, the
product of the *lpxB* gene.[11,16] In this reaction, the 2,3-diacylglucosamine
moiety of UDP-2,3-diacylglucosamine is transferred to the 6 position of lipid
X, generating the β-1-6 linkage, characteristic of lipid A.[11,16]

The disaccharide synthase was the first enzyme of the pathway to be
identified.[16] It has recently been overexpressed and purified to homogeneity.[11]
It is a versatile reagent with which to synthesize lipid A substructures and
analogs.[12] Its mechanism of action has not been studied in detail, but its lack
of a requirement for a nonionic detergent[11] is unusual, suggesting that lipid
X and UDP-2,3-diacylglucosamine may have different physical properties in
water than do classical glycerophospholipids. One possibility is that the monosaccharide precursors of lipid A form micelles rather than bilayers in water.[1]

A specific kinase[10] next incorporates the 4'-monophosphate, generating
lipid IV$_A$ (Figure 2). Large amounts of lipid IV$_A$ can be isolated from mutants
that are temperature sensitive in the generation of KDO.[28] Lipid IV$_A$ (also
referred to as Ia by some authors)[13] has some endotoxin-like properties.[29] In
mouse cells lipid IV$_A$ is a weak agonist,[29,55] but in human cells it appears to
be an endotoxin antagonist.[30,31]

Radiolabeled lipid IV$_A$ can be prepared with the 4'-kinase[10] and is extremely useful for detecting endotoxin-binding proteins on animal cells.[32-34]

We have recently shown that the macrophage scavenger receptor, first identified by its ability to bind damaged LDL,[35] also effectively binds lipid IV_A.[34] Subsequent to binding, the lipid is dephosphorylated at position 1 in a reaction (probably localized in lysosomes) that is sensitive to inhibition by chloroquine.[34]

D. ADDITION OF KDO

Prior to the completion of lipid A biosynthesis, the two stereochemically distinct KDO residues of Re LPS are incorporated by an unusual bifunctional enzyme (Figure 2), encoded by the *kdtA* gene.[9,17] Using colony autoradiography, this gene has recently been mutated, mapped, cloned, and sequenced.[9] Overproduction of a single polypeptide of 425 amino acids causes both the first and second KDO transferases to be overexpressed, consistent with the early observations of Munson et al.,[36] who were unable to separate the two KDO transferases from each other.

The KDO transferase[9] displays a specificity for its lipid A disaccharide bis-phosphate substrates resembling that of lipid A-responsive animal cells or the LPS-binding protein of acute phase serum.[37] A general clustering of basic amino acid residues within the N-terminal half of *kdtA* is reminiscent of the structure of the neutrophil bactericidal permeability-increasing protein (BPI).[9,38] While there is no obvious sequence homology, these functional similarities may reflect a common lipid A binding motif. The overexpressed KDO transferase affords a convenient reagent for the addition of KDO disaccharides to lipid A-like acceptors.

E. FORMATION OF ACYLOXYACYL MOIETIES

In the last stage of Re LPS formation, laurate and myristate residues are incorporated into the nonreducing end of lipid A (Figure 2) to generate the characteristic acyloxyacyl units.[18] Like the UDP-GlcNAc acyltransferases (Figure 2), the "late" acyltransferases function only with thioesters of ACP.[18] The late acyltransferases of *E. coli* display an intriguing specificity for KDO,[18] since they do not function with lipid IV_A as the substrate.[18] This is consistent with the behavior of KDO-deficient mutants of *Salmonella*, in which lipid IV_A (not lipid A) accumulates.[28] However, *Pseudomonas* appears to be different in this regard, since it accumulates fully acylated lipid A when CMP-KDO formation is blocked.[39]

An alternative mechanism for the formation of acyloxyacyl moieties is encountered in the case of palmitate addition to lipid X, generating lipid Y (Figure 3).[40] The same reaction probably accounts for the addition of palmitate to lipid IV_A.[40] In this case, a palmitate residue from the 1 position of any common glycerophospholipid is transferred to *R*-3-hydroxy function of the N-linked *R*-3-hydroxymyristoyl moiety next to the reducing end phosphate.[40] The function of this transesterification reaction is uncertain, since very little palmitate is actually found in mature *E. coli* lipid A.[1] The gene coding for

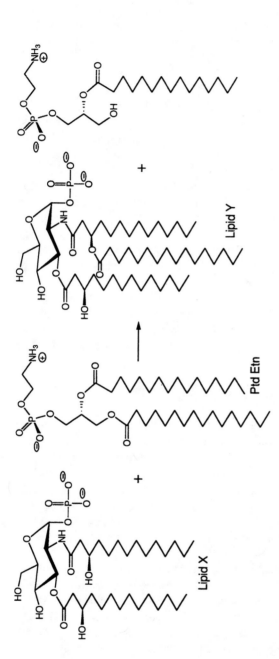

FIGURE 3. Formation of an acyloxyacyl moiety by a transesterification reaction. In the conversion of lipid X to lipid Y, any glycerophospholipid bearing a palmitoyl residue at the *sn*-1 position may be the acyl donor.

the palmitate-incorporating enzyme has not been identified, and mutants lacking it are not yet available.

F. UNEXPLORED ASPECTS OF LIPID A BIOSYNTHESIS

In some organisms, lipid A may be further modified with additional phosphate, phosphoethanolamine, and/or aminoarabinose moieties.[1] The biosynthetic origins and enzymatic incorporation of these "decorations" have not been firmly established.[1] There remains significant controversy about the locations of the polar decorations on the lipid A molecule.[1,13,41,42] Their functions are unknown, but in general they reduce the net negative charge of lipid A.[1]

Preliminary efforts to establish *in vitro* systems for the addition of heptose to $(KDO)_2$-lipid IV_A appear promising (Figure 4).[1,18,43] Heptose addition may occur before or after incorporation of laurate and myristate.[1,18,43] The *rfaC* gene likely encodes for the first heptosyl transferase, while the *rfaF* gene encodes the second.[1,43] The *rfaE* and *D* genes are required for the formation of the ADP-heptose.[1,43,44] These studies would be facilitated by the availability of synthetic ADP-heptose epimers. As explained elsewhere,[1] the biosynthesis of ADP-heptose deserves considerable further investigation.

IV. GENERALITY OF THE BIOSYNTHETIC SCHEME

Three approaches have been used to validate the generality of biosynthetic scheme described above. (1) Mutants defective in UDP-GlcNAc acyltransferase have been constructed and are found to be temperature sensitive for growth.[6] With the available alleles, it is possible to inhibit incorporation of $^{32}P_i$ into lipid A about tenfold relative to glycerophospholipids after 30 min at 42°C.[6] Cells begin to lose viability when the lipid A content is reduced by a factor of two.[6] These findings demonstrate that at least 90% of the lipid A found in *E. coli* is derived by the pathway shown in Figure 2. (2) Several Gram-negative pathogens (including strains of *Salmonella, Enterobacter, Serratia, Klebsiella, Proteus*, and *Pseudomonas*) have been examined for the presence or absence of the enzymes shown in Figure 2.[5] Particular attention has been paid to the UDP-GlcNAc acyltransferases.[5] Both enzymes are found in all organisms examined at specific activities that are within an order of magnitude of those of the *E. coli* enzymes.[5] However, in *Pseudomonas*, the UDP-GlcNAc O-acyltransferase is optimally active with *R*-3-hydroxydecanoate and the N-acyltransferase is optimally active with *R*-3-hydroxylaurate.[5] These findings are consistent with the fatty acyl composition of *Pseudomonas* lipid A.[1,39] (3) Inhibitors of CMP-KDO biosynthesis[45,46] inhibit the growth of the above Gram-negative bacteria and, with the exception of *Pseudomonas* (see above), cause lipid IV_A to accumulate.[39,47]

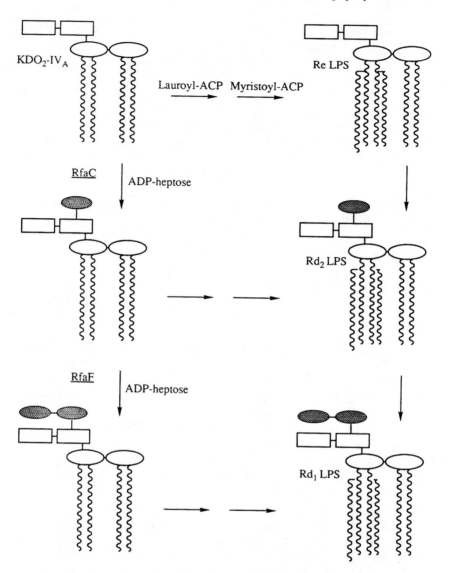

FIGURE 4. Working hypothesis to account for the complex metabolism of $(KDO)_2\text{-IV}_A$ observed in cytosolic extracts of *E. coli* and *S. typhimurium*. Acylation may precede or follow heptose addition,[18,43] but under ordinary circumstances very little lipid A is exported without attachment of laurate and myristate.[1] White ovals, glucosamine; white rectangles, KDO; shaded ovals, heptose.

V. PERSPECTIVE

During the past 5 years, much has been learned about the chemistry and biology of lipid A. The key enzymatic steps in the biosynthesis of lipid A are now well established, and their relevance has been validated by genetic studies.[1] Major issues that require further exploration include the following: (1) the function of lipid A in the bacterial outer membrane; (2) the mechanism by which LPS is exported from its site of biosynthesis on the inner surface of the inner membrane to the outer surface of the outer membrane; and (3) the regulation of lipid A biosynthesis and its coordination with macromolecular synthesis.

In the immediate future, certain straightforward approaches could provide additional insights. (1) The physical properties of lipid A and its precursors could be examined by NMR spectroscopy and X-ray diffraction. Complexes of glucosamine-derived phospholipids with bacterial proteins would be especially informative, since their structures might shed light on the question of how lipid A interacts with proteins on the surfaces of animal cells. (2) The biosynthesis of the core domain should be investigated further at the enzymatic and genetic levels. It is possible that subtle features of core structure have been overlooked,[1] given that synthetic standards of the complete core are not available. Even Re endotoxin might contain additional subtle modifications that have escaped detection. The discovery of an enzyme that cleaves the KDO-lipid A linkage would be extremely helpful in future studies of core assembly. (3) The enzymology and topography of the O-antigen domain[1] deserve further attention. Antibodies to key enzymes, such as the O-antigen ligase and polymerase,[1,48] would help to determine their topographies and might have implications for the mechanism of LPS export.[1] (4) The molecular biology of lipid A assembly must be studied further. Many of the structural genes coding for the enzymes shown in Figure 2 remain to be identified. The function(s) of the complex operon that includes the *lpxA* and *lpxB* genes[7,24,25] must be determined. The DNA sequences of the *rfa* and *rfb* clusters must be completed and analyzed.[49,50] (5) The biosynthesis of lipid A in diverse microbial systems remain to investigated. For instance, the enzymology of lipid A synthesis in organisms that contain diaminoglucose analogs,[51] very long acyl chains,[52] or N-linked 3-ketomyristate[53] has not been studied and could provide unexpected insights.

A broader view of endotoxin biology is certain to emerge.

REFERENCES

1. **Raetz, C. R. H.**, Biochemistry of endotoxins, *Annu. Rev. Biochem.*, 59, 129, 1990.
2. **Nikaido, H. and Vaara, M.**, Outer membrane, in *Escherichia coli and Salmonella typhimurium*, Vol. 1, Neidhardt, F. C., Ed., ASM Publications, American Society of Microbiology, Washington, D.C., 1987, 7.
3. **Nishijima, M. and Raetz, C. R. H.**, Membrane lipid biogenesis in *Escherichia coli*: identification of genes for phosphatidylglycerophosphate synthase and construction of mutants lacking phosphatidylglycerol, *J. Biol. Chem.*, 254, 7837, 1979.
4. **Takayama, K., Qureshi, N., Mascagni, P., Nashed, M. A., Anderson, L., and Raetz, C. R. H.**, Fatty acyl derivatives of glucosamine 1-phosphate in *Escherichia coli* and their relation to lipid A: complete structure of a diacyl GlcN-1-P found in a phosphatidylglycerol-deficient mutant, *J. Biol. Chem.*, 258, 7379, 1983.
5. **Williamson, J. M., Anderson, M. S., and Raetz, C. R. H.**, Acyl-ACP specificity of UDP-GlcNAc acyltransferases from Gram-negative bacteria: relationship to lipid A structure, *J. Bacteriol.*, 173, 3591, 1991.
6. **Galloway, S. M. and Raetz, C. R. H.**, A mutant of *Escherichia coli* defective in the first step of endotoxin biosynthesis, *J. Biol. Chem.*, 265, 6394, 1990.
7. **Crowell, D. N., Reznikoff, W. S., and Raetz, C. R. H.**, Nucleotide sequence of the *Escherichia coli* gene for lipid A disaccharide synthase, *J. Bacteriol.*, 169, 5727, 1987.
8. **Coleman, J. and Raetz, C. R. H.**, First committed step of lipid A biosynthesis in *Escherichia coli*: sequence of the *lpxA* gene, *J. Bacteriol.*, 170, 1268, 1988.
9. **Clementz, T. and Raetz, C. R. H.**, A gene coding for KDO transferase in *Escherichia coli*: identification, mapping, cloning, and sequencing, *J. Biol. Chem.*, 266, 9687, 1991.
10. **Ray, B. L. and Raetz, C. R. H.**, The biosynthesis of Gram-negative endotoxin: a novel kinase in *Escherichia coli* membranes that incorporates the 4' phosphate moiety, *J. Biol. Chem.*, 262, 1122, 1987.
11. **Radika, K. and Raetz, C. R. H.**, Purification and properties of lipid A disaccharide synthase of *Escherichia coli*, *J. Biol. Chem.*, 263, 14859, 1988.
12. **Haselberger, A., Hildebrandt, J., Lam, C., Liehl, E., Loibner, H., Macher, I., Rosenwirth, B., Schütze, E., Vyplel, H., and Unger, F. M.**, Immunopharmacology of lipopolysaccharides (endotoxins) from gram-negative bacteria, *Triangle, Sandoz J. Med. Sci.*, 26, 33, 1987.
13. **Rietschel, E. Th., Brade, L., Schade, U., Seydel, U., Zähringer, U., Kusumoto, S., and Brade, H.**, Bacterial endotoxins: properties and structure of biologically active domains, in *Surface Structures of Microorganisms and their Interactions with the Mammalian Host*, Schrinner, E., Richmond, M. H., Seibert, G., and Schwarz, U., Eds., VCH, New York, 1988, 1.
14. **Anderson, M. S. and Raetz, C. R. H.**, Biosynthesis of lipid A precursors in *Escherichia coli*: a cytoplasmic enzyme that converts UDP-GlcNAc to UDP-3-*O*-(*R*-3-hydroxymyristoyl)-GlcNAc, *J. Biol. Chem.*, 262, 5159, 1987.
15. **Anderson, M. S., Robertson, A. D., Macher, I., and Raetz, C. R. H.**, Biosynthesis of lipid A in *Escherichia coli*: identification of UDP-3-*O*-(*R*-3-hydroxymyristoyl)-α-D-glucosamine as a precursor of UDP-N^2-O^3-bis(*R*-3-hydroxymyristoyl)-α-D-glucosamine, *Biochemistry*, 27, 1908, 1988.
16. **Ray, B. L., Painter, G., and Raetz, C. R. H.**, The biosynthesis of gram-negative endotoxin: formation of lipid A disaccharides from monosaccharide precursors in extracts of *Escherichia coli*, *J. Biol. Chem.*, 259, 4852, 1984.
17. **Brozek, K. A., Hosaka, K., Robertson, A. D., and Raetz, C. R. H.**, Biosynthesis of lipopolysaccharide in *Escherichia coli*: cytoplasmic enzymes that attach 3-deoxy-D-*manno*-octulosonic acid to lipid A, *J. Biol. Chem.*, 264, 6956, 1989.
18. **Brozek, K. A. and Raetz, C. R. H.**, Biosynthesis of lipid A in *Escherichia coli*: acyl carrier protein dependent incorporation of laurate and myristate, *J. Biol. Chem.*, 265, 15410, 1990.

19. **Raetz, C. R. H.**, Molecular genetics of membrane phospholipid synthesis, *Annu. Rev. Genet.*, 20, 253, 1986.
20. **Bulawa, C. E. and Raetz, C. R. H.**, The biosynthesis of Gram-negative endotoxin: identification and function of UDP-diacylglucosamine in *Escherichia coli*, *J. Biol. Chem.*, 259, 4846, 1984.
21. **Anderson, M. S., Bulawa, C. E., and Raetz, C. R. H.**, The biosynthesis of Gram-negative endotoxin: formation of lipid A precursors from UDP-GlcNAc in extracts of *Escherichia coli*, *J. Biol. Chem.*, 260, 15536, 1985.
22. **Park, J. T.**, Murein synthesis, in *Escherichia coli and Salmonella typhimurium*, Vol. 1, Neidhardt, F. C., Ed., ASM Publications, American Society of Microbiology, Washington, DC, 1987, 663.
23. **Cronan, J. E., Jr. and Rock, C. O.**, Biosynthesis of membrane lipids, in *Escherichia coli and Salmonella typhimurium*, Vol. 1, Neidhardt, F. C., Ed., ASM Publications, American Society of Microbiology, Washington, DC, 1987, 474.
24. **Tomasiewicz, H. G. and McHenry, C. S.**, Sequence analysis of the *Escherichia coli dnaE* gene, *J. Bacteriol.*, 169, 5735, 1987.
25. **Tomasiewicz, H. G.**, The Macromolecular Synthesis II Operon in *Escherichia coli*, Doctoral thesis, University of Colorado, Denver, 1990.
26. **Koski, P., Rhen, M., Kantele, J., and Vaara, M.**, Isolation, cloning, and primary structure of a cationic 16-kDa outer membrane protein of *Salmonella typhimurium*, *J. Biol. Chem.*, 264, 18973, 1990.
27. **Icho, T., Sparrow, C. P., and Raetz, C. R. H.**, Molecular cloning and sequencing of the gene for CDP-diglyceride synthetase of *Escherichia coli*, *J. Biol. Chem.*, 260, 12078, 1985.
28. **Raetz, C. R. H., Purcell, S., Meyer, M. V., Qureshi, N., and Takayama, K.**, Isolation and characterization of eight lipid A precursors from a 3-deoxy-D-*manno*-octulosonic acid deficient mutant of *Salmonella typhimurium*, *J. Biol. Chem.*, 260, 16080, 1985.
29. **Sibley, C. H., Terry, A., and Raetz, C. R. H.**, Induction of κ light chain synthesis in 70Z/3 B cell lymphoma cells by chemically defined lipid A precursors, *J. Biol. Chem.*, 263, 5098, 1988.
30. **Loppnow, H., Brade, H., Dürrbaum, I., Dinarello, C. A., Kusumoto, S., Rietschel, E. Th., and Flad, H.-D.**, IL-1 induction capacity of defined lipopolysaccharide partial structures, *J. Immunol.*, 142, 3229, 1989.
31. **Kovach, N. L., Yee, E., Munford, R. S., Raetz, C. R. H., and Harlan, J. M.**, Lipid IV$_A$ inhibits synthesis and release of tumor necrosis factor induced by lipopolysaccharide in whole human blood *ex vivo*, *J. Exp. Med.*, 172, 78, 1990.
32. **Hampton, R. Y., Golenbock, D. T., and Raetz, C. R. H.**, Lipid A binding sites in membranes of macrophage tumor cells, *J. Biol. Chem.*, 263, 14802, 1988.
33. **Raetz, C. R. H., Brozek, K. A., Clementz, T., Coleman, J. D., Galloway, S. M., Golenbock, D. T., and Hampton, R. Y.**, Gram-negative endotoxin: a biologically active lipid, *Cold Spring Harbor Symp. Quant. Biol.*, 53, 973, 1988.
34. **Hampton, R. Y., Golenbock, D. T., Penman, M., Krieger, M., and Raetz, C. R. H.**, Recognition and plasma clearance of endotoxin by scavenger receptors, *Nature*, 352, 342, 1991.
35. **Brown, M. S. and Goldstein, J. L.**, Lipoprotein metabolism in the macrophage: implications for cholesterol deposition in atherosclerosis, *Annu. Rev. Biochem.*, 52, 223, 1983.
36. **Munson, R. S., Jr., Rasmussen, N. S., and Osborn, M. J.**, Biosynthesis of lipid A: enzymatic incorporation of 3-deoxy-D-*manno*-octulosonic acid into a precursor of lipid A in *Salmonella typhimurium*, *J. Biol. Chem.*, 253, 1503, 1978.
37. **Tobias, P. S., Soldau, K., and Ulevitch, R. J.**, Identification of a lipid A binding site in the acute phase reactant lipopolysaccharide binding protein, *J. Biol. Chem.*, 264, 10867, 1989.

38. **Shumann, R. R., Leong, S. R., Flaggs, G. W., Gray, P. W., Wright, S. D., Mathison, J. C., Tobias, P. S., and Ulevitch, R. J.**, Structure and function of lipopolysaccharide binding protein, *Science*, 249, 1429, 1990.

39. **Goldman, R. C., Doran, C. C., Kadam, S. K., and Capobianco, J. O.**, Lipid A precursor from *Pseudomonas aeruginosa* is completely acylated prior to addition of 3-deoxy-D-*manno*-octulosonate, *J. Biol. Chem.*, 263, 5217, 1988.

40. **Brozek, K. A., Bulawa, C. E., and Raetz, C. R. H.**, Biosynthesis of lipid A precursors in *Escherichia coli*: a membrane bound enzyme that transfers a palmitoyl residue from a glycerophospholipid to lipid X, *J. Biol. Chem.*, 262, 5170, 1987.

41. **Strain, S. M., Armitage, I. M., Anderson, L., Takayama, K., Qureshi, N., and Raetz, C. R. H.**, Location of polar substituents and fatty acyl chains on lipid A precursors from 3-deoxy-D-*manno*-octulosonic acid deficient mutants of *Salmonella typhimurium*, *J. Biol. Chem.*, 260, 16089, 1985.

42. **Rietschel, E. Th., Ed.**, *Handbook of Endotoxin: Chemistry of Endotoxin*, Vol. 1, Elsevier/North-Holland Biomedical Press, Amsterdam, 1984.

43. **Sirisena, D. M., Brozek, K. A., MacLachlan, P. R., Sanderson, K. A., and Raetz, C. R. H.**, The *rfaC* gene of *Salmonella typhimurium*: cloning, sequencing, and enzymatic function in heptose transfer to lipopolysaccharide, *J. Biol. Chem.*, in press.

44. **Coleman, W. G., Jr.**, The *rfaD* gene encodes for ADP-L-glycero-D-mannoheptose 6-epimerase, an enzyme required for lipopolysaccharide core biosynthesis, *J. Biol. Chem.*, 258, 1985, 1983.

45. **Hammond, S. M., Claesson, A., Jansson, A. M., Larsson, L.-G., Pring, B. G., Town, C. M., and Ekström, B.**, A new class of synthetic antibacterials acting on lipopolysaccharide biosynthesis, *Nature*, 327, 730, 1987.

46. **Goldman, R., Kohlbrenner, W., Lartey, P., and Pernet, A.**, Antibacterial agents specifically inhibiting lipopolysaccharide synthesis, *Nature*, 329, 162, 1987.

47. **Goldman, R. C., Doran, C. C., and Capobianco, J. O.**, Analysis of lipopolysaccharide synthesis in *Salmonella typhimurium* and *Escherichia coli* using agents which block incorporation of KDO, *J. Bacteriol.*, 170, 2185, 1988.

48. **Rick, P. D.**, Lipopolysaccharide biosynthesis, in *Escherichia coli and Salmonella typhimurium*, Vol. 1, Neidhardt, F. C., Ed., ASM Publications, American Society of Microbiology, Washington, DC, 1987, 648.

49. **Sanderson, K. E. and Hurley, J. A.**, Linkage map of *Salmonella typhimurium*, in *Escherichia coli and Salmonella typhimurium*, Vol. 2, Neidhardt, F. C., Ed., ASM Publications, American Society of Microbiology, Washington, DC, 1987, 877.

50. **Tiang, X.-M., Neal, B., Santiago, F., Lee, S. T., Romana, L. K., and Reeves, P. R.**, Structure and sequence of the *rfb* (O antigen) gene cluster of *Salmonella* serovar typhimurium (strain LT2) *Mol. Microbiol.*, 5, 695, 1991.

51. **Weckesser, J. and Mayer, H.**, Different lipid A types in lipopolysaccharides of phototrophic and related non-phototrophic bacteria, *FEMS Microbiol. Rev.*, 54, 143, 1988.

52. **Hollingsworth, R. I. and Carlson, R. W.**, 27-Hydroxyoctacosanoic acid is a major structural fatty acyl component of the lipopolysaccharide of *Rhizobium trifolii* ANU 843, *J. Biol. Chem.*, 264, 9300, 1989.

53. **Qureshi, N., Honovich, J. P., Hara, H., Cotter, R. J., and Takayama, K.**, Location of fatty acids in lipid A obtained from lipopolysaccharide of *Rhodopseudomonas sphaeroides* ATCC 17023, *J. Biol. Chem.*, 263, 5502, 1988.

54. **Bull, and Raetz, C. R. H.**, unpublished.

55. **Golenbock, D. T., Hampton, R. Y., Qureshi, N., Takayama, K., and Raetz, C. R. H.**, Lipid A-like molecules that antagonize the effects of endotoxins on human monocytes, *J. Biol. Chem.*, 266, 19490, 1991.

Chapter 4

Chemical Synthesis of Lipid A

Shoichi Kusumoto

TABLE OF CONTENTS

I. INTRODUCTION

Lipid A was first described in 1954 by Westphal and Lüderitz, who isolated this fascinating substance as a hydrophobic precipitate after mild acid hydrolysis of lipopolysaccharide (LPS).[1] They soon recognized that lipid A is responsible for most of the so-called endotoxic activity. During the period shortly after the first description of lipid A, there had been much confusion as to the understanding of its biological activity. The poor solubility of lipid A preparations, which lack the hydrophilic polysaccharide portion, caused serious problems, particularly in the reproduction of its biological action. The intrinsic heterogeneity of lipid A and the difficulty in its purification served as additional sources of confusion. Nevertheless, scientific efforts from many investigators gradually led to the generally accepted concept that lipid A is the active site of LPS. The use of triethylammonium salt of lipid A solved many of the solubility problems. Investigations using LPS obtained from rough-type bacteria gave clearer evidence supporting the importance of lipid A for endotoxic activity, since rough-type LPS is devoid of most polysaccharide, but remains endotoxic.[2,3]

By the end of the 1970s, the basic chemical structure of lipid A from enterobacteria was elucidated by the intensive work of Westphal and his group.[3] According to these investigators, lipid A consists of a common backbone structure of a 1,4'-bisphosphorylated β(1-6)disaccharide of D-glucosamine. Many fatty acyl groups bound at two amino as well as hydroxyl groups of the backbone endow the molecule with an overall lipophilic nature. The determination of the fatty acid composition for many bacterial species revealed that 3-hydroxytetradecanoic acid (β-hydroxymyristic acid) of *R*-configuration was the major and characteristic component acid of most lipid A. Some of the 3-hydroxytetradecanoic acids bound to the backbone proved to be further acylated at their 3-hydroxyl groups, forming 3-acyloxytetradecanoyl structures.

Since bacterial cells were used as the source of isolation, however, lipid A could hardly be obtained as a homogeneous compound. The possible influence of trace amounts of unknown contaminants on biological activity could, therefore, not be excluded. Accordingly, lipid A was an attractive target for the challenge of synthetic chemists because of the interesting and diverse biological activities attributed to it and its complicated structure. Many research groups began synthetic work on lipid A around 1980, both in order to confirm the proposed structure and to determine the minimum structure required for the manifesfation of biological activity.

The overall early synthetic work by investigators aimed at the same target structure **1** (Figure 1) in which all possible positions of the backbone were acylated. The 3'-hydroxyl group in **1** remained free, because this position had been proposed to be the linkage site of the polysaccharide in LPS. This target was chosen because no positional distribution of acyl groups had yet

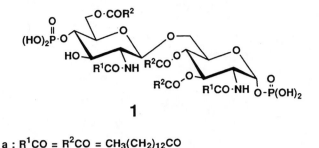

$$a : R^1CO = R^2CO = CH_3(CH_2)_{12}CO$$

$$b : R^1CO = CH_3(CH_2)_{10} \cdot CHCH_2CO \qquad R^2CO = CH_3(CH_2)_{12}CO$$
$$\phantom{b : R^1CO = CH_3(CH_2)_{10} \cdot} OH$$

FIGURE 1. The target structure of early synthetic studies on lipid A.

been determined unequivocally. Comparison of natural lipid A with synthetically obtained compounds **1** containing various combinations of acyl groups revealed that the synthetic and natural materials were significantly different both in the solubility in aqueous systems and in biological activity. The obvious conclusion reached from these observations was that the formula **1** did not represent the true structure of bacterial lipid A. This conclusion dictated the reinvestigation of the structure of lipid A.

In 1985 the revised structure **2** was proposed for the main constituent of lipid A from *Escherichia coli* Re mutant and this structure was subsequently chemically synthesized. The synthetic homogeneous compound exhibited identical biological activity with that of a natural *E. coli* lipid A preparation extracted from bacterial cells.[4,5] This seminal accomplishment provided unequivocal proof that a single pure compound with a definite, though rather complicated, chemical structure is actually responsible for the multiple biological activity described for LPS-endotoxin, a natural amphiphile produced by Gram-negative bacteria. These studies, in this respect, completely excluded the hypothesis that proteins, polysaccharides, or any other bacterial constituents might be responsible for some of the endotoxic activity of LPS. Thereafter, many naturally occurring lipid A analogs have also been synthesized, and their structures unequivocally determined. The endotoxic activity of LPS can, therefore, be discussed at the molecular level in relation to the precise chemical structures of the synthetic homogeneous compounds. We can now also anticipate the utilization of chemical synthesis to prepare any desired lipid A-like compounds which do not exist in nature and thus are never obtained from natural sources. Several research groups are already attempting to modify the chemical structure of natural lipid A, aiming at the creation of less toxic compounds which still retain the beneficial biological activity of the original natural compounds (Figure 2).

In this chapter the result of the synthetic efforts for lipid A and related structural analogs will be described without going into details of their

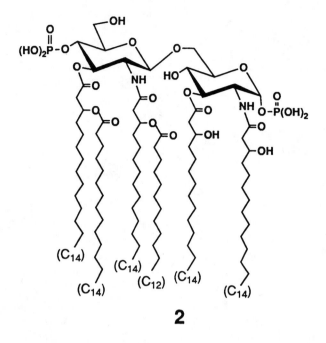

2

FIGURE 2.   The chemical structure of lipid A of *E. coli.*

biological activity, which is a subject of another chapter of this book. The
earlier work aimed at the preparation of the incorrect structure are also explained briefly because many of the important problems for the successful
synthesis of lipid A were solved during these efforts.[6]

II. EARLY SYNTHETIC APPROACHES TOWARD LIPID A

A. PREPARATION OF OPTICALLY PURE (R)-3-HYDROXYTETRADECANOIC ACID

Among the fatty acid components of lipid A of enterobacteria, *(R)*-3-
hydroxytetradecanoic acid is the dominant one.[3] Since an optically and chemically pure form of this hydroxy acid was not available in sufficient amounts
from natural sources, it had to be prepared chemically.

A procedure was described for the optical resolution of racemic 3-hy-
droxytetradecanoic acid with ephedrine for the preparation of one of the
enantiomers of the hydroxy acid.[7] The method was utilized in some of synthetic work on lipid A analogs.

Tai et al. reported an efficient and versatile method for the preparation
of *(R)*-3-hydroxytetradecanoic acid (**3**) by enantioselective hydrogenation of
the corresponding β-keto ester (**4**). They used a Raney Ni catalyst asymmetrically modified by treatment with *(R,R)*-tartaric acid.[8] A large quantity

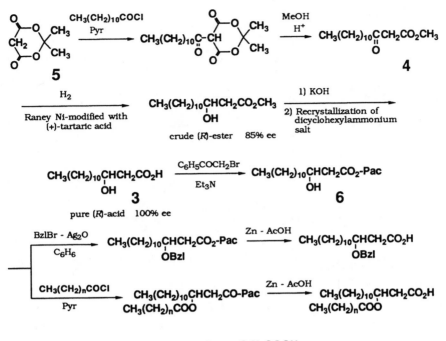

FIGURE 3. Asymmetric synthesis of (R)-3-hydroxytetradecanoic acid and its conversion to the 3-O-benzyl and 3-O-acyl derivatives.

of the optically active hydroxytetradecanoic acid could be prepared by this method and was utilized for much of the synthetic work described in this chapter. The keto ester (4) was readily prepared from Meldrum's acid (5) by a previously reported method.[9] Recently, an alternative efficient method has been described for the enantioselective reduction of 3-keto esters with ruthenium catalyst.[10]

The synthetic hydroxy acid was converted via its phenacyl ester (6) into a protected form, 3-benzyloxytetradecanoic acid, or 3-acyloxytetradecanoic acid, which was then used in the synthesis of lipid A (Figure 3).[11,12] Direct asymmetric synthesis of (R)-3-acyloxytetradecanoic acid with an optical purity of over 95% was also achieved. The key step in this synthesis was an asymmetric allyl boration of 1-dodecanal.[13]

B. SYNTHESIS OF THE BASIC PROPOSED STRUCTURE OF LIPID A

The first successful synthesis of the basic structure was achieved by the group of Shiba and Kusumoto.[14] They synthesized compound 1a (Figure 4) which contains 5 mol of tetradecanoic acid, myristic acid. The product thus corresponds to the proposed structure of lipid A except that all 3-hydroxy-

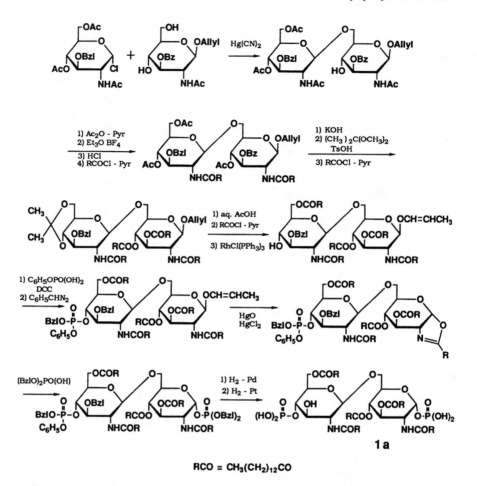

RCO = CH₃(CH₂)₁₂CO

FIGURE 4. Synthesis of pentakis(tetradecanoyl) β(1-6) disaccharide of D-glucosamine 1,4'-bisphosphate (**1a**) corresponding to the early proposed structure of lipid A.

and 3-acyloxytetradecanoic acids are replaced with this nonhydroxylated acid.

In that synthesis, the β(1-6)disaccharide with neither acyl nor phosphate groups was first prepared and then acylated and phosphorylated in a stepwise fashion (Figure 4). Acylation could be effected by a conventional method with the acid chloride. The presence of the phosphate at position 4' did not cause any serious problem. It could be introduced readily with dicyclo-hexylcarbodiimide-phenyl phosphate and was stable in a protected triester form under the various reaction conditions of further transformations. The key step in the synthesis was the formation of the α-phosphate at the glycosidic position. This particular phosphate was by contrast with the 4'-phosphate, chemically unstable, so that it had to be introduced at the latest stage of the entire synthesis just before the final deprotection, which was effected by mild catalytic hydrogenolysis.

The glycosidic phosphate was introduced by a reaction of the corresponding oxazoline, which was formed from the β-(1-propenyl) glycoside. The phosphorylation reaction of an oxazoline was found to proceed by the initial formation of the β-phosphate followed by its isomerization into the α-anomer after sufficiently long reaction times.[14] In the synthesis of **1a**, only the α-phosphate was isolated which has the natural configuration.[15,16]

Among the many other research groups which attempted the synthesis of **1a** and related compounds,[17] van Boeckel et al. succeeded in the synthesis of the same bisphosphate **1a** and its anomer which has the β-glycosyl phosphate.[18] These authors obtained an anomeric mixture of phosphates from a similar oxazoline by a slight modification of the reaction conditions and separation of the anomers before deprotection. They also examined many reagents with various protecting groups for the formation of the 4′-phosphate groups.

C. A NEW SYNTHETIC METHOD FOR GLYCOSYL PHOSPHATES

The basic proposed structure containing glycosyl phosphate with the same α-configuration as the natural lipid A and, if necessary, the corresponding anomeric β-phosphate, could be prepared as described above. However, the oxazoline procedure proved not to be applicable for the synthesis of a glycosyl phosphate of glucosamine, the 2-amino function of which is already acylated with any protected form of 3-hydroxymyristic acid, as in the case of natural lipid A or its synthetic precursor. In that case, an eliminative side reaction inevitably occurred at the step of phosphorylation, with the result that no desired product was obtained.[19,20]

A novel procedure for the preparation of the glycosyl phosphate was then exploited. This method consists of lithiation of the glycosyl hydroxyl group with butyllithium and phosphorylation of the lithiated hydroxyl group with phosphorochloridate.[19] The use of tetrabenzyl pyrophosphate in place of the phosphorochloridate was then recommended.[21] This has proved to be by far the most efficient and practical procedure to date to prepare the desired α-phosphates of *N*-(3-hydroxyacyl)- or *N*-(3-acyloxyacyl)glucosamine related to lipid A, being satisfactorily applied not only in the synthesis of **1** but also for the later synthesis of *E. coli* lipid A, its structural analogs, and lipid X (Figure 5, 6, and 13).

Compound **1b** and several analogous compounds were synthesized in a manner similar to that for **1a**, except that the phosphates were introduced by the butyllithium procedure described above.[22] Surprisingly, the synthetic **1a**, **1b**, and their analogs behaved quite differently from natural lipid A, being much less soluble in aqueous media and far less endotoxic.[23] This completely unexpected observation necessarily prompted an intensive reinvestigation of the chemical structure of natural lipid A.

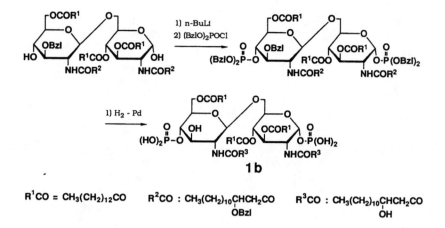

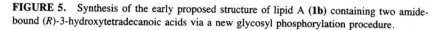

FIGURE 5. Synthesis of the early proposed structure of lipid A (**1b**) containing two amide-bound (*R*)-3-hydroxytetradecanoic acids via a new glycosyl phosphorylation procedure.

III. TOTAL SYNTHESIS OF LIPID A

A. SYNTHESIS OF LIPID A OF *ESCHERICHIA COLI*

In the meantime, the exact chemical structure of the main component of lipid A isolated from *E. coli* Re mutant was elucidated as **2** by use of modern analytical methods such as high field NMR and FD or FAB mass spectrometry.[24-26] Since natural lipid A from *E. coli* Re mutant was known to exhibit full endotoxic activity, it was quite natural that **2** became the next urgent target of chemical synthesis in order not only to confirm this proposed structure but also to verify the principal idea that lipid A is the active site of endotoxin.

E. coli lipid A shares the common backbone of glucosamine β(1-6)disaccharide bisphosphate present in the target **1** of the previous synthesis. Importantly, however, the positions of acylation are not identical. Both glucosamine residues in *E. coli* lipid A are acylated at the same position, namely, at 2-amino and 3-hydroxyl groups. However, the nature of the acyl groups on the individual residues is different. Therefore, a more favorable synthetic scheme could be constructed when two glucosamine residues with the acyl substituents were prepared separately and then coupled to form a disaccharide already containing the acyl functions at the required positions. The other possibility, i.e., to introduce the acyl groups after the formation of the disaccharide, requires more tedious steps of protection and deprotection. In fact, a total synthesis according to the former idea was realized by Imoto et al. as shown in Figure 6.[12,27] In that work, however, a 3-dodecanoyloxytetradecanoyl group bound to the amino group of the nonreducing glucosamine residue was introduced after the formation of the disaccharide, because an elimination reaction leading to the α,β-unsaturated acyl group occurred during

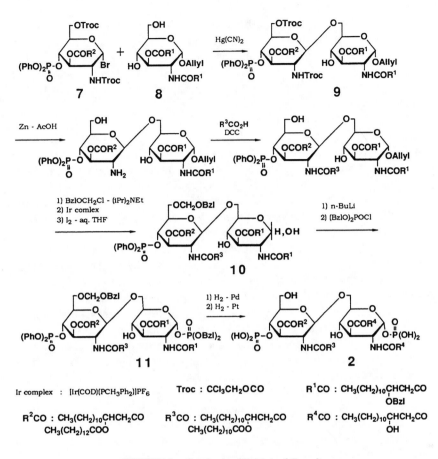

FIGURE 6.   Synthesis of lipid A of *E. coli*.

the glycosidation reaction if an acyloxyacyl group was present on the 2-amino function of the glycosyl donor. This elimination was similar to the reaction observed during the attempted glycosyl phosphorylation via oxazoline (see above). For this reason, the particular 2-amino function was protected with 2,2,2-trichloroethoxycarboyl (Troc) group. A 4-O-phosphoryl group of the nonreducing glucosamine was also introduced in a stable protected form of the diphenyl ester prior to the disaccharide formation. Diphenylphosphorylation could be effected by the method of Szabó et al. with the corresponding chloride and dimethylaminopyridine.[20] The reducing glucosamine component **8** already contained both 3-benzyloxytetradecanoyl groups at the 2-amino and 3-hydroxyl groups, the glycosidic position being protected with allyl group. Coupling of components **7** and **8** gave the disaccharide **9**, the Troc groups of which were then removed and the 2'-amino function then acylated with the above-mentioned 3-dodecanoyloxytetradecanoic acid. Subsequently, the protecting allyl group was removed and the resulting glycosidic hydroxyl group converted into the α-phosphate by the above-described procedure using butyllithium and dibenzylphosphorochloridate. All protecting groups present in **11** were so designed as to be removed hydrogenolytically at this stage. The benzyl-type groups were first split off with palladium black, followed by phenyl groups on the 4'-phosphate with platinum as the catalyst of hydrogenolysis, respectively.

The deprotected product was then purified by preparative thin-layer chromatography (TLC) on silica gel followed by acid precipitation. The latter procedure was indispensable in order to obtain a free acidic form of the final product by removing contaminating cations. Inorganic cations in lipid A preparations are known to disturb the solubilization and it is essential that they be removed.[3] The presence of the acid-labile glycosyl phosphate in the molecule requires careful operation at low temperature during this precipitation process. The free acid form of the product was converted into the soluble triethylammonium salt subsequently used for the biological tests.[12]

Detailed comparison allowed the conclusion that the synthetic and natural lipid A were indeed identical. The synthetic **2** comigrated on TLC with the main component of natural *E. coli* lipid A. Furthermore, the corresponding 4'-monophosphate **12** was prepared by deprotection of the synthetic intermediate **10** prior to the introduction of 1-phosphate.[12] The dimethyl ester of **12** produced by the action of diazomethane was identified by direct comparison of its ^{1}H NMR spectrum with that of the bacterial counterpart isolated in a pure form from *E. coli*.[11] The biological activity of synthetic **2** was also shown to be identical to that of natural lipid A in every respect.[4,5] The chemical synthesis of lipid A therefore provides unequivocal confirmation of the structure which is responsible for the endotoxic activity of bacterial LPS.

The 1-monophosphate **13** as well as the dephospho derivative **14** which lacks both phosphates of **2** was synthesized by modifications of the synthetic procedure described above for **2** (Figure 7).[12] They were used for the elucidation of the role of each phosphate group in manifestation of lipid A biological activities (see Chapter 5).

B. SYNTHESIS OF DISACCHARIDE BIOSYNTHETIC PRECURSORS OF LIPID A

During the course of the structural reinvestigation of *E. coli* lipid A, the positions of direct acylation on the disaccharide backbone were identified to be 2- and 2'-amino as well as 3- and 3'-hydroxyl groups.[28] This result suggested the chemical structure of a biosynthetic precursor of lipid A, designated "precursor Ia", to be represented as **15**. The precursor, which was initially isolated from a temperature-sensitive mutant of *Salmonella*,[29] was shown to contain 4 mol of 3-hydroxytetradecanoic acid as the sole fatty acyl components. Later, Raetz et al. isolated the same compound (named "lipid IVA") and identified the position of this precursor in the biosynthetic pathway (see also Chapter 3).[30,31] Although the structure of precursor Ia had not been

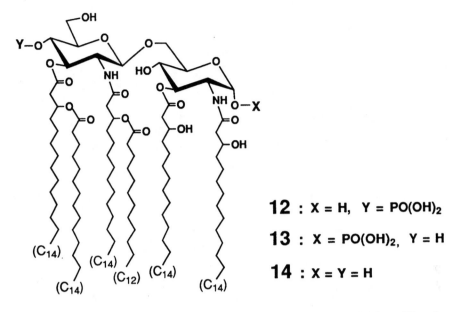

12 : X = H, Y = PO(OH)$_2$

13 : X = PO(OH)$_2$, Y = H

14 : X = Y = H

FIGURE 7.   The chemical structures of the synthetic dephosphorylated derivatives of *E. coli* lipid A.

elucidated at that time, one could assign the structure **15** to it on the basis of the above-mentioned knowledge at that time concerning the positions of acylation on the backbone. Takayama et al. also suggested the same location of the acyl groups in the disaccharide of lipid A on the basis of the structure of lipid X as determined by them (Figure 8).[32]

This assumption was then confirmed by the total synthesis of precursor Ia shown in Figure 9.[11,33] The synthetic strategy was essentially similar to that employed in the early work for the synthesis of **1** described above. Thus, two glucosamine derivatives, appropriately protected as reducing and non-reducing components, were prepared and coupled to form a β(1-6) disaccharide, on which 4 mol of 3-benzyloxytetradecanoic acid was introduced stepwise. The hydroxyl groups at positions 4′ and 1 were phosphorylated with the diphenylphosphorochloridate-dimethylaminopyridine and butyllithium-dibenzylphosphorochloridate procedures, respectively. After hydrogenolysis in two steps and purification as described above in the synthesis of *E. coli* lipid A (**2**), the product obtained was proven as identical to natural precursor Ia. The corresponding 4′-monophosphate, 1-monophosphate, and dephospho derivatives were prepared as well. The dimethyl ester of 4′-monophosphate was identified with the corresponding compound derived from natural precursor Ia, the structure of the particular biosynthetic precursor being thus established for the first time.[33]

Another biosynthetic precursor of lipid A called "precursor Ib" was also isolated from a *S. typhimurium* mutant, which accumulated precursor Ia.[34]

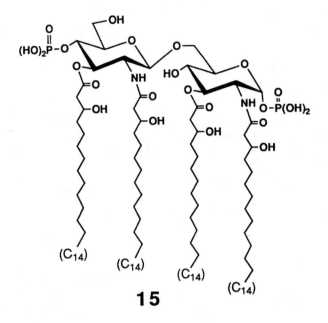

15

FIGURE 8. The chemical structure of ''precursor Ia'' (lipid IVA), a biosynthetic precursor of lipid A.

The compound with the structure **16** proposed for precursor Ib and its isomer **17** were synthesized by an essentially identical procedure to that described in the above synthesis of *E. coli* lipid A.[35,36] Though the two synthetic compounds and natural precursor Ib were indistinguishable on TLC, comparison of 360-MHz ^{1}H NMR spectra of the dimethyl esters of the 4'-monophosphates clearly demonstrated that natural Ib was identical with **16** but not with **17** (Figure 10).

C. SYNTHESIS OF LIPID A OF *SALMONELLA MINNESOTA* AND OTHER BACTERIA

Once the synthetic route for lipid A has been established as described above, many structural analogs which differ in the nature and the distribution of acyl groups could be synthesized by similar methodologies. The above synthesis of precursor Ib and its isomer provides one such example of the methods.

The main component of *S. minnesota* R595 lipid A **(18)** and another analog **(19)** whose pattern of acylation is the same as that observed in lipid A of *Chromobacterium* were synthesized by Kusumoto et al. (Figure 11).[35,37] A synthesis of lipid A of *Proteus mirabilis* by the same principle was reported by Ikeda et al.[38]

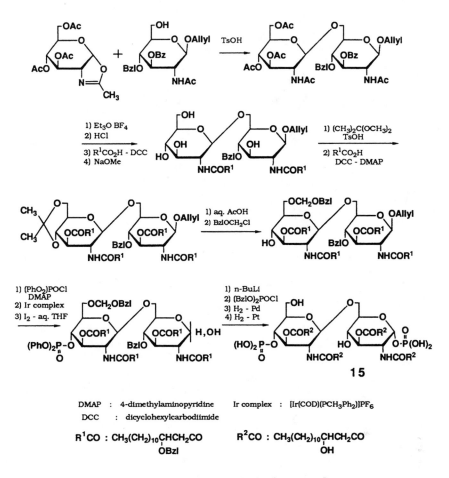

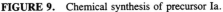

DMAP : 4-dimethylaminopyridine   Ir complex : [Ir(COD)(PCH₃Ph₂)]PF₆
DCC : dicyclohexylcarbodiimide

R¹CO : CH₃(CH₂)₁₀CHCH₂CO
                     |
                    OBzl

R²CO : CH₃(CH₂)₁₀CHCH₂CO
                     |
                    OH

FIGURE 9. Chemical synthesis of precursor Ia.

IV. SYNTHESIS OF MONOSACCHARIDE ANALOGS OF LIPID A

A. SYNTHESIS OF ACYLATED GLUCOSAMINE 1-PHOSPHATES INCLUDING LIPID X AND LIPID Y

2-*N*,3-*O*-Di(3-hydroxytetradecanoyl)-D-glucosamine 1-α-phosphate (**20**) was first isolated from an *E. coli* mutant deficient in phosphatidylglycerol synthesis by Nishijima and Raetz and designated "lipid X".[39] The structure of lipid X corresponds to the reducing half of *E. coli* lipid A, and this monosaccharide derivative was identified as an early biosynthetic precursor of lipid A.[31,32] A closely related monosaccharide derivative **21**, named "lipid Y", was also isolated from the same source.[40] The structure of lipid Y corresponds to the reducing half of *Salmonella* and *Proteus* lipid A (Figure 12).

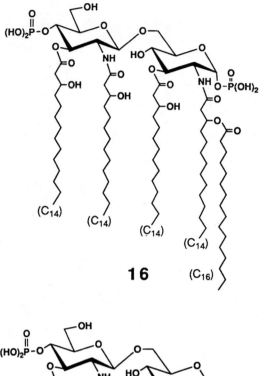

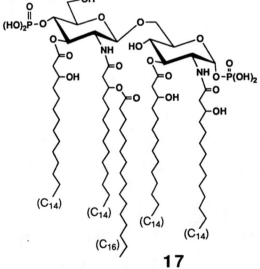

FIGURE 10. The chemical structures of "precursor Ib" (16), another biosynthetic precursor of lipid A, and its structural isomer (17).

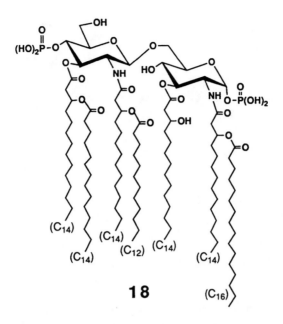

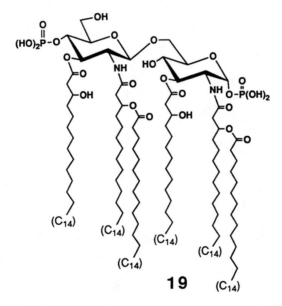

FIGURE 11. The chemical structures of lipid A of *S. minnesota* (**18**) and *Chromobacterium* type (**19**).

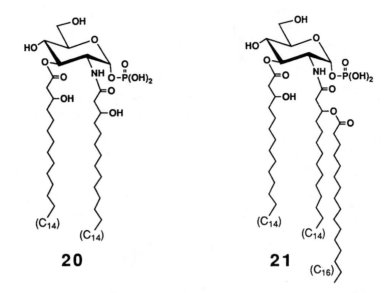

FIGURE 12. The chemical structures of "lipid X" (**20**), a monosaccharide biosynthetic precursor of lipid A, and its analog, "lipid Y" (**21**).

Chemical synthesis of lipid X and lipid Y as well as their dephospho derivatives was reported by Kusumoto et al.[41] The research group of Achiwa also reported the synthesis of lipids X and Y.[42-44] Both groups used the butyllithium procedure for the crucial glycosyl phosphorylation. It should be noted that the latter authors succeeded in removing 4,6-*O*-isopropylidene group in the presence of the glycosyl phosphate, which is usually labile under acidic conditions. The reason for the stability of the phosphate in their preparation is not clear, but might possibly be related to the salt form since the authors did not describe the method by which they removed the inorganic cations which might have been derived from silica gel.

A more improved method of lipid X synthesis was recently described by Macher (Figure 13).[45] He obtained a crystalline tris(hydroxymethyl)-aminomethane salt of lipid X in about 50% overall yield starting from D-glucosamine hydrochloride. He also reported that the protected dibenzyl phosphate intermediate **22** was sufficiently stable to be purified by silica gel column chromatography in spite of the previous observation that the dibenzyl ester of glycosyl phosphate is instable and had to be deprotected immediately.[19] Such divergent stability might be due to the difference in the protecting groups, but it is also possible that an unknown substance such as an acid or excess base present in the mixture might cause the decomposition of the product.

Recently a notable contribution which dealt with the importance of pu-

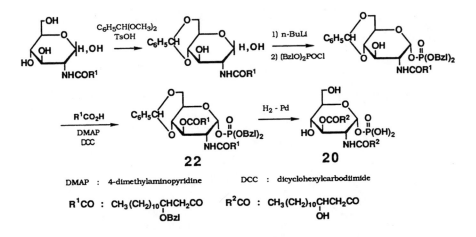

FIGURE 13.   An improved synthetic scheme for lipid X.

rification of synthetic preparations was published by Aschauer et al.[46] In view of the conflicting biological activities of even synthetic lipid X preparations, these authors undertook intensive purification of their synthetic lipid X by means of gel filtration and reversed-phase high-performance liquid chromatography (HPLC), and found that even a crystalline preparation which was assumed to be pure by conventional criteria of purity (e.g., TLC, HPLC, NMR, and elemental analysis) could contain tiny amounts of artifacts **23, 24,** and **25**. Interestingly, **23** and **25** proved to have definite endotoxic activity and have been responsible for the varying biological activity reported for individual preparations, whereas **24** and lipid X **(20)** itself were totally devoid of biological activity. Aschauer et al. stated that the artifacts were formed from **22** or its partly deprotected product by spontaneous glycosidation to form the disaccharide skeleton followed by loss of acyl groups (Figure 14). The reaction was assumed to be induced by the elimination of the glycosyl phosphate residue. Even though such an extreme influence of artifacts on the biological activity could not be always expected, these results underscore the importance of paying enough attention to the purity of even synthetically obtained preparations, in order to obtain unequivocal knowledge on the structure-activity relationships.

Chemical synthesis of the α- and β-1-phosphates and the α-1-pyrophosphate of N-[3-hydroxytetradecanoyl]-D-glucosamine was described recently, as well as related derivatives which have ethanolamine linked to the phosphate part of the compounds. These structures have been identified as being present in bacterial lipid A, though usually not in a stoichiometric amount. The synthetic compounds will be useful in the analysis of the behavior of these structural elements during isolation and degradation studies of bacterial lipid A.[47]

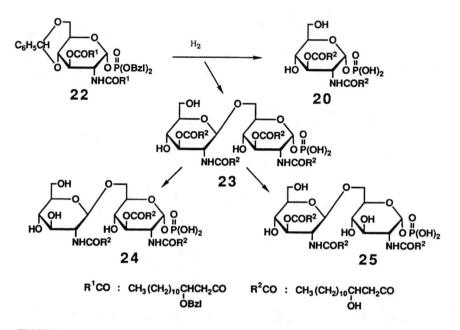

FIGURE 14. The chemical structures of the disaccharide contaminants isolated from a preparation of lipid X and their possible route of formation during the final synthetic steps of lipid X.

B. SYNTHESIS OF ACYLATED GLUCOSAMINE 4-PHOSPHATES

A diacylated glucosamine 4-phosphate **26** which corresponds to the non-reducing half of *E. coli*, *Salmonella*, and *Proteus* lipid A and some of its analogs (**27** to **29**) were synthesized by the group of Achiwa.[48,49] The latter two analogs differ from the natural type (**26**) in the chain length of nonhydroxylated fatty acids bound to the 3-hydroxytetradecanoic acid residues (Figure 15).

Similar acylated glucosamine 4-phosphates were also synthesized by the group of Hasegawa.[50] In spite of its very simple structure and unnatural combination of the acyl groups, compound **30** prepared first was found to manifest some of the beneficial activities of lipid A. This observation, together with the relatively low toxicity of **30**, prompted further synthesis of a series of analogous compounds such as **31**, containing acyl groups of various chain lengths with or without hydroxyl groups.[51]

V. SYNTHESIS OF UNNATURAL ANALOGS OF LIPID A

One of the prominent features of chemical synthesis lies in the point that synthesis can open a way to such compounds never obtained from natural

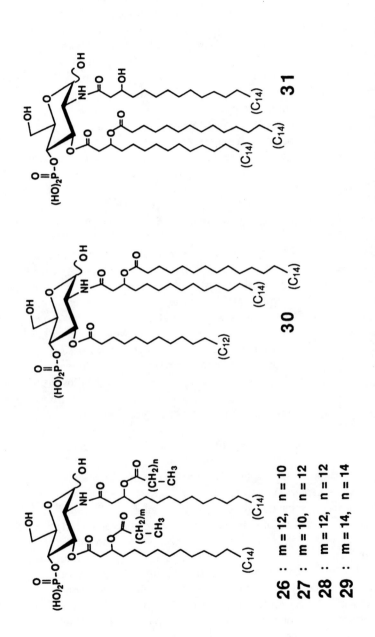

FIGURE 15. The chemical structures of acylated glucosamine 4-phosphates corresponding to the nonreducing half of lipid A and related synthetic analogs.

sources. Through the analysis of structure-activity relationships of synthetic lipid A analogs, we have already been able to conclude that some of the beneficial and detrimental activities of lipid A could be separated depending on the chemical structures. The details of such analyses will be described in another chapter of this book. This conclusion encouraged attempts to obtain synthetic analogs of lipid A which have low or hopefully no toxic activity but still retain beneficial ones such as antitumor activity and so on. Although no ideal compounds have yet been described in this direction, some of the reported attempts will be summarized here.

The synthesis of glucosamine 4-phosphates by Hasegawa's group, described in the preceding paragraph, is one typical example along this line of approach. Those monosaccharides have an obvious advantage that they can be prepared in many fewer steps than a disaccharide derivative corresponding to lipid A.[52-54] These authors also prepared several glycoside derivatives and glycosyl phosphates of the acylated glucosamine 4-phosphates.[55] Similar 1,4-bisphosphorylated and 1,4-bisphosphonylated diacyl glucosamines were described by others.[56]

Substitution of glycosyl phosphate with other polar groups was also attempted. Kusama et al. prepared several lipid A analogs (such as **32**) which have a phosphonooxyethyl group at the glycosyl position in place of the phosphate.[57] Owing to the chemical stability of the phosphonooxyethyl glycosides, the synthesis of these analogs was more straightforward. A monosaccharide (**33**) with a protected phosphonooxyethyl glycoside was first prepared and coupled with the previously described glycosyl donor **7** used in the synthesis of *E. coli* lipid A to give directly a bisphosphorylated disaccharide analog **34**. Deprotection and acylation of the 2′-amino function followed by deprotection completed the synthesis (Figure 16). The final purification of the products was likewise more easy than the natural-type derivatives with a glycosyl phosphate moiety. For example, the phosphonooxyethyl analogs could be converted into free acid form by simple ion exchange chromatography.

Another highly unique approach is the enzymatic synthesis of lipid A analogs. Raetz (see Chapter 3) characterized an enzyme, lipid A synthase, which participates in the biosynthesis of lipid A in an *E. coli* strain.[58] The enzyme catalyzes the formation of a disaccharide 1-monophosphate (**23**) from lipid X and UDP-lipid X (**35**). The specificity of the enzyme toward the lipid X component in this reaction is not very high so that some artificial analogs such as **36** could be incorporated to form a disaccharide **37**.[59] This procedure is quite unique in the point that two components could be coupled at the desired position without any tedious protection of hydroxyl functions. On the other hand, there is an obvious restriction in that disaccharide analogs with 4′-phosphate could not be prepared. This is owing to the specificity of the enzyme, which does not accept UDP-lipid X bearing an additional phosphate group at position 4 (Figure 17).

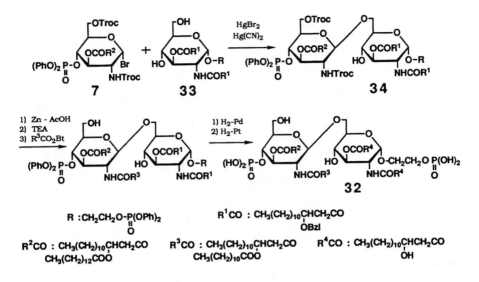

FIGURE 16. Synthesis of a phosphonooxyethyl analog of lipid A of *E. coli*.

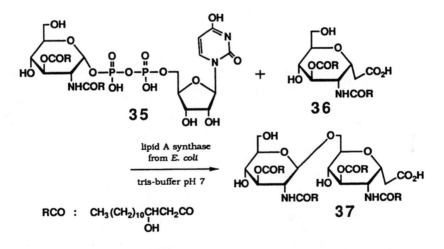

FIGURE 17. Enzymatic synthesis of lipid A analog with lipid A synthase.

VI. CONCLUDING REMARKS

Chemical synthesis of lipid A certainly contributed not only to the determination of its chemical structure but also in many aspects of research related to the biological activities of this complicated molecule. Not being influenced by other contaminants from bacterial cells such as proteins and polysaccharides, synthetic compounds assure a high reproducibility and thus high reliability of results in biological studies. However, it must be emphasized here that the successful contribution of chemistry became possible only on the basis of the available knowledge accumulated by many biological and analytical studies during a long period before the start of the synthetic work. After the synthetic work was initiated, harmonious collaboration was inevitable as well with many research groups of various fields.

It is also evident that completion of the synthesis of lipid A was not a final goal for synthetic chemists. In view of the many new directions of research in this area, synthesis will certainly find many changes for new contributions. Even in a limited research area aiming at the elucidation of the mechanism of action and the metabolic fate of lipid A, for example, one could immediately think of a preparation of radiolabeled lipid A derivatives which will be of great value. In fact, several attempts have already started either by pure chemical synthesis or by combined use of chemical and enzymatic reactions. Synthetic identification of metabolic products may also be important and the preparation of low toxic and clinically applicable lipid A analogs is another dream. It may not be an author's optimistic view to expect that some of these will be realized in the not-so-distant future.

REFERENCES

1. **Westphal, O. and Lüderitz, O.,** Chemische Erforschung von Lipopolysacchariden gramnegativer Bakterien, *Angew. Chem.,* 66, 407, 1954.
2. **Westphal, O. and Lüderitz, O.,** Die Bedeutung von Mutanten bei Enterobakteriaceen für die chemische Erforschung ihrer zellwand Polysaccharide, *Angew. Chem.,* 78, 172, 1966.
3. **Galanos, C., Lüderitz, O., Rietschel, E. Th., and Westphal, O.,** Newer aspects of the chemistry and biology of bacterial lipopolysaccharides with special references to their lipid A component, in *International Review of Biochemistry,* Vol. 14, Goodwin, T., Ed., University Park Press, Baltimore, 1977, 239.
4. **Galanos, C., Lüderitz, O., Rietschel, E. Th., Westphal, O., Brade, H., Brade, L., Freudenberg, M., Schade, U., Imoto, M., Yoshimura, H., Kusumoto, S., and Shiba, T.,** Synthetic and natural *Escherichia coli* free lipid A express identical endotoxic activities, *Eur. J. Biochem.,* 148, 1, 1985.
5. **Kotani, S., Takada, H., Tsujimoto, M., Ogawa, T., Takahashi, I., Ikeda, T., Otsuka, K., Shimauchi, H., Kasai, N., Mashimo, J., Nagao, N., Tanaka, A., Harada, K., Nagai, K., Kutamura, H., Shiba, T., Kusumoto, S., Imoto, M., and Yoshimura, H.,** Synthetic lipid A with endotoxic and related biological activities comparable to those of a natural lipid A from an *Escherichia coli* Re-mutant, *Infect. Immun.,* 49, 225, 1985.
6. **Shiba, T. and Kusumoto, S.,** Chemical synthesis and biological activity of lipid A analogs, in *Handbook of Endotoxin,* Vol. 1, Rietschel, E. Th., Ed., Elsevier, Amsterdam, 1984, 284.
7. **Hiramoto, M., Okada, K., Nagai, S., and Kawamoto, H.,** Structure of viscosin, a peptide antibiotic. I. Syntheses of D- and L-3-hydroxyacyl-L-leucine hydrazides related to viscosin, *Chem. Pharm. Bull.,* 19, 1308, 1971.
8. **Tai, A., Nakahata, M., Harada, T., Izumi, Y., Kusumoto, S., Inage, M., and Shiba, T.,** A facile method for preparation of the optically pure 3-hydroxytetradecanoic acid by an application of asymmetrically modified nickel catalyst, *Chem. Lett.,* p. 1125, 1980.

9. **Oikawa, Y., Sugano, K., and Yonemitsu, O.,** Meldrum's acid in organic synthesis. II. A general and versatile synthesis of β-keto esters, *J. Org. Chem.*, 43, 2087, 1978.

10. **Noyori, R., Ohkuma, T., Kitamura, M., Takaya, T., Sayo, N., Kumobayashi, H., and Akutagawa, S.,** Asymmetric hydrogenation of β-keto carboxylic esters. A practical, purely chemical access to β-hydroxy esters in high enantiomeric purity, *J. Am. Chem. Soc.*, 109, 5856, 1987.

11. **Imoto, M., Yoshimura, H., Yamamoto, M., Shimamoto, T., Kusumoto, S., and Shiba, T.,** Chemical synthesis of a biosynthetic precursor of lipid A with a phosphorylated tetraacyl disaccharide structure, *Bull. Chem. Soc. Jpn.*, 60, 2169, 1987.

12. **Imoto, M., Yoshimura, H., Sakaguchi, N., Shimamoto, T., Kusumoto, S., and Shiba, T.,** Total synthesis of *Escherichia coli* lipid A, the endotoxically active principle of cell-surface lipopolysaccharide, *Bull. Chem. Soc. Jpn.*, 60, 2205, 1987.

13. **Jadhav, P. K.,** Asymmetric synthesis of (3R)-alkanoyloxytetradecanoic acid-components of bacterial lipopolysaccharides, *Tetrahedron Lett.*, 30, 4763, 1989.

14. **Inage, M., Chaki, H., Kusumoto, S., and Shiba, T.,** Chemical synthesis of phosphorylated fundamental structure of lipid A, *Tetrahedron Lett.*, 22, 2281, 1981.

15. **Batley, M., Packer, N. H., and Redmond, J. W.,** Configuration of glycosidic phosphates of lipopolysaccharide from *Salmonella* R595, *Biochemistry*, 21, 6580, 1982.

16. **Zähringer, U., Lindner, B., Seydel, U., Rietschel, E. Th., Naoki, H., Unger, F. M., Imoto, M., Kusumoto, S., and Shiba, T.,** Structure of de-O-acylated lipopolysaccharide from the *Escherichia coli* Re-mutant strain F515, *Tetrahedron Lett.*, 26, 6321, 1985.

17. **Anderson, L. and Unger, F. M., Eds.,** *Bacterial Lipopolysaccharides*, American Chemical Society, Washington, DC, 1983.

18. **van Boeckel, C. A. A., Hermans, J. P. G., Westerduin, P., Oltvoort, J. J., van der Marcel, G. A., and van Boom, J. H.,** Synthesis of two diphosphorylated derivatives containing α- and β-anomeric phosphates, *Recl. Trav. Chim. Pays-Bas*, 102, 438, 1983.

19. **Inage, M., Chaki, H., Kusumoto, S., and Shiba, T.,** A convenient preparative method of carbohydrate phosphates with butyllithium and phosphorochloridate, *Chem. Lett.*, p. 1281, 1982.

20. **Szabó, P., Sarfati, S. R., Diolez, C., and Szabó, L.,** Synthesis of O-[2-deoxy-2-[(3R)-3-hydroxytetradecanamido]-β-D-glucopyranosyl 4-phosphate]-(1-6)-2-deoxy-2-[3R]-3-hydroxytetradecanamido]-D-glucose. The monosaccharide route, *Carbohydr. Res.*, 111, C9, 1983.

21. **Chouinard, P. M. and Bartlett, P. A.,** Conversion of shikimic acid to 5-enolpyruvylshikimate 3-phosphate, *J. Org. Chem.*, 51, 75, 1986.

22. **Inage, M., Chaki, H., Imoto, M., Shimamoto, T., Kusumoto, S., and Shiba, T.,** Synthetic approach to lipid A: preparation of phosphorylated disaccharides containing (R)-3-hydroxyacyl and (R)-3-acyloxyacyl groups, *Tetrahedron Lett.*, 24, 2011, 1983.

23. **Tanamoto, K., Zähringer, U., Mckenzie, G. R., Galanos, C., Rietschel, E. Th., Lüderitz, O., Kusumoto, S., and Shiba, T.,** Biological activities of synthetic lipid A analogs: pyrogenicity, lethal toxicity, anticomplement activity, and induction of gelatin of *Limulus* amoebocyte lysate, *Infect. Immun.*, 44, 421, 1984.

24. **Seydel, U., Lindner, B., Wollenweber, H.-W., and Rietschel, E. Th.,** Structural studies on the lipid A component of enterobacterial lipopolysaccharides by laser desorption mass spectroscopy. Location of acyl groups at the lipid A backbone, *Eur. J. Biochem.*, 145, 505, 1984.

25. **Imoto, M., Kusumoto, S., Shiba, T., Rietschel, E. Th., Galanos, C., and Lüderitz, O.,** Chemical structure of *Escherichia coli* lipid A, *Tetrahedron Lett.*, 26, 907, 1985.

26. **Qureshi, N., Takayama, K., Mascagni, P., Honovich, J., Wong, R., and Cotter, R. J.,** Complete structural determination of lipopolysaccharide obtained from deep rough mutant of *Escherichia coli*, *J. Biol. Chem.*, 263, 11971, 1988.

27. **Imoto, M., Yoshimura, H., Sakaguchi, N., Kusumoto, S., and Shiba, T.,** Total synthesis of *Escherichia coli* lipid A, *Tetrahedron Lett.*, 26, 1545, 1985.

28. **Imoto, M., Kusumoto, S., Shiba, T., Naoki, H., Iwashita, T., Rietschel, E. Th., Wollenweber, H.-W., Galanos, C., and Lüderitz, O.,** Chemical structure of *E. coli* lipid A: linkage site of acyl groups of the disaccharide backbone, *Tetrahedron Lett.,* 24, 4017, 1983.

29. **Lehmann, V.,** Isolation, purification and properties of an intermediate in 3-deoxy-D-*manno*-octulosonic acid-lipid A biosynthesis, *Eur. J. Biochem.,* 75, 257, 1977.

30. **Raetz, C. R. H., Purcel, S., Meyer, M. V., Qureshi, N., and Takayama, K.,** Isolation and characterization of eight lipid A precursors from a 3-deoxy-D-manno-octulosonic acid-deficient mutant of *Salmonella typhimurium, J. Biol. Chem.,* 260, 16080, 1985.

31. **Brozek, K. A. and Raetz, C. R. H.,** Biosynthesis of lipid A in *Escherichia coli, J. Biol. Chem.,* 265, 15410, 1990.

32. **Takayama, K., Qureshi, N., Mascagni, P., Nashed, M. A., Anderson, L., and Raetz, C. R. H.,** Fatty acyl derivatives of glucosamine 1-phosphate in *Escherichia coli* and their relation to lipid A, *J. Biol. Chem.,* 258, 7379, 1983.

33. **Imoto, M., Yoshimura, H., Yamamoto, M., Shimamoto, T., Kusumoto, S., and Shiba, T.,** Chemical synthesis of phosphorylated tetraacyl disaccharide corresponding to a biosynthetic precursor of lipid A, *Tetrahedron Lett.,* 25, 2667, 1984.

34. **Hansen-Hagge, T., Lehmann, V., Seydel, U., Lindner, B., and Zähringer, U.,** Isolation and structural analysis of two lipid A precursors from a KDO-deficient mutant of *Salmonella typhimurium* differing in their hexadecanoic acid content, *Arch. Microbiol.,* 141, 353, 1985.

35. **Kusumoto, S., Kusunose, N., Imoto, M., Shimamoto, T., Kamikawa, T., Takada, H., Kotani, S., Rietschel, E. Th., and Shiba, T.,** Synthesis and biological function of bacterial endotoxin, *Pure Appl. Chem.,* 61, 461, 1989.

36. **Rietschel, E. Th., Brade, L., Schade, U., Galanos, C., Freudenberg, M., Lüderitz, O., Kusumoto, S., and Shiba, T.,** Endotoxic properties of synthetic pentaacyl lipid A precursor Ib and a structural isomer, *Eur. J. Biochem.,* 169, 27, 1987.

37. **Kotani, S., Takada, H., Takahashi, I., Tsujimoto, M., Ogawa, T., Ikeda, T., Harada, K., Okamura, H., Tamura, T., Tanaka, S., Shiba, T., Kusumoto, S., Imoto, M., Yoshimura, H., and Kasai, N.,** Low endotoxic activities of synthetic *Salmonella*-type lipid A with an additional acyloxyacyl group on the 2-amino group of β(1-6)glucosamine disaccharide 1,4′-bisphosphate, *Infect. Immun.,* 52, 872, 1986.

38. **Ikeda, K., Takahashi, T., Kondo, H., and Achiwa, K.,** Total synthesis of *Proteus mirabilis* lipid A, *Chem. Pharm. Bull.,* 35, 1311, 1987.

39. **Nishijima, M. and Raetz, C. R. H.,** Characterization of two membrane-associated glycolipids from an *Escherichia coli* mutant deficient in phosphatidylglycerol, *J. Biol. Chem.,* 256, 10690, 1981.

40. **Takayama, K., Qureshi, N., Mascagni, P., Anderson, L., and Raetz, C. R. H.,** Glucosamine-derived phospholipid in *Escherichia coli, J. Biol. Chem.,* 258, 14245, 1983.

41. **Kusumoto, S., Yamamoto, M., and Shiba, T.,** Chemical synthesis of lipid X and lipid Y, acyl glucosamine 1-phosphates isolated from *Escherichia coli* mutant, *Tetrahedron Lett.,* 25, 3727, 1984.

42. **Takahashi, T., Shimizu, C., Nakamoto, S., Ikeda, K., and Achiwa, K.,** A new methodology for chemoselection of one amino and four hydroxyl groups of glucosamine derivatives and its use for synthesis of lipid X, *Chem. Pharm. Bull.,* 33, 1760, 1985.

43. **Ikeda, K., Takahashi, T., Shimizu, C., Nakamoto, S., and Achiwa, K.,** Lipid A and related compounds. XI. New, efficient synthesis of lipid X, *Chem. Pharm. Bull.,* 35, 1383, 1987.

44. **Ikeda, K., Nakamoto, S., Takahashi, T., and Achiwa, K.,** Lipid A and related compounds. XIV. A new synthesis of lipid Y, the reducing sugar moiety of *Salmonella*- and *Proteus*-type lipid A, *Chem. Pharm. Bull.,* 35, 4436, 1987.

45. **Macher, I.,** A convenient synthesis of 2-deoxy-2-[(*R*)-3-hydroxytetradecanamido]-3-*O*-[(*R*)-3-hydroxytetradecanoyl]-α-D-glucopyranose 1-phosphate (lipid X), *Carbohydr. Res.,* 162, 79, 1987.

46. **Aschauer, H., Grob, A., Hildebrandt, J., Schuetze, E., and Stuetz, P.,** Highly purified lipid X is devoid of immunostimulatory activity, *J. Biol. Chem.*, 265, 9159, 1990.
47. **Trigalo, F., Charon, D., and Szabó, L.,** Chemistry of bacterial endotoxin. IV. Synthesis of anomeric 2-deoxy-2-[(3R)-3-hydroxytetradecanamido]-D-glucopyranosyl phosphate and pyrophosphate derivatives related to 'lipid A', *J. Chem. Soc. Perkin Trans. 1*, 2243, 1988.
48. **Nakamoto, S., Takahashi, T., Ikeda, A., and Achiwa, K.,** Efficient synthesis of novel monosaccharide analogs of lipid A, *Chem. Pharm. Bull.*, 33, 4098, 1985.
49. **Nakamoto, S., Takahashi, T., Ikeda, K., and Achiwa, K.,** Lipid A and related compounds. XV. Efficient synthesis of novel analogs of glucosamine 4-phosphate, the nonreducing sugar moiety of lipid A, *Chem. Pharm. Bull.*, 35, 4517, 1987.
50. **Kiso, M., Tanaka, S., Takahashi, T., Fujishima, Y., Ogawa, Y., and Hasegawa, A.,** Synthesis of 2-deoxy-4-O-phosphono-3-O-tetradecanoyl-2-[(3R)- and (3S)-3-tetradecanoyloxytetradecanamido]-D-glucose: a diastereomeric pair of 4-O-phosphono-D-glucosamine derivative (GLA-27) related to bacterial lipid A, *Carbohydr. Res.*, 148, 221, 1986.
51. **Kiso, M., Tanaka, S., Fujita, M., Fujishima, Y., Ogawa, Y., and Hasegawa, A.,** Synthesis of nonreducing-sugar subunit analogs of bacterial lipid A carrying an amide-bound (3R)-3-acyloxytetradecanoyl group, *Carbohydr. Res.*, 162, 247, 1987.
52. **Kiso, M., Ogawa, Y., Fujishima, Y., Fujita, M., Tanaka, S., and Hasegawa, A.,** Nonreducing-sugar subunit analogs of bacterial lipid A carrying the ester-bond (3R)-3-(acyloxy)tetradecanoyl group, *J. Carbohydr. Chem.*, 6, 625, 1987.
53. **Ogawa, Y., Fujishima, Y., Ishida, H., Kiso, M., and Hasegawa, A.,** Synthesis of 2-deoxy-2-[(3R)-3-hydroxyacyl]amino-4-O-phosphono-3-O-[(3R)-3-tetradecanoyloxytetradecanoyl]-D-glucopranose derivatives related to bacterial lipid A, *J. Carbohydr. Chem.*, 9, 643, 1990.
54. **Ogawa, Y., Fujishima, Y., Ishida, H., Kiso, M., and Hasegawa, A.,** Synthesis of 3-O-[(3R)-3-acyloxytetradecanoyl]-2-deoxy-2-[(3R)-3-hydroxytetradecanamido]-4-O-phosphono-D-glucose derivatives related to bacterial lipid A, *Carbohydr. Res.*, 197, 281, 1990.
55. **Ogawa, Y., Fujishima, Y., Konishi, I., Kiso, M., and Hasegawa, A.,** The chemical modification of the C-1 substituent of a 4-O-phosphono-D-glucosamine derivative (GLA-27) related to bacterial lipid A, *J. Carbohydr. Chem.*, 6, 399, 1987.
56. **Westerduin, P., Veeneman, G. H., and van Boom, J. H.,** Synthesis of diphosphorylated and diphosphonylated lipid A monosaccharide analogues *via* phosphite intermediates, *Rec. Trav. Chim. Pays-Bas*, 106, 601, 1987.
57. **Kusama, T., Soga, T., Shioya, E., Nakayama, K., Nakajima, H., Osada, Y., Ono, Y., Kusumoto, S., and Shiba, T.,** Synthesis and antitumor activity of lipid A analogs having a phosphonooxyethyl group with α- and β-configuration at position 1, *Chem. Pharm. Bull.*, 38, 3366, 1990.
58. **Radica, K. and Raetz, C. R. H.,** Purification and properties of lipid A disaccharides synthase of *Escherichia coli*, *J. Biol. Chem.*, 263, 14859, 1988.
59. **Vyplel, H., Scholz, D., Schindlmayer, K., Bednarik, K., Lam, C., and Schtütze, E.,** C-Glycosidic analogues of lipid A and lipid X: synthesis and biological activities, *J. Med. Chem.*, 34, 2759, 1991.

Chapter 5

STRUCTURE-FUNCTION RELATIONSHIPS OF LIPID A

Haruhiko Takada and Shozo Kotani

TABLE OF CONTENTS

I. INTRODUCTION

In 1985, Japanese and German investigators demonstrated that the *Escherichia coli*-type lipid A (LA-15-PP) (Figure 1) chemically synthesized by Shiba-Kusumoto's group[1,2] reproduced practically all the endotoxic and related bioactivities of bacterial endotoxic lipopolysaccharides (LPS) or lipid A.[3-5] A large number of lipid A structures and related analogs has so far been synthesized by Japanese (Shiba-Kusumoto's group, Hasegawa-Kiso's group, and Achiwa's group), French (Charon's group), and Austrian (Macher's group) scientists (see Chapter 4). Using these homogeneous and chemically well-defined compounds, the structure-function relationships of lipid A have been extensively investigated.

In this chapter we will first review the structural requirements of lipid A for endotoxic and related biological activities (Section II). In Section III we will describe the bioactivities of synthetic lipid A preparations and disaccharide type precursors in relation to different chemical structures. These compounds have a 1,4'-bisphospho β(1-6)-linked D-glucosamine disaccharide backbone. In Section IV we will discuss so-called less (or low) toxic lipid A analogs, either disaccharide or monosaccharide (subunit) type. One of the topics in this section will be their effects as lipid A antagonists. In the concluding section (V), we will discuss possible contributions brought about by elucidation of structure-function relationships to the future development of endotoxicology.

II. STRUCTURAL REQUIREMENTS OF LIPID A FOR VARIOUS BIOLOGICAL ACTIVITIES

We have studied the structure and function relationships of lipid A using a variety of synthetic lipid A and related compounds prepared mainly by Shiba-Kusumoto's group.[3,6-12] We concluded that the structual requirements of lipid A were extremely strict for some bioactivities, but were less so for others. Thus we proposed to discriminate the bioactivities of lipid A and related compounds in five categories as listed in Table 1.[12,13]

The term "endotoxic activities" is sometimes used to indicate all of the collective bioactivities exhibited by endotoxic LPS and its bioactive center, lipid A. In the present description, however, this term refers only to the toxic or harmful effects worthy of the name of "(endo)toxin". In this context, the lethal toxicity in normal animals (exemplified by lethality in the chick embryo), pyrogenicity, and the potency to prepare and provoke the Shwartzman reaction are classified in category I. For these typical endotoxic bioactivities, structural requirements, which are very strict, are as follows. (1) a β(1-6)-linked D-glucosamine disaccharide backbone; (2) bisphosphorylation at positions 1 and 4' of the disaccharide backbone; (3) a suitable number of 3-acyloxyacyl groups per disaccharide unit; and (4) acyl groups of a suitable

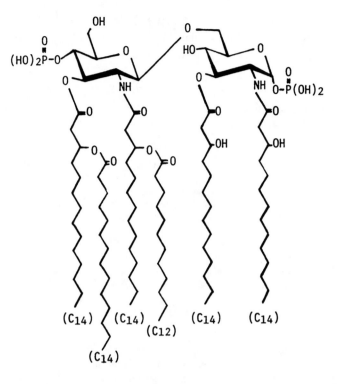

FIGURE 1. Chemical structure of synthetic *E. coli*-type lipid A, LA-15-PP.

length as indicated by Kumazawa et al.[14] and Nakatsuka et al.[15] Thus *E. coli*-type lipid A and closely related disaccharide compounds were most active. On the other hand, the compounds lacking at least one of the above requisites do not exhibit all the endotoxic activities of category I-1, although the structural requirements for pyrogenicity seem to be less strict (category I-2).

Structural requirements for the bioactivities listed in category II are less rigorous as compared with those of category I. The bioactivities classified in category II-1 require the structure of β(1-6)-D-glucosamine disaccharide, but do not need either the presence of 3-acyloxyacyl groups on the backbone structure or a bisphosphate backbone. Thus, such compounds as lipid A precursor Ia (IV$_A$) (LA-14-PP) and monophosphoryl lipid A analogs (for example, LA-15-PH) were active in lethal toxicity assay in galactosamine-loaded mice, the induction of cytokines, immunoadjuvancy, and *Limulus* activity (by the Endospecy test,[16] a colorimetric method by use of a *Tachypleus tridentatus* clotting enzyme preparation from which factor G sensitive to β(1-3)-D-glucan was removed, and a chromogenic substance). Bioactivities included in category II-2, all of which were detected by assays *in vitro*, do not require the disaccharide backbone structure since some monosaccharide lipid A (lipid A-subunit) analogs are definitely active. This may mean that

TABLE 1
Classification of Bioactivities of Lipid A on the Basis of Structural Requirements

Category I

I-1   Lethal toxicity in chick embryos
        Shwartzman reaction
I-2   Pyrogenicity

Category II

II-1   Lethal toxicity in galactosamine-loaded mice
        IFN-α/β-inducing activity in *P. acnes*-primed mice
        TNF-inducing activity in *P. acnes*- or BCG-primed mice
        Immunoadjuvant activity *(in vivo)*
        Limulus activity (by Endospecy test)
II-2   Murine splenocyte stimulating effects; mitogenicity; PBA activity *(in vitro)*
        Murine macrophage stimulating effects; enhancement of PGE_2 production *(in vitro)*
        Guinea pig macrophage-stimulating effects; enhancement of O_2^- production and glucosamine incorporation *(in vitro)*
        Limulus activity (by gelation method)

Category III

        Induction of IL-1 in murine macrophage cultures *(in vitro)*
        Activation of complement cascade in human serum

less toxic lipid A analogs retaining powerful beneficial immunostimulatory but lacking the endotoxic activities may exist. In fact, many investigators have pursued the creation of nontoxic, immunostimulatory lipid A analogs with a view to utilization in clinical and preventive medicine.

III. BISPHOSPHORYL DISACCHARIDE COMPOUNDS — ROLE OF 3-ACYLOXYACYL GROUPS IN MANIFESTATION OF ENDOTOXIC AND RELATED BIOACTIVITIES

In this section we will discuss the structure-function relationships of synthetic lipid A preparations and disaccharide-type lipid A precursors with differing numbers or location of 3-acyloxyacyl groups, a characteristic component of lipid A, on the glucosamine disaccharide backbone (Figure 2). Among them, *E. coli*-type lipid A (LA-15-PP), having two 3-acyloxyacyl groups on the nonreducing moiety of disaccharide unit, *Salmonella minnesota*-type lipid A (LA-16-PP) carrying three 3-acyloxyacyl groups, two on the nonreducing moiety, and one on the reducing part and precursor Ia (IV_A)-type compound (LA-14-PP) with none, have been extensively studied by Japanese and German investigators. Regrettably, information regarding

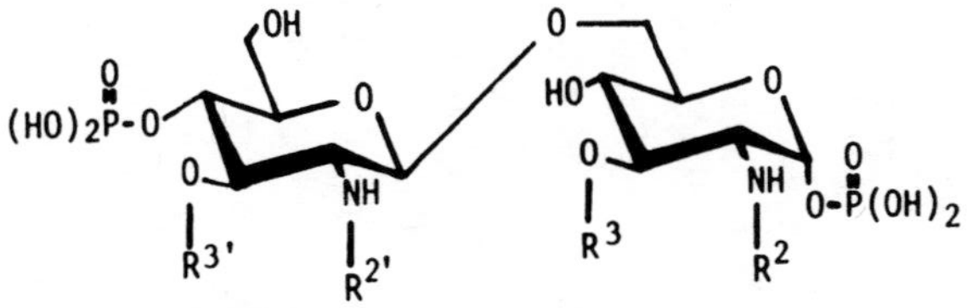

Compound	$R^{3'}$	$R^{2'}$	R^3	R^2	Remarks	Acyl groups per molecule Total	3-Acyloxyacyl-
LA-16-PP	C_{14}-O-(C_{14})	C_{14}-O-(C_{12})	C_{14}-OH	C_{14}-O-(C_{16})	S. minnesota-type	7	3
LA-15-PP	C_{14}-O-(C_{14})	C_{14}-O-(C_{12})	C_{14}-OH	C_{14}-OH	E. coli-type	6	2
LA-22-PP	C_{14}-OH	C_{14}-O-(C_{14})	C_{14}-OH	C_{14}-O-(C_{14})	C. violaceum-type	6	2
LA-20-PP	C_{14}-OH	C_{14}-OH	C_{14}-OH	C_{14}-O-(C_{16})	Precursor Ib (IV_B)-type	5	1
LA-21-PP	C_{14}-OH	C_{14}-O-(C_{16})	C_{14}-OH	C_{14}-OH	Isomer of Ib (IV_B)	5	1
LA-14-PP	C_{14}-OH	C_{14}-OH	C_{14}-OH	C_{14}-OH	Precursor Ia (IV_A)-type	4	0

FIGURE 2. Structures of synthetic lipid A and disaccharide-type lipid A precursors. C_{14}-OH, (R)-3-hydroxytetradecanoyl-; C_{14}-O-(C_{12}), (R)-3-dodecanoyloxytetradecanoyl; C_{14}-O-(C_{14}), (R)-3-tetradecanoyloxytetradecanoyl; C_{14}-O-(C_{16}), (R)-3-hexadecanoyloxytetradecanoyl.

mobacterium violaceum-type lipid A (LA-22-PP), which has two 3-acyloxy-acyl groups, one on the nonreducing and the other on the reducing moieties, and precursor Ib (IV_B)-type compound (LA-20-PP) and its isomer (LA-21-PP), which has one acyloxyacyl group on the reducing and nonreducing moiety, respectively, has so far been limited. As a consequent it remains difficult at this time to draw conclusions on the effects of location of 3-acyloxyacyl groups on the bioactivities of lipid A molecules.

A. ENDOTOXIC ACTIVITIES
1. Lethal Toxicity

Among the synthetic lipid A preparations and related compounds so far tested, only LA-15-PP exhibited distinct lethal toxicity in the chick embryo lethality test; its LD_{50} value was 74 ng per embryo (Table 2). Precursor Ib (IV_B)-type compound (LA-20-PP) may also be highly toxic in this assay, in consideration of its strong endotoxic activities in other assays; however, no data are available at present. Compounds LA-16-PP and LA-14-PP manifest minimal lethal toxicity in chick embryos. Noteworthy findings are that in galactosamine-treated mice, which were rendered highly susceptible to the lethal toxicity of endotoxin (see Chapter 4 in Volume II), LA-14-PP and LA-16-PP, as well as LA-15-PP, were highly lethal: their LD_{50} values were 5 to 10 ng per mouse (Table 2). Thus, structural requirements for lethal toxicity as determined by the above two assay systems were remarkably different.

Although no data on lethal potency in normal mice have so far been available, due primarily to limited amounts of test compounds, lethal toxicity of LA-15-PP and LA-14-PP was compared in two differently primed mice systems. One was an assay in which C57BL/6 were primed by administration of *N*-acetylmuramyl-L-alanyl-D-isoglutamine 4 h before the challenge with test compounds[16a] and the other was the Meth A bearing and *Propionibacterium acnes*-primed BALB/c mouse system.[17] In both systems, LA-15-PP exhibited strong lethal toxicity, whereas LA-14-PP did not. Thus, similar structural requirements were noted between the above tests and the chick embryo lethality assay. In this context, Galanos et al. demonstrated that the lethal toxicity of endotoxin in galactosamine-treated mice was mediated by tumor necrosis factor (TNF) (see Chapter 4 in Volume II). In fact, the structural requirements of lipid A and related compounds for lethal toxicity in galactosamine-loaded mice are closely correlated with those for TNF induction detected by administration to mice (see Section B).

2. Pyrogenicity

Rabbit pyrogenicity tests revealed that a suitable number of 3-acyloxyacyl groups on the disaccharide backbone was required. Compound LA-15-PP with two 3-acyloxyacyl groups was the most pyrogenic, followed by LA-20-PP and LA-21-PP, both of which have one acyloxyacyl group, LA-16-PP, with three, and LA-14-PP with none. LA-14-PP was 100 to 1000 times less pyrogenic than LA-15-PP (Table 2).

TABLE 2
Endotoxic Activities of Synthetic Lipid A and Disaccharide-Type Lipid A Precursors

Compound	Lethal toxicity		Pyrogenicity (minimum effective dose, μg/kg)	Local Shwartzman reaction (preparatory activity)
	Chick embryos	Galactosamine-loaded mice		
LA-16-PP	−	+ + +	0.1—0.4	+
LA-15-PP	+ + +	+ + +	0.001—0.01	+ + +
LA-20-PP	ND	ND	0.01	+ +
LA-21-PP	ND	ND	0.05	+ +
LA-14-PP	−	+ + +	0.4—3.0	−

Note: Based on the reports of Kotani et al.,[3,6,8] Takada et al.,[7,13] Takahashi et al.,[10] Galanos et al.,[4,33] Rietschel et al.,[18] Homma et al.,[5] Kanegasaki et al.,[26,32] ND, not determined.

3. Shwartzman Reaction

The structural requirements of lipid A to prepare rabbit skin for the local Shwartzman reaction were also as strict as those for lethal toxicity in chick embryos. Compound LA-15-PP exhibited potent activity, while LA-16-PP was weakly active and LA-14-PP was totally inactive in this assay (Table 2). Compounds LA-20-PP and LA-21-PP exhibited similar definite preparatory activity.[18] Galanos et al.[19] compared the preparation and provocation of the local Shwartzman reaction by the bacterial precursor Ib (IV$_B$), which structurally corresponds to synthetic LA-20-PP, with that of *E. coli* lipid A, and suggested that LA-20-PP was less active than LA-15-PP. Similar structure-function relationships were reported with the provocation of the Shwartzman reaction, except that LA-21-PP was less active than LA-20-PP in this respect.

B. OTHER BIOACTIVITY DETECTED BY ADMINISTRATION TO ANIMALS (BY ASSAYS *IN VIVO*)

The ability to induce various cytokines (interferon, IFN-α/β, TNF, and others) in properly primed mice and immunoadjuvancy *in vivo* were included in category II-1. Generally speaking, 3-acyloxyacyl groups were not a prerequisite for manifestation of these activities, because LA-14-PP, which lacks a 3-acyloxyacyl group, was as effective as LA-15-PP. On the other hand, compound LA-16-PP with three of the groups was less active in some of these *in vivo* activities than LA-15-PP, indicating that an excess of 3-acyloxyacyl groups generally decreased the bioactivity in this category of lipid A.

Adjuvant activity, manifested by the enhancement of the humoral immune response, was determined in terms of the hemolytic plaque-forming-cell (PFC) response in the spleen of mice immunized with sheep red blood cells (SRBC) together with test compounds. In this assay, both LA-14-PP and LA-16-PP exhibited distinct adjuvant activity roughly comparable to that of LA-15-PP (Table 3). Ukei et al.,[17] on the other hand, examined the adjuvant activity to induce delayed-type hypersensitivity to azobenzen-arsonate-*N*-acetyl-L-tyrosine (ABA-Tyr) in guinea pigs, and found that LA-14-PP exhibited definite adjuvant activity in this system, but with considerably less potency than LA-15-PP (Table 3).

Antitumor activities of synthetic lipid A preparations and related compounds have been examined primarily in mouse systems. Tsujimoto et al.[20] demonstrated that the administration of a mixture of LA-15-PP and Nα-(*N*-acetylmuramyl-L-alanyl-D-isoglutaminyl)-Nε-stearoyl-L-lysine, MDP-Lys(L18), to Meth A fibrosarcoma-bearing BALB/c mice caused marked hemorrhagic necrosis of the tumor at a very high incidence with significant tumor regressive effects. In this system, the activities of LA-14-PP and LA-15-PP were similar (Table 3). Ukei et al.[17] also reported that LA-14-PP induced tumor necrosis in Meth A tumor-bearing BALB/c mice in combination with *P. acnes* pretreatment. In their assay system, LA-15-PP was lethal to all the primed mice within 24 h, probably by heightened susceptibility of the

TABLE 3
Comparison of Immunostimulatory Activities of LA-16-PP,
LA-15-PP, and LA-14-PP

Bioactivity (test animal)	Compound		
	LA-16-PP	**LA-15-PP**	**LA-14-PP**
Assays *in vivo*			
Adjuvant activity			
Antibody response (mouse)[10]	+ +	+ + +	+ + +
DTH response (guinea pig)[17]	ND	+ + +	+ +
Antitumor (mouse)[20]	ND	+ + +	+ +
Analgesic (mouse)[25]	+ + +	+ + +	+ + +
IFN induction (mouse)[8,10]	+	+ + +	+ +
TNF induction (mouse)[3,8]	+	+ + +	+ + +
Assays *in vitro*			
Mitogenicity (mouse)[10]	+ +	+ + +	+ +
PBA (mouse)[10]	+ +	+ + +	+ +
Mϕ-PGE$_2$ (mouse)[10]	+ +	+ + +	+ +
Mϕ-IL-1 (mouse)[11]	+ + + +	+ + +	+ +
Mϕ-O$_2^-$ (guinea pig)[10]	+ +	+ + +	+
Mϕ-GlcN (guinea pig)[10]	+ + +	+ + +	+
Mϕ-IL-1 (human)[27]	+ +	+ + +	−
Complement (human)[12]	+ + + +	+ + +	−

mice to the lethal toxicity effected by *P. acnes*. Nevertheless, LA-15-PP exhibited definite antitumor effects in Meth A-bearing mice that did not receive *P. acnes* pretreatment.[4]

Another noteworthy bioactivity of endotoxins is that directed to the central nervous system (pyrogenicity has already been described). Krueger's group, who discovered the somnogenic potency of bacterial peptidoglycan fragments,[21] examined the somnogenicity of LPS, as well as natural and synthetic lipid A preparations in rabbits.[22,23] Compound LA-15-PP was the most potent among test synthetic lipid A preparations. It significantly increased slow-wave sleep, δ-electroencephalographic amplitudes, and brain-colonic temperatures while reducing rapid-eye-movement (REM) sleep. The data (described later) regarding the somnogenic effects of bisphosphorylated disaccharide analogs without 3-acyloxyacyl groups, monophosphoryl disaccharide compounds, and acylated monosaccharide phosphate, collectively, suggest that the somnogenic activity of lipid A may be classified in category II-1.

Soon after demonstrating structure-dependent analgesic effects of MDP in terms of inhibition of writhing movements in mice induced by intraperitoneal injection of acetic acid,[24] Ogawa et al.[25] examined similar effects of synthetic lipid A and showed that compounds LA-16-PP, LA-15-PP, and LA-14-PP were equally active (Table 3). The analgesic effects may be

TABLE 4
TNF-Inducing Ability and Lethal Toxicity of Synthetic
Lipid A and Analogs

| Compound | TNF induction | Lethal toxicity | |
		Galactosamine-loaded mice	Chick embryos
LA-16-PP	+ +	+ + +	−
LA-15-PP	+ + +	+ + +	+ + +
LA-14-PP	+ +	+ + +	−
LA-18-PP	+ +	+ +	−
LA-17-PP	+ +	+ +	−
401 (lipid X)	−	−	−
408 (lipid Y)	+	+	−
410	−	−	−

Note: Based on the reports of Kotani et al.,[3,8,9] Takada et al.,[7] and Takahashi et al.,[10]

included in category II-1, although no data are so fat available with lipid A-subunit analogs.

Many investigators suggested that endotoxin exerts a variety of bioactivities via the heightened production (generation or liberation) of chemical mediators such as cytokines. This view led us to compare the structural requirements of synthetic lipid A specimens and their analogs for cytokine induction with those for endotoxicity. We examined IFN-α/β induction in *P. acnes*-primed mice and TNF induction in BCG-primed mice. Compound LA-14-PP exhibited activity which, though somewhat weaker than that of LA-15-PP, was definite, whereas LA-16-PP exhibited only marginal effects (Table 3). Homma's group,[26] on the other hand, reported that both LA-14-PP and LA-16-PP induced TNF at a similar level in *P. acnes*-primed mice, although the level was lower than that induced by LA-15-PP. There is some discrepancy between the results of the above two groups, but the structural requirements of lipid A for cytokine induction do not necessarily correlate with those for endotoxic activities except for the lethal toxicity in galactosamine-treated mice. As described above, there is a close correlation between the structural requirements of the test compounds for lethality in galactosamine-sensitized mice and TNF-inducing activity in BCG-primed mice (Table 4).

C. BIOACTIVITY DETECTED BY ASSAYS *IN VITRO*

Structural requirements for lymphocyte and macrophage activation were examined in a variety of *in vitro* assay systems. Generally speaking, LA-14-PP and LA-16-PP as well as LA-15-PP were powerful activators of mouse B cells as well as peritoneal macrophages of mice and guinea pigs. Compound LA-15-PP was the most potent among the three (Table 3). The potency was compared in terms of the following parameters: mitogenicity and polyclonal B cell-activating effects on murine splenocytes, PGE$_2$ production by murine resident peritoneal macrophages, O$_2^-$ production and glucosamine uptake, and spreading by liquid paraffin-irritated guinea pig peritoneal macrophages. These activities are definitely observed also with some monosaccharide-type lipid A analogs (see Section IV.C). Loppnow et al.[27] compared the interleukin-1 (IL-1) induction of human peripheral mononuclear cells by various bacterial and synthetic lipid A preparations and related compounds in terms of the minimum effective dose to induce IL-1 release into culture supernatant. They found that LA-15-PP was the most potent IL-1 inducer, followed by LA-16-PP, then LA-22-PP, whereas synthetic lipid A precursors LA-20-PP and LA-14-PP were inactive. Thus, relative IL-1-inducing effects of lipid A on human blood mononuclear cells were much more structure dependent, and it is pertinent to classify this activity as I-1 or I-2. On the other hand, the structural requirements for IL-1 release from murine peritoneal macrophages were different from those for IL-1 production by human mononuclear cells: the order of potency was LA-16-PP > LA-15-PP > LA-14-PP.[11]

Structural requirements for the activation of the human complement cascade were quite different from those of other bioactivities. Among bisphosphoryl disaccharide compounds, the order of potency was as follows: LA-22-PP > LA-16-PP > LA-15-PP > LA-20-PP. Compound LA-14-PP was practically inactive.[12]

Limulus activity was evaluated by a conventional gelation method (PreGel test) and a colorimetric assay (Endospecy test). In both assays, all the test bisphosphoryl disaccharide compounds powerfully activated the clotting enzyme cascade in horseshoe crab amoebocyte lysates. Taking the activity of lipid A-subunit analogs (see Section IV.C) into consideration, we classified the *Limulus* activity determined by PreGel and Endospecy tests into categories II-1 and II-2, respectively (Table 5).

D. OPTIMAL NUMBER OF ACYLOXYACYL GROUPS

As described above, 3-acyloxyacyl groups on the β(1-6)-D-glucosamine disaccharide backbone are required to exert a whole range of the bioactivities of lipid A. Generally, the optimal number of 3-acyloxyacyl groups was two, and the presence of excess acyloxyacyl groups reduced the bioactivities of the disaccharide molecules in most bioassays. This general rule was also noted by Shimizu et al.,[28] who compared the bioactivities of LA-15-PP with those of *Proteus*-type lipid A, which has a chemical structure similar to LA-16-PP but carries a tetradecanoyloxytetradecanoyl group in place of the dodecanoyloxytetradecanoyl group at position 2′ of the disaccharide backbone. These authors also found that *Proteus*-type lipid A was less active than LA-15-PP in lethal toxicity in galactosamine-treated mice, local Shwartzman reaction in rabbits, and mitogenicity for murine splenocytes.

The position of 3-acyloxyacyl groups attached to the backbone structure is also important. Thus, distinct differences in bioactivity were noted between

TABLE 5
Activation of the Clotting Enzyme
Cascade of *T. tridentatus* by Synthetic
Lipid A and Analogs

Compound	Equivalent of reference LPS (mg/mg) by	
	PreGel test	Endospecy test
LA-16-PP	10	1.41
LA-15-PP	10	5.18
LA-22-PP	10	5.10
LA-20-PP	10	4.33
LA-21-PP	10	8.00
LA-14-PP	10	1.56
LA-18-PP	10	0.25
LA-24-PP	1.0	0.56
LA-17-PP	1.0	0.00071
LA-23-PP	0.1	0.000040
401 (lipid X)	0.1	0.0032
408 (lipid Y)	1.0	0.00030
410	0.1	0.00571
GLA-27	1.0	0.000040

Modified from Takada et al.[12]

disaccharide bisphosphoryl compounds carrying the same number of 3-acyl-oxyacyl groups but at different positions, namely, between LA-15-PP (two at positions 2′ and 3′) and LA-22-PP (two at positions 2 and 2′) and between LA-20-PP (one at position 2) and LA-21-PP (one at position 2′). Nevertheless, further comparative studies are needed before final conclusions can be reached on this issue. In this connection, the findings by Aschauer et al.[29] are note-worthy (the details will be mentioned in the following section). In brief, they described bioactive and inhibitory isomers of 1-monophospho-lipid A analogs carrying three fatty acids but no 3-acyloxyacyl groups.

In summary, the role of 3-acyloxyacyl groups in manifesting bioactivities differs among assays. Thus, discrimination can be expected between beneficial immunostimulation and harmful endotoxic activities. For instance, compounds without acyloxyacyl groups such as LA-14-PP exhibit definite immunostimulation without endotoxicity.

IV. LOW-TOXIC LIPID A-RELATED COMPOUNDS

As described above, whereas the structural requirements of lipid A for endotoxicity are absolute, those for immunostimulation are not. This would suggest that low-toxicity lipid A analogs, which would preserve the beneficial bioactivities of the parent lipid A and could be utilized as a medicinal remedy, could be created by modifying the chemical structure of lipid A. There are

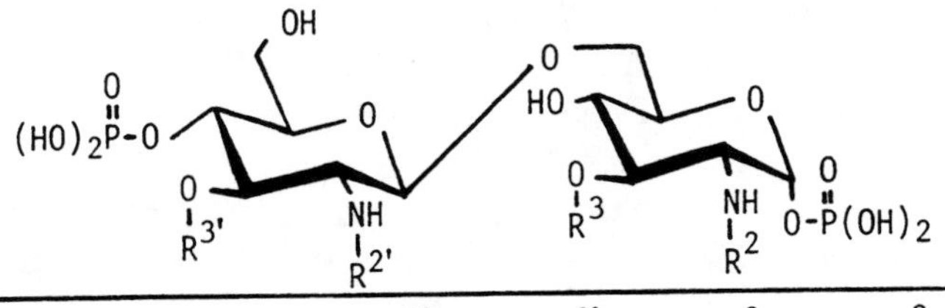

Compound	$R^{3'}$	$R^{2'}$	R^3	R^2
LA-14-PP	C_{14}-OH	C_{14}-OH	C_{14}-OH	C_{14}-OH
LA-18-PP (PH)	C_{14}	C_{14}-OH	C_{14}	C_{14}-OH
LA-24-PP	C_{10}	C_{14}-OH	C_{10}	C_{14}-OH
LA-17-PP (PH)	C_{14}	C_{14}	C_{14}	C_{14}
LA-23-PP	C_{10}	C_{14}	C_{10}	C_{14}

FIGURE 3. Structures of low-toxic analogs of LA-14-PP. Abbreviations other than those in Figure 2: C_{14}, tetradecanoyl; C_{10}, decanoyl.

three approaches by which to achieve this objective: (1) modification of acyl groups on $\beta(1\text{-}6)$ D-glucosamine disaccharide backbone, (2) modification of phosphorylation pattern of propertly acylated disaccharide compounds, and (3) preparation of D-glucosamine phosphates adequately acylated (lipid A-subunit analogs).

A. MODIFICATION OF ACYL GROUPS

Among the bisphosphoryl disaccharide compounds so far described, LA-14-PP seemed the most promising for clinical use. In order to create a compound that preserves beneficial bioactivities inherent to lipid A but shows decreased harmful endotoxicities, six analogs of LA-14-PP have so far been synthesized (Figure 3). Compound LA-18-PP has two tetradecanoyl groups instead of two 3-hydroxytetradecanoyl groups at positions of 3 and 3′ on the disaccharide backbone, whereas all four acyl moieties of LA-17-PP were tetradecanoyl groups. Compounds LA-24-PP and LA-23-PP carried two and four decanoyl groups in place of the tetradecanoyl groups of LA-18-PP and LA-17-PP, respectively. LA-18-PH and LA-17-PH were 4′-monophospho analogs of LA-18-PP and LA-17-PP, respectively. As anticipated, these compounds have so far proven less toxic than the parent LA-14-PP.[9,26] For example, in galactosamine-treated mice, LA-18-PP and LA-17-PP were an order of magnitude less toxic to mice than LA-14-PP. Compounds LA-18-PH and LA-17-PH exhibited no lethal toxicity. LA-17-PP was about tenfold less pyrogenic than LA-14-PP (Table 6). However, there was a decrease in beneficial immunopotentiating activities in these compounds. In particular, LA-18-PH and LA-17-PH were barely bioactive as far as examined. On the other hand, LA-18-PP and LA-17-PP still exhibited some beneficial activities determined by an *in vivo* assay and adjuvant,[9,30] antitumor,[20] and analgesic

TABLE 6
Bioactivities of LA-14-PP, LA-18-PP, and LA-17-PP as Compared with Those of LA-15-PP

Bioactivity	Compound			
	LA-14-PP	LA-18-PP	LA-17-PP	LA-15-PP
Lethal toxicity (LD_{50})				
Chick embryos	—[a]	—	—	0.074 μg
Galactosamine-loaded mice	10 ng	≧126 ng	≧100 ng	7.9 ng
Local Shwartzman reaction (rabbit)	—	—	—	1.25 μg
Pyrogenicity (rabbit)	1.0 μg	1.0 μg	10 μg	0.01 μg
IFN-α/β induction (*P. acnes*-primed mice)[b]	63	13	28	100
TNF induction (BCG-primed mice)[b]	82	4	4	100
Adjuvanticity, humoral (mice)[b]	83	77	83	100
Antitumor activity (mice)[b,c]	85	67	80	100
Analgesic effect (mice)[d]	35	39	55	38

Note: Based on the reports of Kotani et al.,[9] Tsujimoto et al.,[20] and Ogawa et al.[25]

[a] —, not detected at the highest test done (10 or 100 μg per embryo).
[b] Relative potency.
[c] In combination with MDP-Lys(L18).
[d] Percent inhibition of writhing movements induced by intraperitoneal injection of acetic acid.

activity in mice.[25] However, dose-response determinations indicated that about 25-fold greater amount of LA-18-PP and LA-17-PP were required relative to lipid A to induce similar levels of serum TNF in BCG-primed mice[9] (Table 6). Similarly, Cady et al. showed that LA-18-PP induced increases in slow-wave sleep and δ-amplitudes by intravenous administration in rabbits, but a 100-fold higher dose than that of LA-15-PP was required.[23] Compounds LA-24-PP and LA-23-PP are reported to be weakly pyrogenic (to a similar extent to that of LA-14-PP) and to manifest antitumor activity.[13] Compound LA-17-PP, in combination with MDP-Lys(L18), caused significant tumor regression accompanied by marked hemorrhagic necrosis in Meth A tumor-bearing BALB/c mice. Its antitumor activity was comparable to that of LA-14-PP. In contrast, LA-24-PP was scarcely active in this assay.[20]

Compounds LA-18-PP and LA-17-PP manifest significant activity *in vitro*; the potency to induce PGE_2 production and IL-1 release in murine peritoneal macrophage cultures as well as the mitogenicity for murine splenocytes were comparable to those of LA-14-PP, but the potency was less than that of LA-15-PP. Compounds LA-18-PH and LA-17-PH were also definitely active.[9] In the *Limulus* tests, especially in the Endospecy test, it was found that 3-hydroxy groups were important. LA-18-PP and LA-24-PP, which are dehydroxylated at two acyl groups at C-3 and C-3', were three to ten times less active than LA-14-PP in the Endospecy test. LA-17-PP and LA-23-PP, which are dehydroxylated at all 4 acyl groups, were at least 1000 times less active than LA-14-PP[12] (Table 5). The unique structural requirements of lipid

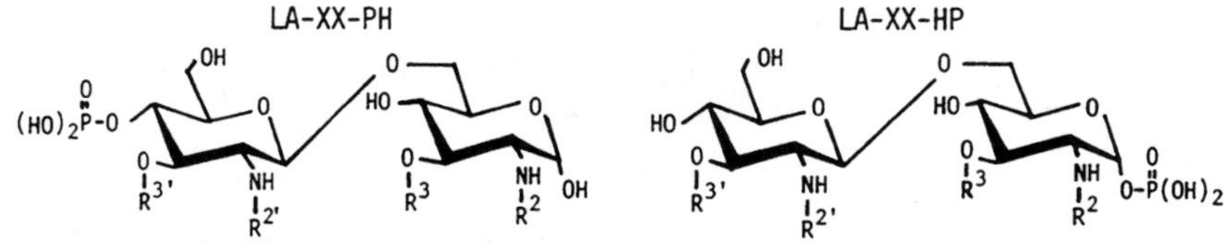

Compound	$R^{3'}$	$R^{2'}$	R^3	R^2
LA-16-PH	C_{14}-O-(C_{14})	C_{14}-O-(C_{12})	C_{14}-OH	C_{14}-O-(C_{16})
LA-15-PH (HP)	C_{14}-O-(C_{14})	C_{14}-O-(C_{12})	C_{14}-OH	C_{14}-OH
LA-14-PH (HP)	C_{14}-OH	C_{14}-OH	C_{14}-OH	C_{14}-OH
LA-18-PH	C_{14}	C_{14}-OH	C_{14}	C_{14}-OH
LA-17-PH	C_{14}	C_{14}	C_{14}	C_{14}

FIGURE 4. Chemical structures of synthetic monophospho-lipid A analogs.

A analogs in human complement activation were partly described above. In this assay, LA-14-PP was inactive, while LA-17-PP, LA-24-PP, LA-18-PP, and LA-23-PP exhibit distinct potency in this order, implying the need of adequate balance between hydrophilicity and hydrophobicity in the test molecule for complement activation.[12]

B. MONOPHOSPHORYL LIPID A ANALOGS

Removal of a 1- or 4'-phosphate group from 1,4'-bisphospho-disaccharide compounds significantly reduces the bioactivity, especially the endotoxicity, of the parent molecule. In the following description, LA-XX-PH and LA-XX-HP denote 4'- and 1-phospho-disaccharide lipid A analogs, respectively. Compound LA-XX-HH denotes dephospho-disaccharide analogs. Synthetic monophosphoryl lipid A analogs so far prepared were as follows: LA-16-PH (an analog of *S. minnesota*-type lipid A), LA-15-PH and LA-15-HP (analogs of *E. coli*-type lipid A), LA-20-PH and LA-21-PH (analogs of precursor Ib, IV_B, and its isomer), LA-14-PH and LA-14-HP (analogs of precursor Ia, IV_A), and LA-18-PH and LA-17-PH (analogs of LA-18-PP and LA-17-PH) (Figure 4). Among these compounds, LA-18-PH and LA-17-PH were essentially inactive, as shown by any of the *in vivo* assays described above. In *in vitro* assays, Loppnow et al.[27] noted slight IL-1 induction by stimulation of human peripheral blood mononuclear cells with LA-20-PH and LA-20-HP, while LA-20-PP and LA-21-PP were completely inactive in this assay. Compound LA-16-PH was less toxic than LA-16-PP; LA-16-PH was devoid of either lethal toxicity in chick embryos, pyrogenicity in rabbits, or preparatory activity for the Shwartzman reaction, and was several times less toxic than LA-16-PP in galactosamine-treated mice.[8] In assays *in vivo*, LA-16-PH was less active than LA-16-PP with regard to adjuvant activity and

TABLE 7
Bioactivities of LA-15-PH and LA-15-HP as Compared with Those of LA-15-PP

Bioactivity	Compound		
	LA-15-PP	LA-15-PH	LA-15-HP
Lethal toxicity (LD_{50})			
Chick embryos	0.074 μg	±[a]	—[b]
Galactosamine-loaded mice	7.9 ng	20.0 ng	50.1 ng
Local Shwartzman reaction (rabbit)	1.25 μg	40.0 μg	±
Pyrogenicity (rabbit)	0.01 μg	1.0 μg	±
TNF induction (BCG-primed mice)[c]	100	61	21
Adjuvanticity			
Antibody response (mouse)[c]	100	84	84
DTH induction (guinea pig)[c]	100	62	29
Antitumor activity (mouse)[c,d]	100	78	66
Analgesic effect (mouse)[e]	38	42	ND[f]
Limulus activity (mg/mg)			
PreGel test	10	10	1.0
Endospecy test	5.52	0.83	0.044

Note: Based on the reports of Kotani et al.,[3] Ukei et al.,[16] Tsujimoto et al.,[20] and Ogawa et al.[25]

[a]   Some animals exhibited positive responses when high dosages (10.000 and 2.000 ng) of test compound were administered.
[b]   Not detected at the highest test dose (10 μg per embryo).
[c]   Relative potency.
[d]   In combination with MDP-Lys(L18).
[e]   Percent inhibition of writhing movements induced by intraperitoneal injection of acetic acid.
[f]   Not determined.

TNF- and IFN-α/β-inducing potency in mice.[8] In keeping with these findings, compound LA-16-PH lacked somnogenicity[23] and exhibited weaker antitumor activity than LA-14-PP.[20] LA-16-PH was also less active than LA-16-PP in other assays *in vitro*.[8]

The importance of phosphate groups is most clearly observed in the LA-15 series of compounds:[3] the potency decreases from LA-15-PP through LA-15-PH and LA-15-HP to practically inactive LA-15-HH, as follows. Comparison of endotoxic activities of LA-15-PH and LA-15-HP with the parent compound LA-15-PP is shown in Table 7. Under the assay conditions where the LD_{50} value for LA-15-PP was 0.074 μg per embryo, compound LA-15-PH manifests lethality in some chick embryos at higher doses (2 to 10 μg per embryo), whereas LA-15-PH was not lethal at similar doses. In galactosamine-treated mice, the LD_{50} values of LA-15-PP, LA-15-PH, and LA-15-HP were 7.9, 20.0, and 50.1 ng per mouse, respectively. In the body weight loss assay, LA-15-PP was the most active, followed by LA-15-PH and LA-15-HP. LA-15-PH exhibited preparatory activity for the Shwartzman reaction at a 16-fold higher dose than LA-15-PP, while LA-15-HP was practically inactive. Compound LA-15-PH required a 100-fold higher dose than that of LA-15-PP to induce a significant fever response in rabbits, and LA-15-HP was scarcely pyrogenic. LA-15-PP and LA-15-PH induced leukopenia in rabbits at doses of 0.01 and 1.0 μg/kg, respectively, but LA-15-HP was totally inactive even at the highest test dose, 1.0 μg/kg. A similar tendency was also noted in other *in vivo* bioactivities. The order of potency was LA-15-PP > LA-15-PH > LA-15-HP with regard to immunoadjuvant activities in guinea pigs,[17] ability to regress Meth A tumors established in BALB/c mice in combination with MDP-Lys (L18),[20] and TNF induction in BCG-primed ICR mice. We also found a similar pattern using several assays *in vitro* except for human complement activation, in which all three compounds exhibited similar activities. A different pattern of structure-function relationships of the LA-15 series was reported by Loppnow et al.[27,31] with regard to IL-1 induction in human peripheral blood mononuclear cell cultures. LA-15-HP was a more potent IL-1 inducer than LA-15-PH in dose-response experiments, and LA-15-HP induced much higher levels of IL-1 than LA-15-PP and LA-15-PH.

Similar effects of the phosphorylation pattern on bioactivity were found with the LA-14 series, although information available is limited by the fact that the parent molecule LA-14-PP lacked some of the bioactivities, especially endotoxicity.[6,7] In addition, there are some discrepancies between investigators.[6,7,32,33] No significant differences due to phosphorylation pattern were found with pyrogenicity. In contrast, remarkable differences were noted in lethal toxicity in galactosamine-treated mice: LD_{50} values of LA-14-PP, LA-14-PH, and LA-14-HP were approximately 10.0, 300, and 3000 ng per mouse, respectively. This order (-PP > -PH > -HP) was generally found in most of the other bioactivities: adjuvant capability in guinea pigs,[17] mitogenicity and polyclonal B cell activating (PBA) activity in murine splenocytes, modulation of phagocytic activity of murine peritoneal macrophages,[7] and *Limulus* activity.[12] LA-14-PH and LA-14-HP caused a similar level of activation of the human complement cascade, despite the fact that the parent LA-14-PP molecule was only weakly active. This suggests that physicochemical properties such as moderate hydrophobicity of test molecules have greater influences than chemical structure itself on the ability to activate the complement cascade.

Bacterial monophosphoryl lipid A (MPL) has been extensively studied with a view to utilizing it as an antitumor/antiinfectious agent, an immunoadjuvant, etc. A major component of this compound is comparable in structure to LA-15-PH. These studies will be reviewed in Chapter 17 in Volume II.

In this connection, Aschauer et al.[29] isolated some disaccharide lipid A analogs from a specimen of synthetic monosaccharide lipid A precursor, lipid

X, that was prepared by the method of Macher.[34] The isolated materials were 1-monophospho β(1-6)-linked D-glucosamine disaccharides carrying four (at positions 2, 3, 2', and 3') or three (at positions 2, 3, and 2' or positions 2, 2', and 3')3-hydroxytetradecanoyl groups. The former corresponds to synthetic LA-14-HP. They found that the compound carrying three acyl groups at positions 2, 2', and 3' was a more potent inducer of TNF in murine bone marrow-derived macrophage cultures than the compound corresponding to LA-14-HP, whereas the compound carrying the same acyl groups at positions 2, 3, and 2' was totally inactive. The latter compound, however, was capable of powerfully inhibiting TNF release by the former compound. Aschauer et al. proposed from these results that some structure-function relationships of the cited lipid A derivatives may require revision. However, no data are available at present on endotoxic activities of the above compounds.

C. MONOSACCHARIDE LIPID A ANALOGS (LIPID A-SUBUNIT ANALOGS)

Representatives of monosaccharide lipid A precursors, lipid X and lipid Y, were chemically synthesized by Kusumoto et al.[35] (compounds 401 and 408, respectively). The former is 1-phospho-D-glucosamine carrying two 3-hydroxytetradecanoyl groups at positions 2 and 3, and the latter is a compound carrying a 3-hexadecanoyloxytetradecanoyl group in place of 3-hydroxy-tetradecanoyl group of the former at position 2 (Figure 5). These synthetic compounds were devoid of endotoxic activities except that lipid X killed some galactosamine-treated mice at very high doses (e.g., 1 μg per mouse) and both compounds were pyrogenic in rabbits at doses higher than 10 μg/kg. In other assays *in vivo*, lipid Y exhibited weak adjuvant activity as well as IFN-α/β and TNF induction in mice. Lipid X was less active than lipid Y in these assays. Cady et al.[23] reported that intracerebroventricular, but not intravenous, injections of a synthetic lipid X supplied by Sandoz Research Institute induced a small, but significant, increase in both slow-wave and REM sleep without affecting δ-amplitudes or brain-colonic temperatures. They cautioned, however, that some other, as-yet undefined, mechanism or other substance may be responsible for the observed somnogenic actions of lipid X in those studies.

Both lipid X and lipid Y exhibited distinct activity when assayed *in vitro*, and lipid X was, in general, more active than lipid Y. In our own studies, the bioactivities of synthetic lipids X and Y were considerably weaker than those of the parent disaccharide compounds such as LA-14-PP.[10] However, many investigators reported that bacterial lipid X and lipid Y exerted definite activities *in vitro*, and their potency was sometimes comparable to those of lipid A itself.[36-38]

The discrepancies in bioactivity between synthetic and bacterial lipid X may derive from a difference in purity. Our study with "ultrapurified"

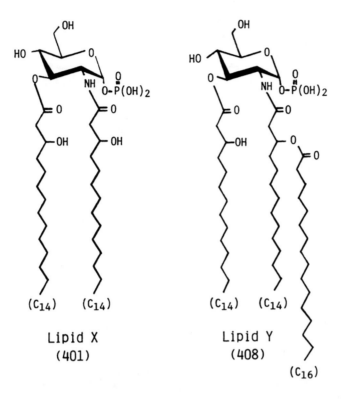

FIGURE 5. Chemical structures of synthetic lipid A subunits, lipid X and lipid Y.

bacterial lipid X revealed no essential differences in bioactivities between bacterial and synthetic compounds.[10] Nevertheless, it is important to acknowledge the possibility that even synthetic lipid X may be contaminated with immunostimulatory or immunoinhabitory impurities, as pointed out by Aschauer et al.[29] Some reported bioactivities of synthetic lipid X may, therefore, be the result of contribution from such impurities. Achiwa's group, who chemically synthesized lipids X and Y, found that both compounds were not lethal when administered to galactosamine-treated mice, and that lipid X was not mitogenic for murine splenocytes.[39]

With a view to creating monosaccharide lipid A analogs which might enhance host defense mechanisms without harmful side-effects, Hasegawa-Kiso's groups have synthesized a large number of lipid A-subunit analogs (see Chapter 4) the bioactivities of which were tested mainly by Homma's group. First they synthesized various analogs of the nonreducing sugar moiety of lipid A. Compound GLA-27, a 4-phospho-D-glucosamine carrying a te-tradecanoyloxytetradecanoyl and a tetradecanoyl group at positions 2 and 3, respectively, was a promising compound[40-42] (Figure 6). It exhibited distinct immunoadjuvant activity and TNF-inducing ability by assays *in vivo* and significant mitogenicity, macrophage-activating, and IFN-inducing activity

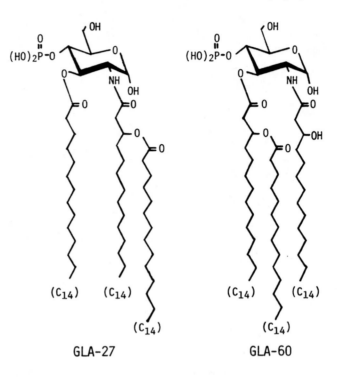

FIGURE 6.   Chemical structures of lipid A-subunit analogs, GLA-27 and GLA-60.

by tests *in vitro*. We showed that these activities of GLA-27 were more potent that those of synthetic lipid X and lipid Y, although nevertheless considerably weaker than those of LA-15-PP.[10] GLA-27 was devoid of pyrogenicity (though seemingly inconsistent its TNF-inducing ability), preparator activity for the Shwartzman reaction, and lethal toxicity in chick embryos and was 100-fold less lethal in galactosamine-treated mice than LA-15-PP[10,40] (Table 8).

Thus, GLA-27 was proven to have a relatively good balance between endotoxic and beneficial activities. However, since its beneficial immuno-stimulatory potency was considerably weaker than those of disaccharide lipid A analogs, Hasegawa-Kiso's group attempted to modify the structure of GLA-27 in various ways. The structure-function relationships of these compounds were extensively studied by Homma's group. The results are summarized as follows. (1) Removal of the tetradecanoyl group at position 3 or transfer of the group from position 3 to 6 diminished the bioactivities of GLA-27 listed above. (2) 6-Monophospho, 1,4-, or 4,6-bisphospho compounds were less active than GLA-27. (3) 1-Deoxy, 1-thio, 1-thio-acetyl, and 2,3-diamino derivatives of GLA-27 showed either increased or decreased immunostimulatory potency as compared with that of the parent structure. For example, by 1-thiolation, GLA-27 acquired the ability to enhance the non-specific resistance to *Pseudomonas aeruginosa* infection in mice, but it had

TABLE 8
Bioactivities of GLA-27 as Compared with Those of LA-14-PP and LA-15-PP

Bioactivity	Compound		
	GLA-27	LA-14-PP	LA-15-PP
Lethal toxicity (LD$_{50}$)			
Chick embryos	—[a]	—	0.074 μg
Galactosamine-loaded mice	≧1260 ng[b]	10 ng	7.9 ng
Shwartzman reaction (rabbit)	—	—	1.25 μg
Pyrogenicity (rabbit)	—	1.0 μg	0.01 μg
TNF induction (BCG-primed mouse)[c]	0.1	82	100
Adjuvanticity, humoral (mouse)[c]	36	82	100
Antitumor activity (mouse)[c,d]	55	85	100
Limulus activity (mg/mg)			
PreGel test	1.0	10	10
Endospecy test	0.000040	1.56	5.18

Note: Based on the reports of Takahashi et al.[10] and Tsujimoto et al.[20]

[a]   —, not detected at the highest test dose (10 μg per embryo).
[b]   100% lethality was not observed even at the highest test dose (10 μg per mouse).
[c]   Relative potency.
[d]   In combination with MDP-Lys(L18).

remarkably lowered IFN-inducing activity.[45] In relation to (2), 4-monophospho isomer of lipid X (GLA-46 in Hasegawa-Kiso's preparation and compound 410 synthesized by Shiba-Kusumoto's group) was devoid of any bioactivity so far examined.[10]

Hasegawa-Kiso's group has continued synthetic studies on lipid A-subunit analogs to obtain improved compounds. Compound GLA-60, 4-phospho-D-glucosamine carrying 3-hydroxy-tetradecanoyl and 3-tetradecanoyloxy-tetradecanoyl groups at positions 2 and 3, respectively (Figure 6), was the compound of choice.[14,15] GLA-60 was practically nontoxic, exhibited adjuvant activity in mice, activated phagocytosis of murine peritoneal macrophages *in vitro*, induced serum IFN and TNF in *P. acnes*-primed mice and serum colony-stimulating factor (CSF) in normal mice, respectively, increased the antiviral activity against vaccinia virus infection in mice, and caused tumor regression of Meth A fibrosarcoma established in BALB/c mice. In each of these activities, GLA-60 was, on a relative dose-response evaluation, more active than GLA-27. Administration of GLA-60 induced serum TNF in IFN-γ-primed mice and resulted in inhibition of lung metastases of B10-B16 melanoma in C57BL/6 mice.[46] In addition, GLA-60 was a powerful mitogen and a PBA for murine splenocytes, and induced both cell-free and cell-associated IL-1 in human mononuclear cell cultures *in vitro*.[47] The latter finding is not necessarily in conflict with the report of Loppnow et al.,[27] that the disaccharide backbone was not necessarily required, but more than one 3-acyloxyacyl group per glucosamine residue were required for IL-1 induction in human peripheral blood mononuclear cell cultures.

In view of the fact that fatty acids are linked to the D-glucosamine residues by the (S) configuration in bacterial lipid A, Hasegawa-Kiso's group synthesized several pairs of 4-*phospho*-glucosamine derivatives with either the (S) or (R) configuration and compared their bioactivities. With one GLA-27 series, the (R) type was a stronger mitogen than the (S) type, while the (S) type exhibited higher TNF- and IFN-inducing ability than the (R) type.[48] Of the GLA-60 series, however, the (R) type was more bioactive than the (S) type in evaluations carried out to date.[14] Shimizu et al.[49] also compared bioactivities of (R)- and (S)-type 4-*O*-phosphonoglucosamine acyl derivatives and found no essential differences among them.

Modification of the chain length of acyl groups at C-2 and C-3 positions of GLA-27 and GLA-60 also influenced their bioactivities. As a whole, the optimal chain length was around the tetradecanoyl structure (C14), although there are minor fluctuations depending on various assays.[14,15]

D. LOW-TOXIC LIPID ANALOGS AS ENDOTOXIN ANTAGONISTS

Proctor et al.[50] found that mice given bacterial lipid X were partially protected from a 100% lethal dose of endotoxin, even when the lipid X was given as late as 6 h after endotoxin challenge. Bacterial lipid X also ameliorated pulmonary hypertension caused by administration of endotoxins in sheep and protected the animals from lethal endotoxicity.[51] These antiendotoxin effects may be mediated by competitive interactions at the binding sites of the key cells involved in endotoxin lethality as suggested by Danner et al.,[52,53] and thus this phenomenon is different from early-phase endotoxin tolerance induced by nontoxic lipid A analogs.[33,54] In fact, the tissue distribution of lipid X resembled that of the endotoxins.[55] The competitive interactions between endotoxin and low-toxic lipid A analogs at cellular binding sites have been suggested by many investigators.[27,29,56,57] These studies may provide a new approach to the clinical use of low-toxic lipid A analogs as endotoxin antagonists in the treatment of Gram-negative sepsis and septic shock. In fact, lipid X protected mice against a fatal *E. coli* infection, although the mechanism responsible for this observation is not clear at present.[58]

V. FUTURE PROSPECTS

Bioactivity studies of synthetic lipid A preparations and their related compounds, either disaccharide or monosaccharide type, have revealed that the structural requirements for endotoxicity and other biological activities can be at least partially differentiated. Many investigators have made efforts to construct lipid A analogs with a better balance between beneficial and toxic activities. Low-toxicity lipid A analogs preserving the immunostimulatory capabilities inherent in endotoxin and lipid A can increase antigen-specific immunity induced by vaccines. They will augment nonspecific resistance against microbial infections and malignant tumors as immunostimulators and will also be useful as therapeutic agents against Gram-negative sepsis and septic shock as endotoxin antagonists. Furthermore, bacterial monophosphoryl lipid A may inactivate suppressor T-cell activity, which may result in immunoadjuvant activity.[59]

Mammalian phagocytes contain lipid A-degrading enzymes. These are capable of dephosphorylating or deacylating the lipid A moieties of endotoxins and possibly produce detoxified "endotoxins" carrying low-toxic lipid A moieties such as precursor Ia (IV$_A$)-type and monophosphoryl lipid A-type structures. These detoxified "endotoxins" may work as "endogenous immunomodulators" or as "endogenous endotoxin antagonists" to regulate physiological as well as host defense mechanisms (see Chapter 18).

It should be noted here that most of the studies on endotoxin and lipid A have so far been concerned with those of Enterobacteriaceae, and inevitably synthetic studies on lipid A have also been focused on mimicking and modifying Enterobacteriaceae lipid A. Nevertheless, it is important to underscore the fact that Gram-negative bacteria other than Enterobacteriaceae are abundant residents of all mammals. For example, *Bacteroides* species are the most prevalent normal flora in the gut and are abundant in periodontal pockets. The lipid A of *Bacteroides* has a quite different structure from that of Enterobacteriaceae and has unique biological activities including low endotoxicity.[60,61]

Natural lipid A, a component of endotoxin of indigenous bacteria, should play important roles in development, modification, and maintenace of defense and physiological mechanisms of the host via interactions with host cellular binding sites for endotoxin. Elucidation of these mechanisms using synthetic lipid A and lipid A-subunit analogs should be an important target of future studies.

NOTE ADDED IN PROOF

Following Loppnow's finding[27] (Section IV.D) that LA-14-PP inhibited IL-1 production and release in human monocyte cultures by lipid A or LPS, similar antagonistic effects of lipid A precursor Ia (IV$_A$), bacterial or synthetic, were reported on TNF and IL-6 production by human monocyte of whole blood cultures[62-64] and β2 integrin CD11b/CD18 (CR3) expression by human neutrophils.[65] Thus compound Ia (IV$_A$) acts as an LPS antagonist with humans, while the same compound can work as an LPS agonist with laboratory animals. So far there have been no reports on antiendotoxic effects of the Ia (IV$_A$) by *in vivo* assays, but there is a good possibility that LA-14-PP and its non-acyloxyacyl derivatives could be useful endotoxin inhibitors in clinical trials.

REFERENCES

1. **Imoto, M., Yoshimura, H., Sakaguchi, N., Kusumoto, S., and Shiba, T.,** Total synthesis of *Escherichia coli* lipid A, *Tetrahedron Lett.,* 26, 1545, 1985.
2. **Imoto, M., Yoshimura, H., Shimamoto, T., Sakaguchi, N., Kusumoto, S., and Shiba, T.,** Total synthesis of *Escherichia coli* lipid A, the endotoxically active principle of cell-surface lipopolysaccharide, *Bull. Chem. Soc. Jpn.,* 60, 2205, 1987.
3. **Kotani, S., Takada, H., Tsujimoto, M., Ogawa, T., Takahashi, I., Ikeda, T., Otsuka, K., Shimauchi, H., Kasai, N., Mashimo, J., Nagao, S., Tanaka, A., Tanaka, S., Harada, K., Nagaki, K., Kitamura, H., Shiba, T., Kusumoto, S., Imoto, M., and Yoshimura, H.,** Synthetic lipid A with endotoxic and related biological activities comparable to those of a natural lipid A from an *Escherichia coli* Re-mutant, *Infect. Immun.,* 49, 225, 1985.
4. **Galanos, C., Lüderitz, O., Rietschel, E. T., Westphal, O., Brade, H., Brade, L., Freudenberg, M., Schade, U., Imoto, M., Yoshimura, H., Kusumoto, S., and Shiba, T.,** Synthetic and natural *Escherichia coli* free lipid A express identical endotoxic activities, *Eur. J. Biochem.,* 148, 1, 1985.
5. **Homma, J. Y., Matsuura, M., Kanegasaki, S., Kawakubo, Y., Kojima, Y., Shibukawa, N., Kumazawa, Y., Yamamoto, A., Tanamoto, K., Yasuda, T., Imoto, M., Yoshimura, H., Kusumoto, S., and Shiba, T.,** Structural requirements of lipid A responsible for the functions: a study with chemically synthesized lipid A and its analogues, *J. Biochem. Tokyo,* 98, 395, 1985.
6. **Kotani, S., Takada, H., Tsujimoto, M., Ogawa, T., Harada, K., Mori, Y., Kawasaki, A., Tanaka, A., Nagao, S., Tanakaka, S., Shiba, T., Kusumoto, S., Imoto, M., Yoshimura, H., Yamamoto, M., and Shimamoto, T.,** Immunobiologically active lipid A analogs synthesized according to a revised structural model of natural lipid A, *Infect. Immun.,* 45, 293, 1984.
7. **Takada, H., Kotani, S., Tsujimoto, M., Ogawa, T., Takahashi, I., Harada, K., Katsukawa, C., Tanaka, S., Shiba, T., Kusumoto, S., Imoto, M., Yoshimura, H., Yamamoto, M., and Shimamoto, T.,** Immunopharmacological activities of a synthetic counterpart of a biosynthetic lipid A precursor molecule and of its analogs, *Infect. Immun.,* 48, 219, 1985.
8. **Kotani, S., Takada, H., Takahashi, I., Tsujimoto, M., Ogawa, T., Ikeda, T., Harada, K., Okamura, H., Tamura, T., Tanaka, S., Shiba, T., Kusumoto, S., Imoto, M., Yoshimura, H., and Kasai, N.,** Low endotoxic activities of synthetic *Salmonella*-type lipid A with an additional acyloxyacyl group on the 2-amino group of β(1-6)glucosamine disaccharide 1,4'-bisphosphate, *Infect. Immun.,* 52, 872, 1986.
9. **Kotani, S., Takada, H., Takahashi, I., Ogawa, T., Tsujimoto, M., Shimauchi, H., Ikeda, T., Okamura, H., Tamura, T., Harada, K., Tanaka, S., Shiba, T., Kusumoto, S., and Shimamoto, T.,** Immunobiological activities of synthetic lipid A analogs with low endotoxicity, *Infect. Immun.,* 54, 673, 1986.
10. **Takahashi, I., Kotani, S., Takada, H., Tsujimoto, M., Ogawa, T., Shiba, T., Kusumoto, S., Yamamoto, M., Hasegawa, A., Kiso, M., Nishijima, M., Amano, F., Akamatsu, Y., Harada, K., Takada, S., Okamura, H., and Tamura, T.,** Requirement of a properly acylated β(1-6)-D-glucosamine disaccharide bisphosphate structure for efficient manifestation of full endotoxic and associated bioactivities of lipid A, *Infect. Immun.,* 55, 57, 1987.
11. **Takahashi, I., Kotani, S., Takada, H., Shiba, T., and Kusumoto, S.,** Structural requirements of endotoxic lipopolysaccharides and bacterial cell walls in induction of interleukin-1, *Blood Purification,* 6, 188, 1988.
12. **Takada, H., Kotani, S., Tanaka, S., Ogawa, T., Takahashi, I., Tsujimoto, M., Komuro, T., Shiba, T., Kusumoto, S., Kusunose, N., Hasegawa, A., and Kiso, M.,** Structural requirements of lipid A species in activation of clotting enzymes from the horseshoe crab, and the human complement cascade, *Eur. J. Biochem.,* 175, 573, 1988.

13. **Takada, H. and Kotani, S.,** Structural requirements of lipid A for endotoxicity and other biological activities, *Crit. Rev. Microbiol.,* 16, 477, 1989.
14. **Kumazawa, Y., Nakatsuka, M., Takimoto, H., Furuya, T., Nagumo, T., Yamamoto, A., Homma, J. Y., Inada, K., Yoshida, M., Kiso, M., and Hasegawa, A.,** Importance of fatty acid substituents of chemically synthesized lipid A-subunit analogs in the expression of immunopharmacological activity, *Infect. Immun.,* 56, 149, 1988.
15. **Nakatsuka, M., Kumazawa, Y., Matsuura, M., Homma, J. Y., Kiso, M., and Hasegawa, A.,** Enhancement of nonspecific resistance to bacterial infections and tumor regressions by treatment with synthetic lipid A-subunit analogs. Critical role of *N*- and 3-*O*-linked acyl groups in 4-*O*-phosphono-D-glucosamine derivatives, *Int. J. Immunopharmacol.,* 11, 349, 1989.
16. **Obayashi, T., Tamura, H., Tanaka, S., Ohki, M., Takahashi, S., Arai, M., Masuda, M., and Kawai, T.,** A new chromogenic endotoxin-specific assay using recombind limulus coagulation enzymes and its clinical applications, *Clin. Chim. Acta,* 149, 55, 1985.
16a. **Takada, H.,** unpublished.
17. **Ukei, S., Iida, J., Shiba, T., Kusumoto, S., and Azuma, I.,** Adjuvant and antitumour activities of synthetic lipid A analogues, *Vaccine,* 4, 21, 1986.
18. **Rietschel, E. T., Brade, L., Schade, U., Galanos, C., Freudenberg, M., Lüderitz, O., Kusumoto, S., and Shiba, T.,** Endotoxic properties of synthetic pentaacyl lipid A precursor Ib and a structural isomer, *Eur. J. Biochem.,* 169, 27, 1987.
19. **Galanos, C., Hansen-Hagge, T., Lehmann, V., and Lüderitz, O.,** Comparison of the capacity of two lipid A precursor molecules to express the local Shwartzman phenomenon, *Infect. Immun.,* 48, 355, 1985.
20. **Tsujimoto, M., Kotani, S., Shiba, T., Kusumoto, S., Hasegawa, A., Kiso, M., and Ono, Y.,** Regressive action with induction of hemorrhagic necrosis of mixtures of acylated muramylpeptides and synthetic, low toxic lipid A analogs on Meth A fibrosarcoma, *Adv. Biosci.,* 68, 151, 1988.
21. **Krueger, J. M., Pappenheimer, J. R., and Karnovsky, M. L.,** Sleep-promoting effects of muramyl peptides, *Proc. Natl. Acad. Sci. U.S.A.,* 79, 6102, 1982.
22. **Krueger, J. M., Kubillus, S., Shoham, S., and Davenne, D.,** Enhancement of slow-wave sleep by endotoxin and lipid A, *Am. J. Physiol.,* 251, R591, 1986.
23. **Cady, A. B., Kotani, S., Shiba, T., Kusumoto, S., and Krueger, J. M.,** Somnogenic activities of synthetic lipid A, *Infect. Immun.,* 57, 396, 1989.
24. **Ogawa, T. and Kotani, S.,** Analgesic effects of *N*-acetylmuramyl-L-alanyl-D-isoglutamine in decreasing the acetic acid-induced abdominal-writhing response, *Infect. Immun.,* 55, 494, 1987.
25. **Ogawa, T., Kotani, S., Kusumoto, S., and Shiba, T.,** Analgesic action of endotoxic lipopolysaccharides, bacterial and synthetic lipid A's and their low toxic analogs in decreasing acetic acid-induced abdominal-writhing response in mice, in *Abstracts of International Symposium on Pyrogen,* Haijun, Z., Ed., Chinese Pharmaceutical Association, Amoy, 1987, 63.
26. **Kanegasaki, S., Tanamoto, K., Yasuda, T., Homma, J. Y., Matsuura, M., Nakatsuka, M., Kumazawa, Y., Yamamoto, A., Shiba, T., Kusumoto, S., Imoto, M., Yoshimura, H., and Shimamoto, T.,** Structure-activity relationship of lipid A: comparison of biological activities of natural and synthetic lipid A's with different fatty acid compositions, *J. Biochem. Tokyo,* 99, 1203, 1986.
27. **Loppnow, H., Brade, H., Dürrbaum, I., Dinarello, C. A., Kusumoto, S., Rietschel, E. T., and Flad, H.-D.,** IL-1 induction-capacity of defined lipopolysaccharide partial structures, *J. Immunol.,* 142, 3229, 1989.
28. **Shimizu, T., Akiyama, S., Masuzawa, T., Yanagihara, Y., Ikeda, K., Takahashi, T., Kondo, H., and Achiwa, K.,** Biological activities of chemically synthesized *Proteus*-type lipid A, *Microbiol. Immunol.,* 31, 381, 1987.

29. **Aschauer, H., Grob, A., Hildebrandt, J., Schuetze, E., and Stuetz, P.,** Highly purified lipid X is devoid of immunostimulatory activity. Isolation and characterization of immunostimulating contaminants in a batch of synthetic lipid X, *J. Biol. Chem.*, 265, 9159, 1990.

30. **Tsujimoto, M., Kotani, S., Okunaga, T., Kubo, T., Takada, H., Kubo, T., Shiba, T., Kusumoto, S., Takahashi, T., Goto, Y., and Kinoshita, F.,** Enhancement of humoral immune responses against viral vaccines by a non-pyrogenic 6-*O*-acetylmura-myldipeptide and synthetic low toxicity analogues of lipid A, *Vaccine*, 7, 39, 1989.

31. **Loppnow, H., Brade, L., Brade, H., Rietschel, E. T., Kusumoto, S., Shiba, T., and Flad, H.-D.,** Induction of human interleukin 1 by bacterial and synthetic lipid A, *Eur. J. Biochem.*, 16, 1263, 1986.

32. **Kanegasaki, S., Kojima, Y., Matsuura, M., Homma, J. Y., Yamamoto, A., Kumazawa, Y., Tanamoto, K., Yasuda, T., Tsumita, T., Imoto, M., Yoshimura, H., Yamamoto, M., Shimamoto, T., Kusumoto, S., and Shiba, T.,** Biological activities of analogues of lipid A based chemically on the revised structural model. Comparison of mediator-inducing, immunomodulating and endotoxic activities, *Eur. J. Biochem.*, 143, 237, 1984.

33. **Galanos, C., Lehmann, V., Lüderitz, O., Rietschel, E. T., Westphal, O., Brade, H., Brade, L., Freudenberg, M. A., Hansen-Hagge, T., Lüderitz, T., McKenzie, G., Schade, U., Strittmatter, W., Tanamoto, K., Zähringer, U., Imoto, M., Yoshimura, H., Yamamoto, M., Shimamoto, T., Kusumoto, S., and Shiba, T.,** Endotoxic properties of chemically synthesized lipid A part structures. Comparison of synthetic lipid A precursor and synthetic analogues with biosynthetic lipid A precursor and free lipid A, *Eur. J. Biochem.*, 140, 221, 1984.

34. **Macher, I.,** A convenient synthesis of 2-deoxy-2-[(*R*)-3-hydroxytetradecanamido]-3-*O*[(*R*)-3-hydroxytetradecanoyl]-α-D-glucopyranose 1-phosphate (lipid X), *Carbohydr. Res.*, 162, 79, 1987.

35. **Kusumoto, S., Yamamoto, M., and Shiba, T.,** Chemical syntheses of lipid X and lipid Y, acyl glucosamine 1-phosphates isolated from *Escherichia coli* mutants, *Tetrahedron Lett.*, 25, 3727, 1984.

36. **Nishijima, M., Amano, F., Akamatsu, Y., Akagawa, K., Tokunaga, T., and Raetz, C. R. H.,** Macrophage activation by monosaccharide precursors of *Escherichia coli* lipid A, *Proc. Natl. Acad. Sci. U.S.A.*, 82, 282, 1985.

37. **Amano, F., Nishijima, M., Akagawa, K., and Akamatsu, Y.,** Enhancement of O_2^- generation and tumoricidal activity of murine macrophages by a monosaccharide precursor of *Escherichia coli* lipid A, *FEBS Lett.*, 192, 263, 1985.

38. **Amano, F., Nishijima, M., and Akamatsu, Y.,** A monosaccharide precursor of *Escherichia coli* lipid A has the ability to induce tumor-cytotoxic factor production by a murine macrophage-like cell line, J774.1, *J. Immunol.*, 136, 4122, 1986.

39. **Shimizu, T., Akiyama, S., Masuzawa, T., Yanagihara, Y., Nakamoto, S., Takahashi, T., Ikeda, K., and Achiwa, K.,** Comparison of biological activities of chemically synthesized monosaccharide analogues of reducing and nonreducing sugar moieties of lipid A, *Chem. Pharm. Bull.*, 34, 5169, 1986.

40. **Matsuura, M., Kojima, Y., Homma, J. Y., Kubota, Y., Yamamoto, A., Kiso, M., and Hasegawa, A.,** Biological activities of chemically synthesized analogues of the nonreducing sugar moiety of lipid A, *FEBS Lett.*, 167, 226, 1984.

41. **Kumazawa, Y., Matsuura, M., Homma, J. Y., Nakatsuru, Y., Kiso, M., and Hasegawa, A.,** B cell activation and adjuvant activities of chemically synthesized analogues of the nonreducing surgar moiety of lipid A, *Eur. J. Immunol.*, 15, 199, 1985.

42. **Kumazawa, Y., Matsuura, M., Maruyama, T., Homma, J. Y., Kiso, M., and Hasegawa, A.,** Structural requirements for inducing *in vitro* B lymphocyte activation by chemically synthesized derivatives related to the nonreducing D-glucosamine subunit of lipid A, *Eur. J. Immunol.*, 16, 1099, 1986.

43. **Matsuura, M., Yamamoto, A., Kojima, Y., Homma, J. Y., Kiso, M., and Hasegawa, A.,** Biological activities of chemically synthesized partial structure analogues of lipid A, *J. Biochem. Tokyo,* 98, 1229, 1985.

44. **Kumazawa, Y., Takimoto, H., Yamamoto, A., Homma, J. Y., Agawa, Y., Kiso, M., and Hasegawa, A.,** Effect of chemical modification at C1 of the glucosamine backbone of lipid A-subunit analog GLA-27 manifestation of immunopharmacological activity, *FEBS Lett.,* 239, 117, 1988.

45. **Nakatsuka, M., Ikeda, S., Kumazawa, Y., Matsuura, M., Nishimura, C., Homma, J. Y., Kiso, M., and Hasegawa, A.,** Enhancement of nonspecific resistance to microbial infections of synthetic lipid A-subunit analogues of GLA-27 modified at the C1 position of the glucosamine backbone, *Int. J. Immunopharmacol.,* 12, 599, 1990.

46. **Saiki, I., Maeda, H., Murata, J., Yamamoto, N., Kiso, M., Hasegawa, A., and Azuma, I.,** Antimetastatic effect of endogenous tumor necrosis factor induced by the treatment of recombinant interferon γ followed by an analogue (GLA-60) to synthetic lipid A subunit, *Cancer Immunol. Immunother.,* 30, 151, 1989.

47. **Saiki, I., Maeda, H., Murata, J., Takahashi, T., Sekiguchi, S., Kiso, M., Hasegawa, A., and Azuma, I.,** Production of interleukin 1 from human monocytes stimulated by synthetic lipid A subunit analogues, *Int. J. Immunopharmacol.,* 12, 297, 1990.

48. **Kumazawa, Y., Ikeda, S., Takimoto, H., Nishimura, C., Nakatsuka, M., Homma, J. Y., Yamamoto, K., Kiso, M., and Hasegawa, A.,** Effect of stereospecificity of chemically synthesized lipid A-subunit analogues GLA-27 and GLA-40 on the expression of immunopharmacological activities, *Eur. J. Biochem.,* 17, 663, 1987.

49. **Shimizu, T., Masuzawa, T., Yanagihara, Y., Nakamoto, S., Itoh, H., and Achiwa, K.,** Antitumor activity, mitogenicity, and lethal toxicity of chemically synthesized monosaccharide analogs of lipid A, *J. Pharmacobio-Dyn.,* 11, 512, 1988.

50. **Proctor, R. A., Will, J. A., Burhop, K. E., and Raetz, C. R. H.,** Protection of mice against lethal endotoxemia by a lipid A precursor, *Infect. Immun.,* 52, 905, 1986.

51. **Golenbock, D. T., Will, J. A., Raetz, C. R. H., and Proctor, R. A.,** Lipid X ameliorates pulmonary hypertension and protects sheep from death due to endotoxin, *Infect. Immun.,* 55, 2471, 1987.

52. **Danner, R. L., Joiner, K. A., and Parrillo, J. E.,** Inhibition of endotoxin-induced priming of human neutrophils by lipid X and 3-aza-lipid X, *J. Clin. Invest.,* 80, 605, 1987.

53. **Danner, R. L., van Dervort, A. L., Doerfler, M. E., Stuetz, P., and Parrillo, J. E.,** Antiendotoxin activity of lipid A analogues: requirements of the chemical structure, *Pharm. Res.,* 7, 260, 1990.

54. **Madonna, G. S., Peterson, J. E., Ribi, E. E., and Vogel, S. N.,** Early-phase endotoxin tolerance: induction by a detoxified lipid A derivative, monophosphoryl lipid A, *Infect. Immun.,* 52, 6, 1986.

55. **Golenbock, D. T., Ebert, S., Will, J. A., and Proctor, R. A.,** Elimination and tissue distribution of the monosaccharide lipid A precursor, lipid X, in mice and sheep, *Antimicrob. Agents Chemother.,* 32, 37, 1988.

56. **Schwartz, B. S., Monroe, M. C., and Bradshaw, J. D.,** Endotoxin-induced production of plasminogen activator inhibitor by human monocytes is autonomous and can be inhibited by lipid X, *Blood,* 8, 2188, 1989.

57. **Romano, M. and Hawiger, J.,** Interaction of endotoxic lipid A and lipid X with purified human platelet protein kinase C, *J. Biol. Chem.,* 265, 1765, 1990.

58. **Golenbock, D. T., Leggett, J. E., Pasmussen, P., Craig, W. A., Raetz, C. R. H., and Proctor, R. A.,** Lipid X protects mice against fatal *Escherichia coli* infection, *Infect. Immun.,* 56, 779, 1988.

59. **Baker, P. J., Hiernaux, J. R., Fauntleroy, M. B., Prescott, B., Cantrell, J. L., and Rudbach, J. A.,** Inactivation of suppressor T-cell activity by nontoxic monophosphoryl lipid A, *Infect. Immun.,* 56, 1076, 1988.

60. **Lindberg, A. A., Weintraub, A., Zähringer, U., and Rietschel, E. T.**, Structure-activity relationships in lipopolysaccharides of *Bacteroides fragilis*, *Rev. Infect. Dis.*, 12 (Suppl.), 133, 1990.
61. **Hamada, S., Takada, H., Ogawa, T., Fujiwara, T., and Mihara, J.**, Lipopolysaccharides of oral anaerobes associated with chronic inflammation: chemical and immunomodulating properties, *Int. Rev. Immunol.*, 6, 247, 1990.
62. **Kovach, N. L., Yee, E., Munford, R. S., Raetz, C. R. H., and Harlan, J. M.**, Lipid IV_A inhibits synthesis and release of tumor necrosis factor induced by lipopolysaccharide in human whole blood ex vivo, *J. Exp. Med.*, 172, 77, 1990.
63. **Golenbock, D. T., Hampton, R. Y., Qureshi, N., Takayama, K., and Raetz, C. R. H.**, Lipid A-like molecules that antagonize the effects of endotoxins on human monocytes, *J. Biol. Chem.*, 266, 19490, 1991.
64. **Feist, W., Ulmer, A. J., Wang, M.-H., Musehold, J., Schlüter, C., Gerdes, J., Herzbeck, H., Brade, H., Kusumoto, S., Diamanstein, T., Rietschel, E. T., and Flad, H.-D.**, Modulation of lipopolysaccharide-induced production of tumor necrosis factor, interleukin 1, and interleukin 6 by synthetic precursor Ia of lipid A, *FEMS Microbiol. Immunol.*, 89, 73, 1992.
65. **Lynn, W. A., Raetz, C. R. H., Qureshi, N., and Golenbock, D. T.**, Lipopolysaccharide-induced stimulation of CD11b/CD18 expression on neutrophils. Evidence of specific receptor-based response andinhibition by lipid A-based antagonists, *J. Immunol.*, 147, 3072–3079, 1991.

Chapter 6

CHEMICAL STRUCTURE OF THE CORE REGION OF LIPOPOLYSACCHARIDES

Otto Holst and Helmut Brade

TABLE OF CONTENTS

I. INTRODUCTION[1]

Lipopolysaccharides (LPS) are integral constituents of the Gram-negative bacterial cell wall. They display essential physiological functions for the bacterium and exhibit a variety of biological activities in higher organisms, including humans. LPS of all Gram-negative bacteria share a common architecture, being made up by a lipid, termed lipid A, to which a saccharide moiety is covalently linked. In many bacteria, e.g., the Enterobacteriaceae, Pseudomonadaceae, Vibrionaceae, Legionellaceae, and Brucellaceae, the saccharide portion consists of a heterooligosaccharide of less than 15 sugar residues (core oligosaccharide) to which a polysaccharide of repeating oligosaccharide units (O-specific chain) is linked. This type of LPS is the "classical" one which is found in wild-type (smooth- or S-form) strains of the afore-mentioned bacteria. The two saccharide portions, O-chain and core oligosaccharide, can be distinguished by their genetic determination, biosynthesis, chemical structure, and biological features. Defects in the gene cluster responsible for the biosynthesis of the O-chain (*rfb* locus) lead to a truncated LPS consisting of lipid A and the core oligosaccharide only. Such mutants are named rough (R-) mutants, according to the rough appearance of their colony morphology; they have a reduced pathogenicity compared to their parent S-form strain. However, R-type LPS is also observed in other wild-type bacteria including those being highly pathogenic for man, such as *Bordetella pertussis*, *Bacteroides fragilis*, *Haemophilus influenzae*, *Neisseria meningitidis*, and *N. gonorrhoeae*, *Acinetobacter calcoaceticus*, and Chlamydiaceae. It is not known whether these bacteria are generally unable to make S-type LPS or whether the genes for the synthesis of O-side chains are deleted or suppressed by regulatory mechanisms. Nevertheless, the occurrence of such R-type LPS in pathogenic wild-type bacteria clearly shows that the

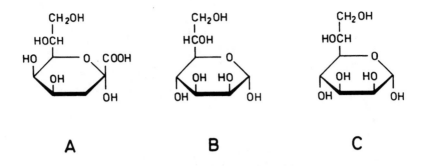

FIGURE 1. Formulas of (A) 3-deoxy-D-*manno*-octulopyranosonic acid, (B) L-*glycero*-D-*manno*-heptopyranose, and (C) D-*glycero*-D-*manno*-heptopyranose.

O-chain cannot be regarded as a pathogenicity factor per se and that it is dispensible without deleterious effects for the bacterium. Thus, the core oligosaccharide together with the lipid A component are the structural principle which is present in all Gram-negative bacteria investigated so far.

To understand the common and unique features of the different core oligosaccharides, we investigated during recent years the molecular structure of the core region of LPS from different bacteria in comparison to their biological activity, in particular to their immunoreactive properties. Here we review the chemical structure of core regions from bacteria of remote genetic origin.

A chemical characteristic of the core region is the presence of two sugars, i.e., 3-deoxy-D-*manno*-octulopyranosonic acid (Kdo) and heptopyranose, occurring in both the L-*glycero*-D-*manno* (L,D-Hep*p*) and D-*glycero*-D-*manno* configuration (D,D-Hep*p*) (Figure 1). D,D-Hep*p* was proposed to be a biosynthetic precursor of L,D-Hep*p*, and the conversion to the latter by epimerization of C-6 of ADP-D,D-Hep*p* catalyzed by ADP-D-*glycero*-D-*manno*-heptopyranose-6-epimerase was reported.[2-4] These glycosyl residues are found in the lipid A-proximal part of LPS which, particularly in enterobacterial LPS, is also termed the *inner core region*, in contrast to the *outer core region* which contains hexoses. However, as will be shown below, this differentiation based on composition is not very strict.

The linkages of Kdo, a polyfunctional ketose[5] with eight carbon atoms, are extremely sensitive to acid-catalyzed hydrolysis. Kdo was found in all LPS, and structural analyses of core oligosaccharides identified at least one Kdo unit as being located proximal to lipid A in the inner core region. Heptoses do not occur in all LPS;[6-14] however, when present, they are usually located adjacent to the Kdo region, although their occurrence in more distal parts of the core (see below) or even in the O-antigen[15] has been reported.

In the following, core structures of LPS of various bacterial families and genera are described, beginning with the enterobacterial core types. To facilitate this discussion, we have numbered the frequently occurring hexose,

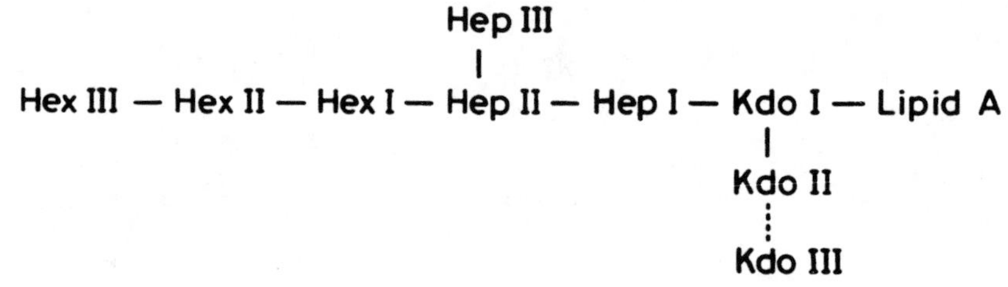

```
                        Hep III
                          |
Hex III — Hex II — Hex I — Hep II — Hep I — Kdo I — Lipid A
                                              |
                                            Kdo II
                                              ⋮
                                            Kdo III
```

FIGURE 2.   Scheme of enumerating frequently occurring sugar constituents of the core region of LPS. Dotted line indicates nonstoichiometric substitution.

heptose, and Kdo residues as schematically illustrated in Figure 2, beginning from the reducing end. With few exceptions, the substitution with phosphate groups is not shown in tables and figures and, where not stated otherwise, sugars are present as pyranosides.

II. ENTEROBACTERIACEAE

The structural elucidation of the core region was initiated by the discovery of enterobacterial rough (R) mutants which were first isolated from *Salmonella* species and later also from all enterobacterial genera.[16] In all cases the core consisted of a heterooligosaccharide, the structural variability of which was limited. Only one core type (Ra-core) was identified for the whole genus *Salmonella*, and in *Escherichia coli* five core types designated R1, R2, R3, R4, and K-12 have been described.[1]

A. *SALMONELLA* (TABLE 1)

The structure of the outer core region of *Salmonella* LPS (Ra-core) had already been elucidated more than 2 decades ago (reviewed in Reference 16). It comprised a pentasaccharide of D-glucopyranose (D-Glc*p*), D-galactopyranose (D-Gal*p*), and 2-acetamido-2-deoxy-D-glucopyranose (Glc*p*NAc). This outer core was also found in the LPS of *Arizona* sp.[16] Part of the inner core of *Salmonella* LPS was also structurally analyzed. A branched trisaccharide of L,D-Hep*p* with the sequence L,D-Hep*p*1.7-L,D-Hep*p*1.3-L,D-Hep*p* was identified to which the outer core was linked to position 3 of HepII. The heptose trisaccharide was bound to HO-5 of Kdo. All anomeric configurations of the linkages in the *Salmonella* Ra-core were later determined to be α.[17] The definite structural analysis of the Kdo region was hampered for a long time by the difficult chemistry of Kdo. Since this polyfunctional sugar was found to be sensitive to different chemicals used to degrade LPS for analysis, leading to artifacts,[5,18] new methodologies for the structural analysis of the

TABLE 1
Structure of the Core Region of *Salmonella* LPS
(Ra core type)[17,20,21,23,24]

```
O–chain–1.4αD–Glc1.2αD–Gal1.3αD–Glc1.3LαD–Hep1.3LαD–Hep1.5αKdo
              2                6            7            4
              |                |            |            |
              1                1            1            2
          αD–GlcNAc        αD–Gal      LαD–Hep     αKdo2.4αKdo
```

Note: KdoIII is present in nonstoichiometric amounts in most LPS analyzed.

Kdo region had to be developed. This topic has been reviewed in more detail elsewhere.[19] Structural analysis was facilitated by the isolation of core-defective R mutants which, due to a genetic defect in the *rfa* locus (encoding the enzymes for the core biosynthesis), made an incomplete core oligosaccharide. The Re and Rd chemotypes, which contain only the Kdo or Hep-Kdo region, respectively, were particularly useful in determining the structure of the inner core region. An α2.4-linked Kdo disaccharide was isolated from enterobacterial LPS by mild hydrolysis in acetate buffer, and characterized by combined gas-liquid chromatography/mass spectrometry (GLC-MS) and ^{1}H- and ^{13}C-nuclear magnetic resonance (NMR) spectroscopy.[20,21] This compound was identified in the LPS of many bacterial strains.[19] The development of an appropriate methylation analysis, which identified the linkages between the Kdo residues and that of the adjacent neutral sugars, allowed the structural analysis of the inner core region as shown in Table 1.[22-24] The substitution of KdoII by KdoIII was nonstoichiometric in all the LPS studied so far, except in *S. minnesota* chemotype Rb$_2$ strain R345. The occurrence of HepIII at HO-7 of HepII in the complete core (Ra) or the Rb chemotype was shown by methylation analyses of *S. minnesota* R345 (unpublished) and by identification of the methylated trisaccharide LαD-Hep*p*1.7LαD-Hep*p*-1.3L,D-Hep*p* in comparison with a synthetic standard.[25]

The substitution of the inner core region by phosphate at an unknown position of HepII and 2-aminoethyl pyrophosphate at HO-4 of HepI has been described.[26] Furthermore, Kdo-7-(2-aminoethyl phosphate) was isolated from *S. minnesota* R5 and R7 and characterized.[27] This compound was obtained after hydrolysis of these LPS in acetate buffer, which cleaves ketosidic but not glycosidic linkages. Therefore, it was suggested that KdoII was substituted by 2-aminoethyl phosphate. It should be noted that all the described charged substituents were present in nonstoichiometric proportions. We also would like to point out here that the structural determination of phosphate substituents of the core region is difficult for technical reasons. Therefore, the structural proposals concerning the phosphorylation of the inner core should be regarded as tentative.

TABLE 2
Structures of the Core Oligosaccharides from *E. coli* R1-R4 and K-12 Core Types[17,29-36,41]

Chemotype	Structure[a]

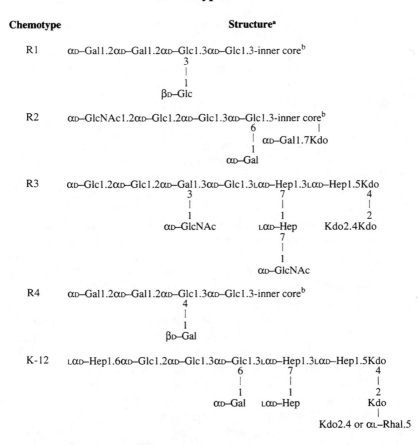

R1   αD–Gal1.2αD–Gal1.2αD–Glc1.3αD–Glc1.3-inner core[b]
                                    3
                                    |
                                    1
                                  βD–Glc

R2   αD–GlcNAc1.2αD–Glc1.2αD–Glc1.3αD–Glc1.3-inner core[b]
                                        6 |
                                        | αD–Gal1.7Kdo
                                        1
                                     αD–Gal

R3   αD–Glc1.2αD–Glc1.2αD–Gal1.3αD–Glc1.3LαD–Hep1.3LαD–Hep1.5Kdo
                    3   7   4
                    | | |
                    1   1   2
                αD–GlcNAc   LαD–Hep   Kdo2.4Kdo
                                7
                                |
                                1
                            αD–GlcNAc

R4   αD–Gal1.2αD–Gal1.2αD–Glc1.3αD–Glc1.3-inner core[b]
                                    4
                                    |
                                    1
                                  βD–Gal

K-12   LαD–Hep1.6αD–Glc1.2αD–Glc1.3αD–Glc1.3LαD–Hep1.3LαD–Hep1.5Kdo
                            6   7   4
                            | | |
                            1   1   2
                        αD–Gal   LαD–Hep   Kdo
                                                   |
                                        Kdo2.4 or αL–Rhal.5

[a] Nonstoichiometric substitutions: R3, D-GlcN linked to HepIII, KdoIII linked to KdoII, and NAc; K-12, LαD-Hep linked to GlcIII, and all substituents at KdoII.
[b] The Hep/Kdo region has not been fully determined.

B. *ESCHERICHIA COLI* (TABLE 2)

Five redundant core types have been identified for the LPS of *E. coli* by chemical, serological, and genetic investigations, termed R1, R2, R3, R4, and K-12.[1,28] The structure of the outer region of each of these core types has been published;[17,29-33] however, that of the K-12 core has recently required revision.[34,35]

The common partial structure of the outer core region of the *E. coli* and *Salmonella* core types was the αD-HexIII(1.2)-αD-HexII(1.3)αD-HexI tri-saccharide, which was substituted at position HO-2 of HexIII by either another hexose or by D-GlcNAc in α-pyranosidic linkage. An exception is the K-12 core where HexIII was substituted at position HO-6 by another α-linked L,D-Hep*p* residue. The respective hexose trisaccharide was found to be further substituted either at HO-6 of HexI by an αD-Gal*p* residue (Ra, R2, K-12) or at HO-3 or HO-4 of HexII (R1, R3, R4) by another hexose or hexosamine.

The structures of the inner core regions of the core types R1, R2, and R4 have not yet been completely elucidated. In the case of the R1 core, the presence of the trisaccharide L,D-Hep*p*1.7L,D-Hep*p*1.3L,D-Hep*p*, to which the outer core was linked at HO-3 of HepII, was indicated.[29] Preliminary experiments identified an αD-Glc*p*N1.7αD-Hep*p*-1.7αD-Hep*p* trisaccharide,[29a] which was also isolated from other core oligosaccharides (see below). NMR investigations of the core oligosaccharide indicated the linkages of the inner core region to be α.[17] The R1 core was proposed to be substituted at HO-4 of HepI by 2-aminoethyl pyrophosphate and at position HO-4 of HepII by a phosphate residue.[26]

The data obtained from methylation analysis of the R2 core oligosaccharide showed the presence of three L,D-Hep*p* units, presumably present as L,D-Hep*p*1.7L,D-Hep*p*1.3L,D-Hep*p*, the linkages of which are α as indicated by NMR studies of the core oligosaccharide.[17,30] The outer core was found to be linked to HO-3 of HepII. An αD-Gal*p*1.7Kdo disaccharide was isolated from the inner core region and characterized.[36] Since this compound, which was identified only in the R2 core, was isolated after hydrolysis of LPS in acetate buffer, D-Gal*p* is likely to be linked to KdoII.

The inner core region of *E. coli* R3 has been investigated only recently.[37] An αD-GlcN*p*1.7LαD-Hep*p*1.7LαD-Hep*p* trisaccharide and an αD-GlcN*p*1.7LαD-Hep*p* disaccharide were isolated and their structures identified. The same oligosaccharides were isolated from *E. coli* J-5,[38] an Rc-mutant derived from the wild-type strain *E. coli* O111 which harbors the R3 core type. Chemical, methylation, and NMR analyses of the isolated dephosphorylated core oligosaccharide and its product after Smith degradation, and methylation analysis of the Kdo region identified the structure as given in Table 2.

The structure of the Kdo region in *E. coli* R4 is not known. However, some information was provided on the heptose part of the inner core, indicating the presence of the trisaccharide LαD-Hep*p*1.7LαD-Hep*p*1.3L,D-Hep*p*, to which the outer core is linked at HO-3 of HepII.[32] Data of methylation analysis indicated that this trisaccharide was substituted by 2-aminoethyl pyrophosphate groups at positions HO-4 of each HepI and II.

From earlier publications it was not clear whether the core oligosaccharide of LPS from *E. coli* K-12 contained three or four L,D-Hep*p* residues.[17,39,40] Therefore, we reinvestigated this core and proved by chemical and methylation analyses the presence of four L,D-Hep*p* units.[34] We then isolated and characterized the disaccharide LαD-Hep*p*1.6αD-Glc*p*, the analytical data of which were identical to those of the synthetic counterpart.[35] Furthermore, LαD-Hep*p*1.6αD-Glc*p*1.2αD-Glc*p* trisaccharide and the tetrasaccharide

LαD-Hep*p*1.6αD-Glc*p*1.2αD-Glc*p*1.3αD-Glc*p* were isolated in their reduced and methylated forms. These data proved unequivocally that an L,D-Hep*p* residue was linked to HO-6 of GlcIII. The substitution was nonstoichiometric. The structure of the heptose region, includingthe linkage to KdoI, was shown by analysis of the product from Smith degradation and by identification of methylated LαD-Hep*p*1.7LαD-Hep*p*1.3LαD-Hep*p*. No Glc*p*NAc-hexose-containing oligosaccharide(s) could be identified. The identification of a D-GlcN-L,D-Hep disaccharide[35a] showed that the D-Glc*p*NAc residue was not 1.6-linked to GlcIII but rather to heptose. In NMR analyses only α-linkages were identified, which is in contrast to an earlier reported β-linkage of the D-GlcNAc residue.[17] Since it is not yet known to which heptosyl residue the D-GlcNAc is bound, the position of the latter is not shown in Table 2.

Different partial structures were isolated from the inner core region of the LPS from *E. coli* K-12, i.e., αL-Rha*p*1.5Kdo,[41] Kdo-7-(2-aminomethyl phosphate), and small amounts of Kdo phosphate.[27] All these compounds were obtained after hydrolysis of the LPS in acetate buffer, indicating that the identified substituents were present most likely at KdoII. These results, and the additional methylation analysis of the Kdo region, collectively allow us to propose the structure shown in Table 2.[42] Despite the identification of Kdo-7-(2-aminoethyl phosphate) and Kdo phosphate, the substitution of the inner core region by phosphate derivatives is not yet known.

C. *SHIGELLA* (TABLE 3)

The core structures of LPS from *S. sonnei* and *S. flexneri* have been extensively investigated.[43,44] In both species, LPS of the (*E. coli*) R1 core type was identified, which was substituted by 2-aminoethyl pyrophosphate at HO-4 of HepI and, in some R mutants, by a second 2-aminoethyl pyro-phosphate residue at HO-4 of HepII. Interestingly, HepIII carried an α1.7-linked D-GlcN unit, similar to that found in *E. coli* R1 and R3. The R4 core type was proposed to occur in *S. dysenteriae*.[16] The structure of the Kdo region of the inner core has so far not been investigated thoroughly; however, an α2.4-linked Kdo disaccharide was detected in *S. sonnei*.[19]

D. *PROTEUS* (TABLE 4)

The genus *Proteus* consists of four species: *P. vulgaris*, *P. mirabilis*, *P. myxofaciens*, and *P. penneri*.[45,46] Most of the structural investigations on the core region were performed with LPS of *P. mirabilis*.[47-50] These and other more fragmentary studies[51] on different Proteae such as *Morganella* and *Providencia* showed that the core structures, in which D-GalA was a characteristic constituent, are quite different from those of *Salmonella* and *E. coli*. The core structures of *Proteus* 027, *Providencia*, and *Morganella* contained additional D-GalN.[51-53]

The structure of the core from an Rc-type mutant (strain R4) from *Proteus mirabilis* 028 differed from that of the Rc mutant of *Salmonella*; for example,

TABLE 3
Structures of the Core Oligosaccharides from LPS of *Shigella sonnei* and *Shigella flexneri* Serotype 6

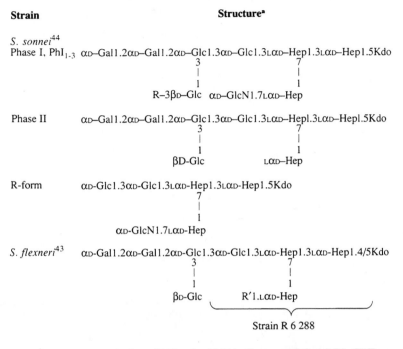

Strain	Structure[a]

S. sonnei[44]

Phase I, PhI$_{1-3}$ αD–Gal1.2αD–Gal1.2αD–Glc1.3αD–Glc1.3LαD–Hep1.3LαD–Hep1.5Kdo

Phase II   αD–Gal1.2αD–Gal1.2αD–Glc1.3αD–Glc1.3LαD–Hep1.3LαD–Hep1.5Kdo

R-form   αD-Glc1.3αD-Glc1.3LαD-Hep1.3LαD-Hep1.5Kdo

S. flexneri[43]   αD-Gal1.2αD-Gal1.2αD-Glc1.3αD-Glc1.3LαD-Hep1.3LαD-Hep1.4/5Kdo

[a]   Nonstoichiometric substitutions: GlcII and D-GlcN in *Sh. sonnei* R-form LPS; GlcII was not substituted by D-Glc in *Sh. flexneri* strains R V 6 975 and R 6 551 LPS. R = linkage point of the repeating unit(s) in PhI$_1$ and PhI$_2$ LPS, R′ = αD-GlcN in strains S 6 standard, S 6 975, R V 6 975, R 6 551, and R 6 288, and H in strain R 6 488.

HepI (which was α1.5-linked to Kdo) carried D-Glc*p* at HO-3 and HepII at HO-4.[48] HepII was substituted at HO-7 by either 2-aminoethyl phosphate or by phosphate in nonstoichiometric amounts. The complete core of the LPS from *P. mirabilis* R110/1959 was much more complex.[49] The tetrasaccharide L,D-Hep*p*1.7L,D-Hep*p*1.3L,D-Hep*p*1.5Kdo was present; however, HepIII was β-linked. HepI was substituted at position HO-4 by an αD-Glc*p*1.6-D-Glc*p* disaccharide, and HepII at position HO-6 by 2-aminoethyl phosphate in nonstoichiometric amounts. A second 2-aminoethyl phosphate residue was suggested to be linked to Kdo. The negative charge of the inner core region was provided by the presence of Kdo and two D-GalA residues linked α1.3 to HepII and β1.7 to HepIII. D,D-Hep*p* in addition to L,D-Hep*p* was present. Investigations on the cores of *P. mirabilis* 027 and 028 indicated basically the same structures as in R110/1959, but D-GlcN was exchanged by D-GalN in 027.[52,53] In serotype 028, D,D-Hep*p* was replaced by L,D-Hep*p*, which represents another example of the presence of L,D-Hep*p* in the outer core

TABLE 4
Structures of the Core Oligosaccharides from LPS of
***Proteus mirabilis* Strains R4/028 and R110/1959**

Strain	Structure

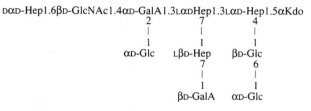

---

[a] The substitution of KdoI by KdoII was proposed tentatively.

region. Some information was available on the structure of the Kdo region. Investigations on the Re mutant LPS of *P. mirabilis* 1959 (strain R45) and the Rc mutant of wild-type 028 (strain R4) indicated that Kdo was α2.6-linked to the nonreducing D-GlcN of lipid A.[48,54] Furthermore, in strain R45 an α2.4-linked Kdo disaccharide was identified.[21] Small proportions of a third Kdo residue may be present in R4/028, but its position could not be determined.[48] From the Re mutant, R45, an 8-O-(4-amino-4-deoxy-βL-arabinopyranosyl)-Kdo disaccharide was isolated, but it is not yet clear whether the 4-amino-4-deoxy-βL-arabinopyranose is linked to KdoI or to KdoII.[47] A (2-aminoethyl phosphate)-Kdo unit was proposed for strain R110/1959.[49]

E. *CITROBACTER* (TABLE 5)

Characteristic for the genus *Citrobacter* is the simultaneous occurrence of complete and incomplete core structures in wild-type LPS, which is not known for other enterobacterial genera. The identified core structures from *C. freundii* LPS are summarized in Table 5.[55-59] The structures of the outer core regions of the investigated strains were different from those of *Salmonella* and *E. coli*. The typical hexose pentasaccharide present in the Ra and R1-R4 core types was not found except in strain PCM 1487, which possessed a structure very similar to that of *S. flexneri* and the R3 core of *E. coli*.[55,56] The only difference was that D-Gal*p*NAc instead of D-GlcNAc was α1.3-linked to Hex II (D-Gal*p*). The core structures of serotypes 04 and 036 were nearly identical;[58,59] the terminal D-Glc*p* in the side chain of the outer core of 04

TABLE 5
Structure of Core Oligosaccharides of *Citrobacter* LPS[55-59]

Serotype   **Structure**

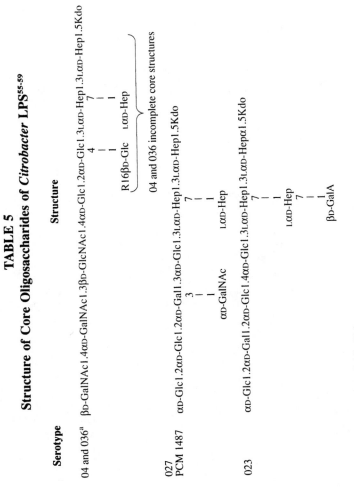

04 and 036[a]   βD-GalNAc1.4αD-GalNAc1.3βD-GlcNAc1.4αD-Glc1.2αD-Glc1.3Lαp-Hep1.3Lαp-Hep1.5Kdo

                          4   7
                          ↑   ↑
                  R16βD-Glc   Lαp-Hep

04 and 036 incomplete core structures

027
PCM 1487   αD-Glc1.2αD-Glc1.2αD-Gal1.3αD-Glc1.3Lαp-Hep1.3Lαp-Hep1.5Kdo

                                 3   7
                                 ↑   ↑
                           αD-GalNAc   Lαp-Hep

023   αD-Glc1.2αD-Gal1.2αD-Glc1.4αD-Glc1.3Lαp-Hep1.3Lαp-Hepα1.5Kdo

                                        7
                                        ↑
                                Lαp-Hep
                                        7
                                        ↑
                                βD-GalA

[a] R = αD-Glc in 04 and αD-Gal in 036.

TABLE 6
Structure of Core Oligosaccharides from LPS of
Yersinia enterocolitica and *Erwinia carotovora*

Strain	Structure

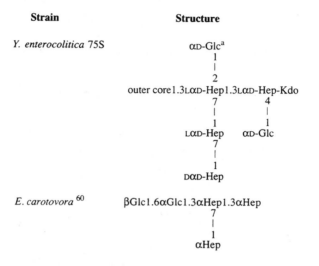

$Y.$ *enterocolitica* 75S

αD-Glc^a
|
|
2
outer core1.3LαD-Hep1.3LαD-Hep-Kdo
7   4
| |
1   1
LαD-Hep   αD-Glc
7
|
1
DαD-Hep

$E.$ *carotovora* [60]

βGlc1.6αGlc1.3αHep1.3αHep
7
|
1
αHep

---

^{a} Nonstoichiometric substitution.

was positionally replaced by D-Gal$p$ in 036. As in *Salmonella* and *E. coli*, the characteristic structural element of the inner core region was the LαD-Hep$p$-1.7-LαD-Hep$p$1.3LαD-Hep$p$1.5Kdo tetrasaccharide, which is substituted at HO-3 of HepII by D-Glc$p$ and at HO-4 of HepI by 2-aminoethyl phosphate.[57] Position HO-7 of HepIII was substituted in most cases by 2-aminoethyl pyrophosphate; however, in serotype 023, a D-Gal$p$A residue was identified at this position.[59] The structure of the Kdo region has not yet been elucidated.

F. *YERSINIA* (TABLE 6)

The structure of the heptose region of the Rc-type LPS from *Y. enterocolitica* strain 75R has been elucidated only very recently.[59a] The trisaccharide LαD-Hep$p$1.7LαD-Hep$p$1.3LαD-Hep$p$, which is linked to Kdo, was identified. HepI was substituted at HO-4 by αD-Glc$p$; HepII partially carried another αD-Glc$p$ residue at HO-2. In strain 75S (having the complete core) the outer core region extended from HO-3 and HepIII was 7-substituted by an α-linked D,D-Hep$p$. Thus, the linkages within the heptose region are unique among enterobacterial core types.

G. *ERWINIA* (TABLE 6)

Only the core structure of the LPS from *E. carotovora*, a plant pathogen, has been investigated to date.[60] This LPS contains the characteristic core

TABLE 7
Structure of Core Oligosaccharides from LPS of
***Bacteroides fragilis* NCTC 9343[8]**

PSI   βD-Gal1.6βD-Gal1.6βD-Gal1.6βD-Gal1.4/6αD-Glc1.2αL-Rha
                                                    4/6
                                                     |
                                                     1
                                                  βD-Gal

PSII   βD-Gal1.6αD-Glc1.2αL-Rha
              4
              |
              1
           βD-Gal

trisaccharide Hep1.7Hep1.3Hep in the inner core region, to which at HO-3 of HepII a β1.6-linked D-Glcp disaccharide is α-linked. The substitution of the core by phosphoryl residues and the structure of the Kdo region were not reported.

III. BACTEROIDEACEAE (*BACTEROIDES*, TABLE 7)

B. fragilis is found in the normal human colonic flora. Two polysaccharides (PS I and PS II) were isolated from its LPS after weak acid hydrolysis, and their structures were characterized.[8] Neither heptose nor Kdo were reported as constituents of the LPS. PS II is a partial structure of PS I. Speculatively, PS II could represent a very unusual and small core structure, which carried one repeating unit (the trisaccharide of D-Galp) in PS I. On the other hand, the sugar linking the polysaccharide moiety to lipid A was not determined. The genus *Bacteroides* had been reported for a long time to possess an LPS devoid of Kdo. However, recently Kdo phosphate was detected in LPS of *B. gingivalis*[61] and *B. fragilis*,[61a] suggesting that this compound, which cannot be identified under the hydrolysis conditions used in the standard assays for the determination of Kdo, could also be present in the LPS from other species.

IV. VIBRIONACEAE

A. *VIBRIO* (TABLE 8)

V. cholerae cause severe epidemics of diarrhea and are divided into the O1 (cholera vibrios) and the non-O1 serogroups. The O1 group is further subdivided into two major serological types, Ogawa and Inaba.[62] *V. metschnikovii* is a separate species that had formerly been included into the species *V. cholerae* as biotype "proteus". *V. parahaemolyticus* is a halophilic, marine bacterium, which is responsible for food poisoning. *V. ordalii* and *V. anguillarum* are fish pathogens.[63]

TABLE 8
Structures of Core Oligosaccharides Obtained from LPS of Different *Vibrio* Species

Species   Structure

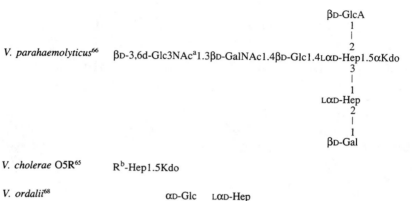

Species	Structure
V. parahaemolyticus[66]	βD-3,6d-Glc3NAc[a]1.3βD-GalNAc1.4βD-Glc1.4LαD-Hep1.5αKdo (with βD-GlcA1→2 branch, and 3→1 LαD-Hep2→1 βD-Gal branch)
V. cholerae O5R[65]	R[b]-Hep1.5Kdo

V. ordalii[68]

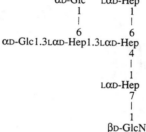

[a]   D-3,6d-Glc3NAc = 3-acetamido-3,6-dideoxy-D-glucopyranose.
[b]   R = unknown substituent.

The LPS from the genus *Vibrio* had for a long time been known as another example of LPS lacking Kdo.[26] However, later studies have identified phosphorylated Kdo residues in various species, e.g., Kdo-5-P in *V. cholerae* Ogawa and Inaba,[64] Kdo-4-P in *V. parahaemolyticus* and *V. cholerae* O5R,[65,66] and Kdo-P of unknown structure in *V. metschnikovii*.[67] In three different strains of *V. cholerae*, 7-O-(2-amino-2-deoxy-αD-glucopyranosyl)-L-*glycero*-D-*manno*-heptopyranose was identified as another LPS core constituent.[38]

The structural analyses of three core oligosaccharides from *Vibrio* LPS have been reported. The core of LPS from *V. parahaemolyticus* serotype O12 contains the trisaccharide LαD-Hep*p*1.3LαD-Hep*p*1.5Kdo in the inner core region.[66] A third residue of L,D-Hep*p* was missing. HepI was substituted at HO-4 by βD-Glc*p*, which carried the disaccharide 3-acetamido-3,6-dideoxy-βD-Glc*p*-1.6-D-Gal*p*NAc. Negative charges were provided by Kdo-4-P, a D-Glc*p*A residue β-linked to HepI, 3-deoxy-D-*threo*-hexulosonic acid,

D-Gal*p*A, and two additional phosphate residues, the positions of which were not determined. A second structure published was that of the Kdo region of LPS from *V. cholerae* O5R.[65] As in *V. parahaemolyticus*, only one Kdo residue was present, which was phosphorylated at HO-4 and substituted by L,D-Hep*p* at HO-5.

A different core structure was reported for the LPS of *V. ordalii*.[68] Four units of L,D-Hep*p* were clustered. HepI was substituted at HO-3 by HepII and at HO-4 by a β1.7-linked D-Glc*p*N-L,D-Hep*p* disaccharide, which (in α-linkage) was also identified in other cores, e.g., from *E. coli* R1, R3, *S. flexneri*, *V. cholerae*, *Aeromonas salmonicida*, and *Bordetella pertussis* LPS (see below). Furthermore, HepI was substituted at HO-6 by a fourth L,D-Hep*p* residue, and HepII carried two D-Glc*p* units at HO-3 and HO-6, respectively.

B. *AEROMONAS* (TABLE 9)

The genus *Aeromonas* comprises mostly fish pathogens and is associated with diseases in freshwater fish.[69] Three major chemotypes (I, II, and III) of the LPS from *A. hydrophila* are known.[70] Chemotypes I and II are characterized by the presence of an α1.7-linked D-Glc*p*N-L,D-Hep*p* disaccharide, which was not found in chemotype III.[71] Chemotype II possesses a cluster of four L,D-Hep*p* residues, in which the central HepII residue is α1.2-linked to HepI and substituted at HO-4,HO-6, and HO-3 by D-Glc*p*, HepIII, and αD-Glc*p*N1.7LαD-Hep*p*, respectively.[72] The same substitution pattern was found at HepII of chemotype III.[73] Only two L,D-Hep*p* units were present in the latter chemotype; HepII was substituted by two D-Glc*p* residues at HO-4 and HO-6, respectively, and by 3-O-(3-acetamido-3,6-dideoxy-βL-glucopyranosyl)-D-galactopyranose at HO-3. The core region of a human isolate of *A. hydrophila* was similar to that of chemotype II, but additionally carried at HO-6 of the D-Glc*p* residue a trisaccharide consisting of one D-Gal*p* and two D,D-Hep*p* units.[74]

The most prominent feature of the core structures from LPS of *A. salmonicida* and *A. salmonicida salmonicida* is the occurrence of a furanosidic Kdo substituted at HO-6 by HepI,[75,76] which so far is the only example of furanoid Kdo in LPS. Wheras the core of LPS from *A. salmonicida* contained three L,D-Hep*p* residues,[75] that of *A. salmonicida salmonicida* was comprised of five such units,[76] of which HepI (and not HepII as in *A. hydrophila*) was substituted at positions HO-3, HO-4, and HO-6.

V. NEISSERIACEAE

Most of the members of this family possess a rough-type LPS.

A. *NEISSERIA* (TABLE 10)

N. meningitidis is an important human pathogen causing severe meningitis and fulminant sepsis.[77] Eleven serotypes (L1 to L11) of the LPS have been

TABLE 9
Core Structures from LPS of *Aeromonas* Species

Species   **Structure**

A. hydrophila

    Chemotype II[72]

LαD-Hep
1
|
6
βD-Glc1.4LαD-Hep1.2LαD-Hep
3
|
1
αD-GlcN1.7LαD-Hep

    Chemotype III[73]

αD-Glc
1·
|
6
βL-3,6d-Glc3NAc[a]1.3αD-Gal1.3LαD-Hep1.2LαD-Hep
4
|
1
αD-Glc

    A6 (human isolate)[74]

DαD-Hep   LαD-Hep
1   1
| |
6   6
βD-Gal1.4DαD-Hep1.6βD-Glc1.4LαD-Hep1.2LαD-Hep
3
|
1
αD-GlcN1.7LαD-Hep

A. salmonicida[75]   LαD-Hep1.2LαD-Hep1.3LαD-Hep1.6Kdo*f*

A. salmonicida salmonicida[76]

LαD-Hep
1
|
6
αGal1.6αGlc1.4αGalNAc1.4LαD-Hep1.2LαD-Hep1.3LαD-Hep1.6Kdo*f*
4
|
1
αD-GlcN1.7LβD-Hep

[a] L-3,6d-Glc3NAc = 3-acetamido-3,6-dideoxy-L-glucopyranose.

identified, the epitopes of which are located in the oligosaccharide part of the LPS.[78-80] The structures of the core oligosaccharides of LPS from *N. meningitidis*, which carry the determinants for the serotypes L1 to L7 and L9, were identified.[81-84] The LαD-Hep*p*1.3LαD-Hep*p*1.5Kdo trisaccharide was present without being substituted by a third L,D-Hep*p* residue. Instead, HepII carried a D-Glc*p*NAc unit at HO-2. A D-Glc*p* residue, which represented the link of a neutral oligosaccharide to the inner core region, was bound to position HO-4 of HepI. In contrast to most enterobacterial core structures most of the

TABLE 10
Structures of Core Oligosaccharides Isolated from LPS of *Neisseria meningitidis* Serotypes, of *N. gonorrhoeae*, and of *Acinetobacter calcoaceticus*

Serotype	Structure
N. meningitidis[81-84]	(Basic structure)

$$R1.4\beta D\text{-}Glc1.4L\alpha D\text{-}Hep1.5Kdo$$
$$3$$
$$|$$
$$1$$
$$R'1.3L\alpha DHep$$
$$2$$
$$|$$
$$1$$
$$\alpha D\text{-}GlcNAc^{a}$$

Serotype	Structure
L1	R = αD-Gal1.4βD-Gal
L2	R = βD-Gal1.4βD-GlcNAc1.3βD-Gal, R' = αD-Glc
L3, L7, L9	R = βD-Gal1.4βD-GlcNAc1.3βD-Gal, R' = H
L4	R = βD-Gal1.4βD-GlcNAc1.3βD-Gal,
L5	R = βD-Gal1.4βD-GlcNAc1.3βD-Gal1.4βD-Glc, R' = αD-Glc
L6	R = βD-GlcNAc1.3βD-Gal

N. gonorrhoeae[85-87]

F62

$$Glc^{b}$$
$$|$$
$$\beta GalNAc1.3\beta Gal1.4\beta GlcNAc^{b}1.4Gal\text{-}L\alpha D\text{-}Hep\text{-}Kdo$$
$$|$$
$$\beta GlcNAc\text{-}L\alpha D\text{-}Hep$$

JW31R

$$Gal\text{-}Gal\text{-}Glc\text{-}Hep\text{-}Kdo$$
$$|$$
$$GalNAc\text{-}Hex\text{-}Hex\text{-}Hex\text{-}Hep$$
$$|$$
$$GlcNAc$$

A. calcoaceticus

Serotype	Structure
NCTC 10305[92]	D-Glc1.6D-Glc1.6D-Glc1.4/5D-Glc1.5Ko^{c}
NCTC 10303[94]	αL-Rha1.3αL-Rha1.8Kdo

[a] Nonstoichiometrically O-acetylated in serotypes L4 and L5.
[b] Assignment exchangeable.
[c] Ko = octulosonic acid.

hexoses are present in β-linkage. The core oligosaccharides of the LPS from the serotypes L3, L7, and L9 differed in substitution by 2-aminoethyl phosphate, which was suggested to be responsible for the induction of an immune response against the serotypes. Fast atom bombardment mass spectrometry investigations on serotypes L4 and L5 revealed differences in the main core structure, which were generated by additional 2-aminoethyl phosphate or O-acetyl groups, and by various proportions of hexoses.[83] The structure of the L2 core oligosaccharide was also basically the same as that in L3, L7, and L9, but carried an additional D-Glc*p* residue at HO-3 of HepII as in L5.[84] In the core oligosaccharides from serotypes L1 and L6 only disaccharide residues consisting of two D-Gal units (L1) and of D-GlcNAc and D-Gal (L6) were found to be linked to the D-Glc residue of the basic structure.[81] Both core oligosaccharides were partially O-acetylated and carried 2-aminoethyl phosphate at HO-3 (L1) and HO-7 (L6) of HepII, respectively. Many LPS of *N. meningitidis* contain the terminal tetrasaccharide βD-Gal1.4βD-Glc-NAc1.3βD-Gal1.4βD-Glc, which is identical to the glycosyl moiety of lacto-neotetraglycosylceramide, a glycosphingolipid of human erythrocytes, and may play a role in *Neisseria* avoidance of immunologic host defense by antigen mimicry.[86]

Some work was performed on core structures of LPS from *N. gonor-rhoeae*, an important sexually transmitted pathogen.[85-87] Most of the linkages of the core oligosaccharide from LPS of *N. gonorrhoeae* F62[85] remained undetermined, but, roughly, the general structure resembles that of the core from *N. meningitidis* LPS. The composition of the terminal tetrasaccharide at HepI was different from that in the *N. meningitidis* core, and HepI was additionally substituted by another D-Glc or D-GlcNAc residue. The partial structure of four core oligosaccharides from LPS of *N. gonorrhoeae* JW31R were determined,[87] the general sequence of which is shown in Table 10. The identified oligosaccharides differed in substitution by one or two 2-aminoethyl phosphate groups and by the presence of the terminal Gal of the Gal-Gal-Glc residue at HepI.

The core oligosaccharides of neisserial LPS are thought to be linked via Kdo to lipid A. The structure of the Kdo region of the LPS from *Neisseria* has not yet been determined.

An interesting observation of great relevance for the host defense against gonococci was made with gonococcal LPS. When the bacteria are grown under *in vivo* conditions, i.e., in serum-containing media, they are resistant to the bactericidal effects of normal serum. However, when grown on artificial media they become serum sensitive. This phenomenon was understood when it was found that *N*-acetyl-neuraminic acid (5-NeuNAc), which inhibits alternative complement activation, was transferred to LPS if its precursor CMP-5-NeuNAc was present in the growth medium.[88]

B. *ACINETOBACTER* (TABLE 10)

A. calcoaceticus causes opportunistic infections in hospitalized patients.[89,90] The LPS of strain 10305 NCTC contained Kdo, but neither L,D-nor D,D-Hep*p*.[91] After degradation of the LPS with acid (0.1 *M* HCl, 100°C, 1 h), the product obtained consisted of lipid A and a phosphorylated core oligosaccharide composed of D-Glc*p* and octulosonic acid (Ko).[92] Ko was substituted at position HO-5 (by D-Glc*p*), as Kdo in other LPS. Ko was shown to be isosteric to Kdo with respect to the configuration of the hydroxyl groups (D-*glycero*-D-*talo*-configuration[92a] and linked the core oligosaccharide to HO-6′ of the nonreducing D-GlcN of lipid A. This is the first example where a sugar other than Kdo forms the link between core and lipid A. In addition, 2-keto-3-deoxy-1,7-dicarboxyheptonic acid of unknown configuration was detected.[93] Furthermore, the disaccharide D-Glc*p*1.5Kdo was identified.[93a] In LPS of strain 10303 NCTC, the trisaccharide αL-Rha*p*1.3αL-Rha*p*1.8Kdo was isolated and characterized by GLC-MS and NMR.[94]

VI. CHLAMYDIACEAE

The family Chlamydiaceae comprises only one genus with three species; (1) *Chlamydia trachomatis*, a human pathogen causing sexually transmitted genital infections in adults and pneumonia in newborns, (2) *C. psittaci*, a pathogen for animals causing organ-specific infections, and *C. pneumoniae* (formerly called TWAR) causing broncho-pulmonary infections in humans.[95,96]

The LPS represents one of the major surface antigens of *Chlamydia*, and was shown to be chemically and serologically related to LPS from rough mutants of the Re type.[9,10,97,98] In addition, it carried a genus-specific epitope which was defined by a monoclonal antibody.[99] Using the tools of molecular genetics, it was possible to express this epitope in *E. coli* and *S. minnesota* strains which contained a recombinant plasmid carrying an insert of *C. trachomatis* DNA.[100,101] These recombinants produced an LPS containing the *Chlamydia*-specific epitope in addition to that of the parent strain.[101] Since *Chlamydia* are hazardous pathogens and difficult to grow in large quantities, the availability of such recombinants was highly advantageous. Transformation of Re-mutants of *S. minnesota* and *S. typhimurium* with the recombinant plasmid still resulted in the expression of the genus-specific LPS epitope in addition to those of the parent strain.[101] Thus, the chemical structure of the recombinant and the parent Re LPS could be compared. The former contained, in addition to the α2.4-linked Kdo disaccharide also present in the parent LPS, a 2.8-linked Kdo disaccharide and a trisaccharide with the sequence Kdo2.8Kdo2.4Kdo. This was shown by gas-liquid chromatography-mass spectrometry (GLC-MS) of appropriately derivatized and purified samples obtained after mild acid hydrolysis of recombinant LPS.[102] These experiments did not, however, allow the determination of the anomeric configurations of Kdo and of the linkage to lipid A. Therefore, we synthesized the terminal disaccharide in α2.8- and β2.8-linkage. These compounds could be separated by GLC, and their comparison with the natural compound indicated that an α2.8-linked Kdo disaccharide is present as the terminal part of the trisaccharide.[103] Next, we determined the linkage of the Kdo trisaccharide to

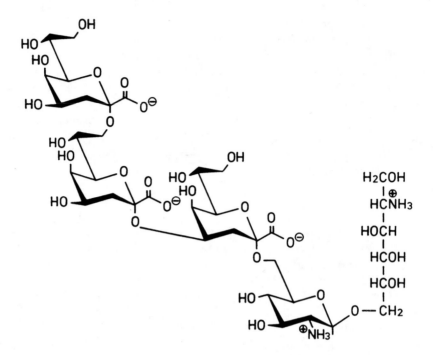

FIGURE 3. Structure of the isolated pentasaccharide from the recombinant LPS of *Salmonella minnesota* r595-207, carrying the *Chlamydia*-specific LPS epitope.

the lipid A backbone.[104] Recombinant LPS was de-O-acylated with hydrazine at 37°C, dephosphorylated with hydrogen fluoride, reduced with sodium borohydride, and finally de-N-acylated with hydrazine at 85°C. Following purification by high-voltage paper electrophoresis, a pentasaccharide was obtained, the ^{13}C-NMR spectrum of which was assigned to the structure shown in Figure 3. It shows that all three Kdo residues are present as α-pyranosides and that the trisaccharide is linked to position HO-6′ of the nonreducing glucosamine residue of lipid A.

The pentasaccharide was then coupled to bovine serum albumin and its immunoreactive properties were investigated. The artificial glycoconjugate still contained the genus-specific epitope, the minimal structure of which was found to be the Kdo trisaccharide.[104] The α2.8-linkage between two Kdo units is unique among bacterial LPS and, therefore, most likely represents the immunodominant part of the *Chlamydia*-specific LPS epitope as recognized by monoclonal antibodies.[99,104]

VII. BRUCELLACEAE (*BORDETELLA*, TABLE 11)

B. pertussis causes whooping cough in humans. Two types of rough-like LPS (LPS I and LPS II) were isolated, which differed at least by the presence

TABLE 11
Structure of the Core Oligosaccharide from LPS of
Bordetella pertussis[106,107]

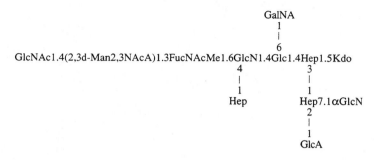

Note: 2,3d-Man2,3NAcA = 2,3-diacetamido-2,3-dideoxy-mannosaminuronic acid, FucNAcMe = methylacetamidofucosamin, GalNA = galactosaminuronic acid.

of a phosphate group at HO-4 of Kdo in LPS II.[105,106] A second type of heterogeneity of the LPS from this organism was identified in sodium dodecyl sulfate-polyacrylamide gel electrophoresis, where two LPS bands, A and B, were found, which did not correlate with LPS I or LPS II.[107] The structural difference in the polysaccharide portions was shown to be located terminally, where in LPS A an additional trisaccharide D-GLcNAc1.4(2,3-diacetamido-2,3-dideoxymannosaminuronic acid)1.3(2,6-dideoxy-2-methylacetamidogalactose) was found.[107,107a]

The polysaccharide portions of LPS I and II were identical to the polysaccharide from LPS A.[107a] It is noted that the disaccharide 7-*O*-(2-amino-2-deoxy-αD-glucopyranosyl)-L-*glycero*-D-*manno*-heptopyranose, which is also present in other LPS (see above), was for the first time completely analyzed in the LPS of *B. pertussis*.[108]

As mentioned above, Kdo was found to be substituted at position HO-4 by a phosphate group. Additionally, Kdo-5-phosphate was identified in LPS from *B. pertussis*.[109] However, no structural analysis of the entire Kdo region has been performed.

VIII. PASTEURELLACEAE (*HAEMOPHILUS*)

H. influenzae causes invasive and bacteremic infections, e.g., meningitis in young children.[110] Only R-type LPS was detected in wild strains of *H. influenzae*.[111-114] The core oligosaccharide consisted of Glc, Gal, Hep, and Kdo, and contained in the case of strain EAG additional GalN.[113,114] The sequence of the core oligosaccharides was not identified. A Kdo-phosphate residue was detected in the LPS of the deep-rough mutant I-69 Rd⁻b⁺,[115] and later the structure of the LPS was elucidated.[116] Only one Kdo residue

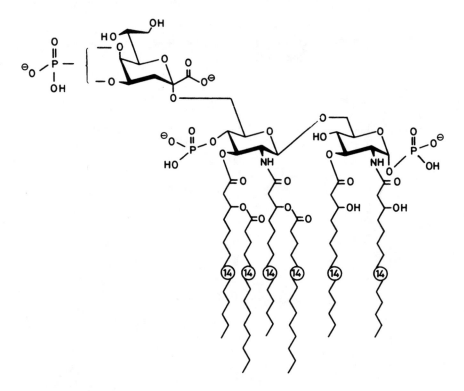

FIGURE 4. Structure of the LPS isolated from *Haemophilus influenzae* strain I-69 Rd⁻/b⁺. Two LPS populations were found, differing in the substitution of Kdo with phosphate at either position HO-4 or HO-5.

was present, which was linked to position HO-6′ of the nonreducing D-GlcN of the lipid A and which was phosphorylated at either HO-4 or HO-5 (Figure 4). The existence of this mutant suggests, parenthetically, that a single (phosphorylated) Kdo residue linked to lipid A is sufficient for the survival of Gram-negative bacteria.

IX. PSEUDOMONADACEAE

A. *PSEUDOMONAS* (TABLE 12)

A partial structure of the core oligosaccharide of *P. aeruginosa* was proposed for strain NCTC 1999 which consists of a L,D-Hep*p* disaccharide bound to Kdo and substituted by a branched hexasaccharide, to which alanine is linked.[117] This was also the case in the LPS of strain PAC1R.[118] Here, only one Hep*p* residue was identified, which was linked to Kdo and substituted by the outer core at an unknown position and by D-Glc at HO-7. The linkage of the heptose to Kdo was either at HO-4 or at HO-5. The LPS of *Ps. maltophila* (NCTC 10257) was reported not to contain any heptose at all.[7] A

TABLE 12
Partial Structures of Core Oligosaccharides from LPS of *Pseudomonas aeruginosa, Ps. maltophila, Xanthomonas sinensis,* and *Rhizobium trifolii*

Strain	Structure

P. aeruginosa[117,118]
NCTC 1999

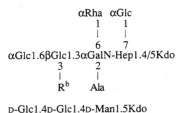

Ala[a]
|
βD-Glc1.2αL-Rha1.6D-Glc-GalN-3Hep1.3Hep1.4/5Kdo
|
αD-Glc1.6D-Glc

PAC1R

αRha αGlc
1   1
| |
| |
6   7
αGlc1.6βGlc1.3αGalN-Hep1.4/5Kdo
3   2
| |
R[b]   Ala

P. maltophila[7]

D-Glc1.4D-Glc1.4D-Man1.5Kdo
6
|
1
D-GalN

X. sinensis[119]

Man-Man-Glc-Man-Kdo
|
P
|
GalA-amide[c]

R. trifolii[121,122]

αD-Gal1.6αD-Man1.5Kdo   αD-GalA1.5Kdo
4   4
| |
1   1
αD-GalA   αD-GalA

[a]   Partial substitution with alanine.
[b]   R = linkage point of the O-antigen.
[c]   GalA-amide = galacturonamide.

partial core structure was proposed, which consisted of a branched hexose-tetrasaccharide linked to HO-5 of Kdo via D-Man*p*. The mannose residue was suggested to be substituted by either D-Gal*p*A phosphate or 2-aminoethyl pyrophosphate at HO-2 or at HO-3. The occurrence of a phosphorylated Kdo was also discussed.

B. *XANTHOMONAS* (TABLE 12)

Some fragmentary studies on the core of the LPS from *X. sinensis* were reported.[18,119] Only one unit of Kdo was identified and this was the first example at that time that a single Kdo residue is present in the main chain. As in *P. maltophila*, the LPS was free of heptose, and a mannose residue, substituted by a hexose tetrasaccharide, is linked to Kdo. Furthermore, the mannose residue carried a galacturonamide phosphate.

X. RHIZOBIACEAE (*RHIZOBIUM*, TABLE 12)

Rhizobia are nonpathogenic bacteria living in symbiosis with legume plants, where they participate in the process of nitrogen assimilation.[120] LPS and capsular polysaccharides were identified on their cell surface, and both were suggested to play a role in the specific adhesion of the bacteria to their leguminoses host and, therefore, in symbiosis.

Two core oligosaccharides have been isolated from the LPS of *R. trifolii* ANU843 which were negatively charged.[121,122] A tetrasaccharide comprised a D-Man*p* residue α1.5-linked to Kdo: mannose was substituted at HO-4 by D-Gal*p*A and at HO-6 by D-Gal.[122] The other oligosaccharide was a trisaccharide consisting of two D-Gal*p*A residues and Kdo. The structure originally proposed in Reference 121 has been recently revised in that the uronic acids are now thought to be linked to positions HO-4 and HO-5 of Kdo.[122a]

Two oligosaccharides of identical composition to the above-described were isolated from *R. leguminosarum* biovar *phaseoli*, and the structures were reported as (1) a tetrasaccharide, in which D-Man*p* was linked to HO-4 of Kdo and substituted at HO-4 by D-Gal*p* and at HO-6 by D-Gal*p*A and (2) a trisaccharide, in which Kdo was substituted at HO-4 and HO-7 by D-Gal*p*A, respectively.[123] These structures were stated to be identical to those of the oligosaccharides isolated from LPS of *R. trifolii* ANU843, and may therefore also have to be revised.

XI. PHOTOTROPHIC BACTERIA

A. GENERAL

Phototrophic bacteria comprise the nonsulfur purple bacteria (formerly named Rhodospirillaceae), the sulfur-containing purple bacteria (Chromatiaceae), the green sulfur bacteria (Chlorobiaceae), and the Chloroflexaceae, a family of thermophilic, gliding bacteria. Except for the latter, the cell wall of which is composed quite differently as compared to that of the other families, LPS has been isolated from members of each of the first three families.[124] One characteristic feature of the LPS from phototrophic bacteria is the occurrence of methylated sugars. Several LPS were found to be free of heptose,[13,14,125] others contained only D,D-Hep*p*,[126-128] and some contained both D,D- and L,D-Hep*p*.[129,130] To date, only core structures of the LPS from nonsulfur purple bacteria have been elucidated, which will be discussed below.

B. NON-SULFUR PURPLE BACTERIA (TABLE 13)

This family consists of the genera *Rhodopseudomonas*, *Rhodocyclus*, *Rhodomicrobium*, and *Rhodobacter*. It should be noted that, in the core of

TABLE 13
Core Structures of LPS from Members of the
Non-Sulfur Purple Bacteria

Strain	Structure
R. tenue[132]	(Basic structure)

$$\begin{array}{cc}
\text{L-D-Hep1.3L,D-Hep-Kdo} \\
7 \qquad\quad 4 \\
| \qquad\quad | \\
1 \qquad\quad 1 \\
\text{L,D-Hep} \quad\ \ \text{R}
\end{array}$$

Strain	Structure
EU1	R = L,D-Hep
2761	R = D,D-Hep1.2L,D-Hep
3661	R = D,D-Hep1.2D,D-Hep
G FUy	R = y1.2D-Man (y = unknown substituent)
R. sphaeroides[14]	
ATCC 17023	αD-GlcA1.4αD-GlcA1.4αD-GlcA1.4Kdo

$$\begin{array}{c}
6 \\
| \\
\text{Thr}^a
\end{array}$$

Strain	Structure
R. gelatinosus[134]	
Dr2	αGalA1.4αGalA1.4Kdo

$$\begin{array}{c}
5 \\
| \\
1 \\
\text{D,D-Hep}
\end{array}$$

^a Thr = threonine.

the LPS from *Rhodobacter*, 5-NeuNAc is present.[131] Structural investigations were performed on the heptose region of *Rhodocyclus tenuis* (formerly *Rhodospirillum tenue*).[132] The trisaccharide L,D-Hep*p*.7L,D-Hep*p*1.3L,D-Hep*p* was proposed to be present, in which HepII is partially substituted by phosphate at HO-3. Different structures of strain-specific regions were identified, containing L,D- and/or D,D-Hep*p*, and D-Man*p*. However, the complete core structures were not identified.

The structure of the oligosaccharide moiety of the LPS from *Rhodobacter sphaeroides* (formerly *Rhodopseudomonas sphaeroides*) has been elucidated.[14] Only one unit of Kdo was present, which linked the oligosaccharide to lipid A. This oligosaccharide represented a cluster of negatively charged residues, since Kdo was phosphorylated at HO-8 and further substituted at HO-4 by a D-Glc*p*A trisaccharide, the terminal residue of which carried threonine at the carboxyl group. No heptose was found in this LPS.

A very similar structure was identified in the core of *Rhodocyclus gelatinosus* (formerly *Rhodopseudomonas gelatinosa*) Dr2 LPS, which contains

D,D-Hep$p$.[133,134] The core oligosaccharide comprises one Kdo residue, which is substituted at HO-4 by a disaccharide of D-Gal$p$A. The D,D-Hep$p$ is linked to HO-5 of Kdo and is partially phosphorylated at HO-7.

XII. CONCLUDING REMARKS

Although there is a considerable structural variability in the different core oligosaccharides, some general principles can be formulated which will be discussed in the following. The lipid A-proximal inner core region of enterobacterial LPS is characterized by the occurrence of heptose and Kdo and, therefore, the terms inner core and heptose-Kdo region are often used synonymously. This is correct for most enterobacterial and many other bacterial species, but one should keep in mind that heptose may be absent from LPS or may occur also in the outer core region. In addition, as shown for the first time in *A. calcoaceticus*, Kdo may be replaced positionally by Ko. Many LPS which have been reported to lack Kdo were later shown, by the improved analytical technology developed in recent years, still to contain this sugar. Therefore, the occurrence of at least one Kdo residue can be regarded as a general principle of the inner core region of bacterial LPS. Another common feature of the inner core region of different LPS is the accumulation of negatively charged groups, contributed by carboxyl and phosphoryl groups. Since these charged groups are located in the proximity of lipid A which also carries negatively charged phosphoryl groups, their neutralization by divalent cations or polyamines under physiological conditions may influence the conformation of LPS and could be essential for the function of LPS to maintain the integrity of the outer membrane. In this context, it is important to define the minimal structure of the core oligosaccharide still allowing a bacterial viability. For a long time, a Kdo disaccharide was considered to represent this minimal structure; however, the recent report on a rough mutant from *H. influenzae* has shown that a single Kdo residue obviously fulfills the structural requirements for this important physiological function of LPS. Interestingly, this single Kdo is phosphorylated, which may indicate the importance of a critical number of negative charges.

Another highly conserved partial structure being found in all enterobacterial and many nonenterobacterial species, is the tetrasaccharide LαD-Hep$p$1.7LαD-Hep$p$1.3LαD-Hep$p$1.5Kdo in which the second heptose is substituted at HO-3 with either glucose or galactose. An exception is the core structure of *P. mirabilis* R4/028, where HepII is linked to position HO-4 of HepI. In LPS lacking heptose, HO-5 of Kdo or Ko is substituted by another sugar (D-Man$p$ in *Rhizobium trifolii*, *Pseudomonas maltophila*, and *X. sinensis* or D-Glc$p$ in *A. calcoaceticus*). Therefore, HO-5 of KdoI (or Ko) represents the favorite linkage point of the core oligosaccharide in most of the LPS investigated. There are some characteristic positions at which the above tetrasaccharide is substituted. KdoI is often substituted at HO-4 by a second Kdo residue or other negatively charged groups such as phosphate in *H.*

influenzae, B. pertussis, V. cholerae, and *V. parahaemolyticus* or GalA in *Rhodocyclus gelatinosus* and *Rhizobium trifolii.* In addition, substitution of Kdo was reported (1) at HO-5 with D-Glc*p*, L-Rha*p*, D-Gal*p*A, and phosphate; (2) at HO-7 with D-Gal*p*, 2-aminoethyl phosphate; and (3) at HO-8 with Kdo, L-Rha*p*, phosphate, and 4-amino-4-deoxy-L-arabinopyranose. However, in many cases it is not clear whether the substitution is on KdoI, KdoII, or KdoIII.

The characteristic point of substitution at HepI is obviously HO-4, where phosphate, 2-aminoethyl phosphate, 2-aminoethyl pyrophosphate, D-Glc*p*, and HepII were identified. HepII is most often substituted at HO-3 by the first unit of the outer core (usually a Glc or Gal residue) and at HO-7 by HepIII which in turn often carries substituents at HO-7 (Glc*p*N, Gal*p*A, D,D-Hep*p*).

Although there exist reports on the substitution of core structures with phosphate, this still appears to be one of the major analytical tasks of the future. Chemical methods are often not very reliable, since phosphate groups tend to migrate under different experimental conditions and phosphodiesters and pyrophosphate groups may be cleaved during the degradation process required for the particular analytical method. Most likely, NMR spectroscopy will help to solve this problem. However, at present the complexity of NMR spectra obtained with native LPS does not allow an unambiguous interpretation of the data.

NOTE ADDED IN PROOF

The structure of the core oligosaccharide from *Neisseria gonorrhoeae* F62 LPS (compare Table 10) has in the meantime been accomplished:[135]

βGalNAc1.3βGal1.4βGlcNAc1.3βGal1.4βGlc1.4αHep-Kdo

$$3$$

$$|$$

$$1$$

GlcNAc1.2αHep

The structures of the isolated oligosaccharides from *Rhizobium leguminosarum* biovar *phaseoli* LPS (compare Chapter 10) have been revised and are identical to those shown for *R. trifolii* in Table 12.[136] Finally, neuraminic acids containing oligosaccharides have recently been isolated from LPS of *Campylobacter jejuni* and *Rhodobacter capsulatus* 37b4, and their structures have been elucidated.[137,138]

ACKNOWLEDGMENTS

We are indebted to our colleagues U. Zähringer (Forschungsinstitut Borstel, Germany), H. Mayer (Max-Planck-Institut für Immunbiologie, Freiburg

i. Br., Germany), J. Radziejewska-Lebrecht (University of Lodz, Poland), M. Caroff (Université de Paris-Sud, France), and R. W. Carlson (Complex Carbohydrate Research Center, Athens, GA) for providing unpublished data, and to E. T. Rietschel for reading this manuscript. The research of our laboratory is supported by the Bundesministerium für Forschung und Technologie (grants 01ZR8604 and 01Ki8818) and the Deutsche Forschungsgemeinschaft (grant Br. 731/9-1).

REFERENCES

1. **Rietschel, E. T., Brade, L., Holst, O., Kulshin, V. A., Lindner, B., Moran, A. P., Schade, U. F., Zähringer, U., and Brade, H.**, Molecular structure of bacterial endotoxin in relation to bioactivity, in *Cellular and Molecular Aspects of Endotoxin Reactions,* (Endotoxin Research Series, Vol. 1), Nowotny, A., Spitzer, J. J., and Ziegler, E. J., Eds., Excerpta Medica, Amsterdam, 1990, 15.

2. **Raetz, C. H. R.**, Biochemistry of endotoxins, *Annu. Rev. Biochem.,* 59, 129, 1990.

3. **Coleman, W. G., Jr.**, The *rfaD* gene codes for ADP-L-*glycero*-D-*manno*-heptose-6-epimerase: an enzyme required for lipopolysaccharide core biosynthesis, *J. Biol. Chem.,* 258, 1985, 1983.

4. **Kocsis, B. and Kontrohr, T.**, Isolation of adenosine 5'-diphosphate-L-*glycero*-D-*mann*noheptose, the assumed substrate of heptose transferase(s), from *Salmonella minnesota* R595 and *Shigella sonnei* Re mutants, *J. Biol. Chem.,* 259, 11858, 1984.

5. **Unger, F. M.**, The chemistry and biological significance of 3-deoxy-D-*manno*-2-octulosonic acid (KDO), *Adv. Carbohydr. Chem. Biochem.,* 38, 323, 1981.

6. **Brade, H., and Galanos, C.**, Isolation, purification, and chemical analysis of the lipopolysaccharide and lipid A of *Acinetobacter calcoaceticus* NCTC 10305, *Eur. J. Biochem.,* 122, 233, 1982.

7. **Neal, D. J. and Wilkinson, S. G.**, Lipopolysaccharides from *Pseudomonas maltophila.* Structural studies of the side-chain, core, and the lipid A regions of the lipopolysaccharide from strain NCTC 10257, *Eur. J. Biochem.,* 128, 143, 1982.

8. **Weintraub, A., Zähringer, U., and Lindberg, A. A.**, Structural studies on the polysaccharide part of the cell wall lipopolysaccharide from *Bacteroides fragilis* NCTC 9343, *Eur. J. Biochem.,* 151, 657, 1985.

9. **Nurminen, M., Rietschel, E. T., and Brade, H.**, Chemical characterization of *Chlamydia trachomatis* lipopolysaccharide, *Infect. Immun.,* 48, 573, 1985.

10. **Brade, L., Schramek, S., Schade, U., and Brade, H.**, Chemical, biological and immunochemical properties of the *Chlamydia psittaci* lipopolysaccharide, *Infect. Immun.,* 54, 568, 1986.

11. **Sonesson, A., Jantzen, E., Bryn, K., Larsson, L., and Eng, J.**, Chemical composition of a lipopolysaccharide from *Legionella pneumophila*, *Arch. Microbiol.,* 153, 72, 1989.

12. **Yokota, A., Rodriguez, M., Yamada, Y., Imai, K., Borowiak, D., and Mayer, H.**, Lipopolysaccharides of *Thiobacillus* species containing lipid A with 2,3-diamino-2,3-dideoxyglucose, *Arch. Microbiol.,* 149, 106, 1987.

13. **Tegtmeyer, B., Weckesser, J., Mayer, H., and Imhoff, J. F.**, Chemical composition of the lipopolysaccharides of *Rhodobacter sulfidophilus, Rhodopseudomonas acidophila,* and *Rhodopseudomonas blastica, Arch. Microbiol.,* 143, 32, 1985.

14. **Salimath, P. V., Tharanathan, R. N., Weckesser, J., and Mayer, H.**, The structure of the polysaccharide moiety of *Rhodopseudomonas sphaeroides* ATCC 17023 lipopolysaccharide, *Eur. J. Biochem.,* 144, 227, 1984.

15. **Ansari, A. A., Kenne, L., Lindberg, B., Gustafsson, B., and Holme, T.,** Structural studies of the O-antigen from *Vibrio cholerae* 0:21, *Carbohydr. Res.,* 150, 213, 1986.
16. **Galanos, C., Lüderitz, O., Rietschel, E. T., and Westphal, O.,** Newer aspects of the chemistry and biology of bacterial lipopolysaccharides, with special reference to their lipid A component, in *International Review of Biochemistry, Biochemistry of Lipids II,* Vol. 14, Goodwin, T. W., Ed., University Park Press, Baltimore, 1977, 239.
17. **Jansson, P. E., Lindberg, A. A., Lindberg, B., and Wollin, R.,** Structural studies on the hexose region of the core in lipopolysaccharides from *Enterobacteriaceae, Eur. J. Biochem.,* 115, 571, 1981.
18. **Volk, W. A., Salomonsky, N. L., and Hunt, D.,** *Xanthomonas sinensis* cell wall lipopolysaccharide. I. Isolation of 4,7-anhydro- and 4,8-anhydro-3-deoxy-octulosonic acid following acid hydrolysis of *Xanthomonas sinensis* lipopolysaccharide, *J. Biol. Chem.,* 247, 3881, 1972.
19. **Brade, H., Brade, L., and Rietschel, E. T.,** Structure-activity relationships of bacterial lipopolysaccharides (endotoxins), *Zentralbl. Bakteriol. Hyg.,* A268, 151, 1988.
20. **Brade, H., Zähringer, U., Rietschel, E. T., Christian, R., Schulz, G., and Unger, F. M.,** Spectroscopic analysis of a 3-deoxy-D-*manno*-2-octulosonic acid(KDO)-disaccharide from the lipopolysaccharide of a *Salmonella godesberg* Re mutant, *Carbohydr. Res.,* 134, 157, 1984.
21. **Brade, H. and Rietschel, E. T.,** α-2→4-Interlinked 3-deoxy-D-*manno*-octulosonic acid disaccharide. A common constituent of enterobacterial lipopolysaccharides, *Eur. J. Biochem.,* 145, 231, 1984.
22. **Tacken, A., Brade, H., Unger, F. M., and Charon, D.,** GLC-MS of partially methylated and acetylated derivatives of 3-deoxyoctitols, *Carbohydr. Res.,* 149, 263, 1986.
23. **Brade, H., Moll, H., and Rietschel, E. T.,** Structural investigations on the inner core region of lipopolysaccharides from *Salmonella minnesota* rough mutants, *Biomed. Mass Spectrom.,* 12, 602, 1985.
24. **Tacken, A., Rietschel, E. T., and Brade, H.,** Methylation analysis of the heptose/3-deoxy-D-*manno*-2-octulosonic acid region (inner core) of the lipopolysaccharide from *Salmonella minnesota* rough mutants, *Carbohydr. Res.,* 149, 279, 1986.
25. **Holst, O., Brade, H., Dziewiszek, K., and Zamojski, A.,** GLC-MS of partially methylated and acetylated derivatives of L-*glycero*-D-*manno*- and D-*glycero*-D-*manno*-heptopyranoses and -heptitols, *Carbohydr. Res.,* 204, 1, 1990.
26. **Rietschel, E. T., Galanos, C., Lüderitz, O., and Westphal, O.,** Chemical structure, physiological function and biological activity of lipopolysaccharides and their lipid A component, in *Immunopharmacology and Regulation of Leucocyte Function,* Webb, D., Ed., Marcel Dekker, New York, 1982, 183.
27. **Holst, O., Röhrscheidt-Andrzejewski, E., Brade, H., and Charon, D.,** Isolation and characterisation of 3-deoxy-D-*manno*-2-octulopyranosonate 7-(2-aminoethyl phosphate) from the inner core region of *Escherichia coli* K-12 and *Salmonella minnesota* lipopolysaccharides, *Carbohydr. Res.,* 204, 93, 1990.
28. **Schmidt, G., Jann, B., and Jann, K.,** Immunochemistry of R lipopolysaccharides of *Escherichia coli.* Studies on R mutants with an incomplete core, derived from *E. coli* O8:K27, *Eur. J. Biochem.,* 16, 382, 1970.
29. **Feige, U. and Stirm, S.,** On the structures of the *Escherichia coli* C cell wall lipopolysaccharide core and on its X174 region, *Biochem. Biophys. Res. Commun.,* 71, 566, 1976.
29a. **Haishima, Y., Holst, O., and Brade, H.,** unpublished.
30. **Hämmerling, G., Lüderitz, O., Westphal, O., and Mäkelä, P. H.,** Structural investigations on the core polysaccharide of *Escherichia coli* O100, *Eur. J. Biochem.,* 22, 331, 1971.
31. **Jansson, P. E., Lindberg, B., Lindberg, A. A., and Wollin, R.,** Structural studies on the hexose region of the Enterobacteriaceae type R3 core polysaccharide, *Carbohydr. Res.,* 68, 385, 1979.

164   *Bacterial Endotoxic Lipopolysaccharides*

32. **Feige, U., Jann, B., Jann, K., Schmidt, G., and Stirm, S.**, On the primary structure of the *Escherichia coli* R4 cell wall lipopolysaccharide core, *Biochem. Biophys. Res. Commun.*, 79, 88, 1977.
33. **Jansson, P. E., Lindberg, B., Bruse, G., Lindberg, A. A., and Wollin, R.**, Structural studies on the hexose region of the lipopolysaccharide from *Escherichia coli* C, *Carbohydr. Res.*, 54, 261, 1977.
34. **Holst, O., Zähringer, U., Brade, H., and Zamojski, A.**, Structural analysis of the heptose/hexose region of the lipopolysaccharide from *Escherichia coli* K-12, *Carbohydr. Res.*, 215, 323, 1991.
35. **Pakulski, Z., Zamojski, A., Holst, O., and Zähringer, U.**, The synthesis and characterisation of 6-O-L-*glycero*-D-*manno*-heptopyranosyl-D-*gluco*-pyranose, *Carbohydr. Res.*, 215, 337, 1991.
35a. **Holst, O.**, unpublished.
36. **Holst, O., Röhrscheidt-Andrzejewski, E., Cordes, H.-P., and Brade, H.**, Isolation and identification of 3-deoxy-7-O-α-D-galactopyranosyl-D-*manno*-2-octulopyranosonate from the inner core region of the lipopolysaccharide of *Escherichia coli* E 100, *Carbohydr. Res.*, 188, 212, 1989.
37. **Haishima, Y., Holst, O., and Brade, H.**, Structural investigation on the lipopolysaccharide of *Escherichia coli* rough mutant F653 representing the R3 core type, *Eur. J. Biochem.*, 203, 127, 1992.
38. **Kaca, W., de Jongh-Leuvenink, J., Zähringer, U., Rietschel, E. T., Brade, H., Verhoef, J., and Sinnwell, V.**, Isolation and chemical analysis of 7-O-(2-amino-2-deoxy-α-D-glucopyranosyl)-L-*glycero*-D-*manno*-heptose as a constituent of the lipopolysaccharides of the UDP-galactose epimerase-less mutant J-5 of *Escherichia coli* and *Vibrio cholerae*, *Carbohydr. Res.*, 179, 289, 1988.
39. **Prehm, P., Stirm, S., Jann, B., Jann, K., and Boman, H. G.**, Cell-wall lipopolysaccharide of ampicillin-resistant mutants of *Escherichia coli* K-12, *Eur. J. Biochem.*, 66, 369, 1976.
40. **Prehm, P., Schmidt, G., Jann, B., and Jann, K.**, The cell-wall lipopolysaccharide of *Escherichia coli* K-12: structure and acceptor site for O-antigen and other substituents, *Eur. J. Biochem.*, 70, 171, 1976.
41. **Holst, O. and Brade, H.**, Isolation and identification of 3-deoxy-5-O-α-L-rhamnopyranosyl-D-*manno*-2-octulopyranosonate from the inner core region of the lipopolysaccharide of *Escherichia coli* K-12, *Carbohydr. Res.*, 207, 327, 1990.
42. **Holst, O., Röhrscheidt-Andrzejewski, E., Zähringer, U., and Brade, H.**, Structural analysis of the lipid A-Kdo region of the lipopolysaccharide from *Escherichia coli* K-12 strain W3100, presented at 5th Eur. Carbohydrate Symp., Prague, August 21 to 25, 1989, B20.
43. **Katzenellenbogen, E. and Romanowska, E.**, Structural studies on *Shigella flexneri* serotype 6 core region, *Eur. J. Biochem.*, 113, 205, 1980.
44. **Gamian, A. and Romanowska, E.**, The core structure of *Shigella sonnei* lipopolysaccharide and the linkage between O-specific polysaccharide and the core region, *Eur. J. Biochem.*, 129, 105, 1982.
45. **Kotelko, K.**, *Proteus mirabilis*: taxonomic position, peculiarities of growth, components of the cell envelope, *Curr. Top. Microbiol. Immunol.*, 129, 181, 1986.
46. **Hickman, F. W., Steigerwalt, A. G., Farmer, J. J., III, and Brenner, D. J.**, Identification of *Proteus penneri* sp. nov., formerly known as *Proteus vulgaris* Indole negative or as *Proteus vulgaris* Biogroup 1, *J. Clin. Microbiol.*, 15, 1097, 1982.
47. **Sidorczyk, Z., Kaca, W., Brade, H., Rietschel, E. T., Sinnwell, V., and Zähringer, U.**, Isolation and structural characterization of an 8-O-(4-amino-4-deoxy-β-L-arabinosyl)-3-deoxy-D-*manno*-octulosonic acid disaccharide in the lipopolysaccharide of a *Proteus mirabilis* deep rough mutant, *Eur. J. Biochem.*, 168, 269, 1987.
48. **Radziejewska-Lebrecht, J., Bhat, U. R., Brade, H., and Mayer, H.**, Structural studies on the core and lipid A region of a 4-amino-L-arabinose-lacking Rc-type mutant of *Proteus mirabilis*, *Eur. J. Biochem.*, 172, 535, 1988.

49. **Radziejewska-Lebrecht, J. and Mayer, H.**, The core region of *Proteus mirabilis* R110/ 1959 lipopolysaccharide, *Eur. J. Biochem.*, 183, 573, 1989.
50. **Radziejewska-Lebrecht, J., Krajewska-Pietrasik, D., and Mayer, H.**, Terminal and chain-linked residues of D-galacturonic acid: characteristic constituents of the R-core regions of *Proteeae* and of *Serratia marcescens*, *Syst. Appl. Microbiol.*, 13, 214, 1990.
51. **Basu, S., Radziejewska-Lebrecht, J., and Mayer, H.**, Lipopolysaccharide of *Providencia rettgeri*. Chemical studies and taxonomical implication, *Arch. Microbiol.*, 144, 213, 1986.
52. **Krajewska, D. and Gromska, W.**, Heterogeneity of the lipopolysaccharide from *Proteus mirabilis* O27, *Arch. Immunol. Ther. Exp.*, 29, 581, 1981.
53. **Radziejewska-Lebrecht, J., Kotelko, K., Krajewska, D., and Mayer, H.**, The R-core region of *Proteus mirabilis* lipopolysaccharides, *EOS Immunol. Immunopharm.*, 6, 167, 1986.
54. **Sidorczyk, Z., Zähringer, U., and Rietschel, E. T.**, Chemical structure of the lipid A component of the lipopolysaccharide from *Proteus mirabilis* Re mutant, *Eur. J. Biochem.*, 137, 15, 1983.
55. **Romanowska, E., Gamian, A., and Dabrowski, J.**, Core region of *Citrobacter* lipopolysaccharide from strain PCM 1487: structure elucidation by two dimensional ^{1}H-NMR spectroscopy at 500 MHz and methylation analysis/mass spectrometry, *Eur. J. Biochem.*, 161, 557, 1986.
56. **Dabrowski, J., Hauck, M., Romanowska, E., and Gamian, A.**, Structure elucidation of the core octasaccharide from *Citrobacter* PCM 1487, with the aid of 500 MHz, two-dimensional phase sensitive correlated, relayed-coherence transfer, double-quantum, triple-quantum filtered, and N.O.E. ^{1}H-NMR spectra, *Carbohydr. Res.*, 180, 163, 1988.
57. **Gamian, A. and Romanowska, E.**, Structure of the heptose-3-deoxy-octulosonic acid region of *Citrobacter* lipopolysaccharide core, *Carbohydr. Res.*, 198, 381, 1990.
58. **Romanowska, E., Gamian, A., Lugowski, C., Romanowska, A., Dabrowski, J., Hauck, M., Opferkuch, H. J., and von der Lieth, C.-W.**, Structure elucidation of the core regions from *Citrobacter* O4 and O36 lipopolysaccharides by chemical and enzymatic methods, gas chromatography/mass spectrometry, and NMR spectroscopy at 500 MHz, *Biochemistry*, 27, 4153, 1988.
59. **Romanowska, E., Gamian, A., Katzenellenbogen, E., Romanowska, A., Lugowski, C., Kulakowska, M., Dabrowski, J., and Dabrowski, U.**, Lipopolysaccharide core regions of *Citrobacter*: structure and serology, in *Cellular and Molecular Aspects of Endotoxin Reactions*, (Endotoxin Research Series, Vol. 1), Nowotny, A., Spitzer, J. J., and Ziegler, E. J., Eds., Excerpta Medica, Amsterdam, 1990, 103.
59a. **Radziejewska-Lebrecht, J. and Mayer, H.**, personal communication.
60. **Sandulache, R. and Prehm, P.**, Structure of the core oligosaccharide from lipopolysaccharide of *Erwinia carotovora*, *J. Bacteriol.*, 161, 1226, 1985.
61. **Kumada, H., Watanabe, K., Umemoto, T., Haishima, Y., Kondo, S., and Hisatsune, K.**, Occurrence of O-phosphorylated 2-keto-3-deoxy-octonate in the lipopolysaccharide of *Bacteroides gingivalis*, *FEMS Microbiol. Lett.*, 51, 77, 1988.
61a. **Brade, H., Weintraub, , and Zähringer, U.**, unpublished.
62. **Finkelstein, R. A.**, *Cholera*, *Crit. Rev. Microbiol.*, 2, 553, 1973.
63. **Baumann, P., Furniss, A. L., and Lee, J. V.**, Genus I. *Vibrio*, in *Bergey's Manual of Systematic Bacteriology*, Vol. 1, Krieg, N. R., Ed., Williams & Wilkins, Baltimore, 1984, 518.
64. **Brade, H.**, Occurrence of 2-keto-3-deoxyoctonic acid 5-phosphate in lipopolysaccharides of *Vibrio cholerae* Ogawa and Inaba, *J. Bacteriol.*, 161, 795, 1985.
65. **Kondo, S., Haishima, Y., and Hisatsune, K.**, Analysis of the 2-keto-3-deoxyoctonate (KDO) region of lipopolysaccharides isolated from non-O1 *Vibrio cholerae* O5R, *FEMS Microbiol. Lett.*, 68, 155, 1990.

66. **Kondo, S., Zähringer, U., Seydel, U., Sinnwell, V., Hisatsune, K., and Rietschel, E. T.**, Chemical structure of the carbohydrate backbone of *Vibrio parahaemolyticus* serotype O12 lipopolysaccharide, *Eur. J. Biochem.*, 200, 689, 1991.

67. **Kondo, S., Iguchi, T., and Hisatsune, K.**, A comparative study of the sugar composition of lipopolysaccharides isolated from *Vibrio cholerae*, '*Vibrio albensis*', and *Vibrio metschnikovii*, *J. Gen. Microbiol.*, 134, 1699, 1988.

68. **Banoub, J. H. and Hodder, H. J.**, Structural investigation of the lipopolysaccharide core isolated from a virulent strain of *Vibrio ordalii*, *Can. J. Biochem. Cell Biol.*, 63, 1199, 1985.

69. **Popoff, M.**, Genus III. *Aeromonas*, in *Bergey's Manual of Systematic Bacteriology*, Vol. 1, Krieg, N. R., Ed., Williams & Wilkins, Baltimore, 1984, 545.

70. **Shaw, D. H. and Hodder, H. J.**, Lipopolysaccharides of the motile aeromonads; core oligosaccharide analysis as an aid to taxonomic classification, *Can. J. Microbiol.*, 24, 864, 1978.

71. **Banoub, J. H., Michon, F., Shaw, D. H., and Roy, R.**, E.i. and c.i. mass-spectral identification of some derivatives of 7-*O*-(2-amino-2-deoxy-α-D-glucopyranosyl)-L-*glycero*-D-*manno*-heptose, obtained from lipopolysaccharide representative of the *Vibrionaceae* family, *Carbohydr. Res.*, 128, 203, 1984.

72. **Banoub, J. H., Choy, Y.-M., Michon, F., and Shaw, D. H.**, Structural investigations on the core oligosaccharides of *Aeromonas hydrophila* (chemotype II) lipopolysaccharide, *Carbohydr. Res.*, 114, 267, 1983.

73. **Banoub, J. H. and Shaw, D. H.**, Structural investigations on the core oligosaccharide of *Aeromonas hydrophila* (chemotype III) lipopolysaccharide, *Carbohydr. Res.*, 98, 93, 1981.

74. **Michon, F., Shaw, D. H., and Banoub, J. H.**, Structure of the lipopolysaccharide core isolated from a human strain of *Aeromonas hydrophila*, *Eur. J. Biochem.*, 145, 107, 1984.

75. **Shaw, D. H., Squires, M. J., Ishiguro, E. E., and Trust, T. J.**, The structure of the heptose-3-deoxy-D-*manno*-octulosonic-acid region in a mutant form of *Aeromonas salmonicida* lipopolysaccharide, *Eur. J. Biochem.*, 161, 309, 1986.

76. **Shaw, D. H. and Hart, M. J.**, Structural studies on the core oligosaccharide isolated from a smooth lipopolysaccharide of *Aeromonas salmonicida*, presented at 15th Int. Carbohydrate Symp., Yokohama, August 12 to 17, 1990, 352.

77. **Vedros, N. A.**, Genus I. *Neisseria*, in *Bergey's Manual of Systematic Bacteriology*, Vol. 1, Krieg, N. R., Ed., Williams & Wilkins, Baltimore, 1984, 290.

78. **Zollinger, W. D. and Mandrell, R. E.**, Outer membrane protein and lipopolysaccharide serotyping of *Neisseria meningitidis* by inhibition of a solid-phase radioimmunoassay, *Infect. Immun.*, 18, 424, 1977.

79. **Zollinger, W. D. and Mandrell, R. E.**, Type specific antigens of group A *Neisseria meningitidis*: lipopolysaccharide and heat-modifiable outer membrane proteins, *Infect. Immun.*, 28, 451, 1980.

80. **Jennings, H. J., Lugowski, C., and Ashton, F. E.**, Conjugation of meningococcal lipopolysaccharide R-type oligosaccharides to tetanus toxoid as route to a potential vaccine against group B *Neisseria meningitidis*, *Infect. Immun.*, 43, 407, 1984.

81. **Di Fabio, J. L., Michon, F., Brisson, J.-R., and Jennings, H. J.**, Structure of the L1 and L6 core oligosaccharide epitopes of *Neisseria meningitidis*, *Can. J. Chem.*, 68, 1029, 1990.

82. **Jennings, H. J., Johnson, K. G., and Kenne, L.**, The structure of an R-type oligosaccharide core obtained from some lipopolysaccharides of *Neisseria meningitidis*, *Carbohydr. Res.*, 121, 233, 1983.

83. **Dell, A., Azadi, P., Tiller, P., Thomas-Oates, J., Jennings, H. J., Beurret, M., and Michon, F.**, Analysis of oligosaccharide epitopes of meningococcal lipopolysaccharides by fast-atom-bombardment mass spectrometry, *Carbohydr. Res.*, 200, 59, 1990.

84. **Jennings, H. J., Beurret, M., Gamian, A., and Michon, F.,** Structure and immunochemistry of meningococcal lipopolysaccharides, *Antonie van Leeuwenhoek; J. Microbiol. Serol.,* 53, 519, 1987.

85. **Yamasaki, R., Bacon, B., Nasholds, W., and Schneider, H.,** Structural determination of oligosaccharides of lipooligosaccharides (LOS) F62 of *Neisseria gonorrhoeae,* presented at 15th Int. Carbohydr. Symp., Yokohama, 1990, 335.

86. **Griffiss, J. McL., Schneider, H., Mandrell, R. E., Yamasaki, R., Jarvis, G. A., Kim, J. J., Gibson, B. W., Hamadeh, R., and Apicella, M. A.,** Lipooligosaccharides: the principal glycolipids of the neisserial outer membrane, *Rev. Infect. Dis.,* 10, S287, 1988.

87. **Gibson, B. W., Webb, J. W., Yamasaki, R., Fisher, S. J., Burlingame, A. L., Mandrell, R. E., Schneider, H., and Griffiss, J. McL.,** Structure and heterogeneity of the oligosaccharides from the lipopolysaccharides of a pyocin-resistant *Neisseria gonorrhoeae, Proc. Natl. Acad. Sci. U.S.A.,* 86, 17, 1989.

88. **Parsons, N. J., Andrade, J. R. C., Patel, P. V., Cole, J. A., and Smith, H.,** Sialylation of lipopolysaccharide and loss of absorption of bactericidal antibody during conversion of gonococci to serum resistance by cytidine 5'-monophospho-N-acetyl neuraminic acid, *Microb. Pathogenesis,* 7, 63, 1989.

89. **Retailliau, H. F., Hightower, A. W., Dixon, R. E., and Allan, J. R.,** *Acinetobacter calcoaceticus:* a nocosomial pathogen with an unusual seasonal pattern, *J. Infect. Dis.,* 139, 371, 1979.

90. **Sherertz, R. J. and Sullivan, M. L.,** An outbreak of infections with *Acinetobacter calcoaceticus* in burn patients: contamination of patients mattresses, *J. Infect. Dis.,* 151, 252, 1985.

91. **Brade, H. and Galanos, C.,** Isolation, purification, and chemical analysis of the lipopolysaccharide and lipid A of *Acinetobacter calcoaceticus, Eur. J. Biochem.,* 122, 233, 1982.

92. **Kawahara, K., Brade, H., Rietschel, E. T., and Zähringer, U.,** Studies on the chemical structure of the core-lipid A region of the lipopolysaccharide of *Acinetobacter calcoaceticus* NCTC 10305: detection of a new 2-octulosonic acid interlinking the core oligosaccharide and lipid A, *Eur. J. Biochem.,* 163, 489, 1987.

92a. **Zähringer, U.,** personal communication.

93. **Brade, H. and Rietschel, E. T.,** Identification of a 2-keto-3-deoxy-1,7-dicarboxyheptonic acid as a constituent of the lipopolysaccharide of *Acinetobacter calcoaceticus* NCTC 10305, *Eur. J. Biochem.,* 153, 249, 1985.

93a. **Brade, H.,** unpublished.

94. **Brade, H., Tacken, A., and Christian, R.,** Isolation and identification of a rhamnosyl-rhamnosyl-3-deoxy-D-*manno*-octulosonic acid trisaccharide from the lipopolysaccharide of *Acinetobacter calcoaceticus* (10303 NCTC London), *Carbohydr. Res.,* 167, 295, 1987.

95. **Schachter, J. and Caldwell, H. D.,** Chlamydiae, *Annu. Rev. Microbiol.,* 34, 285, 1980.

96. **Grayston, J. T., Campbell, L. A., Kuo, C.-C., Mordhorst, C. H., Saikku, P., Thom, D. H., and Wang, S.-P.,** A new respiratory tract pathogen: *Chlamydia pneumoniae* strain TWAR, *J. Infect. Dis.,* 161, 618, 1990.

97. **Brade, L., Nurminen, M., Mäkelä, P. H., and Brade, H.,** Antigenic properties of *Chlamydia trachomatis* lipopolysaccharide, *Infect. Immun.,* 48, 569, 1985.

98. **Nurminen, M., Leinonen, M., Saikku, P., and Mäkelä, P. H.,** The genus-specific antigen of *Chlamydia*: resemblance to the lipopolysaccharide of enteric bacteria, *Science,* 220, 1279, 1983.

99. **Caldwell, H. D. and Hitchcock, P. J.,** Monoclonal antibody against a genus-specific antigen of *Chlamydia* species: location of the epitope on chlamydial lipopolysaccharide, *Infect. Immun.,* 44, 306, 1984.

100. **Nano, F. E. and Caldwell, H. D.,** Expression of the chlamydial genus-specific lipopolysaccharide epitope in *Escherichia coli, Science,* 228, 742, 1985.

101. **Brade, L., Nano, F. E., Schlecht, S., Schramek, S., and Brade, H.**, Antigenic and immunogenic properties of recombinants from *Salmonella typhimurium* and *Salmonella minnesota* rough mutants expressing in their lipopolysaccharides a genus-specific chlamydial epitope, *Infect. Immun.*, 55, 482, 1987.

102. **Brade, H., Brade, L., and Nano, F. E.**, Chemical and serological investigations on the genus-specific lipopolysaccharide epitope of *Chlamydia*, *Proc. Natl. Acad. Sci. U.S.A.*, 84, 2508, 1987.

103. **Kosma, P., Schulz, G., and Brade, H.**, Synthesis of a trisaccharide of 3-deoxy-D-*manno*-2-octulopyranosylonic acid (KDO) residues related to the genus-specific lipopolysaccharide epitope of *Chlamydia*, *Carbohydr. Res.*, 183, 183, 1988.

104. **Holst, O., Brade, L., Kosma, P., and Brade, H.**, Structure, serological specificity, and synthesis of the genus-specific lipopolysaccharide epitope of *Chlamydia*, *J. Bacteriol.*, 173, 1862, 1991.

105. **LeDur, A., Chaby, R., and Szabo, L.**, Isolation of two protein-free and chemically different lipopolysaccharides from *Bordetella pertussis* phenol-extracted endotoxin, *J. Bacteriol.*, 143, 78, 1980.

106. **Caroff, M., Lebbar, S., and Szabo, L.**, Detection of 3-deoxy-2-octulosonic acid in thiobarbiturate-negative endotoxins, *Carbohydr. Res.*, 161, C4, 1987.

107. **Caroff, M., Chaby, R., Karibian, D., Perry, J., Deprun, C., and Szabo, L.**, Variations in the carbohydrate regions of *Bordetella pertussis* lipopolysaccharides: electrophoretic, serological, and structural features, *J. Bacteriol.*, 172, 1121, 1990.

107a. **Caroff, M.**, personal communication.

108. **Chaby, R. and Szabo, L.**, 7-O-(2-Amino-2-deoxy-α-D-glycopyranosyl)-L-*glycero*-D-*manno*-heptose: a constituent of the endotoxin of *Bordetella pertussis*, *Eur. J. Biochem.*, 70, 115, 1976.

109. **Chaby, R. and Szabo, L.**, 3-Deoxy-2-octulosonic acid 5-phosphate: a component of the endotoxin of *Bordetella pertussis*, *Eur. J. Biochem.*, 59, 277, 1975.

110. **Kilian, M. and Biberstein, E. L.**, Genus III. *Haemophilus*, in *Bergey's Manual of Systematic Bacteriology*, Vol. 1, Krieg, N. R., Ed., Williams & Wilkins, Baltimore, 1984, 558.

111. **Flesher, A. R. and Insel, R. A.**, Characterization of lipopolysaccharide of *Haemophilus influenzae*, *J. Infect. Dis.*, 138, 719, 1978.

112. **Inzana, T. J.**, Electrophoresis heterogeneity and interstrain variation of the lipopolysaccharide of *Haemophilus influenzae*, *J. Infect. Dis.*, 148, 492, 1983.

113. **Zamze, S. E. and Moxon, E. R.**, Composition of the lipopolysaccharide from different capsular serotype strains of *Haemophilus influenzae*, *J. Gen. Microbiol.*, 133, 1443, 1987.

114. **Inzana, T. J., Seifert, W. E., Jr., and Williams, R. P.**, Composition and antigenic activity of the oligosaccharide moiety of *Haemophilus influenzae* type b lipopolysaccharide, *Infect. Immun.*, 48, 324, 1985.

115. **Zamze, S. E., Ferguson, M. A. J., Moxon, E. R., Dwek, R. A., and Rademacher, T. W.**, Identification of phosphorylated 3-deoxy-*manno*-octulosonic acid as a component of *Haemophilus influenzae* lipopolysaccharide, *Biochem. J.*, 245, 583, 1987.

116. **Helander, I. M., Lindner, B., Brade, H., Altmann, K., Lindberg, A. A., Rietschel, E. T., and Zähringer, U.**, Chemical structure of the lipopolysaccharide of *Haemophilus influenzae* strain I-69 Rd⁻/b⁺: description of a novel deep-rough chemotype, *Eur. J. Biochem.*, 177, 483, 1988.

117. **Drewry, D. T., Symes, K. C., Gray, G. W., and Wilkinson, S. G.**, Studies of polysaccharide fractions from the lipopolysaccharide of *Pseudomonas aeruginosa* NCTC 199, *Biochem. J.*, 149, 93, 1975.

118. **Rowe, P. S. N. and Meadow, P. M.**, Structure of the core oligosaccharide from the lipopolysaccharide of *Pseudomonas aeruginosa* PAC1R and its defective mutants, *Eur. J. Biochem.*, 132, 329, 1983.

119. **Lüderitz, O., Freudenberg, M. A., Galanos, C., Lehmann, V., Rietschel, E. T., and Shaw, D. H.**, Lipopolysaccharides of gram-negative bacteria, *Curr. Top. Membr. Transp.*, 17, 79, 1982.

120. **Jordan, D. C.**, Genus I. *Rhizobium*, in *Bergey's Manual of Systematic Bacteriology*, Vol. 1, Krieg, N. R., Ed., Williams & Wilkins, Baltimore, 1984, 235.

121. **Carlson, R. W., Hollingsworth, R. I., and Dazzo, F. B.**, A core oligosaccharide component from the lipopolysaccharide of *Rhizobium trifolii* ANU843, *Carbohydr. Res.*, 176, 127, 1988.

122. **Hollingsworth, R. I., Carlson, R. W., Garcia, F., and Gage, D. A.**, A new core tetrasaccharide component from the lipopolysaccharide of *Rhizobium trifolii* ANU 843, *J. Biol. Chem.*, 265, 12752, 1990.

122a. **Carlson, R. W.**, personal communciation.

123. **Carlson, R. W., Garcia, F., Noel, D., and Hollingsworth, R.**, The structures of the lipopolysaccharide core components from *Rhizobium leguminosarum* BIOVAR *phaseoli* CE3 and two of its symbiotic mutants, CE109 and CE309, *Carbohydr. Res.*, 195, 101, 1989.

124. **Weckesser, J. and Mayer, H.**, Lipopolysaccharide aus phototrophen Bakterien: ein Beitrag zur Phylogenie und zur Endotoxin-Forschung, *Forum Mikrobiol.*, 108, 242, 1987.

125. **Holst, O., Hunger, U., Gerstner, E., and Weckesser, J.**, Lipophilic lipopolysaccharide (O-antigen) in *Rhodomicrobium vannielii* ATCC 17100, *FEMS Microbiol. Lett.*, 10, 165, 1981.

126. **Weckesser, J., Mayer, H., Drews, G., and Fromme, I.**, Lipophilic O-antigens containing D-*glycero*-D-*manno*-heptose as the sole neutral sugar in *Rhodopseudomonas gelatinosa*, *J. Bacteriol.*, 123, 449, 1975.

127. **Meißner, J., Borowiak, D., Fischer, U., and Weckesser, J.**, The lipopolysaccharide of the phototrophic bacterium *Ectothiorhodospira vacuolata*, *Arch. Microbiol.*, 149, 245, 1988.

128. **Meißner, J., Pfennig, N., Krauss, J. H., Mayer, H., and Weckesser, J.**, Lipopolysaccharides of *Thiocystic violacea*, *Thiocystic pfennigii*, and *Chromatium tepidum*, species of the family *Chromatiaceae*, *J. Bacteriol.*, 170, 3217, 1988.

129. **Weckesser, J., Mayer, H., Metz, E., and Biebl, H.**, Lipopolysaccharide of *Rhodocyclus purpureus*: taxonomic implication, *Int. J. Syst. Bacteriol.*, 33, 53, 1983.

130. **Meißner, J., Fischer, U., and Weckesser, J.**, The lipopolysaccharide of the green-sulfur bacterium *Chlorobium vibrioforme* f. *thiosulfatophilum*, *Arch. Microbiol.*, 149, 125, 1987.

131. **Krauss, J. H., Reuter, G., Schauer, R., Weckesser, J., and Mayer, H.**, Sialic-acid containing lipopolysaccharides in purple nonsulfur bacteria, *Arch. Microbiol.*, 150, 584, 1988.

132. **Radziejewska-Lebrecht, J., Feige, U., Mayer, H., and Weckesser, J.**, Structure of the heptose region of lipopolysaccharides from *Rhodospirillum tenue*, *J. Bacteriol.*, 145, 138, 1981.

133. **Masoud, H.**, Chemical and Biological Studies on the Lipopolysaccharides of *Rhodocyclus gelatinosus* and *Sphaerotilus natans*, and on Two "Haptenic" Polysaccharides of *Rhodocyclus gelatinosus*, Thesis, University of Freiburg, 1989.

134. **Masoud, H., Mayer, H., Kontrohr, T., Holst, O., and Weckesser, J.**, The structure of the core region of the lipopolysaccharide from *Rhodocyclus gelatinosus* Dr2, *Syst. Arch. Microbiol.*, 14, 222, 1991.

135. **Yamasaki, R., Bacon, B. E., Nasholds, W., Schneider, H., and Griffiss, J. M.**, Structural determination of oligosaccharides derived from lipooligosaccharide of *Neisseria gonorrhoeae* F62 by chemical, enzymatic, and two-dimensional NMR methods, *Biochemistry*, 30, 10566, 1991.

136. **Bhat, U. R., Krishnaiah, B. S., and Carlson, R. W.**, Reexamination of the structures of the lipopolysaccharide core oligosaccharides from *Rhizobium leguminosarum* biovar phaseoli, *Carbohydr. Res.*, 220, 219, 1991.

137. **Aspinall, G. O., McDonald, A. G., Raju, T. S., Pang, H., Mills, S. D., Kurjanczyk, L. A., and Penner, J. L.,** Serological diversity and chemical structures of *Campylobacter jejuni* low-molecular weight lipopolysaccharides, *J. Bacteriol.*, 174, 1324, 1992.

138. **Krauss, J. H., Himmelspach, K., Reuter, G., Schauer, R., and Mayer, H.,** Structural analysis of a novel sialic-acid-containing trisaccharide from *Rhodobacter capsulatus* 37b4 lipopolysaccharide, *Eur. J. Biochem.*, 204, 217, 1992.

Chapter 7

CHEMICAL SYNTHESIS OF CORE STRUCTURES

Peter L. Stütz and Frank M. Unger

TABLE OF CONTENTS

I. INTRODUCTION

Lipopolysaccharides (LPS) are common cell wall constituents of Gram-negative bacteria. The LPS of the family of Enterobacteriaceae have been most intensively investigated. In contrast to the extreme diversity of the "O-specific" antigenic chains, the core region shows higher taxonomical restriction. In enterobacterial strains, the so-called inner core exhibits a rather typical architecture, as depicted below for *Salmonella* strains. This or similar schemes have served as a guideline for objectives to synthetic organic chemists:

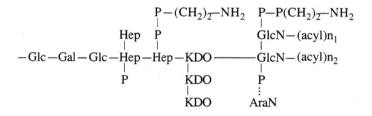

The preceding chapter deals with the "state of the art" of the structural elucidation of LPS core structures. Previous reviews have summarized our knowledge about the chemical structures of lipopolysaccharides.[1] The present chapter is intended to provide an overview of recent synthetic achievements in the field of core structures, as found in the LPS of Enterobacteriaceae (e.g., *Salmonella* and *Acinetobacter*), of Neisseriae, and in the LPS-like glycolipid of the membrane of Chlamydiae, a family not related to Gram-negative bacteria and lacking a typical cell wall. The literature from 1985 onwards is discussed in some detail.

In addition to pursuing the purely synthetic challenges for obtaining access to genuine structures occurring in bacteria, the scientists involved had also more interdisciplinary objectives in mind:

- Unequivocal proof of spectroscopically deduced structures from fragments of bacterial "rough" (R) strains. While oligosaccharides of the O-antigen or outer core regions are amenable to classical analysis, e.g., by methylation and mass spectroscopy, the structures of the inner core region, mostly due to their content of KDO, have been much more difficult to determine. (For example, it was believed for a long time that the first KDO of the inner core was attached at position 3' of the lipid A moiety. Only after spectroscopic reevaluation by NMR and MS spectroscopy and by comparison with synthetic model disaccharides could the correct α-KDOp-(2→6)-GlcN linkage between the inner core sugars and the lipid A component be unequally established.[2-5])

- Epitope analysis of monoclonal antibodies against Re-type LPS of *Salmonella minnesota*, *Proteus mirabilis*, and *Escherichia coli*.[6] For this purpose, partial structures of the oligosaccharide backbone were synthesized for binding studies to various Re LPS-specific antibodies. High molecular weight antigens were obtained by copolymerization of corresponding allyl-glycosides with acrylamide.[7] It was concluded that the immunodominant region of Re LPS comprised both KDO and lipid A domains.

- Production of vaccines against pathogenic bacteria, by synthesizing oligosaccharides that contain suitable spacers for coupling to immunoreactive carriers.
- Establishment of structure-activity relationships with certain immunopharmacological parameters as readouts (see also Chapter 8). Such investigations are much more plentiful with lipid A analogs, but synthetic α(2→6)KDO-linked monosaccharide analogs of lipid A were tested for their potential to induce mitogenicity or to elicit endotoxicity.[8] The minimal inner core substructure having the ability to induce interleukin-1 in mononuclear cells has been claimed to be L-α-Hep$p$ (1→5)-α-KDO.[9]

During recent years, efficient and highly diastereoselective syntheses for L-*glycero*-D-*manno*-heptose (LD-Hep$p$), suitably protected for the correct construction of inner core oligosaccharides of various bacteria, have been developed. The art of preparing KDO-containing oligosaccharides of native regio- and stereochemistry has almost been brought to perfection. This also holds true for the construction of the acid-labile α-glycosidic (2→6)linkage to acylated glucosamines representing the lipid A part. Thus, in principle, methods are now available for the total synthesis of bacterial (inner) core LPS.

II. SYNTHESIS OF OUTER CORE OLIGOSACCHARIDES

A. TRISACCHARIDE OF *SALMONELLA*

$$\begin{array}{c} \alpha-\text{D}-\text{Gal}p-1 \\ \uparrow \\ 6 \\ \alpha-\text{D}-\text{Gal}p-(1\to3)-\alpha-\text{D}-\text{Glc}- \end{array}$$

This trisaccharide is part of the LPS core of *Salmonella*, recognized as a binding site by bacteriophages in the initial step of infection.[10] Mutants defective in certain structural elements of the core region are known to be resistant to phage infection.

The synthetic strategy of Norberg et al.[10] relied on appropriate methyl 1-thioglycosides as intermediates. Thus, 2,3,4,6-tetra-*O*-benzyl-α-D-galactopyranosyl bromide, prepared from the corresponding 1-thioglycoside, was

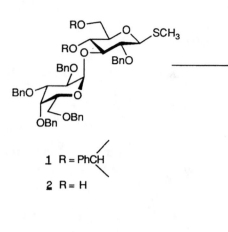

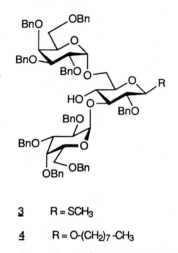

1 R = PhCH

2 R = H

<u>3</u>   R = SCH₃

<u>4</u>   R = O-(CH₂)₇-CH₃

<u>5</u>   R = O-(CH₂)₂-Ph-p-NHCOCF₃

FIGURE 1.

reacted with methyl-4,6-*O*-benzylidene-2-O-benzyl-1-thio-β-D-glucopyrano-
side in the presence of halide ion to give the disaccharide **1** in 87% yield.
The disaccharide diol **2** was obtained by treatment with aqueous acetic acid
in 89% yield, and was subjected to a second halide-prompted glycosidation
step, using the same glycosyl donor. The protected trisaccharide **3** was obtained in 89% yield. Its structure was proven by NMR after O-acetylation to
give **3a** (formula not shown). The signal for H-4 in **3a** was found shifted
downfield in comparison with the corresponding signal in **2**, indicating that
position 4 was O-acylated and glycosidation had regioselectively occurred in
the 6-position. Treatment of **3** with bromine, then with tetraethylammonium
bromide and octano or *p*-trifluoroacetamidophenylethanol gave the corresponding glycosides **4** and **5** in good yield, so as to enable eventual attachment of the oligosaccharide to liposomes or proteins for vaccination studies
(Figure 1).

Deprotection of **4** and **5** was achieved by hydrogenation with Pd on
charcoal.

B. SYNTHESIS OF THE TETRASACCHARIDE CORE REGION OF *MYCOBACTERIUM KANSASI* (FIGURE 2)

-D-Glc*p*-(1→3)-β-D-Glc*p*-(1→4)-αD-Glc*p*-(1→1)-α-D-Glc*p*

The above tetrasaccharide backbone occurs in all seven species-specific
lipooligosaccharide antigens of *M. kansasi* known so far. Interestingly, tre-

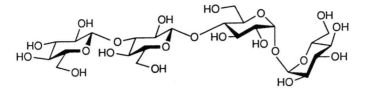

FIGURE 2.

halose derivatives appear to be widespread constituents of mycobacterial surface structures. Infections with *M. kansasii* are chronic pulmonary disorders resembling tuberculosis.[11]

The first synthetic problem that Liptak et al.[11] had to solve was to provide access of nonsymmetrical, partially protected derivatives of trehalose as starting materials. This was achieved by treating the easily accessible 2,3,2′,3′,4′,6′-hexa-*O*-benzyl-α-α-trehalose with acetophenone dimethyl acetal in the presence of *p*-toluene-sulfonic acid at room temperature without any solvent to yield the thermodynamic 4,6-ketalized product **6** exclusively. The postulated axial position of the phenyl ring in **6** was confirmed by NMR spectroscopy. Likewise, treatment with α-α-dimethoxytoluene gave the corresponding 4,6-benzylidene derivative. Hydrogenolysis of **6** with lithium aluminum hydride/aluminum chloride afforded the protected trehalose derivative **7** with a free hydroxy function at C-5, which served as acceptor in the subsequent glycosylation step. After reaction with 2,4,6-tri-*O*-acetyl-3-allyl-α-D-glucopyranosylbromide in the presence of HgBr$_2$, the protected trisaccharide **8** with β-glycosidic linkage was obtained in good yield; no trace of the corresponding α-anomer could be detected.

A 2 + 2 block synthesis expected to lead to the desired trisaccharide directly was unsuccessful because of the low reactivity of the corresponding acetobromo derivative of laminaribiose. Therefore, a step-by-step approach was chosen. The O-acetyl derivative **8** was saponified to the alcohol and perbenzylated. The reaction product was subjected to 3-O-deallylation, using tris(phenylphosphine)-rhodium(I)chloride as catalyst for the double bond isomerization to give an appropriate acceptor (formulas not shown). Glycosylation to the protected tetrasaccharide **9** could be achieved in 71% yield by treatment with methyl-2,3,4,6-tetra-*O*-acetyl-1-thio-β-D-glucopyranoside in the presence of four equivalents of methyl triflate. Again, only the β-anomer could be detected. Catalytic hydrogenation of **9** and subsequent acetylation yielded the crystalline tetradecaacetate, the ^{13}C-NMR spectrum of which was fully consistent with the proposed structure. Zemplén deacetylation gave the amorphous title compound (Figure 3).

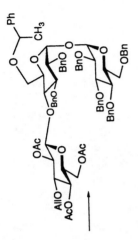

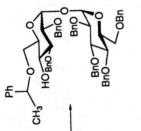

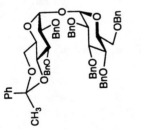

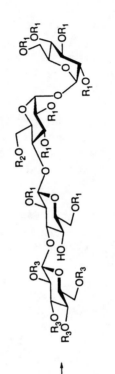

FIGURE 3.

III. SYNTHESES OF OLIGOSACCHARIDES CORRESPONDING TO THE HEPTOSE REGION

A. NEW DIASTEREOSELECTIVE SYNTHESES OF L-*GLYCERO*-D-*MANNO*-HEPTOSE AS CENTRAL BUILDING BLOCK

LD-Hep*p* links the "outer core" oligosaccharides with the "inner core" region in the form of a trisaccharide of three Hep units connected to the "KDO region". Recently described, diasteroselective syntheses make this central building block of the inner core region easily accessible.

The rather lengthy, though intellectually rewarding, synthesis of Paulsen et al.[12] is based on inexpensive diisopropylidene-mannose 10 as starting material. After reaction with 2.5 equivalents of 2-lithio-1,3-dithiane in tetrahydrofurane, the crystalline adduct 11 was obtained in 77% yield as the only diastereoisomer. This heptose derivative has the D-*glycero*-D-*galacto* configuration (with the two hydroxy functions at C-2 and C-3 in the threo configuration). For conversion into the desired L-*glycero*-D-*manno* configuration, a series of steps was necessary. The dithio-ketal alcohol obtained by mild acid hydrolysis was selectively acylated at position 7 with pivaloylchloride at −40°C to give 12 in 74% yield (formula not shown). After ketalization with 2,2-dimethoxypropane, a mixture of protected forms of 12 was obtained which was directly hydrolyzed to the aldehyde in quantitative yield, using CH_3J and 2,4,6-trimethylpyridine as base. Reduction of the aldehyde with $NaCHBH_3$ at pH 3.5 gave alcohol 13. Under these conditions, isomerization reactions and hydrolysis of the pivaloyl ester are avoided. Further reaction with 2,2-dimethoxypropane gave a mixture of triisopropylidene derivatives together with significant amounts of diisopropylidene compounds which were removed by chromatography. Hydrolysis of the pivaloyl ester gave alcohol 14, which was oxidatized to aldehyde 15. On acid hydrolysis, 15 gave the desired LD-Hep*p* as a mixture of anomers. These were characterized in the form of their crystalline hexaacetates.

The aldehyde 15 could not be obtained by the more conventional oxidation methods. Best results were obtained by oxidation of the Mg-alcoholate of 14 with 1,1'-azo(dicarbonyl)-dipiperidine. When crude 14 was directly converted into the LD-Hep*p*-hexa-acetates, the isolated yield of α- and β-anomers 16 and 17 was 44 and 24%, respectively (Figure 4).

Another interesting and highly efficient synthesis of LD-Hep*p* was described by Brimacombe and Kabir.[13] Starting from D-mannose, the (*E*)-α-D-*lyxo*-hept-5-enofuranoside 18 was obtained in six high-yielding steps via a Wittig-type elongation of the side chain. Bishydroxylation of 18 with OsO_4 produced a mixture of benzyl-2,3-*O*-isopropylidene L-*glycero*-D-*manno*-heptofuranoside 19 and its L-*glycero*-L-*gulo* isomer in a ratio of 7:1 in 92% yield. The preponderant isomer was easily purified by crystallization of the corresponding triacetate. This derivative gave, after Zemplén deacylation, hydrogenation and acid hydrolysis, the desired LD-Hep*p* in high yield. Similarly,

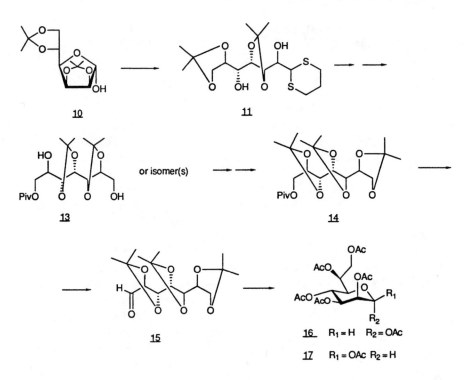

FIGURE 4.

by starting from the (Z)-isomer of **18** the corresponding D-*glycero*-L-*manno*-heptose was obtained.

Compound **18** can serve as starting material for syntheses of both the L-*glycero*- and D-*glycero*-L-*manno*-heptoses as has been elegantly demonstrated by the same authors. The enantiomerically pure crystalline oxiran **20** with the L-*glycero*-D-*manno* configuration was obtained by titanium-catalyzed epoxidation of **18** in the presence of diisopropyl L-(+)-tartrate in 88% yield.[13] Payne rearrangement of **20** afforded the primary oxiran. Hydrolysis of this product with sodium hydroxide in aqueous 1,4-dioxane produced exclusively diol **21** with inverted configuration at C-6 in about 60% yield (Figure 5).

B. SYNTHESIS OF OLIGOSACCHARIDES CONTAINING L-*GLYCERO*-D-*MANNO*-HEPTOSE (LD-HEP*p*)

The intermediate **19** as described by Brimacombe and Kabir[13] was successfully used by Paulsen and associates for the synthesis of both α- and β-D-Glc*p*N-(1→7)-LD-Hep*p*.[14] These structures are core constituents of bacteria whose LPS apparently does not contain lipid A. Thus, α-D-Glc*p*N-(1→7)-L-α-D-Hep*p* occurs exclusively in *Aeromonas hydrophila*, whereas the β-anomer is found in *Vibrio ordalii*.

Transformation of **19** into **22** was easily achieved by successive treatment

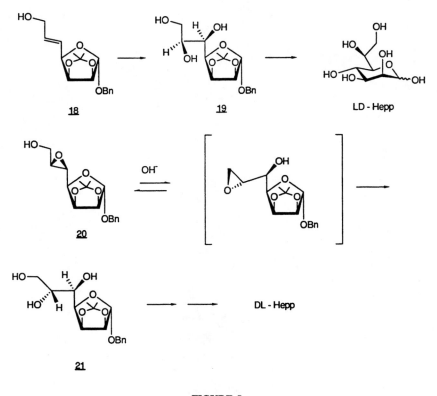

FIGURE 5.

with trifluoro acetic acid, selective protection of the 7-OH function by *tert*-butyl-diphenylsilylation, and perbenzylation of silyl ether **21**. Compound **22** gave the stable glycosyl acceptor **23** upon treatment with tetrabutylammonium fluoride. Reaction with the known donor **24** in the presence of Ag-triflate at −60°C yielded a 7:1 mixture of the corresponding α- and β-linked disaccharides **25** and **26**. When the glycosylation reaction was performed in heterogeneous phase using insoluble Ag-silicate as catalyst at −50°C, the β-glycosidically linked disaccharide **26** was the only product (87%) isolated yield). De-O-acetylation, followed by catalytic hydrogenation of either disaccharide alcohol, gave the corresponding (1→7) glycosidically linked disaccharides **27** and **28** as hydrochlorides in the desired pyranose form (Figure 6).

An elegant synthesis of LD-Hep*p*, suitably protected for the synthesis of an inner core trisaccharide of *Neisseria meningitidis*, was described by van Boom et al.[15] demonstrating that the dimethyl(phenyl)silyl-group can be introduced into precursor molecules as the equivalent of a hydroxy function. Thus, aldehyde **29**, obtained in good yield in a series of conventional steps from allyl-α-D-*manno*-pyranoside, was reacted with the Grignard reagent **30** dimethylphenylsilylmethyl-Mg-chloride. The protected heptose equivalent **31** was obtained in 70% yield in at least 95% diastereomeric purity. The key

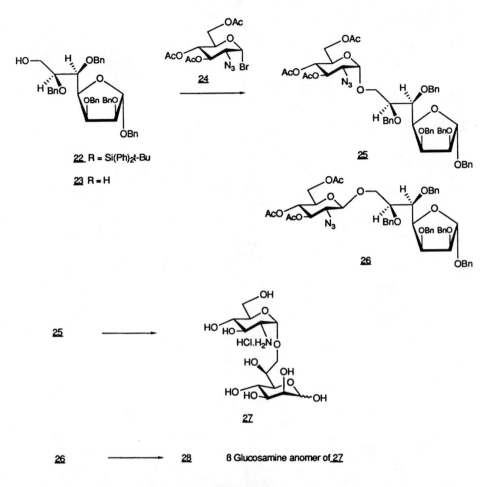

FIGURE 6.

intermediate **32** was obtained in 42% total yield via benzylation, deallylation follwed by acetylation, and regioselective deacetylation. The pyranosyl chloride **33**, synthesized by treating the pyranose with chlorodimethylformamidiniumchloride under Vilsmeier conditions, was subsequently converted into **34** in order to enable coupling to an appropriate carrier for eventual vaccination against meningitis, using the final trisaccharide as the antigen. Using **34** as acceptor and the known glucopyranosyl chloride **35** as donor, glycosylation was achieved in 72% yield with silver triflate, in the absence of base, to give **36**. Zemplén deacetylation of **36** gave **37** which could be coupled to the glycosyl donor **38**, using trimethylsilyl triflate as catalyst, in the presence of 4Å molecular sieve to yield the fully protected trisaccharide **39** in 92% yield. Conversion into the deprotected, α-glycosidically linked trisaccharide **40** was performed in six steps with a total yield of 30%. Removal of the dimethyl(phenyl)silyl group to afford the desired alcohol was effected with KBr and acetyl peroxide in the dark. Obviously, the utilization of the silylation-desilylation strategy to generate an unprotected OH-function at a later stage of the synthesis can be extended to simpler LD-Hep*p* analogs. Consequently, an efficient two-step synthesis of LD-Hep*p* from **29** has also been described (Figure 7).[16]

The first syntheses of α(1→3) and α(1→7)-linked LD-Hep*p*-disaccharides and a trisaccharide linked α(1→7) and α(1→3) as major backbone of enterobacterial inner core LPS were described by Zamojski et al.,[17] starting with a homologization at C-6 of mannose derivative **41** in a Grignard reaction with benzyloxymethyl magnesium chloride. A 28:1 mixture of heptoses diastereomeric at C-6 was obtained in 48% yield, with the desired L-*glycero* isomer **42** as prevailing product.[18] The stereochemistry of this reaction was predicted on the basis of Cram's cyclic model of asymmetric 1,2-induction and confirmed by NMR.[18] After benzylation and deallylation, the acceptor molecule **43** was obtained in good yield.

The trichloroacetimidate method was successfully employed for the glycosylation reactions.[19] Coupling of **43** with **44**, prepared according to Schmidt from the corresponding free anomeric heptose, gave the α-linked heptobiose **45** as a single product in 50% yield. Deallylation of **45** to **46** and glycosylation with **44** gave the single α-linked trisaccharide **47** in 52% yield. All glycosylation reactions were catalyzed by anhydrous *p*-toluene-sulfonic acid. Deprotection of **45** to L-α-D-Hep*p*-(1→3)-L-α-D-Hep*p* and of **47** to L-αβ-D-Hep*p*-(1→7)-L-α-D-Hep*p*-(1→3)-L-α-D-Hep*p* was performed by conventional procedures (Figure 8).

Very recently, the synthesis has been described of the trisaccharide methyl 3-*O*-(α-D-glucopyranosyl)-7-*O*-(L-*glycero*-α-D-*manno*-heptopyranosyl)-L-*glycero*-α-D-*manno*-heptopyranoside, which links the "hexose" and "heptose" regions.[20]

In this investigation the heptose precursor **49** was synthesized from the aldehyde **48** and the Grignard complex of isopropoxydimethylsilylmethylchloride, followed by oxidative cleavage of the C-Si-bond as described by van Boom et al.[15,16] The authors stated that this type of homologization is more convenient than that described by Zamojski et al.[17] who used the rather unstable Grignard reagent benzyloxymethyl magnesium chloride for this purpose.

For the synthesis of the heptobiose L-α-D-Hep*p*-(1→7)-L-α-Hep*p* as first target, the glycosyl donor was obtained by hydrogenolysis of **49**, followed by acetylation and acetolysis, resulting in the crystalline hexa-acetate as intermediate. The desired glycosyl bromide **50** was obtained by treatment with hydrogen bromide in glacial acetic acid.

For the target structures with free hydroxyl groups in positions 3 or 7, the appropriate heptosyl acceptors **52** and **53** were obtained as follows: regioselective silylation of **49** at O-7 with *tert*-butyldimethylsilyl chloride in pyridine, followed by benzoylation of the silylether **50** afforded **51** as central

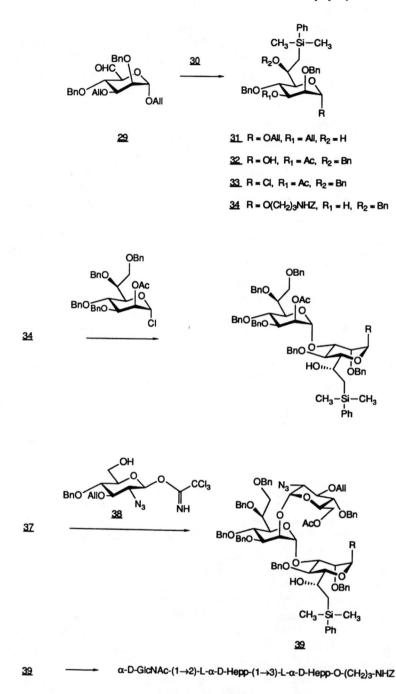

FIGURE 7.

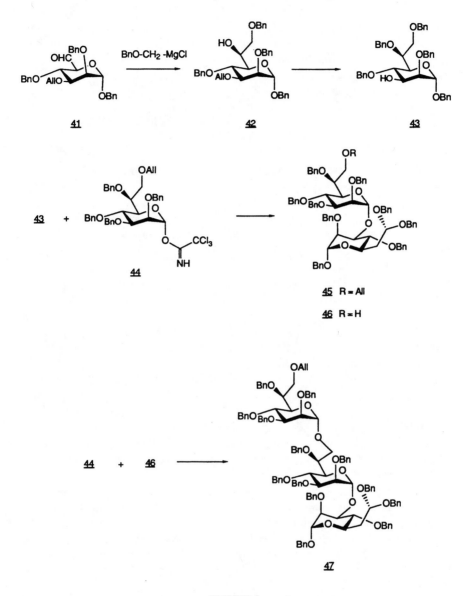

FIGURE 8.

intermediate. Hydrogenation of **51** gave the heptose protected only in positions 6 and 7, which was successively treated with trimethyl orthoacetate, acetic anhydride, and acetic acid to afford **52**. Treatment of **51** with aqueous acetic acid gave **53** in 69% yield.

Coupling of **50** with **53**, promoted by silver triflate in the presence of collidine overnight at room temperature, afforded the heptobiose **54** in 82% yield.

Halide-assisted glycosylation of **52** with 2,3,4,6-tetra-*O*-D-glucopyrano-syl bromide gave the α-linked disaccharide **55** which was subjected to de-silylation; the resulting alcohol was glucosylated with **50** in the presence of silver triflate and collidine to afford the trisaccharide derivative **56** in 70% yield. Again, a prolonged reaction time was required (Figure 9).

IV. SYNTHESIS OF OF KDO-CONTAINING OLIGOSACCHARIDES

A. KDO-KDO CONTAINING OLIGOSACCHARIDES
1. Enterobacterial Type

Besides LD-Hep*p*, 3-deoxy-D-*manno*-2-octulosonic acid (KDO) is the sec-ond unique component of the inner core of LPS in Gram-negative bacteria. It is now firmly established[5a] that this region contains an α-(2→4)-linked KDO-disaccharide as a common constituent in rough-mutant lipopolysac-charides. In addition, the first successful synthesis of the tetrasaccharide α-KDO-(2→4)-α-KDO-KDO(2→6)-β-D-GlcN-(1→6)-D-GlcN by Paulsen et al.[21] has provided an independent structural proof of a degraded LPS from *Salmonella minnesota* Re 595.

A most challenging feature of KDO and its glycosidic derivatives to synthetic organic chemists has been the sensitivity toward acids. Hence, glycosylation reactions have to be performed with extreme care. Furthermore, monitoring of anomeric configuration is rather cumbersome. Finally, uni-versally applicable glycosylation procedures affording good stereocontrol are not available, so that these oligosaccharide syntheses require time-consuming optimization procedures. For instance, when the protected KDO-pyranosyl-bromide **57**, as described by Paulsen et al.,[22] was condensed with allyl alcohol in dichloromethane in the presence of insoluble silver salts and molecular sieve 4 Å at low temperature, the β-allyl-glycoside **58** was obtained in 72% yield, together with 12% of the glycal ester derivative **60** as an elimination product. The latter compound always appears as side product in varying amounts after glycosidation reactions using activated KDO derivatives as donors. Reaction of allyl alcohol with **57** in nitromethane catalyzed by mer-cury(II)-cyanide afforded the α-anomer **59** as the major isomer (3:1) in 81% yield (Figure 10).[22]

For the synthesis of KDO-(2→4)-α-KDO **64** compound **59** was deacylated and subsequently treated with 4-nitrophenyl chloroformate-pyridine to give a mixture of 4,5:7,8-di-O-carbonyl derivatives **61** (22%), the crystalline 7,8-O-carbonyl derivative **62** (30%), and starting material (30%) as major products that were separated by silica gel chromatography. Glycosidation of **62** with **57** under catalysis by mercury(II) cyanide in acetonitrile gave a 1:3 mixture of β-D-(2→4) linked disaccharide **63** and the α-D-(2→4) anomer **64** in 69% yield. The anomers were separated by silica gel chromatography. It is im-portant to note that products corresponding to glycosylation of the axial OH

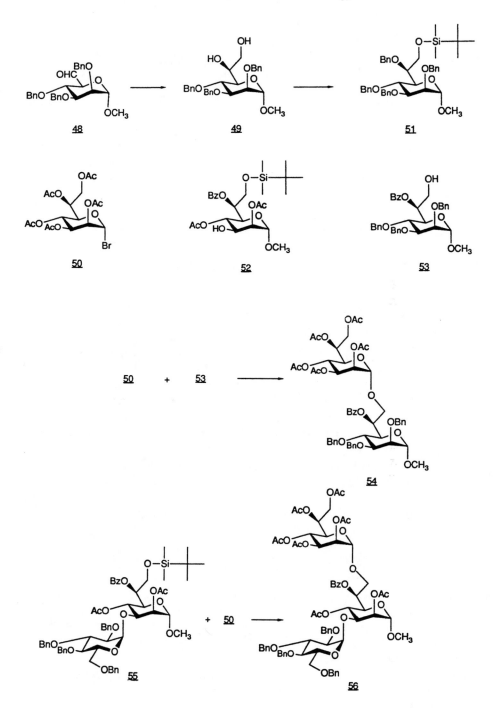

FIGURE 9.

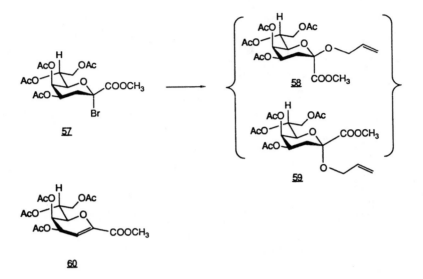

FIGURE 10.

at C-5 of **62** were not detected. Sequential deacetylation with sodium methoxide in methanol and deesterification in aqueous sodium hydroxide gave sodium *O*-(sodium 3-deoxy-α-D-*manno*-2-octulopyranosylonate)-β-(2→4)-β-(allyl-3-deoxy-α-D-*manno*-2-octulopyranoside)onate **65**. This product can then be conveniently copolymerized with four equivalents of acrylamide in the presence of *N*,*N*,*N*′,*N*′-tetramethylethylenediamine and ammonium persulfate to give an antigenic preparation for studies of the epitope specificity of anti-LPS-antibodies[23] (Figure 11).

A similar synthesis with improved yields was reported in parallel by the same group[24] utilizing the double-silylated KDO derivative **66** as a key intermediate (Figure 12).

For the synthesis of the trisaccharide α-KDO*p*-(2→4)-, α-KDO*p*- α-KDO Kosma et al.[25] started with the peracetylated disaccharide **67**, which was deallylated via isomerization into the propenyl glycoside with bis(methyldiphenylphosphine)cycloocta-1,5-diene iridium(I) hexafluorophosphate and subsequent hydrolysis catalyzed by iodine in aqueous oxolane, in 67% yield. After acetylation, a 1:5 mixture of the 2-*O*-acetyl-βα- and αβ-derivatives was obtained, separable by silica gel chromatography. The assignment of the anomeric configuration was possible by comparison of the chemical shift of the signals for equatorial H-3 in both isomers. The crystalline α-anomer **68** was converted into the disaccharide bromide **69** with titanium tetrabromide in quantitative yield and immediately used for the glycosylation of three molar equivalents of the 7,8-*O*-carbonyl derivative **62**, catalyzed by mercury(II)-cyanide in acetonitrile. The glycal corresponding to **67** was formed as elimination product in 50% yield, together with a 1:2 mixture of the

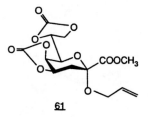

61

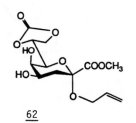

62

57 + 62

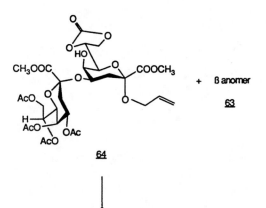

+ ß anomer

63

64

α-KDO-(2→4)-ß-KDO-α-OAll

65

FIGURE 11.

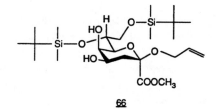

66

FIGURE 12.

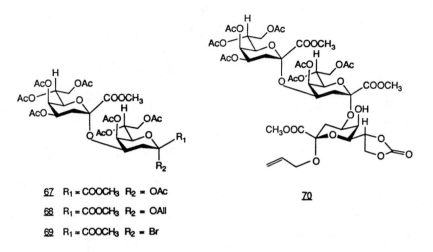

67 R₁ = COOCH₃ R₂ = OAc

68 R₁ = COOCH₃ R₂ = OAll

69 R₁ = COOCH₃ R₂ = Br

70

FIGURE 13.

β-(2′→4)- and α-(2′→4) trisaccharide derivatives, also in 50% yield. The desired α-(2′→4) trisaccharide derivative **70** was isolated by column chromatography on silica gel. This product was also transformed into the deprotected allyl glycoside and copolymerized with acrylamide to form an artificial antigen for immunochemical studies (Figure 13).

2. KDO-KDO Oligosaccharides As Found in Glycolipids from the Genus *Chlamydia*

The glycolipids of *Chlamydia* contain LPS with some of the characteristics of LPS from Enterobacteriaceae. However, in addition, these materials also contain the sequence KDO-(2→8)-KDO-(2→4)-KDO. The anomeric configuration of the (2→8)-linked KDO residue was unknown and was determined by comparative NMR spectroscopy using synthetic anomers as model compounds.[26]

For the synthesis of the (2→8)-linked KDO disaccharides, the allylglycoside **71** was chosen as starting material. Desilylation with 2% hydrogen fluoride in acetonitrile was accompanied by acetyl migration. Therefore, the crude alcohol was immediately glycosylated with an excess of **57** under modified Helferich conditions. The α-anomer **72** was the major product (59% yield) and was crystallized following chromatography on silica gel. By a series of chemical transformations, the α-anomeric configuration of **72** was shown to correspond to the one naturally occurring in LPS from *C. psittaci*.

The synthesis of the trisaccharide was achieved by transformation of **72** into the corresponding glycosyl bromide, essentially as described above for **69**. Glycosylation with four equivalents of **62** in acetonitrile with mercury(II) cyanide as catalyst afforded a 2:5 mixture of β:α anomers in 30% yield which

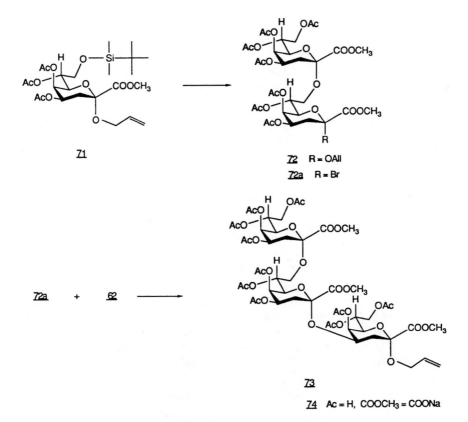

FIGURE 14.

was separated by silica gel chromatography. The isomer **73** having the natural configuration was deacetylated in methanolic sodium methoxide and hydrolyzed in aqueous sodium hydroxide to afford the tris-sodium salt **74** for copolymerization with acrylamide (Figure 14).

B. OLIGOSACCHARIDES CONTAINING L-*GLYCERO*-D-*MANNO*-HEPTOSYL-KDO

Considering the architecture of enterobacterial inner core LPS as illustrated below, one realizes that the joining of the heptose oligosaccharide with the KDO-oligosaccharide part represents a daunting synthetic challenge. Not only is a new strategy of protective groups at the central KDO required; in addition, position 5 of KDO is highly unreactive toward glycosyl donors. The problem becomes even more complicated if the glucosamine moieties of lipid A are to be included.

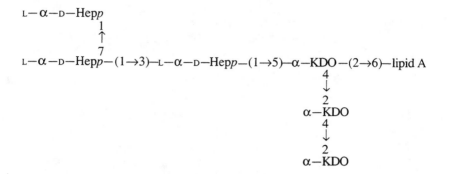

Paulsen and Heitmann reported the synthesis of trisaccharide building blocks consisting of LD-Hep*p* and KDO. These can be deblocked at the anomeric centers for the required condensations, e.g., with lipid A analogs.

The central KDO-building block, **77**, as glycosyl acceptor was prepared by the following steps: first, the KDO-pyranosyl bromide **57** was converted into the β-benzyl-ketoside **75** in near quantitative yield. After deacetylation, selective protection of positions 7,8 was effected with 1,3-dichloro-1,1,3,3-tetra-isopropyldisiloxane to afford **76**, containing the versatile TIPS-protective group. Compound **76** was then reacted with dibutyltin oxide to form the 4,5-O-stannylidene acetal as an intermediate which, upon treatment with allyl bromide in the presence of tetrabutylammonium iodide, gave **77** in a totally regioselective manner (Figure 15).

The known glycosyl donor of the heptosyl portion, **78**, also described by Paulsen and Heitmann, was condensed with **77** in the presence of silver triflate at −40°C to give the corresponding α-disaccharide **79** in 86% yield. Deallylation at C-4 of **79** gave the glycosyl acceptor appropriate for glycosylation to the trisaccharide **80**. Thus, reaction of **79a** with **57** in the presence of mercury(II)bromide exclusively gave the α-trisaccharide **80** in 43% yield, which was deprotected to L-α-D-Hep-(1→5)-KDO-(4→2)-α-KDO in two conventional steps (Figure 16).

C. SYNTHESIS OF KDO ANALOGS

The synthetic strategies of Heath and Galambor,[29a] as refined by Unger[29b] and of Schmidt[30] toward obtaining larger amounts of natural KDO, had involved coupling of suitable arabinose derivatives with either pyruvate or an oxaloacetate derivative by partial synthesis. Recently, a total synthesis of (±)-KDO, using a cycloaddition approach, has also been described.[31] Because the resolution of racemic KDO has not yet been described, the interested reader is referred to the original literature.

In an attempt to determine the configuration of a new octulosonic acid from *Acinetobacter calcoaceticus* containing an additional hydroxyl function at C-3, Paulsen et al.[32] treated the glucal **60** with osmium tetroxide in the presence of excess morpholine-*N*-oxide and obtained the single hydroxylation

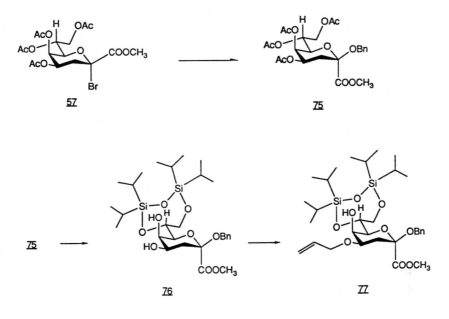

FIGURE 15.

product **81**. The configuration at C-3 became clear upon analysis of its ^{1}H-NMR spectrum because of the large diaxial coupling constant of $J_{3,4} = 10.4$ Hz. This product, however, was not identical with the natural hydroxylated KDO analog. Meanwhile, the synthesis of the latter has been achieved by a different route[33] (Figure 17).

In this context, it can be but briefly mentioned that the known 5-deoxy-KDO*p* moiety was also used as glycosyl acceptor for the synthesis of allyl α-KDO*p*-(2→4)-α-5-deoxy-KDO*p*, which was copolymerized with acrylamide for epitope analysis studies.[34]

Another recent study reported the synthesis of β-KDO analogs having a substituent attached via C-C bond to C-2 as potential inhibitors of the enzyme CMP-KDO synthetase.[35] Inhibitors of this key enzyme for the incorporation of KDO into the lipopolysaccharide of Gram-negative bacteria could, thus, have a potential as antibiotics with specific activity against Gram-negative bacteria.

V. SYNTHESIS OF MIXED OLIGOSACCHARIDES THAT CONTAIN THE GLUCOSAMINE BACKBONE OF LIPID A AND KDO

A. OLIGOSACCHARIDES WITHOUT O-PHOSPHATE GROUPS

The attachment of the KDO moiety to the β-glucosamine-(1→6)-disaccharide backbone of lipid A was again pioneered by the group of Paulsen et al.[36] The first synthetic problem, i.e., the preparation of the disaccharide

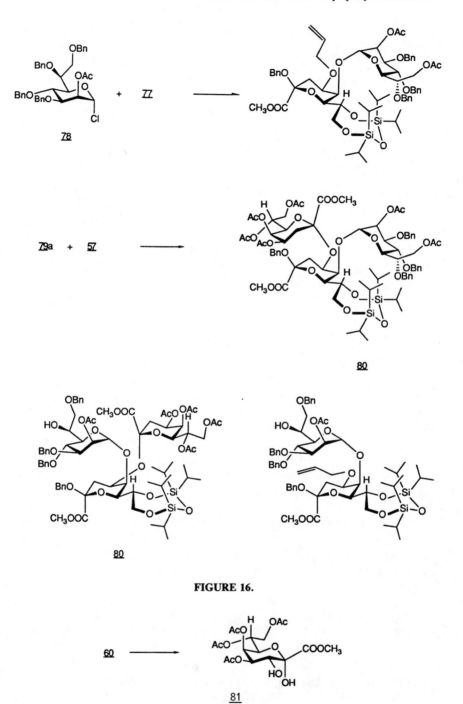

FIGURE 16.

FIGURE 17.

β-D-GlcA-(1→6)-GlcA was solved in an elegant way by taking advantage of the fact that α-pyranosyl bromides of 2-azido-2-deoxy-D-glucose can be converted into β-glycosides by heterogenous catalysis even without neighboring group participation.

The ring opening of the known epoxide **82** with azide, followed by O-benzylation, afforded the derivative of 1,6-anhydro-glucose **33** in good yield which, upon acetolytic ring opening using acetanhydride/trifluoroacetic acid and subsequent treatment with titanium tetrabromide, gave the glycosyl donor **84**. The authors also showed that an appropriately O-protected oxirane **82** could serve as an intermediate in 4-phosphorylated analogs. The corresponding glycosyl acceptor **85** was obtained from **84** by reaction with benzyl alcohol in dichloromethane at −20°C in the presence of silver silicate and subsequent deacetylation.

Dissacharide **86** was obtained by reaction of **84** and **85** in dichloromethane at −20°C in the presence of molecular sieves (4 Å) and silver silicate. The reaction product was directly deacetylated and then purified by column chromatography. The yield of **87** was 66% after two steps.

Owing to the largely different reactivities of the free hydroxyl groups in positions 4′ and 6′, **87** could be directly used as glycosyl acceptor for the condensation with KDO donor **57**. In the presence of a 10:1 ratio of silver carbonate/silver perchlorate in dichloromethane at 0°C, 30% of the desired α-ketosidically linked trisaccharide **88** was isolated. Only 2% of the corresponding β-anomer was formed, together with the product resulting from elimination of hydrogen bromide from **57** as the major side product. A complete stereochemical assignment of the observed proton signals by two-dimensional NMR spectroscopy was performed.

Reduction of the azido groups in **88** was achieved with hydrogen sulfide in aqueous pyridine. Acylation of free amino groups with (R)-3-hydroxymyristic acid with water-soluble carbodiimide led to the expected product, which could be deprotected to **89** in a conventional manner. Compound **89** is tightly adsorbed by erythrocytes, a property which makes it useful in a system for testing the epitope specificity of monoclonal antibodies (Figure 18).

The wealth of available building blocks and glycosylation methods enabled the group of Paulsen to synthesize the trisaccharide α-KDO-(2→4)-α-KDO-(2→6)-D-GlcN as well as higher homologs.[37-40] Thus, Paulsen et al.[38] reported the syntheses of the tetrasaccharide α-KDO-(2–4)-α-KDO-(2→6)-D-GlcN-(1→6)-D-GlcN and the pentasaccharide α-KDO-(2→4)-α-KDO-(2→4)-α-KDO-(2→6)-β-D-GlcN-β-D-(1→6)-GlcN, acylated with (R)-3-hydroxy-myristic acid at both nitrogen amino functions of the glucosamine disaccharide.[39] The synthesis of the latter is outlined to illustrate the general synthetic strategy applied to all of these oligosaccharides.

The silylated β-KDO-glycoside **6** was condensed with the KDO donor **57** to afford the disaccharide **90** in 74% yield. Adduct **90** was transformed

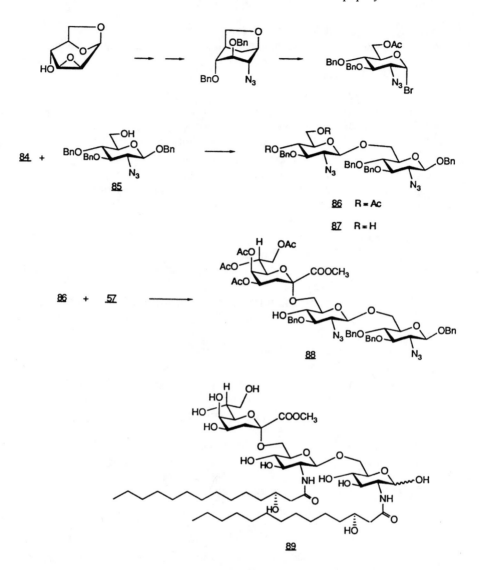

FIGURE 18.

into the KDO disaccharide **91** by a few conventional deprotection-protection steps, as silylated KDO-pyranosyl-halogenides were found to be too reactive for easy handling. Glycosylation of the partially protected trisaccharide **92** with **91** in toluene-nitromethane with mercury-II-bromide as catalyst afforded the expected pentasaccharide **93** in 21% yield without formation of the corresponding β-anomer. Main by-products were the hydrolysis product (40%) and the elimination product of **91** (30%). Characteristically for this class of oligosaccharides, formation of significant amounts of interglycosidic lactone **94** was observed during desilylation of **93** using tetrabutylammonium fluoride.

The synthesis of the trisaccharide **92** has been described by Paulsen and Krogmann.[40a] These authors demonstrated that the 4',6'-bis-silylated disaccharide **95** undergoes smooth rearrangement into the thermodynamically more stable 3',4'-isomer **96** under catalysis by *p*-toluene-sulfonic acid. The primary alcohol was used for condensation with the KDO-donor **57** to form **92** in 40% yield. This molecule was extended by an additional KDO residue upon condensation with **57** to form the corresponding α-(2-5)-linked tetrasaccharide.

The above-mentioned myristoylated but otherwise deprotected KDO-GlcN oligosaccharides were fixed on sheep erythrocytes for the determination of the epitope specificities of monoclonal antibodies directed against inner core LPS determinants (Figure 19).

Most recently, the synthesis of a tetrasaccharide of the genus-specific LPS epitope of *Chlamydia* has been obtained.[41] It differs from the enterobacterial analog in that it contains an α-KDO-(2$\rightarrow$8)-α-KDO-(2$\rightarrow$4)-α-KDO-(2$\rightarrow$6)-GlcN linkage pattern. This synthesis could be achieved, albeit in only 16% yield and only in the form of a 3:1 mixture of α:β anomers, by condensation of the bis-KDO-glycosyl donor **72a** with the disaccharide **97** in the presence of 3:1 mercury(II)cyanide/mercury(II)bromide (Figure 20).

B. KDO-GLUCOSAMINE CONTAINING OLIGOSACCHARIDES BEARING ALSO PHOSPHATE GROUPS

The inner core LPS of Gram-negative bacteria is not only characterized by its relatively conserved oligosaccharide backbone but also by the presence of phosphate- and pyrophosphate ester groups. The latter, together with the long-chain fatty acids of the lipid A part, convey unusual physicochemical properties to the molecules. Indeed, these features represent yet another challenge to the synthetic organic chemist and, in particular, to the analytical chemist. The synthetic target compounds have a high propensity to form micelles and mixed micelles in solution (see Chapter 9) and, hence, purification of the deprotected target structures is difficult. Most probably, this amphiphilic character is a prerequisite for the elicitation of the well-known endotoxic and immunomodulatory effects. There is some evidence that not only lipid A but also KDO, linked to O-phosphorylated O,N-acylated lipid A partial structures, contributes to the endotoxic properties of LPS (see Chapter 8).

The syntheses of N,O-acylated and O-phosphorylated lipid A structure analogs have been dealt with in the preceding article. Here, the state of the art regarding the attachment of KDO to lipid A, culminating in the total synthesis of Re chemotype LPS, will be described.

The first successful synthesis of a disaccharide involving KDO and GlcN with native O,N-substitution patterns has been described by the group of Hasegawa.[42,43] The presence of ester-bound fatty acids in the lipid A part precludes the convenient use of O-acetyl groups for the KDO residues if one aims at native target substructures. These problems were overcome by the

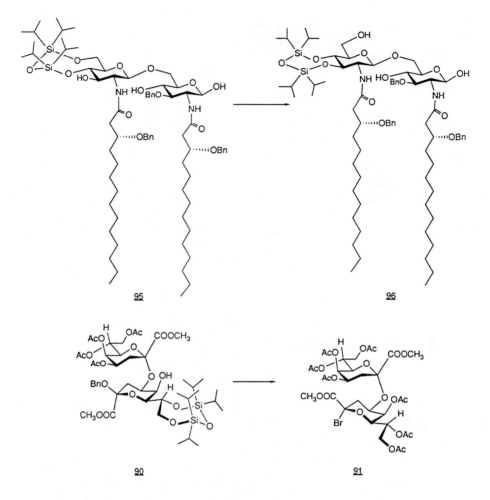

FIGURE 19.

use of chloroacetyl protecting groups which can be removed under very mild conditions. Suitably protected KDO derivatives were activated to the glycosyl donors the form of either of pyranosyl bromides or fluorides. The latter were obtained by reaction of corresponding 2-unsubstituted KDO derivatives with diethylaminosulfur trifluoride (DAST). As glycosyl acceptors, partial structure analogs of the nonreducing end of lipid A, either 4-phosphorylated or with a free 4-OH group, were used. In the latter case, the disaccharide product was phosphorylated at a later stage of the synthesis.[44]

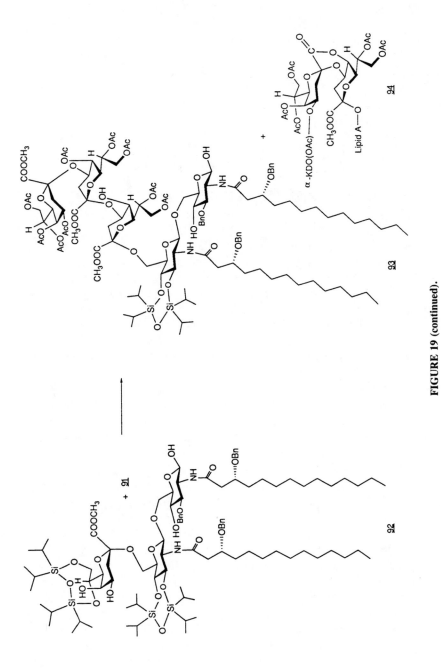

FIGURE 19 (continued).

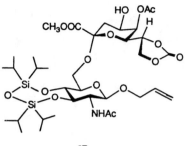

97

FIGURE 20.

VI. TOWARD THE TOTAL SYNTHESIS OF "ROUGH" LPS

Despite the tremendous progress made in recent years with regard to regio- and stereoselective synthesis of oligosaccharides that represent subunits of bacterial core LPS, the total synthesis of complete Re LPS, e.g., α-KDO-(2→4)-α-KDO (2→6)-linked to native lipid A of *Escherichia coli*, has not yet been achieved. However, meticulous and elegant work of Kusumoto et al.[45-47] has resulted in the completion of the total synthesis of the 1-dephospho derivative of *E. coli* Re lipopolysaccharide.

The essential steps in this major undertaking were the following:

- Synthesis of the protected KDO-pyranosyl-fluoride, **98**.[48]
- Condensation of **98** with the protected 1-dephospho-lipid A derivative **99** to the trisaccharide **100** with ethyldiisopropylamine and boron trifluoride etherate in methylene chloride under argon at 0°C in high yield.
- Deketalization and selective monoketalization of position 7,8 of the KDO-part of **100** and subsequent regioselective condensation with **98** to the corresponding tetrasaccharide, isolated in 31% yield as a single product **101**, after cleavage of all isopropylidene groups by treatment with trifluoroacetic acid (Figure 21).

The following sequence of steps led to the deprotected target structure **102**: deallylation, using [Ir(COD)PCH₃(C₆H₅)₂)]PF₆ and I₂ in aqueous THF, and two hydrogenation steps with H₂-Pd black and then H₂-PtO₂ in order to completely cleave the phenylphosphate ester bonds. Purification was achieved by centrifugal partition chromatography of the final product **103**. On thin layer chromatography, it behaved in a manner identical to the product obtained from natural *E. coli* Re LPS after selective cleavage of the glycosidic phosphate by mild acid treatment. This synthesis, therefore, represents the first authentic confirmation of the proposed structure of Re LPS (Figure 22).

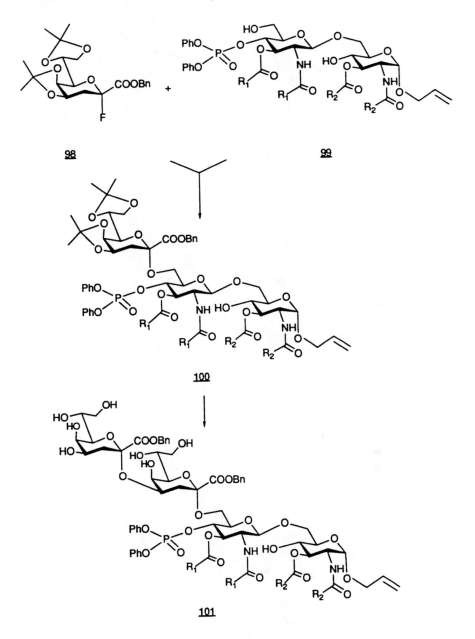

R₁CO = (R)-3- tetradecanoyloxytetradecanoyl

R₂CO = (R)-3-hydroxytetradecanoyl

FIGURE 21.

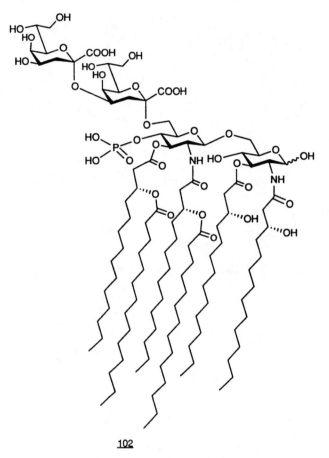

102

FIGURE 22.

VII. CONCLUDING REMARKS

Decisive progress has been made during recent years regarding the total synthesis of core structures of "rough" LPS of Gram-negative bacteria. Building blocks, repeating partial structures up to the size of pentasaccharides, have been prepared under regio- and stereocontrol and and mostly good yields. Further progress may be expected, for example, by applying solid-phase syntheses.

The chemistry associated with the introduction of suitably protected phosphate groups into LPS-related glycolipid structures continues to present considerable obstacles. Problems arise with quantitative O-phosphorylation of sterically hindered hydroxyl groups, exhaustive deprotection of the phosphate esters, and separation and purification of amphiphilic end products.

Presently, techniques are being developed by specialist groups to solve these problems and to improve the synthetic access to a variety of biologically important LPS derivatives. Thus, Szabo's group has reported on a useful O-phosphorylation procedure while Kusumoto's group has developed interesting partition methodology for separation and purification of synthetic, LPS-related phosphoglycolipids.

In due course, a multitude of LPS core structures will be freely accessible by chemical synthesis, contributing importantly to progress in immunopharmacology and antiinfective medicine.

ACKNOWLEDGMENT

The authors are indebted to Dr. R. Ippolito, Chembiomed, for having contributed the chemical formulas with the CHEMDRAW program.

REFERENCES

1. **Mayer, H., Ramadas Bhat, H., Masoud, H., Radziejewska-Lebrecht, J., Wiedemann, C., and Krauss, J. H.,** Bacterial lipopolysaccharides, *Pure Appl. Chem.,* 61, 1271, 1989.
2. **Imoto, M., Kusumoto, S., Shiba, T., Naoki, H., Iwashita, T., Rietschel, E. T., Wollenweber, H. W., Galanos, C., and Lüderitz, O.,** Chemical structure of *E. coli* lipid A: linkage site of acyl groups in the disaccharide backbone, *Tetrahedron Lett.,* 24, 4017, 1983.
3. **Strain, S. M., Fesik, S. W., and Armitage, I. M.,** Characterization of lipopolysaccharide from a heptoseless mutant of *E. coli* by carbon 13 nuclear magnetic resonance, *J. Biol. Chem.,* 258, 2906, 1983.
4. **Qureshi, N., Takayama, K., Heller, D., and Fenselau, C.,** Position of ester groups in the lipid A backbone of lipopolysaccharide obtained from *Salmonella typhimurium, J. Biol. Chem.,* 258, 12947.
5a. **Christian, R., Schulz, G., Waldstätten, P., and Unger, F. M.,** Zur Struktur der 3-Desoxyoctulosonsäure-(KDO)-Region des Lipopolysaccharides von *Salmonella minnesota* Re 595, *Tetrahedron Lett.,* 25, 3433, 1984.
5b. **Paulsen, H., Stiem, M., and Unger, F. M.,** Synthese eines 3-*deoxy*-D-*manno*-2-octulosonsäure(KDO)-haltigen Tetrasaccharides und dessen Strukturvergleich mit einem Abbauprodukt aus Bakterien-Lipopolysaccharided, *Tetrahedron Lett.,* 27, 1135, 1986.
6. **Rozalski, A., Brade, L., Kosma, P., Appelmelk, B. J., Krogmann, C., and Brade, H.,** Epitope specificities of murine monoclonal and rabbit polyclonal antibodies against enterobacterial lipopolysaccharides of the Re chemotype, *Infect. Immun.,* 57, 2645, 1989.
7. **Rozalski, A., Brade, L., Kuhn, H. M., Brade, H., Kosma, P., Appelmelk, B. J., Kusumoto, S., and Paulsen, H.,** Determination of the epitope specificity of monoclonal antibodies against the inner core region of bacterial lipopolysaccharides by use of 3-*deoxy*-D-*manno*-octulosonate-containing synthetic antigens, *Carbohydr. Res.,* 193, 257, 1989.
8. **Shimizu, T., Masuzawa, T., Yanagihara, Y., Shimizu, C., Ikeda, K., and Achiwa, K.,** Biological activities of chemically synthesized N-acetylneuraminic acid-($\alpha 2 \rightarrow 6$)-monosaccharide analogs of lipid A, *FEBS Lett.,* 228, 99, 1988.

9. **Lebbar, S., Cavaillon, J. M., Caroff, M., Ledur, A., Brade, H., Sarfati, R., and Haeffner-Cavaillon, N.**, Molecular requirement for interleukin 1 induction by lipopolysaccharide-stimulated human monocytes: involvement of the heptosyl-2-keto-3-deoxy-octulosonate, *Eur. J. Immunol.*, 16, 87, 1986.

10. **Norberg, T., Walding, M., and Westman, E.**, Synthesis of the methyl, 1-octyl, and *p*-trifluoroacetamidophenethyl α-glycosides of 3,6-di-O-(α-D-galactopyranosyl)-D-glucopyranose and an acyclic analogue thereof, and references listed therein, *J. Carbohydr. Chem.*, 7, 283, 1988.

11. **Liptake, A., Kerekgyarto, J., Szurma, Z., and Duddek, H.**, Synthesis of the tetrasaccharide core region of antigenic lipo-oligosaccharides characteristic of *Mycobacterium kansasii*, *Carbohydr. Res.*, 175, 241, 1988.

12. **Paulsen, H., Schüller, M., Heitmann, A., Nashed, M. A., and Redlich, H.**, Diastereoselektive Synthese von L-*glycero*-D-*manno*-Heptose, einem Baustein der inneren Core-Region von Lipopolysacchariden, *Justus Liebigs Ann. Chem.*, 675, 1986.

13. **Brimacombe, J. S. and Kabir, K. M. S.**, Convenient synthesis of L-*glycero*-D-*manno*-heptose and D-*glycero*-D-*manno*-heptose, *Carbohydr. Res.*, 152, 329, 1986.

14. **Paulsen, H., Wulff, A., and Heitmann, A. C.**, Synthese von Disacchariden aus L-*glycero*-D-*manno*-heptose und 2-Amino-2-desoxy-D-glucose, *Justus Liebigs Ann. Chem.*, 1073, 1988.

15. **Boons, G., Overhand, M., van der Maarel, G. A., and van Boom, J. H.**, Anwendung der Dimethyl(phenyl)silyl-Guppe als Äquivalent einer Hydroxyfunktion bei der Synthese eines L-*glycero*-α-D-*manno*-heptapyranos d-haltigen Trisaccharids aus der dephosphorylierten "inner core" Region von *Neisseria meningitidis*, *Angew. Chem.*, 101, 1538, 1989.

16. **Boons, G., van der Klein, P., van der Maarel, G. A., and van Boom**, Synthesis of L-*glycero*-D-*manno*-heptose, *Recl. Trav. Chim. Pays-Bas*, 107, 507, 1988.

17. **Dziewiszek, K., Banaszek, A., and Zamojski, A.**, The synthesis of the heptose region of the Gram-negative bacterial core oligosaccharides, *Tetrahedron Lett.*, 28, 1569, 1987.

18. **Dziewisek, K. and Zamojski, A.**, New syntheses of D- and L-*glycero*-D-*manno*-heptoses, *Carbohydr. Res.*, 150, 163, 1986.

19. **Schmidt, R. R.**, Neue Methoden zur Glycosid- und Oligosaccharidsynthese — gibt es Alternativen zur Koenigs-Knorr-Methode?, *Angew. Chem.*, 98, 213, 1986.

20. **Garegg, P. J., Oscarson, S., and Szönyi, M.**, Synthesis of methyl 3-O-(α-D-glucopyranosyl)-7-O-(L-*glycero*-α-D-*manno*-heptopyranosyl)-L-*glycero*-α-D-*manno*-heptopyranoside, *Carbohydr. Res.*, 205, 125, 1990.

21. **Paulsen, H., Stiem, M., and Unger, F. M.**, Synthese eines 3-Desoxy-D-*manno*-2-octulosonsäure (KDO)-haltigen Tetrasaccharides und dessen Strukturvergleich mit einem Abbauprodukt aus Bakterien-Lipopolysacchariden, *Tetrahedron Lett.*, 27, 1135, 1986.

22. **Paulsen, H., Hayauchi, Y., and Unger, F. M.**, Synthese von Disacchariden der 3-Desoxy-D-*manno*-2-octulosonsäure (KDO) und D-Glucosamin, *Justus Liebigs Ann. Chem.*, 1270, 1984.

23. **Kosma, P., Gass, J., Schulz, G., Christian, R., and Unger, F. M.**, Artificial antigens. Synthesis of polyacrylamide copolymers containing 3-deoxy-D-*manno*-2-octulo-pyranosylonic acid (KDO) residues, *Carbohydr. Res.*, 167, 39, 1987.

24. **Kosma, P., Schulz, G., and Unger, F. M.**, Synthesis of polyacrylamide copolymers containing α-(2→4)-β-, β(2→4)-β and β-(2→4)-α-linked O-(3-deoxy-D-*manno*-2-octulo-pyranosylono)-(3-deoxy-D-*manno*-2-octulopyranosyono) (KDO) residues, *Carbohydr. Res.*, 180, 19, 1988.

25. **Kosma, P., Schulz, G., Unger, F. M., and Brade, H.**, Synthesis of trisaccharides containing 3-deoxy-D-*manno*-2-octulosonic acid residues related to the KDO-region of enterobacterial lipopolysaccharides, *Carbohydr. Res.*, 190, 191, 1989.

26. **Kosma, P., Schulz, G., and Brade, H.**, Synthesis of a trisaccharide of 3-deoxy-D-*manno*-2-octulopyranosylonic acid (KDO) residues related to the genus-specific lipopolysaccharide epitope of *Chlamydia*, *Carbohydr. Res.*, 183, 183, 1988.

27. **Paulsen, H. and Heitmann, A.**, Synthese von Trisaccharid-Einheiten der inneren Core-Region von Lipopolysacchariden, *Justus Liebigs Ann. Chem.*, 655, 1989.
28. **Paulsen, H. and Heitmann, A.**, Synthese von Strukturen der inneren Core-Region von Lipopolysacchariden, *Justus Liebigs Ann. Chem.*, 1061, 1988.
29a. **Heath, E. and Ghalambor, A.**, 2-Keto-3-deoxy-octonate, a constituent of cell wall lipopolysaccharide preparations obtained from *E. coli, Biochem. Biophys. Res. Commun.*, 10, 340, 1963.
29b. **Unger, F.**, The chemistry and biological significance of 3-deoxy-D-*manno*-2-octulosonic acid (KDO), *Adv. Carbohydr. Chem. Biochem.*, 38, 323, 1981.
30. **Schmidt, R. R.**, Abstract of Papers, 187th Natl. Meeting of the American Chemistry Society, St. Louis, April 1984.
31. **Danishefsky, S., Pearson, W., and Segmuller, B.**, Total Synthesis of (±)-3-deoxy-D-*manno*-2-octulopyranosate, *J. Am. Chem. Soc.*, 107, 1280, 1985.
32. **Paulsen, H., Krogmann, C., and Dessen, U.**, Synthese von α-D-*glycero*-D-*galacto*-2-octuloson-säure und 5-Acetamido-5-desoxy-D-*erythro*-L-*gluco*-2-nonulonsäure, *Liebigs Ann. Chem.*, 277, 1988.
33. **Zähringer, U., Kosma, P., Sinnwell, V., and Rietschel, E. T.**, *Carbohydr. Res.*, in preparation.
34. **Kosma, P., Waldstätten, P., Daoud, L., Schulz, G., and Unger, F.**, Synthesis of poly(acrylamide) copolymers containing 3,5-dideoxy-D-*arabino*-2-octulopyranosylonic acid (5-deoxy-KDO) residues, *Carbohydr. Res.*, 194, 145, 1989.
35. **Waglund, T., Luthman, K., Orbe, M., and Claesson, A.**, Synthesis of C-(β-D-glycosy)analogues of 3-deoxy-D-*manno*-2-octulosonic acid (KDO) as potential inhibitors of CMP-KDO-synthetase, *Carbohydr. Res.*, 206, 269, 1990.
36. **Paulsen, H. and Schüller, M.**, Synthese von KDO-haltigen Lipoid-A-Analoga, *Justus Liebigs Ann. Chem.*, 249, 1987.
37. **Paulsen, H., Stiem, M., and Unger, F. M.**, Synthese der Sequenz α-KDO-(2→4)-α-KDO-(2→6)-D-GlcN der inneren Core-Struktur von Lipopolysacchariden, *Justus Liebigs Ann. Chem.*, 273, 1987.
38. **Paulsen, H., Stiem, M., and Unger, F. M.**, Synthese von O-(3-Desoxy-α-D-*manno*-2-octulopyranosylonsäure)-(2→4)-O-(3-desoxy-α-D-*manno*-2-octulopyranosylonsäure)-(2→6)-O-(2-amino-2-desoxy-β-D-glucopyranosyl)-(1→6)-2-amino-2-desoxy-D-glucopyranose, *Carbohydr. Res.*, 172, 11, 1988.
39. **Paulsen, H. and Krogmann, C.**, Synthese einer KDO-haltigen Pentasaccharidsequenz der inneren Core- und Lipid A-Region von Lipopolysacchariden, *Carbohydr. Res.*, 205, 31, 1990.
40a. **Paulsen, H. and Krogmann, C.**, Synthese von KDO-haltigen Tri- und Tetrasaccharid-Sequenzen der inneren Core- und Lipoid A-Region von Lipopolysacchariden, *Justus Liebigs. Ann. Chem.*, 1203, 1989.
40b. **Paulsen, H., Stiem, M., and Unger, F. M.**, Synthese eines 3-Desoxy-D-*manno*-2-Octulosonsäure(KDO)-haltigen Tetrasaccharides und dessen Strukturvergleich mit einem Abbauprodukt aus Bakterien-Lipopolysacchariden, *Tetrahedron Lett.*, 27, 1135, 1986.
41. **Kosma, P., Bahnmüller, R., Schulz, G., and Brade, H.**, Synthesis of a tetrasaccharide of the genus-specific lipopolysaccharide epitope of *Chlamydia, Carbohydr. Res.*, 208, 37, 1990.
42. **Kiso, M., Fujita, M., Hayashi, E., Hasegawa, A., and Unger, F.**, Novel Disaccharides containing 3-deoxy-D-*manno*-2-octulosonic acid (KDO) and lipid A subunit analogs, *J. Carbohydr. Chem.*, 6, 691, 1987.
43. **Kiso, M., Fujita, M., Tanahashi, M., Fujishima, Y., Ogawa, Y., Hasegawa, A., and Unger, F.**, Conjugates of biologically active lipid A subunit analogs with 3-deoxy-D-*manno*-2-octulosonic acid (KDO) and its methyl ester, *Carbohydr. Res.*, 177, 51, 1988.
44. **Nakamoto, S. and Achiwa, K.**, Lipid A and related compounds. XVI. Synthesis of biologically active tetraacetyl-3-deoxy-D-*manno*-2-octulosonic acid (KDO)-(α2-6)-D-glucosamine-4-phosphates. Novel analogs of the non-reducing sugar moiety of lipid A, *Chem. Pharm. Bull.*, 35, 4537, 1987.

45. **Imoto, M., Kusunose, N., Kusumoto, S., and Shiba, T.,** Synthetic approach to bacterial lipopolysaccharide. Preparation of trisaccharide part structures containing KDO and 1-dephospholipid A, *Tetrahedron Lett.,* 29, 2227, 1988.
46. **Kusumoto, S., Kusunose, N., Kamikawa, T., and Shiba, T.,** Chemical synthesis of 1-dephosphoderivative of *Escherichia coli* Re lipopolysaccharide, *Tetrahedron Lett.,* 29, 6325, 1988.
47. **Kusumoto, S., Kusunose, N., Imoto, M., Shimamoto, T., Kamikawa, T., Takada, H., Kotani, S., Rietschel, E., and Shiba, T.,** Synthesis and biological function of bacterial endotoxin, *Pure Appl. Chem.,* 61, 461, 1989.
48. **Imoto, M., Kusunose, N., Matsuura, Y., Kusumoto, S., and Shiba, T.,** Synthetic approach to bacterial lipopolysaccharide. Preparation of trisaccharide part structures containing KDO and 1-dephospholipid A, *Tetrahedron Lett.,* 28, 2227, 1988.

Chapter 8

STRUCTURE-FUNCTION RELATIONSHIPS TO CORE OLIGOSACCHARIDE

Jean-Marc Cavaillon and Nicole Haeffner-Cavaillon

TABLE OF CONTENTS

I. INTRODUCTION

For a long time, the role played by the core oligosaccharide within the lipopolysaccharide (LPS) molecule in endotoxin-induced biological activities has been widely underestimated. It was commonly concluded from experimental data obtained with LPS from enterobacterial deep rough mutants (Re or Rd LPS) that the lipid A moiety was the only active part of the molecule. Although isolated lipid A is endowed with the endotoxic properties of LPS, the presence of the core moiety has often been recognized to enhance its biological activities. The involvement of the core oligosaccharide, and particularly the residues within the inner core, has been clarified by recent biophysical studies establishing that the fluidity of the hydrocarbon chains, which is an important parameter with respect to the expression of biological activities, is significantly influenced by the presence of oligosaccharides.[1] Other reports have suggested possible lectin-like interactions of endotoxins with different ligands (soluble molecules or receptors), involving residues within the core moiety. A few studies have dealt with isolated core regions or core fragments; however, the more common approach has been to compare the biological activities of lipid A with those of different rough mutants. Such studies have often allowed the conclusion that R LPS had a higher biological activity than the isolated lipid A. Being the most conserved common carbohydrate structural element of all endotoxin, the inner core region has been an attractive target for obtaining cross-reactive antibodies in the hope of protection against Gram-negative sepsis (see Volume II). In this review, we will not discuss experiments which have employed antibodies directed against the core region. Mainly as a result of the steric hindrance caused by a relatively large molecule (generally an IgM) as compared to LPS, we feel that such reagents are not ideal tools to study the structure-function relationships of core oligosaccharides. Rather, we will focus on experiments which demonstrate how residues within the core region can amplify the lipid A activities as well as how those residues can directly interact with different binding sites and thus manifest their own biological activities.

II. BIOCHEMICAL STRUCTURE OF CORE OLIGOSACCHARIDES

Detailed biochemical characteristics of the core oligosaccharides are given in Chapter 4. Briefly, the core oligosaccharide can be subdivided into the lipid A-proximal inner core and the lipid A-distal outer core. Within different enterobacterial species, the structural variability of the core is limited (e.g., one type for the *Salmonella* genus, five types for *Escherichia coli*). The inner core is the most conserved moiety, being composed of two unusual sugars, heptose and 3-deoxy-D-*manno*-2-octulosonic acid or KDO (for the former name 2-*keto*-3-*deoxy*-*o*ctonic acid). KDO, an eight-carbon atom sugar acid, is the obligatory link between the lipid A and the core regions. As few as one or as many as three KDO residues may be present, depending upon the origin of LPS. Charged constituents such as phosphate and ethanolamine are common elements of the inner core region. Some LPS, for example those derived *Vibrio* species, first reported to be devoid of KDO residues, contain substituted KDO molecules which prevent their detection by the classical thiobarbiturate assay.[2] Trifluoroacetic acid treatment of these LPS, however, readily allows detection of KDO.[3] We have found that such treatment greatly enhanced the macrophage activation properties of these LPS.[4] This observation, together with other findings discussed later, suggested that the KDO residue(s) may play a role in the biological activies of LPS. The outer core region contains more common sugars, mainly hexoses and hexosamines, and reveals a higher structural diversity between LPS from different origins. In LPS isolated from Enterobacteriaceae, *Pseudomonas*, and *Vibrio*, the outer core is linked to the repetitive units of oligosaccharide which define the O-antigen specificity. LPS isolated from *Neisseria (meningitidis), Bordetella (pertussis), Bacteroides (fragilis), Haemophilus (influenzae),* and *Acinobacter (calcoaceticus)* lack the O-specific polysaccharide with repetitive units and are therefore composed exclusively of lipid A and a core oligosaccharide. The LPS obtained from *Chlamydia trachomatis* and *Rhodopseudomonas sphaeroides* are composed only of lipid A and an inner core region.

The core region has an important role in the tertiary and/or quaternary conformation of the LPS. It lowers the temperature for which the gel to liquid-crystalline phase transition occurs in the molecule.[1] Furthermore, the presence of KDO allows expression of a specific high-affinity binding site capable of interacting with divalent cations (Ca^{2+}, Mg^{2+}).[5] Ca^{2+}- and Mg^{2+}-containing buffers significantly reduce the solubility of isolated LPS and this is particularly apparent with the rough forms of LPS.[6] The interaction of LPS with divalent cations was pointed out to be required for assembly and maintenance of the normal structural organization of the outer membrane of Gram-negative bacteria. Indeed, as stated by Rietschel et al.,[7] the presence of KDO, together with lipid A, excellently fulfilled the requirement of bacteria for the organization and function of the outer membrane, i.e., for bacterial growth and survival: KDO is an obligatory constituent of LPS and plays a vital role for Gram-negative bacteria. Enterobacterial rough-type LPS have been used to analyze the structure of the core region and are essential tools in further investigating the structure-function relationship of the core residues. Deep rough mutant LPS extracted from *Salmonella minnesota* R595, R4, and R7 and *Escherichia coli* D31m4 contain lipid A and KDO residues (Re mutants) or lipid A, KDO, and heptose residues (Rd mutants). Ra, Rb, and Rc (like *E. coli* J5) LPS mutants contain, in addition, hexose residues from the outer core (Figure 1).

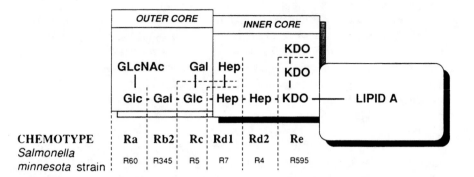

FIGURE 1. Structure of core oligosaccharide. The *S. minnesota* core moiety is given as an example. The outer and inner core are indicated as well as the chemotypes of rough mutants.

III. ROLE PLAYED BY THE CORE REGION IN THE BIOLOGICAL ACTIVITIES OF ENDOTOXINS

A. INTERACTION WITH SOLUBLE MOLECULES

1. Complement Activation

Gram-negative bacteria activate the alternative and classical complement pathways (reviewed in Volume II). Both pathways lead to C3 convertase formation and result in C3 cleavage. The efficiency of complement activation increases markedly with S LPS, Re LPS, and lipid A.[8] The polysaccharide portion allows activation of complement through the alternative pathway.[9] The length of the polysaccharide chain[8,9] as well as its chemical composition[10,11] influence the capacity of the molecule to activate the complement system. Lipid A leads to an antibody-independent activation of the classical pathway.[9] Rough J5 mutants of *E. coli* activate C1 tenfold more than smooth *E. coli* bacteria,[12] suggesting that O-specific polysaccharide may mask determinants within the core and/or the lipid A involved in the interaction with the C1 component. Vukajlovich[13] further reported that only isolated lipid A and Re-chemotype LPS were able to activate the classical pathway. Clas et al.[14] indicated that, to achieve 50% inhibition of C1 activity, tenfold more lipid A than Re LPS was required. This observation suggests a positive influence of KDO residue(s) on lipid A-mediated activation of classical pathway of complement. Although the additional monosaccharide L-glycero-D-manno-heptose in the Rd2 LPS compared to the Re LPS was sufficient to block lipid A-dependent activation of the classical pathway,[13] it was shown that Rd LPS binds the C1q molecule.[14] In contrast to the activation of the classical pathway, Rd LPS initiates significant activation of alternative pathway of complement whereas Re LPS does not.[15] This observation indicates that L-glycero-D-manno-heptose provides a critical recognition role for activation of this pathway.

2. Interaction with Serum Proteins

Many serum components (proteins, lipoproteins) have been shown to interact with LPS.[16-18] At least two different proteins isolated from mouse serum have been reported to interact specifically with core residues on LPS.[19-22] These two serum factors share some properties: both are 28-kDa proteins, labile above 56°C, found in the serum of all mouse strains including nude mice, are not increased in acute serum, and once complexed to LPS allow activation of serum complement. However, these two factors do not interact with similar residues in the core region. For one of these, binding to the LPS molecule is inhibited by the heptose-KDO disaccharide or heptose-heptose-KDO trisaccharide.[22] It is rapidly cleared from serum after an injection of LPS, and reappears within 1 h. The other is specific for N-acetyl-D-glucosamine present in the proximal end of *S. typhimurium* Ra LPS and for L-glycero-D-manno-heptose which is a component of a side branch of Ra LPS.[20] It reacts specifically with Ra LPS and only weakly binds to smooth or Rb through Re chemotype LPS.[21] With other serum compounds known to combine with LPS, such as the recently defined LPS-binding protein (LBP),[23] binding is mediated by lipid A. Although LBP binds to lipid A very efficiently, it has been reported that the presence of KDO residues renders the Re LPS three- to sixfold more effective than isolated lipid A in competing with the binding of LBP to LPS.[24]

3. Interaction with *Limulus*-Derived Factor(s)

The reactivity of LPS with the lysates of amebocytes of the American horseshoe crab *Limulus polyphemus* (LAL) is widely used to detect minute amounts of LPS. However, as a consequence of endotoxin biochemical heterogeneity, the reactivity of LAL differs significantly among LPS preparations, depending upon the origin of the LPS.[25] On a weight basis, it has been shown that *S. typhimurium* lipid A is 19-fold less reactive in the LAL test than the homologous Re LPS,[26] and the rough LPS is 3-fold more reactive than the S-type LPS. Altogether, these results suggest that core residues are essential for optimally efficient reactivity of LPS in the LAL test. An LPS-binding lectin, limulin, has been recovered from the serum of *L. polyphemus*. Qualitative precipitin tests revealed no reactivity of limulin with isolated lipid A, whereas it reacted with Re LPS.[27] This lectin was reported to be specific for N-acetylneuraminic acid and KDO residues. An antigenically distinct lectin has also been isolated from the horseshoe crab *Carcinoscorpius rotunda cauda* found in the Bay of Bengal.[28] Similarly, this lectin (called carcinoscorpin) binds to sialic acid as well as to KDO. Interestingly, this lectin failed to interact with *Vibrio cholerae* LPS, known to possess a highly substituted KDO. These examples clearly indicate that LPS can interact via specific core residues through specific binding of lectin molecules found in *Limulus* amebocyte lysates.

4. Interaction with Antibiotics

The interaction of LPS with hydrophobic antibiotics correlates directly with the length of the polysaccharidic chain from Re LPS to S-type LPS.[29] Polymyxin B is the best-known antibiotic that interacts with LPS molecules. The efficiency of polymyxin B to inhibit endotoxin-induced biological activities greatly depends upon the origin of LPS.[30] Morrison and Jacobs[31] established that polymyxin B interacts with the lipid A-KDO region of the LPS molecule. Little is known about the exact role of KDO or other core residues in the interaction of LPS with polymyxin B. We observed that polymyxin B did not impair the *in vitro* biological activities of the isolated core oligosaccharide.[4] It has been postulated that polymyxin B binds relatively nonselectively to both KDO and/or phosphate groups of LPS.[5] The substitution of an acidic phosphate group in the lipid A was postulated to account for the decreased binding of polymyxin B to LPS isolated from a polymyxin B-resistant mutant of *S. typhimurium*.[32] In addition, it was proposed that loss of hydroxy fatty acid contributed to the polymyxin B resistance.[33] That the KDO-lipid A complex is essential for the structural and functional integrity of the Gram-negative bacteria provided the basis for the development of new antibiotics; these antibiotics are able to interfere with the KDO-cytidylytransferase, an enzyme which converts KDO to cytidine-5'-monophosphate-KDO, the substrate for a series of enzymes able to incorporate KDO into the LPS.[34-36]

B. ACTIVATION OF MAMMALIAN CELLS

Endotoxin can either interact specifically with the cell membrane through different specific receptor(s) (see Section IV) or nonspecifically via lipid-lipid interactions. The nature of these different binding sites on the cell membrane for LPS may generate different signals of activation.

1. Activation of Polymorphonuclear (PMN) Cells

Activation of neutrophils is a good example of how initiation of different cellular functions may require different biochemical structures within the LPS molecule. Lipid A was far more active than any other form of LPS in triggering chemiluminescence[37,38] and in inhibiting random migration[38] (Rc and Rd LPS being the less active, Ra and S-type LPS exerting similar activity). In contrast, chemotaxis experiments in the presence of heat-inactivated serum indicated that lipid A and S-type and Re LPS were devoid of any activity whereas Ra, Rb, Rc, and Rd LPS initiated granulocyte migration.[38] Further, it was reported that lipid A, Rb LPS, and Re LPS all had similar capacities to trigger the release of leukotrienes LTD4 and LTC4 from PMN cells.[39,40] It is noteworthy that, in contrast to PMN responses to LPS, LTC4 production by macrophages was mainly triggered by rough LPS and that isolated natural or synthetic lipid A had a very limited activity.[41] This difference in the induction of leukotriene production by neutrophils and macrophages may reflect the presence of a different binding site on both cell types.

2. Activation of B Lymphocytes

Purified LPS and isolated lipid A are well-known polyclonal activators of B lymphocytes, leading to cell proliferation (mitogenic response) and to differentiation and immunoglobulin secretion. Purified lipid A-free polysaccharides derived from *Bordetella pertussis, Haemophilus influenzae,* and *Bacteroides fragilis* LPS have been shown to be mitogenic and capable of triggering immunoglobulin synthesis.[42-44] Williamson et al.[44] demonstrated that the polysaccharide fraction from *B. fragilis* LPS was active, whereas that obtained from smooth *E. coli* LPS was a poor B-cell activator. The main difference between both polysaccharides is that the former is only composed of a core oligosaccharide, whereas the latter is a more complex structure with repetitive units of O-specific polysaccharides. Similarly, isolated *Bordetella pertussis* and *H. influenzae* polysaccharides, which have been reported to be active on B cells,[42,43] are composed of unique core oligosaccharides. It is generally agreed that S-type *S. minnesota* LPS are less active than rough mutants.[45,46] Contradictory results were reported in studies comparing rough mutants LPS with lipid A extracted from *S. minnesota*: Fleebe et al.[45] illustrated that the activity of rough mutants (1 μg/ml) was higher than that of free lipid A, whereas Tanamoto et al.[46] quoted similar activities. Whether this discrepancy reflects different solubilization procedures is unknown. We have observed that the relative mitogenic activity induced by *E. coli* Re and S LPS and by isolated lipid A was in the order Re LPS > S LPS = lipid A.[47] Futher, it has been reported that, on a weight basis, *Bacteroides fragilis* rough LPS was more active than the lipid A derived therefrom in triggering the proliferation of mouse spleen cells.[44] Polyclonal activation of B cells induced by LPS also leads to the production of auto-antibodies. Bloembergen et al.[26] observed that, on a weight basis, *S. typhimurium* Re LPS was far more active than the homologous lipid A in inducing anti-hemolysin- and anti-DNA-secreting spleen cells.

The low endotoxic responder C3H/HeJ mouse strain model has been widely used as a tool for better understanding the mechanisms of LPS activation. Of interest, some LPS preparations have been reported to be able to trigger the proliferation of spleen B cells from C3H/HeJ mice (e.g., *Brucella abortus, Pseudomonas aeruginosa, Bordetella pertussis, Bacteroides fragilis*).[44,48-51] Although Moreno and Berman[48] showed that smooth *Brucella abortus* LPS was more active than rough LPS in low endotoxin response mice, most investigators have found that the reactivity of B cells was induced only by LPS which lack the O-specific polysaccharide, such as *Bordetella pertussis*[50] or *Bacteroides fragilis*[44,51] LPS. Flebbe et al.[45] established that the refractory state of C3H/HeJ B cells to mitogenic stimulation by *Salmonella* and *E. coli* LPS did not extend to R-chemotype LPS. The mitogenic activity of LPS was dependent upon a structural requirement involving an intact KDO linkage to lipid A within the core oligosaccharide.

Another approach using synthetic subunits of lipid A covalently coupled to KDO has failed to resolve the role of KDO in mitogenic activity of LPS.

According to Shimizu et al.,[52] the addition of KDO enhanced the mitogenic activity of 2,3-diacyloxyacylglucosamine-4-phosphate. In contrast, Kumazawa et al.[53] reported that addition of KDO residues modified neither the mitogenicity, the adjuvanticity, nor the tumor necrosis factor (TNF)-inducing capacities of 4-O-phosphoro-D-glucosamine derivatives carrying N- and 3-O-acyl substituents. Additional studies will (hopefully) resolve these differences.

3. Activation of Macrophages

Previous experiments in our laboratory established that the B cell activation by LPS and its isolated components is macrophage dependent.[42] The role of macrophages in B cell activation by endotoxins has been confirmed macrophages. While it has been established that the lipid A component can trigger monocyte/macrophage activation, several reports have implied that residues within the core oligosaccharide modulate the activities of LPS and are directly responsible for some activation processes. During the last decade, much information has emerged from studies on cytokine production, mainly on interleukin-1 (IL-1), interleukin-6 (IL-6), and tumor necrosis factor alpha (TNFα) produced by mouse macrophages and human monocytes (see Volume 2). The potency of LPS to induce the release of cytokines by mononuclear phagocytes depends on the LPS bacterial origin,[56-58] at least in part reflecting the biochemical diversity among LPS from various bacterial species. Similar findings have been published for IL-6 production by human fibroblasts triggered by endotoxins.[59] Interestingly, *in vivo* LPS-induced activities often parallel the capacities of the different LPS to induce *in vitro* cytokine production. Thus, the potency of IL-1-inducing properties of different *Klebsiella* LPS correlates well with their *in vivo* adjuvanticity.[60] Loppnow et al.[61] documented that *in vivo* toxicity parallels the ability of different LPS to stimulate IL-1, IL-6, and TNFα production *in vitro*. As far as TNFα and IL-1 production are concerned, comparisons between S-type and R LPS have led to contradictory results. Studying human monocytes, Männel and Falk[62] indicated that optimal cytokine production was achieved with S LPS rather than with R LPS. In contrast, Chen et al.[63] described similar activities induced by S-form and R LPS. Further, Newton[57] claimed that *S. minnesota* Re LPS was the most potent activator of IL-1 production among the different tested LPS, and Feist et al.[64] provided evidence that Rb2 LPS was more active than Re and S LPS in inducing TNFα. In experiments using mouse peritoneal macrophages, Flebbe et al.[65] showed that Ra, Rb, Rc, and Rd LPS were far more potent than S-type LPS in inducing IL-1 and TNFα production. Similarly, in an investigation of TNFα production by rabbit macrophages, Mathison et al.[66] pointed out that the R LPS was maximally active at concentrations 10- to 100-fold less than S-type LPS.

Production of cytokines by monocytes/macrophages stimulated with purified natural or synthetic lipid A remains also a controversial issue. Using human peripheral blood mononuclear cells, Kovach et al.[67] observed that Re and Rc LPS were 11- and 18-fold more efficient, respectively, than lipid A in inducing TNFα production. According to Loppnow et al.[68] synthetic lipid A was mainly active on mononuclear cells, whereas purified mononuclear phagocytes were poorly activated. In our studies using human monocytes isolated by adherence, we clearly demonstrated that natural purified lipid A or synthetic lipid A were very poor inducers of IL-1 secretion, despite the fact that both compounds induced intracellular IL-1 accumulation.[69] The relative low activity of synthetic *E. coli* lipid A (compound 506) compared to *E. coli* Re LPS is illustrated in Figure 2, which shows the total amounts of IL-1α and IL-1β (i.e., intracellular and released IL-1) produced by human monocytes *in vitro* in response to these stimuli. Similar findings were obtained when we compared natural *Neisseria meningitidis* LPS and natural homologous lipid A.[47] Our findings were confirmed by Hurme and Serkkola,[70] who pointed out further that synthetic lipid A induced IL-1 mRNA transcription and IL-1 translation but not IL-1 secretion; however, Saiki et al.[71] reported that synthetic *E. coli* lipid A was perfectly active on human monocytes.

In vivo, both LPS and lipid A are potent inducers of macrophage class II major histocompatibility complex (MHC) antigen expression.[72] It is worth noting that neither the biosynthesis nor the expression of class II MHC antigens was increased by direct exposure of macrophages to LPS or lipid A. Rather, the available evidence has implied the requisite participation of accessory cells (T lymphocytes, NK cells) together with the release of mediators including γ-interferon.[72,73] These findings underscore the concept that the results of *in vivo* or *in vitro* experiments can be misinterpreted when using a mixture of cell populations, as this may allow cascade events not reflecting a direct interaction between LPS and the target cell being studied. It may also explain the apparently conflicting results from our group, indicating a lack of IL-1 release by lipid A-stimulated purified monocytes, compared with those reported by Loppnow et al.[68] who did report IL-1 release by human peripheral blood mononuclear leukocytes stimulated by synthetic lipid A.

Furthermore, in *in vitro* studies, *E. coli* Re LPS was significantly more potent than its homologous synthetic lipid A in inducing the release of IL-6 and TNFα from human monocytes (Figure 2). With respect to our findings on TNFα production, it is tempting to compare our data with those reported by Tanamoto et al.,[74] who found that, in an *in vivo* model, *Pseudomonas aeruginosa* lipid A manifested no anti-tumor activity, whereas both homologous LPS and Re *S. typhimurium* LPS were active. The authors pointed out that chemical modifications of the polysaccharide region or deacylation of the LPS diminished antitumor activity. In a study of mouse peritoneal macrophages, both Lasfargues et al.[75] and we[76] found that lipid A was a weak inducer of IL-1 release, although elicited peritoneal macrophages were more reactive to lipid A than resident macrophages. C3H/HeJ macrophages exhibited both cell-associated and membrane IL-1 upon LPS activation[76] and IL-1 release if Ra LPS was employed,[65] whereas they were not stimulated by lipid A. In a recent review of this subject[77] we compiled our results pointing

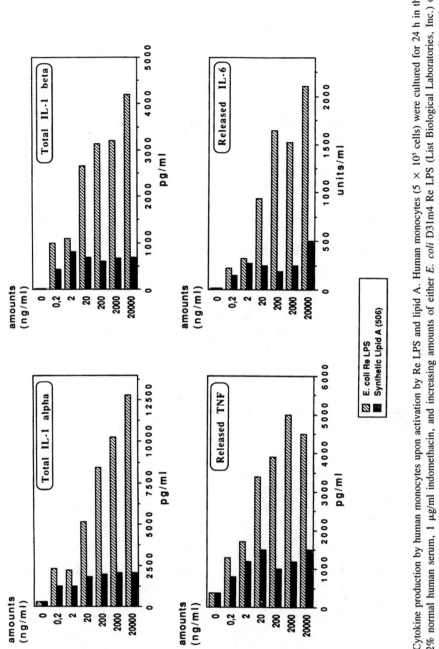

FIGURE 2. Cytokine production by human monocytes upon activation by Re LPS and lipid A. Human monocytes (5 × 10⁵ cells) were cultured for 24 h in the presence of 0.2% normal human serum, 1 μg/ml indomethacin, and increasing amounts of either *E. coli* D31m4 Re LPS (List Biological Laboratories, Inc.) or synthetic *E. coli* lipid A (compound 506, Daiichi, Japan). Cell-associated and released IL-1α was assessed by enzyme-linked immunosorbent assay, cell-associated and released IL-1β and released TNFα were assessed by radioimmunoassay, and released IL-6 was assessed by a biological assay using the 7TD1 cell line.

out that the inner core, and particularly the KDO residue(s), plays a major role within the LPS molecule in triggering the release of IL-1 by human monocytes and mouse macrophages. In addition to experiments demonstrating that LPS is far more active than lipid A in inducing cytokine release,[58,69,76] we established that

1. The isolated core oligosaccharides from *Bordetella pertussis* and *N. meningitidis* endotoxins were able to induce IL-1 production[78,79]
2. A pentasaccharide, isolated from the core of *B. pertussis* LPS, with a terminal methyl ketoside KDO residue, was similarly active to the whole core[4,80]
3. The trisaccharide heptose-heptose-KDO isolated from *S. minnesota* Rd1 mutant was able to induce IL-1 release,[80] although high concentrations of the trisaccharide were required[68,80]
4. The bioactivity of endotoxin correlated with the thiobarbiturate response (specific for KDO), reflecting the degree of substitution of the KDO molecule[4,25]
5. The carboxyl group of the KDO should be unsubstituted in order to manifest biological activity on monocytes[4]

According to our results and those reported by others, uncleaved LPS is the most potent inducer of IL-1 release when compared to fragments derived therefrom. In order to induce IL-1 release, lipid A must be linked to a KDO residue. One could postulate that similar requirements are necessary for IL-6 and TNFα release. In agreement with our conclusion, Lüderitz et al.[41] have recently demonstrated that natural and synthetic lipid A were far less active than Re LPS in inducing the release of LTC4 by mouse peritoneal macrophages. That, in contrast, lipid A and Re LPS were equally potent in inducing LTC4 release by PMN cells[39,40] suggests that probably there exist different binding molecules that are being triggered on neutrophils and monocytes. Interestingly, certain macrophage activities are similarly induced by LPS and lipid A, indicating that oligosaccharides are not always involved in the conformational requirement of lipid A activation. Optimal concentrations of LPS and lipid A for protein phosphorylation are very similar.[81] LPS and lipid A exhibit similar abilities to stimulate the conversion of arginine into ornithine by macrophages[82] or to initiate enhanced hydrolysis of phosphatidylinositol-4,5-bisphosphate.[83] Lipid A was even more active than LPS in inducing elevated intracellular levels of Ca^{2+}.[83] These data suggest that different parts of the LPS model may be involved in the interaction with the monocyte/macrophage membrane and induce different functions depending upon the receptor triggered.

TABLE 1
Specific Binding of LPS to Mouse Macrophages Requires the
Presence of Oligosaccharide Residues

Competing agent (3 µg)	Percent inhibition of [³H]-LPS binding (mean of three individual experiments ± SD)
Neisseria meningitidis LPS	61 ± 8
Escherichia coli S LPS	47 ± 11
E. coli Re LPS	39 ± 4
Natural *N. meningitidis* lipid A	8 ± 5
Natural *E. coli* lipid A	18 ± 5
Synthetic (506) *E. coli* lipid A	8 ± 7

Note: Similar results were obtained with human monocytes. *N. meningitidis* [³H]-LPS
(0.1 µg) was incubated in RPMI medium supplemented with 4% normal human
serum for 1 h at 22°C with Balb/c mouse elicited peritoneal macrophages either
alone or in the presence of an excess of competing agents.

IV. CELLULAR RECEPTORS FOR THE CORE

Once encapsulated within liposomes, LPS is far less potent than free LPS
in stimulating macrophage functions and cytokine release.[84-87] This observation established that maximum activation of macrophages is only achieved
if the LPS molecule is able to interact with the cell membrane rather than be
delivered within the cell via mechanisms which avoid an interaction with the
membrane. Well-defined LPS receptors which implicate the lipid A portion
of the molecule have been demonstrated on macrophages and other cells (see
Chapters 12 and 13). However, several reports implicate other binding sites
involving an interaction with the core region. Thus, Parent[88] identified a
47-kDa glycoprotein present on the surface of hepatocytes, which bound
smooth and rough LPS. Its interaction with LPS can be inhibited by S and
R LPS and heptose residues, but neither by free KDO residues nor by isolated
lipid A. These data suggest strongly the existence of an inner core heptose-
specific receptor. Human monocytes and mouse and rabbit peritoneal macrophages may also bind LPS through a lectin-like receptor. The specific
binding of radiolabeled *N. meningitidis* or *B. pertussis* LPS could be inhibited
by various LPS,[89] by isolated core polysaccharide,[47,90] though not by lipid
A. Although S-type LPS as well as R LPS are able to compete with radiolabeled *N. meningitidis* LPS, natural or synthetic lipid A were inefficient
(Table 1). This specific binding required the presence of serum[89] and cations
(Table 2). Tahri-Jouti and Chaby,[91] who recently identified a lipid A binding
site on mouse macrophages, also indicated similar requirements for LPS-
specific interaction with the cell surface. Compounds that inhibit the specific
interaction of LPS with its lectin-like receptor (e.g., gangliosides, lipoproteins) prevented IL-1 production by monocytes/macrophages.[92-94] A third type
of interaction with human monocytes has been illustrated by Warner et al.,[95]

TABLE 2
Specific Binding of LPS to Human Monocytes
Requires the Presence of Cations

Concentrations of Ca^{2+} and Mg^{2+} added to Ca^{2+}/Mg^{2+}-free Hanks medium (mmol/l)	Specifically bound [³H]-LPS (Δ cpm)
0	787
0.03	739
0.06	768
0.12	1365
0.25	4692
0.5[a]	6677
0.75	7837
1.0	6461

Note: Binding experiments were performed with *N. meningitidis* [³H]-LPS (0.1 μg) in the absence or presence of unlabeled *N. meningitidis* LPS (4 μg) using Hanks medium supplemented with different concentrations of cations.

[a] Equivalent to concentrations of calcium and magnesium present in RPMI medium.

who observed that the O-specific polysaccharide could inhibit the binding of LPS to the cells. T lymphocytes can also bind LPS via the core region.[96] Activated T cells, but not resting T cells, interact with Rb LPS, and isolated Rb and Rc core oligosaccharides can inhibit the generation of cytotoxic T lymphocytes. Finally, it has been suggested that inner core oligosaccharide may enhance the binding of lipid A. Kirikae et al.[97] provided evidence showing that mouse erythrocytes can be hemagglutinated by LPS, especially the R form: the lipid A played the essential role, although the KDO residues also participated since Rc through to Re LPS displayed higher capacities of agglutination than free lipid A.

V. CONCLUSIONS

As stated by Rietschel and colleagues,[98] "the KDO containing inner core and lipid A should not be viewed as separate regions but as one molecular entity." In this respect, it is important to recognize that the minimal LPS structure required for bacterial survival resides in the KDO-lipid A domain. We have pointed out that some of the biological activities claimed to be solely due to the lipid A moiety are significantly enhanced by the presence of the inner core residues. Whether this reflects an adaptive response of the host organism to the most conserved part of the molecule is not known. The probability that, *in vivo*, during bacterial infection free lipid A could be encountered is extremely low compared to the R and S forms of LPS. Because of the continuing synthesis of LPS by the bacteria, it is well known that even so-called smooth bacteria contain significant amounts of R-chemotype LPS. The exact nature of the endotoxins present *in vivo* during infection remains to be fully established. It is likely that, *in vivo* free LPS is rarely, if ever, encountered by host cells. LPS, via its lipid A region, binds to various serum proteins (LBP, lipoproteins, etc.) and lipid A is probably often hidden from the target cells. This would suggest that polysaccharide-dependent interaction of the core region with lectin-like receptors has important *in vivo* relevance. Very probably, the influence of the core oligosaccharide on the tertiary and/or quaternary conformation of the LPS, the presence of charged residues, and the possible interaction with lectin-like ligands are, either individually or in association, responsible for strengthening the interaction of LPS with its different target molecules.

ACKNOWLEDGMENTS

The authors thank Martine Caroff, Catherine Fitting, Carlos Munoz, and Christine Couturier for their contributions.

REFERENCES

1. **Brandenburg, K. and Seydel, U.**, Investigation into the fluidity of lipopolysaccharide and free lipid A membrane systems by Fourier-transform infrared spectroscopy and differential scanning calorimetry, *Eur. J. Biochem.*, 191, 229, 1990.
2. **Brade, H.**, Occurrence of 2-keto-deoxyoctonic acid 5-phosphate in lipopolysaccharide of *Vibrio cholerae* ogawa and inaba, *J. Bacteriol.*, 161, 795, 1985.
3. **Caroff, M., Lebbar, S., and Szabo, L.**, Detection of 3-deoxy-2-octulosonic acid in thiobarbiturate negative endotoxins, *Carbohydr. Res.*, 161, c4, 1987.
4. **Haeffner-Cavaillon, N., Caroff, M., and Cavaillon, J. M.**, Interleukin-1 induction by lipopolysaccharides: structural requirements of the 3-deoxy-D-manno-2-octulosonic acid (KDO), *Mol. Immunol.*, 26, 485, 1989.
5. **Schlinder, M. and Osborn, M. J.**, Interaction of divalent cations and polymyxin B with lipopolysaccharide, *Biochemistry*, 18, 4425, 1979.
6. **Galanos, C. and Lüderitz, O.**, Lipopolysaccharide: properties of an amphipathic molecule, in *Handbook of Endotoxin*, Vol. 1, Rietschel, E. T., Ed., Elsevier Science, Amsterdam, 1984, 46.
7. **Rietschel, E. T., Brade, L., Shade, U., Seydel, U., Zähringer, U., Kusumoto, S., and Brade, H.**, Bacterial endotoxins: properties and structure of biologically active domains, in *Surface Structures of Microorganisms and their Interactions with the Mammalian Host*, Vol. 18, Schrinner, E., Richmond, M. H., Seibert, G., and Schwarz, Eds., VCH, Weinheim, Germany, 1988, 1.
8. **Dierich, M. P., Bitter-Suermann, D., Köning, W., Hadding, U., Galanos, C., and Rietschel, E. T.**, Analysis of bypass activation of C3 by endotoxic LPS and loss of this potency, *Immunology*, 24, 721, 1973.
9. **Morrison, D. C. and Kline, L. F.**, Activation of the classical and properdin pathways of complement by bacterial lipopolysaccharides (LPS), *J. Immunol.*, 118, 362, 1977.

10. **Grossman, N. and Leive, L.**, Complement activation via the alternative pathway by purified salmonella lipopolysaccharide is affected by its structure but not its O-antigen length, *J. Immunol.*, 132, 376, 1984.

11. **Yokochi, T., Inoue, Y., Kimura, Y., and Kato, N.**, Strong interaction of lipopolysaccharide possessing the mannose homopolysaccharides with complement and its relation to adjuvant action, *J. Immunol.*, 144, 3106, 1990.

12. **Tenner, A. J., Ziccardi, R. J., and Cooper, N. R.**, Antibody-independent C1 activation by *E. coli, J. Immunol.*, 133, 886, 1984.

13. **Vukajlovich, S. W.**, Antibody-independent activation of the classical pathway of human serum complement by lipid A is restricted to Re-chemotype lipopolysaccharide and purified lipid A, *Infect. Immun.*, 53, 480, 1986.

14. **Clas, F., Schmidt, G., and Loos, M.**, The role of the classical pathway for the bactericidal effect of normal sera against gram-negative bacteria, in *Current Topics in Microbiology and Immunology*, Vol. 121, Springer-Verlag, Berlin, 1985, 19.

15. **Vukajlovich, S. W., Hoffman, J., and Morrison, D. C.**, Activation of human serum complement by bacterial lipopolysaccharides: structural requirements for antibody independent activation of the classical and alternative pathways, *Mol. Immunol.*, 24, 319, 1987.

16. **Tesh, V. L., Vukajlovich, S. W., and Morrison, D. C.**, Endotoxin interactions with serum proteins relationship to biological activity, in *Bacterial Endotoxins: Pathophysiological Effects, Clinical Significance and Pharmacological Control*, Alan R. Liss, New York, 1988, 47.

17. **Ulevitch, R. J. and Tobias, P. S.**, Interactions of bacterial lipopolysaccharides with serum proteins, in *Bacterial Endotoxins: Pathophysiological Effects, Clinical Significance and Pharmacological Control*, Alan R. Liss, New York, 1988, 309.

18. **Berger, D. and Beger, H. G.**, Quantification of the endotoxin-binding capacity of human transferrin, in *Bacterial Endotoxins: Pathophysiological Effects, Clinical Significance and Pharmacological Control*, Alan R. Liss, New York, 1988, 115.

19. **Ihara, I., Ueda, H., Suzuki, A., and Kawakami, M.**, Physicochemical properties of a new bactericidal factor, Ra-reactive factor, *Biochem. Biophys. Res. Commun.*, 107, 1185, 1982.

20. **Ihara, I., Harada, Y., Ihara, S., and Kawakami, M.**, A new complement dependent bactericidal factor found in nonimmune mouse sera: specific binding to polysaccharide of Ra chemotype salmonella, *J. Immunol.*, 128, 1256, 1982.

21. **Kawakami, M., Ihara, I., Suzuki, A., and Harada, Y.**, Properties of a new complement-dependent bactericidal factor specific for Ra chemotype salmonella in sera of conventional and germ-free mice, *J. Immunol.*, 129, 2198, 1982.

22. **Brade, L. and Brade, H.**, A 28,000 dalton protein of normal mouse serum binds specifically to the inner core region of bacterial lipopolysaccharide, *Infect. Immun.*, 50, 687, 1985.

23. **Schumann, R. R., Leong, S. R., Flaggs, G. W., Gray, P. W., Wright, S. D., Mathison, J. C., Tobias, P. S., and Ulevitch, R. J.**, Structure and function of lipopolysaccharide binding protein, *Science*, 249, 1429, 1990.

24. **Tobias, P. S., Soldau, K., and Ulevitch, R. J.**, Identification of a lipid A binding site in the acute phase reactant lipopolysaccharide binding protein, *J. Biol. Chem.*, 264, 10867, 1989.

25. **Laude-Sharp, M., Haeffner-Cavaillon, N., Caroff, M., Lantreibecq, F., Pusineri, C., and Kazatchkine, M. D.**, Dissociation between the interleukin 1-inducing capacity and limulus reactivity of lipopolysaccharides from gram-negative bacteria, *Cytokine*, 2, 253, 1990.

26. **Bloembergen, P., Hofhuis, F. M. A., Hol, C., and van Dijk, H.**, Endotoxin induced auto-immunity in mice. III. Comparison of different endotoxin preparations, *Int. Arch. Allergy Immunol.*, 92, 124, 1990.

27. **Rostam-Abadi, H. and Pistole, T. G.**, Lipopolysaccharide-binding lectin from the horseshoe crab, *Limulus polyphemus*, with specificity for 2-keto-3-deoxyoctonate (KDO), *Dev. Comp. Immunol.*, 6, 209, 1982.

28. **Dorai, D. T., Srimal, S., Mohan, S., Bachhawat, B. X., and Balganesh, T. S.**, Recognition for 2-keto-3-deoxyoctonate in bacterial cells and lipopolysaccharides by the sialic acid binding lectin from the horseshoe crab *Carcinoscorpius rotunda cauda*, *Biochem. Biophys. Res. Commun.*, 104, 141, 1982.

29. **Schlect, S. and Schmidt, G.**, Möglichkeiten zur Differenzierung von Salmonella-R-Formen mittels Antibiotica und Antibakterieller Farbstoffe, *Zentralbl. Bakteriol. Mikrobiol. Hyg. A*, 212, 505, 1969.

30. **Cavaillon, J. M. and Haeffner-Cavaillon, N.**, Polymyxin-B inhibition of LPS-induced interleukin-1 secretion by human monocytes is dependent upon the LPS origin, *Mol. Immunol.*, 23, 965, 1986.

31. **Morrison, D. C. and Jacobs, D. M.**, Binding of polymyxin B to the lipid A portion of bacterial lipopolysaccharides, *Immunochemistry*, 13, 813, 1976.

32. **Vaara, M., Vaara, T., Jensen, M., Helander, I., Nurminen, M., Rietschel, E. Th., and Mäkela, P. H.**, Characterization of the lipopolysaccharide from the polymyxin-resistant pmrA mutants of *Salmonella typhimurium*, *FEBS Lett.*, 129, 145, 1981.

33. **Conrad, R. S. and Galanos, C.**, Fatty acid alterations and polymyxin B binding by lipopolysaccharides from *Pseudomonas aeruginosa* adapted to polymyxin B resistance, *Antimicrob. Agents Chemother.*, 33, 1724, 1989.

34. **Hammond, S. M., Claesson, A., Jansson, A. M., Larsson, L.-G., Pring, B. G., Town, C. M., and Ekström, B.**, A new class of synthetic antibacterials acting on lipopolysaccharide biosynthesis, *Nature*, 327, 730, 1987.

35. **Goldman, R., Kohlbrenner, W., Lartey, P., and Pernet, A.**, Antibacterial agents specifically inhibiting lipopolysaccharide synthesis, *Nature*, 329, 162, 1987.

36. **Goldman, R. C., Doran, C., and Capobianco, J. O.**, Antibacterial agents which specifically inhibit lipopolysaccharide synthesis, in *Cellular and Molecular Aspects of Endotoxin Reactions*, Nowotny, A., Spitzer, J. J., and Ziegler, E. J., Eds., Elsevier Science, Amsterdam, 1990, 157.

37. **Kapp, A., Freudenberg, M., and Galanos, C.**, Induction of human granulocyte chemiluminescence by bacterial lipopolysaccharides, *Infect. Immun.*, 55, 758, 1987.

38. **Pugliese, C., LaSalle, M. D., and DeBari, V. A.**, Relationships between the structure and function of lipopolysaccharide chemotypes with regard to their effects on the human polymorphonuclear neutrophil, *Mol. Immunol.*, 25, 631, 1988.

39. **Bremm, K. D., Köning, W., Spur, K. B., Crea, A., and Galanos, C.**, Generation of slow-reacting substance (leukotrienes) by endotoxin and lipid A from human polymorphonuclear granulocytes, *Immunology*, 53, 299, 1984.

40. **Bremm, K. D., Köning, W., Thelestam, M., and Alouf, J. E.**, Modulation of granulocyte functions by bacterial endotoxin and endotoxins, *Immunology*, 62, 363, 1987.

41. **Lüderitz, T., Brandenburg, K., Seydel, U., Roth, A., Galanos, C., and Rietschel, E. T.**, Structural and physicochemical requirements of endotoxin for the activation of arachidonic acid metabolism in mouse peritoneal macrophages *in vitro*, *Eur. J. Biochem.*, 179, 11, 1989.

42. **Haeffner-Cavaillon, N., Cavaillon, J. M., and Szabo, L.**, Macrophage dependent polyclonal activation of splenocytes by *Bordetella pertussis* endotoxin and its isolated polysaccharide and lipid A regions, *Cell. Immunol.*, 74, 1, 1982.

43. **Guenounou, M., Raichvarg, D., Hatat, D., Brossard, C., and Agneray, J.**, *In vitro* immunological activities of the polysaccharide fraction from *Haemophilus influenzae* type a endotoxin, *Infect. Immun.*, 36, 603, 1982.

44. **Williamson, S. I., Wannemuehler, M. J., Jirillo, E., Pritchard, D. G., Michalek, S. M., and McGhee, J. R.**, LPS regulation of the immune response: separate mechanisms for murine B cell activation by lipid A (direct) and polysaccharide (macrophage-dependent) derived from *Bacteroides* LPS, *J. Immunol.*, 133, 2294, 1984.

45. **Flebbe, L., Vukajlovich, S. W., and Morrison, D. C.,** Immunostimulation of C3H/HeJ lymphoid cells by R-chemotype lipopolysaccharide preparations, *J. Immunol.*, 142, 642, 1989.

46. **Tanamoto, K.-I., Galanos, C., Lüderitz, O., Kusumoto, S., and Shiba, T.,** Mitogenic activities of synthetic lipid A analogs and suppression of mitogenicity of lipid A, *Infect. Immun.*, 44, 427, 1984.

47. **Cavaillon, J. M., Munoz, C., Fitting, C., Couturier, C., and Haeffner-Cavaillon, N.,** Signals involved in interleukin-1 production induced by endotoxins, in *Cellular and Molecular Aspects of Endotoxin Reactions*, Nowotny, A., Spitzer, J. J., and Ziegler, E. J., Eds., Elsevier Science, Amsterdam, 1990, 257.

48. **Moreno, E. and Berman, D. T.,** *Brucella abortus* lipopolysaccharide is mitogenic for spleen cells of endotoxin-resistant C3H/HeJ mice, *J. Immunol.*, 123, 2915, 1979.

49. **Pier, G. B., Markham, R. B., and Eardley, D.,** Correlation of the biologic responses of C3H/HeJ mice to endotoxin with the chemical and structural properties of the lipopolysaccharides from *Pseudomonas aeruginosa* and *Escherichia coli*, *J. Immunol.*, 127, 184, 1981.

50. **Girard, R., Chaby, R., and Bordenave, G.,** Mitogenic response of C3H/HeJ mouse lymphocytes to polyanionic polysaccharides obtained from *Bordetella pertussis* endotoxin and from other bacterial species, *Infect. Immun.*, 31, 122, 1981.

51. **Joiner, K. A., McAdam, K. P. W. J., and Kasper, D. L.,** Lipopolysaccharides from *Bacteroides fragilis* are mitogenic for spleen cells from endotoxin responder and non-responder mice, *Infect. Immun.*, 36, 1139, 1982.

52. **Shimizu, T., Akiyama, S.-I., Masuzawa, T., Yanagihara, Y., Nakamoto, S.-I., and Achiwa, K.,** Biological activities of chemically synthesized 2-keto-3-deoxyoctonic acid-(α2→6)-D-glucosamine analogs of lipid A, *Infect. Immun.*, 55, 2287, 1987.

53. **Kumazawa, Y., Matsurra, M., Homma, J. Y., Furuya, T., Takimoto, H., Inagaki, K., Nagumo, T., Kiso, M., and Hasegawa, A.,** Immunopharmacological activities of 2-keto-3-deoxyoctonic acid-(α2→6)-linked 4-O-phosphono-D-glucosamine derivatives carrying N- and 3-O-acyl substituents, *Infect. Immun.*, 57, 1845, 1989.

54. **Fernandez, C. and Severinson, E.,** The polyclonal lipopolysaccharide response is accessory-cell-dependent, *Scand. J. Immunol.*, 18, 279, 1983.

55. **Corbel, C. and Melchers, F.,** Requirement for macrophages or for macrophage- or T cell-derived factors in the mitogenic stimulation of murine B lymphocytes by lipopolysaccharides, *Eur. J. Biochem.*, 13, 528, 1983.

56. **Hanazawa, S., Nakada, K., Ohmori, Y., Miyoshi, T., Amano, S., and Kitano, S.,** Functional role of interleukin 1 in periodontal disease: induction interleukin 1 production by *Bacteroides gingivalis* lipopolysaccharide in peritoneal macrophages from C3H/HeN and C3H/HeJ mice, *Infect. Immun.*, 50, 262, 1985.

57. **Newton, R. C.,** Human monocyte production of interleukin-1: parameters of the induction of interleukin-1 secretion by lipopolysaccharides, *J. Leuk. Biol.*, 39, 299, 1986.

58. **Cavaillon, J. M. and Haeffner-Cavaillon, N.,** Characterization of the induction of human interleukin 1 by endotoxins, in *Lipid Mediators in the Immunology of Shock*, (NATO ASI Series, Vol. 139), Paubert-Braquet, M., Ed., Plenum Press, New York, 1978, 395.

59. **Hamada, S., Takada, H., Mihara, J., Nakagawa, I., and Fujiwara, T.,** LPS of oral Bacteroides species: general properties and induction of cytokines in human gingival fibroblast cultures, in *Cellular and Molecular Aspects of Endotoxin Reactions*, Nowotny, A., Spitzer, J. J., and Ziegler, E. J., Eds., Elsevier Science, Amsterdam, 1990, 285.

60. **Kido, N., Nakashima, I., and Kato, N.,** Correlation between strong adjuvanticity of Klebsiella 03 lipopolysaccharide and its ability to induce interleukin-1 secretion, *Cell. Immunol.*, 85, 477, 1984.

61. **Loppnow, H., Libby, P., Freudenberg, M., Krauss, J. H., Weckesser, J., and Mayer, H.,** Cytokine induction by lipopolysaccharide (LPS) corresponds to lethal toxicity and is inhibited by nontoxic *Rhodobacter capsulatus* LPS, *Infect. Immun.*, 58, 3743, 1990.

62. **Männel, D. N. and Falk, W.**, Optimal induction of tumor necrosis factor production in human monocytes requires complete S-form lipopolysaccharide, *Infect. Immun.*, 57, 1953, 1989.
63. **Chen, A. R., McKinnon, K. P., and Koren, H. S.**, Lipopolysaccharide (LPS) stimulates fresh human monocytes to lyse actinomycin D-treated WEHI-164 target cells via increased secretion of a monokine similar to tumor necrosis factor, *J. Immunol.*, 135, 3978, 1985.
64. **Feist, W., Ulmer, A. J., Musehold, J., Brade, H., Kusumoto, S., and Flad, H.-D.**, Induction of tumor necrosis factor-alpha release by lipopolysaccharide and defined lipopolysaccharide partial structures, *Immunobiology*, 179, 293, 1989.
65. **Flebbe, L. M., Chapes, S. K., and Morrison, D. C.**, Activation of C3H/HeJ macrophage tumoricidal activity and cytokine release by R-chemotype lipopolysaccharide preparations. Differential effects of IFN-γ, *J. Immunol.*, 145, 1505, 1990.
66. **Mathison, J. C., Virca, G. D., Wolfson, E., Tobias, P. S., Glaser, K., and Ulevitch, R. J.**, Adaptation to bacterial lipopolysaccharide controls lipopolysaccharide-induced tumor necrosis factor production in rabbit macrophages, *J. Clin. Invest.*, 85, 1108, 1990.
67. **Kovach, N. L., Yee, E., Munford, R. S., Raetz, C. R. H., and Harlan, J. M.**, Lipid IV$_A$ inhibits synthesis and release of tumor necrosis factor induced by lipopolysaccharide in human whole blood *ex vivo*, *J. Exp. Med.*, 172, 77, 1990.
68. **Loppnow, H., Brade, H., Dürrbaum, I., Dinarello, C. A., Kusumoto, S., Rietschel, E. T., and Flad, H.-D.**, IL-1 induction-capacity of defined lipopolysaccharide partial structures, *J. Immunol.*, 142, 3229, 1989.
69. **Caroff, M., Cavaillon, J. M., Fitting, C., and Haeffner-Cavaillon, N.**, Inability of pyrogenic, purified *Bordetella pertussis* lipid A to induce interleukin-1 release by human monocytes, *Infect. Immun.*, 54, 465, 1986.
70. **Hurme, M. and Serkkola, E.**, Comparison of interleukin 1 release and interleukin 1 mRNA expression of human monocytes activated by bacterial lipopolysaccharide or synthetic lipid A, *Scand. J. Immunol.*, 30, 259, 1989.
71. **Saiki, I., Maeda, H., Murata, J., Takahashi, T., Sekiguchi, S., Kiso, M., Hasegawa, A., and Azuma, I.**, Production of interleukin 1 from human monocytes stimulated by synthetic lipid A subunit analogues, *Int. J. Immunopharmacol.*, 12, 297, 1990.
72. **Wentworth, P. A. and Ziegler, H. K.**, Modulation of macrophage Ia expression by lipopolysaccharide: stem cell requirements, accessory lymphocyte involvement, and IA-inducing factor production, *Infect. Immun.*, 57, 2028, 1989.
73. **Marshall, N. E., Skeen, M. J., and Ziegler, H. K.**, Regulation of macrophage Ia expression by bacterial endotoxins, in *Cellular and Molecular Aspects of Endotoxin Reactions*, Nowotny, A., Spitzer, J. J., and Ziegler, E. J., Eds., Elsevier Science, Amsterdam, 1990, 215.
74. **Tanamoto, K.-I., Abe, C., Homma, J. Y., and Kojima, Y.**, Regions of lipopolysaccharide of *Pseudomonas aeruginosa* essential for antitumor and interferon-inducing activities, *Eur. J. Biochem.*, 97, 623, 1979.
75. **Lasfargues, A., Ledur, A., Charon, D., Szabo, L., and Chaby, R.**, Induction by lipopolysaccharide of intracellular and extracellular interleukin 1 production: analysis with synthetic models, *J. Immunol.*, 139, 429, 1987.
76. **Cavaillon, J. M., Fitting, C., Caroff, M., and Haeffner-Cavaillon, N.**, Dissociation of cell-associated interleukin-1 (IL-1) and IL-1 release induced by lipopolysaccharide and lipid A, *Infect. Immun.*, 57, 791, 1989.
77. **Cavaillon, J. M. and Haeffner-Cavaillon, N.**, Signals involved in interleukin 1 synthesis and release by lipopolysaccharide-stimulated monocytes/macrophages, *Cytokine*, 2, 313, 1990.
78. **Haeffner-Cavaillon, N., Cavaillon, J. M., Moreau, M., and Szabo, L.**, Interleukin 1 secretion by human monocytes stimulated by the isolated polysaccharide region of the *Bordetella pertussis* endotoxin, *Mol. Immunol.*, 21, 389, 1984.
79. **Cavaillon, J. M. and Haeffner-Cavaillon, N.**, The role of serum in interleukin 1 production by human monocytes activated by endotoxins and their polysaccharide moieties, *Immunol. Lett.*, 10, 35, 1985.

80. **Lebbar, S., Cavaillon, J. M., Caroff, M., Ledur, A., Brade, H., Sarfati, R., and Haeffner-Cavaillon, N.,** Molecular requirement for interleukin 1 induction by lipopolysaccharide-stimulated human monocytes: involvement of the heptosyl-2-keto 3-deoxyoctulosonate region, *Eur. J. Immunol.,* 16, 87, 1986.

81. **Weiel, J. E., Hamilton, T. A., and Adams, D. O.,** LPS induces altered phosphate labeling of proteins in murine peritoneal macrophages, *J. Immunol.,* 136, 3012, 1986.

82. **Benninghoff, B., Dröge, W., and Lehmann, V.,** The lipopolysaccharide-induced stimulation of peritoneal macrophages involves at least two signal pathways. Partial stimulation by lipid A precursors, *Eur. J. Biochem.,* 179, 589, 1989.

83. **Prpic, V., Weiel, J. E., Somers, S. D., DiGuiseppi, J., Gonias, S. L., Pizzo, S. V., Hamilton, T. A., Herman, B., and Adams, D. O.,** Effects of bacterial lipopolysaccharide on the hydrolysis of phosphatidylinositol-4,5-bisphosphate in murine peritoneal macrophages, *J. Immunol.,* 139, 526, 1987.

84. **Dijkstra, J., Mellors, J. W., Ryan, J. L., and Szoka, F. C.,** Modulation of the biological activity of bacterial endotoxin by incorporation into liposomes, *J. Immunol.,* 138, 2663, 1987.

85. **Bakouche, O., Koff, W. C., Brown, D. C., and Lachman, L. B.,** Interleukin 1 release by human monocytes treated with liposome-encapsulated lipopolysaccharide, *J. Immunol.,* 139, 1120, 1987.

86. **Dijkstra, J., Larrick, J. W., Ryan, J. L., and Szoka, F. C.,** Incorporation of LPS in liposomes diminishes its ability to induce tumoricidal activity and tumor necrosis factor secretion in murine macrophages, *J. Leuk. Biol.,* 43, 436, 1988.

87. **Daemen, T., Veninga, A., Dijkstra, J., and Scherphof, G.,** Differential effects of liposome-incorporation on liver macrophage activating potencies of rough lipopolysaccharide, lipid A, and muramyl dipeptide, *J. Immunol.,* 142, 2469, 1989.

88. **Parent, J. B.,** Membrane receptors on rat hepatocytes for the inner core region of bacterial lipopolysaccharides, *J. Biol. Chem.,* 265, 3455, 1990.

89. **Haeffner-Cavaillon, N., Cavaillon, J. M., Etievant, M., Lebbar, S., and Szabo, L.,** Specific binding of endotoxin to human monocytes and mouse macrophages: serum requirement, *Cell. Immunol.,* 91, 119, 1985.

90. **Haeffner-Cavaillon, N., Chaby, R., Cavaillon, J. M., and Szabo, L.,** Lipopolysaccharide receptor on rabbit peritoneal macrophages. I. Binding characteristics, *J. Immunol.,* 128, 1950, 1982.

91. **Tahri-Jouti, M.-A., and Chaby, R.,** Specific binding of lipopolysaccharides to mouse macrophages. I. Characteristics of the interaction and inefficiency of the polysaccharide region, *Mol. Immunol.,* 27, 751, 1990.

92. **Cavaillon, J. M., Fitting, C., Hauttecoeur, B., and Haeffner-Cavaillon, N.,** Inhibition by gangliosides of the specific binding of lipopolysaccharide (LPS) to human monocytes prevents LPS-induced interleukin-1 production, *Cell. Immunol.,* 106, 293, 1987.

93. **Cavaillon, J. M., Fitting, C., Haeffner-Cavaillon, N., Kirsch, S. J., and Warren, H. S.,** Cytokine response by monocytes and macrophages to free and lipoprotein-bound lipopolysaccharide, *Infect. Immun.,* 58, 2375, 1990.

94. **Haeffner-Cavaillon, N. and Cavaillon, J. M.,** Involvement of the LPS receptor in the induction of interleukin-1 in human monocytes stimulated with endotoxins, *Ann. Inst. Pasteur/Immunol.,* 138, 461, 1987.

95. **Warbner, S. J. C., Savage, N., and Mitchell, D.,** Characteristics of lipopolysaccharide interaction with human peripheral-blood monocytes, *Biochem. J.,* 232, 379, 1985.

96. **Lehmann, V., Streck, H., Minner, I., Krammer, P. H., and Ruschmann, E.,** Selection of bacterial mutants from *Salmonella* specifically recognizing determinants on the cell surface of activated T lymphocytes. A novel system to define cell surface structures, *Eur. J. Immunol.,* 10, 685, 1980.

97. **Kirikae, T., Inada, K., Hirata, M., Yoshida, M., Galanos, C., and Lüderitz, O.,** Hemagglutination induced by lipopolysaccharides and lipid A, *Microbiol. Immunol.,* 30, 269, 1986.

98. **Rietschel, E. T., Brade, L., Holst, O., Kulshin, V. A., Lindner, B., Moran, A. P., Schade, U. F., Zähringer, U., and Brade, H.,** Molecular structure of bacterial endotoxin in relation to bioactivity, in *Cellular and Molecular Aspects of Endotoxin Reactions,* Nowotny, A., Spitzer, J. J., and Ziegler, E. J., Eds., Elsevier Science, Amsterdam, 1990, 15.

Chapter 9

SUPRAMOLECULAR STRUCTURE OF LIPOPOLYSACCHARIDES AND LIPID A

Ulrich Seydel and Klaus Brandenburg

TABLE OF CONTENTS

I. INTRODUCTION

Lipopolysaccharides (LPS) are the major amphiphilic molecules expressed at the surface of Gram-negative bacteria. They are anchored in the membrane by their covalently linked amphiphilic lipid component, lipid A, which is known to harbor the endotoxic principle of LPS.[1] Amphiphilic molecules, in general, consist of a hydrophilic, polar headgroup and hydrophobic, apolar hydrocarbon chains. They aggregate in an aqueous environment above a certain concentration, termed the critical micellar concentration (CMC), into supramolecular structures which depend, in their geometry, on the primary chemical structure of the amphiphile and on its secondary structure, which corresponds to the molecular shape of the individual molecules. Aggregation is promoted by hydrophobic interaction or, more precisely, the minimization of the Gibbs free energy of the water-amphiphile system, which is accomplished by the tendency of free water to increase its entropy. The molecular shape of a given amphiphile is not constant; it depends rather on the phase state of its hydrocarbon moiety. Thus, the same amphiphilic molecules may exist in a highly ordered gel state below and in a less-ordered liquid crystalline (fluid) phase state above a phase transition temperature T_c which depends, among others, on the length and on the degree of saturation of the acyl chains. Changes in the phase state may, therefore, also provoke a change in the supramolecular structure. This interrelationship is further complicated by the influence of other ambient conditions such as water concentration and the presence of ions, negative or positive, which interact with charged amphiphiles.

For a large number of amphiphiles, in particular phospholipids, the major constituents of most biological membranes, the phase states and the supramolecular structures have been investigated in detail and phase diagrams have been established.[2-4] Moreover, the importance of these physical quantities for the function of biological membranes, especially that of the incorporated proteins, and on the interaction between biological membranes is widely accepted.

In this review, we will summarize experimental results obtained for LPS and lipid A including the theoretical background and the applied physical techniques. We will, furthermore, try to elucidate the biological relevance of phase states and supramolecular structures for the function of LPS as major amphiphilic constituents of the outer leaflet of the outer membrane and, with some examples, for the manifold actions of LPS and lipid A in biological systems. An understanding of the physical mechanisms of the biological action of endotoxin is, at present, hampered by some still-unanswered basic questions, for instance, as to the role of binding proteins expressed on the host cell membrane and to the configuration of the active "endotoxic principle", monomer vs. supramolecular aggregate.

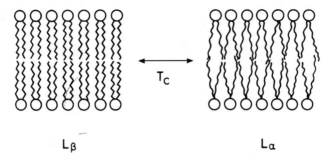

FIGURE 1. Schematic representation of the gel (β) to liquid crystalline (α) chain melting-transition within a lamellar (L) phase at the transition temperature T_c.

II. GENERAL ASPECTS OF STRUCTURAL POLYMORPHISM

A. BASIC PRINCIPLES

The basic principles of the structural polymorphism of lipids, including the phase states, are briefly discussed for phospholipids for which these parameters and their correlation to the chemical structure have been intensively studied.

Phospholipids assume two main phase states, the gel (β) and the liquid crystalline (α) state of their hydrocarbon chains, between which a temperature-induced, reversible first-order transition takes place (Figure 1). In the highly ordered β-phase, the acyl chains are in the *all-trans* configuration and with increasing temperature an increasing number of *gauche*-conformers is introduced, leading to a lowering of the degree of order and, with that, to a fluidization. The temperature T_c of the β↔α phase transition (main transition) depends on the kind of the hydrophilic headgroup (charge, size) and on the number, length, and degree of saturation of the hydrophobic acyl chains. Thus, for example, the T_cs of the negatively charged phospholipids dimyristoyl- and dipalmitoyl-phosphatidic acid are 49 and 67°C, respectively, those of the neutral (zwitterionic) phospholipids dimyristoyl- and dipalmitoylphosphatidylcholine 25 and 41°C, respectively, each at neutral pH and excess water.[5] The value of T_c, for example, for the negatively charged dipalmitoylphosphatidylglycerol, is shifted from 42°C at neutral pH to 46.5°C at pH 3[6] and 50.5°C at pH 2.7.[7] An increase in the ratios of the lipid to water concentration leads, for example, for dipalmitoylphosphatidylcholine to an increase in T_c from 41 to 59°C at 10% water content (w/w).[8] A transition between different phase states is accompanied by changes in the geometry of the involved molecules and has, therefore, an impact on the structure of the supramolecular assemblies.

The relation between the molecular shape of an individual molecule and the three-dimensional supramolecular structure of aggregates formed above the CMC can be described by a simple geometric model which relates the final structure to the ratio of the effective cross-sectional areas of the hydrophilic polar and the hydrophobic apolar regions of the amphiphilic molecules. Israelachvili et al.[9] have introduced a dimensionless shape parameter S which is defined as $S = v/(a_o * l_c) = a_h/a_o$ with v as the volume per molecule of the hydrophobic moiety, a_e and a_h the cross sections of the hydrophilic and hydrophobic moieties, respectively, and l_c the length of the fully extended hydrophobic portion. From the absolute values of the shape parameter, which may be determined, for example, by X-ray small- and wide-angle diffraction experiments, the possible structure of the supramolecular aggregates can be deduced. The correlation between molecular shape and supramolecular structure is schematically outlined in Figure 2. For $S < 1/2$ micellar structures (A), for $S = 1$ lamellar structures (C), and for $S > 1$ inverted micelles (D) are favored. If the amphiphile has a prominent axis originating, for example, from headgroup charges or from an asymmetric distribution of the acyl chains, the spherical, i.e., the micellar and inverted micellar, structures are modified to hexagonal H_I (B) and hexagonal H_{II} (F), respectively. In the latter case, the water component is assumed to form idealized circular rods lined with the hydrated lipid headgroups and with the remaining volume filled by the fluid hydrocarbon chains. For the H_I phase, the situation is opposite, with the circular lipid aggregates being embedded in a water continuum. Newer investigations have shown the importance of other nonlamellar so-called cubic structures (E) also for phospholipids which may occur as stable intermediate phases between the lamellar L and inverted hexagonal H_{II}.[2,10-12] Other than the L and H_{II} phases, the cubic Q phases cannot be described by a unique structure. It rather exists in various structures which may be assigned to different crystallographic space groups.[11] Those space groups relevant for an understanding of the LPS and lipid A structural polymorphism are depicted in Figure 2.

Since the molecular shape varies as a function of ambient conditions, alterations of the latter may lead to modifications of spatial requirements of the molecules and, with that, of the shape parameter S. Thus, for instance, an increase in a_h or a decrease in a_o may induce a transition from a lamellar into a hexagonal H_{II} phase. An increase in a_h might be caused by an increase in temperature; a decrease in a_o might be provoked, particularly for negatively charged lipids, by higher concentrations of mono- and divalent cations or a decrease in pH. The lamellar to hexagonal H_{II} transition occurs mainly in the liquid crystalline phase. The main parameters influencing a transition from lamellar to nonlamellar Q,H_{II} are summarized in Figure 3.

B. BIOLOGICAL RELEVANCE

The importance of the phase state of the hydrocarbon moiety for membrane function is readily intelligible. The lipid composition of biological membranes guarantees a sufficiently high mobility (high fluidity or low state

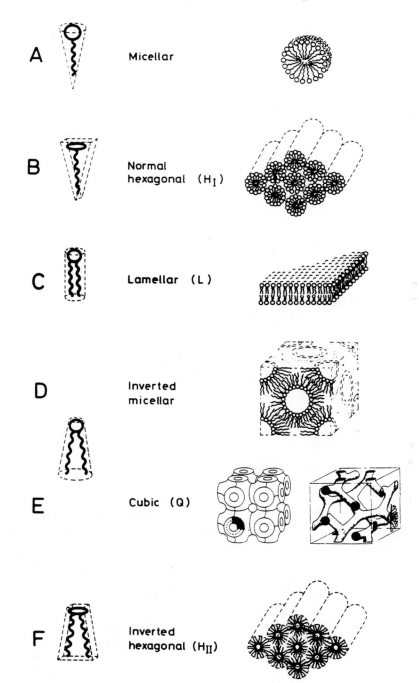

FIGURE 2. Correlation between the molecular shape of lipids and the three-dimensional supramolecular structures they form.

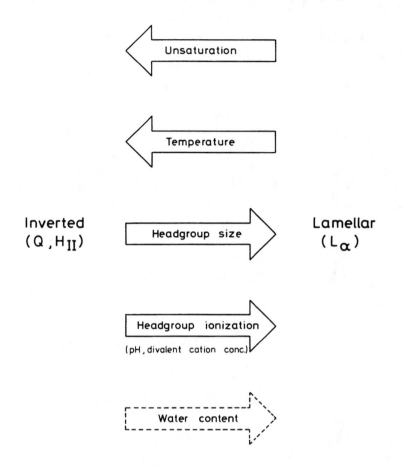

FIGURE 3. Factors influencing the lamellar to inverted phase transition. The dashed arrow indicates the qualitatively different influence of water for phospholipids and LPS.

of order) of the acyl chains which is a prerequisite for transport processes and protein activity. The role of nonlamellar structures is more difficult to understand. In fact, there is little evidence for the existence of nonlamellar structures in biological bilayer membranes. However, as Seddon[10] pointed out, nonlamellar phases are of fundamental relevance for many processes taking place in, on, or through most biomembranes independent of whether such phases actually exist as stable structures in biomembranes. It is known that most membranes contain, beside those amphiphiles forming bilayer structures, lipid constituents also, which tend, as isolated components, to adopt nonlamellar, cubic, or H_{II} structures. In the membrane, this tendency will exert forces or constraints on the other membrane constituents, which are, therefore, influenced in their functional properties. Thus, alterations of the bilayer system by H_{II}-forming lipids may lead to macroscopic changes in membrane thickness and permeability which might then, in turn, influence microscopic properties of individual membrane constituents such as rates of rotation and lateral diffusion.

Some of the processes which might be related to the ability of lipid bilayer constituents to assume nonlamellar structures were outlined approximately 10 years ago by Cullis et al.[13,14] in the so-called metamorphic mosaic model. This model includes various processes such as (1) transbilayer transport of polar molecules, (2) membrane continuity between membrane-bound compartments, (3) budding process of membrane-bound vesicles, (4) transmembrane configuration of a protein without the requirement of apolar amino acids, (5) compartmentalization within a continuous membrane system, and (6) transmembrane transport through an aqueous pore formed by H_{II}-phase lipids.

III. STRUCTURAL POLYMORPHISM OF LIPOPOLYSACCHARIDE AND LIPID A

In principle, the basic characteristics of the phase behavior of phospholipids are valid also for endotoxins. However, some major factors specific for LPS and lipid A have to be considered, which make straightforward interpretations of the respective data more difficult. These factors relate mainly to variations in the primary chemical structure, i.e., to the headgroup conformation and charge state, to the acylation pattern and length of the carbohydrate moiety, even within the same preparation, in the way that the macroscopically averaged measured values represent a microscopically heterogeneous mixture. This was already recognized by Shands[15] in his early review on the physical structure of bacterial LPS as the major impediments to the study of the physicochemical properties of LPS. Also, due to the fact that LPS are isolated from whole bacterial cells, contamination with various amounts of phospholipids and peptides cannot be excluded. Thus, the often insufficiently defined chemical structure might partly explain the controversial or incomparable data published on structural polymorphism of endotoxin.

A. PHASE STATE

The importance of the phase state for the three-dimensional supramolecular structure has been intensively elucidated for phospholipids.[2,11,16] This connection applies in a similar way to LPS and lipid A. However, here the situation is complicated by the extremely strong influence of the water concentration on the phase states and on many properties of the transition between these, such as transition enthalpy ΔH_c, temperature T_c, and broadness of the transition range.[17] This so-called lyotropism depends on the length of the sugar moiety and its influence can hardly be assessed for natural isolates, in particular for smooth-form LPS. It seems reasonable to assume that the strong lyotropic behavior of LPS and lipid A is the reason for partly contradictory statements in the literature as well on the phase behavior as on the structure of endotoxin. Further reasons for controversy may arise from the different physical techniques applied which, in many cases, do not allow a direct comparison of the measured data. This becomes evident, for example, from the measurements of membrane fluidity. With infrared spectroscopy, this parameter is obtained as an average value over the full length of the acyl chains, whereas with electron spin resonance (ESR) or fluorescence polarization spectroscopy the fluidity is determined with a probe molecule adsorbed to a specific site of the hydrocarbon chains. Therefore, for a calibration of the data obtained with different techniques a new parameter, the so-called order parameter, has been introduced as a measure for the state or degree of order and is defined to 1 for perfectly aligned and to 0 for isotropically arranged hydrocarbon chains.

The observation of a $\beta \leftrightarrow \alpha$ phase transition of endotoxin was first published by Emmerling et al.[18] in 1977. They found for LPS of *Escherichia coli* K12, applying fluorescence spectroscopy with *N*-phenylnaphthylamine (NPN), a sharp increase in fluorescence intensity in a narrow transition range around 22°C exhibiting hysteretic behavior within a complete heating and cooling cycle typical for a cooperative first-order phase transition. They confirmed the existence of an order-disorder transition at a similar temperature (20.5°C) in X-ray diffraction experiments evaluating the precise position and intensity of the wide-angle reflections in the range 0.43 to 0.45 nm. The occurrence of a thermotropic phase transition in endotoxin was confirmed in the following years by various groups; however, the measured values for T_c differed considerably even for LPS of comparable chemical structures. Thus, Van Alphen et al.[19] applying light-scattering spectroscopy obtained for LPS from *E. coli* K12 grown at 37 and 12°C, respectively, T_c values of 29.6°C (transition range 3.0°C) and 20.0°C (transition range 7.6°C), respectively. In differential scanning calorimetric (DSC) experiments the LPS from cells grown at 37°C showed a transition in the temperature range 29 to 38°C in a heating and 34 to 24°C in a cooling scan. Coughlin et al.[20] investigated LPS from *E. coli* using ESR probes and found broad "temperature-dependent" ($\beta \leftrightarrow \alpha$) transitions with different T_cs for native ($T_c = 29$°C) and electrodialyzed ($T_c = 24$°C) LPS. The same group also investigated the phase behavior of short and long O-antigen-containing fractions of LPS from *E. coli* O111:B4, again using spin probes.[21] Measurements with the spin probe 5-doxyl stearate (5-DS) showed a relatively slow decrease in polarity with increasing temperature. Thus, a gradual partitioning of 5-DS into LPS should be indicative of a broad phase transition from approximately 25°C to above 50°C. It should be mentioned that the mobility of the probe CAT_{12}, labeling the headgroup region, was greater in the long-chain than in the short-chain fraction, indicating that the sugar portion of the latter was more tightly packed. The mobility of CAT_{12} in the Mg^{2+} salt form of all samples investigated, long-chain, short-chain, and unseparated LPS, was lower than in their respective Na-forms, although the same differences in mobility between the fractions were observed, suggesting that the headgroup packing of the long-chain fraction is inherently more open than that of the short-chain fraction, regardless of the LPS salt form. These measurements gave first evidence for the influence of cations on the phase behavior of LPS and were, in further studies, extended by other groups. In these studies also the dependence of T_c on the length of the sugar moiety covering the full range from lipid A over rough mutant to wild-type LPS was determined. We will review the more recent results on endotoxin phase behavior, including those on the influence of other parameters such as salt form, pH, and water content.

In our own investigations, we have found a very characteristic dependence of T_c on the chemical structure of the sugar moiety of LPS by applying fluorescence- and Fourier-transform infrared-spectroscopy (FT-IR) as well as DSC.[17,22] Thus, for enterobacterial strains (natural salt forms), T_cs are highest for free lipid As (around 45°C), lowest for deep rough mutant LPS (around 30°C), and, with increasing length of the polysaccharide portion toward completion of the O-chain (wild-type LPS), increase again up to 37 to 40°C. For LPS and free lipid A from *Salmonella minnesota*, this dependence of T_c is presented in Figure 4. Naumann et al.,[23] using FT-IR, have arrived at similar results. They found for Na-salt forms of lipid A Re and Rd1 LPS (from *S. minnesota*) and S LPS (from *Citrobacter freundii*) T_cs of 46, 29, 37, and 38°C, respectively. For S-form LPS from *S. abortus equi* (Ca salt form), they found a T_c of approximately 42°C. The deviations between the different wild-type LPS may be related to the heterogeneity of the LPS which are presumed to exist as a mixture of different LPS structures varying in the length of the sugar moiety as well as in the acylation pattern of their lipid A.[24] For synthetic lipid A of *E. coli* and *S. minnesota* (compounds 506 and 516, respectively) T_c values of 44 and 47°C, respectively, were found, values in close agreement with the natural compounds. Furthermore, the entire thermotropic phase behavior of synthetic and natural lipid A preparations showed a very good overall agreement.[25]

The phase transition temperatures of some nonenterobacterial LPS and lipid A were found to lie significantly lower than those of Enterobacteriaceae.[17] This observation correlates with the fact that the former contain shorter acyl chains with a higher degree of unsaturation.[26-28] The enthalpy changes of the $\beta \leftrightarrow \alpha$ acyl chain-melting transition measured with DSC were shown to give typical molar values of 30 kJ for enterobacterial LPS.[17,22,29,30] Interestingly, the comparison with the values measured for C 14:0 phospholipids (24 to 27 kJ/mol),[5] shows a drastic reduction of the calorimetric endotherm with respect to the contribution of a single methylene group. This observation was attributed to a lower state of order of the acyl chains in the gel state or to a decrease of the cooperative association of LPS molecules; however, a satisfactory quantitative explanation could not be given up to now.[17,30,31]

For a given LPS, the values of T_c seem to be strongly dependent on its salt form. It was shown for Re LPS from *E. coli* that T_c increases from 28.5°C

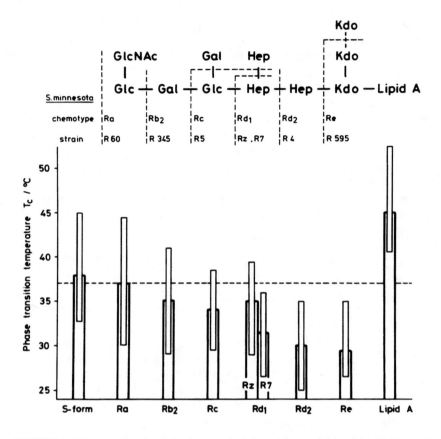

FIGURE 4.   Temperatures T_c of the $\beta \leftrightarrow \alpha$ acyl chain melting transition of free lipid A and various LPS preparations from *S. minnesota*. The inserted bars indicate the temperatures of the onset and the end of the melting transition, respectively, and the dashed line the physiological temperature.

(Na salt) over 33.0 (ethanolamine salt), 35.0 (H salt), and 36.0°C (triethyl-amine salt) to 40.0°C (K salt).[17] The influence of the other parameters mentioned above may be summarized as follows: the addition of divalent cations and a lowering of pH cause a significant rigidification of the acyl chains of free lipid A and LPS preparations and, partly, also lead to an increase in T_c. At basic pH, a fluidization of the LPS and lipid A acyl chains and a decrease of T_c take place. The influence of divalent cations are strongest for lipid A and deep rough mutant LPS, whereas the effect of a change in the pH is similar for all different LPS samples. Free lipid A and LPS exhibit an extremely strong lyotropic behavior, i.e., a strong dependence of the expression of the $\beta \leftrightarrow \alpha$ chain melting transition and of its enthalpy on the water content. Thus, for example, for Re LPS the phase transition appears distinctly only at water concentrations >30 to 50%. This observation may explain why Labischinski et al.,[32] in an earlier investigation with X-ray diffraction, did not detect any phase transition up to 50°C with dry or "hydrated" lipid A and LPS samples from *S. minnesota*.

For reasons not so obvious, Nikaido et al.[33] applying ESR did not obtain evidence for the existence of a $\beta \leftrightarrow \alpha$ transition in preparations of rough mutant LPS (Rc type) from *S. typhimurium* HN202. In a recent paper, Vaara et al.[34] confirmed these findings and stated "that the reported transitions are artifacts perhaps caused by the imperfect packing of LPS molecules". The reasons why these authors did not detect a phase transition might relate to the applied technique and to the fact that they used significant amounts (10 m*M*) of divalent cations in their LPS preparations. It is known that a phase transition may not be detectable with a particular technique even though its existence was proven for an identical sample with other techniques. Possibly, the partitioning of ESR probes or of certain fluorophores into LPS aggregates is inhibited or masked at higher concentrations of divalent cations.

B. SUPRAMOLECULAR STRUCTURE

The early work on the supramolecular structure was reviewed by Shands[15] in his article on "the physical structure of bacterial LPS". It was focused on the characterization of LPS particles with typical molecular weights of 1×10^6 to 20×10^6 Da as estimated using physicochemical data on diffusion coefficients, intrinsic viscosity, and partial specific volumes. From their size, the LPS particles were well suited for electron microscopic visualization, which revealed only a few major morphological forms, mainly droplet-like forms at neutral pH and ribbon-like structures at high pH, even though the investigated LPS were isolated from different bacteria and should therefore differ in chemical composition. From these investigations, Milner et al.[35] suggested that the morphology of LPS particles might be determined by the aggregation of linear molecules of polysaccharides into a "fringe micelle". Rothfield et al.[36] found that LPS particles were surrounded by a bimolecular leaflet-like structure, from which they suggested that LPS formed a phospholipid-like ordered leaflet structure. The width of such leaflets was derived by Burge and Draper[37] from electron microscopy and X-ray diffraction to be 70 Å for the bilayer (giving rise to a "trilaminar" structure in electron micrographs). Shands et al.[38] arrived at values of 60 Å for a heptoseless mutant (Re) and 90 Å for an Ra mutant and wild type, respectively, each from *S. typhimurium*. In later measurements, these values were found to be too high by approximately 20%, from which the authors interpreted the visible LPS structure to consist of a lipid leaflet of about 40 Å with an additional 15 Å on each side occupied by the core polysaccharide.

In the more recent work, the rather descriptive electron microscopy was replaced by analytical physical techniques such as small-angle X-ray and neutron diffraction for the determination of the long-range order (supramolecular conformation) and wide-angle X-ray diffraction for the determination of the short-range order (arrangement of the acyl chains). Among these,

X-ray diffraction is the most direct technique. From the evaluation of the periodicities of Bragg reflections, the crystal structure can be directly determined by mathematical procedures. The evaluation of neutron-scattering data is performed by fitting these to *a priori* model structures, that is, the most probable supramolecular configuration is that which fits best the measured data. The different techniques applied and the different sample conditions, mainly concerning the degree of hydration, might be considered as reasons for the two basically different statements on endotoxin supramolecular structure which is found to be either lamellar or nonlamellar.

Emmerling et al.[18] have also reported on X-ray-scattering data of LPS; however, unfortunately it is not clear from their paper which one of the described LPS, from *E. coli* B/r, *E. coli* strains K2464 (wild type), or K2754 (heptoseless mutant), was used in the X-ray experiments. They found a very simple phase behavior of the investigated LPS with only one ordered phase in the temperature range 0 to 50°C for LPS samples between O and 30% water content. Its structure was interpreted to be lamellar due to the existence of sharp reflections with equidistant spacing. Above 30% water content, the reflections broadened and eventually disappeared. The repeat distance varied from 7 nm for dry LPS to about 10 nm for LPS at about 25% water. According to the authors, dry LPS formed bilayers and water caused swelling of the lamellar phase by hydration of the polar headgroups. They stated that the swelling was limited and that additional water led to a disordering of the lamellar stacking. Wawra et al.[39] also performed X-ray small- and wide-angle diffraction experiments with dried samples of LPS from *S. minnesota* wild-type (SF1111) and deep-rough mutant R595 (SF1167). They obtained results similar to those of Emmerling and co-workers in that they found LPS bilayer stacks with a single bilayer thickness of 5.4 nm for R595 LPS and 10.9 nm for wild-type LPS. In both preparations they observed a wide-angle reflection at 0.41 nm which they interpreted as arising from the hexagonal dense packing of the acyl chains.

In more recent investigations, Labischinski et al.[32] applied X-ray diffraction to elucidate the three-dimensional conformation of S-form, Re-mutant LPS and of free lipid A from *Salmonella* in their Ca-, Na-, and triethylammonium salt forms. They prepared the lipids as dry samples and as aqueous solutions (approximately 5 to 10% LPS:water wt/wt). In all cases, they found no evidence for other than lamellar structures at temperatures up to 50°C. The reflection from the hexagonal packing of the hydrocarbon chains at 0.425 nm observed under wide-angle conditions was not found to be altered significantly in position and sharpness when recorded up to 50°C. Similar experiments with synthetic compounds corresponding in their chemical structures to those of lipid A from *E. coli* (compound 506) and from *S. minnesota* (compound 516), respectively, and to that of the lipid A-precursor Ia (compound 406) led, in each case, to diffraction patterns that were interpreted to result from lamellar structures.[40] These results are surprising when considering the fact that the primary chemical structures of the investigated compounds vary considerably: the number of the hydrocarbon chains attached to an identical bisphosphoryl diglucosamine backbone ranges from four (406) over six (506) to seven (516). However, for an interpretation of these results it should be kept in mind that the samples were analyzed as dry foils. In very recent experiments, Labischinski et al.[41] applied also small-angle neutron scattering to lipid A samples obtained from LPS of *S. minnesota* R595 by acid hydrolysis, which should contain mainly the 4'-monophosphoryl compound. Again, only lamellar structures were found at 10 and 50°C, i.e., supposedly well below and above the β↔α-phase transition temperature T_c and at a lipid concentration of 10 mg/ml. The analysis of the scattering curves and the interpretation of the corresponding thickness distribution functions of the lamellar aggregates led to a model of a lipid A bilayer of about 5 nm in thickness. This thickness was explained by interdigitating fatty acid chains of the lipid A.

Hayter et al.[42] applied the same technique to the structural analysis of Re LPS (Na salt form) from *E. coli* D21f2 at different pH. At pH 6.0 and 7.4, they could best fit their data by randomly coiled tubular micelle structures with a diameter of 10.0 nm. At pH 9.1, the data were compatible with uniform tubular micelles of diameter 11.0 nm and length 144 nm. These structures correspond, at first sight, to those published as electron micrographs by Coughlin et al.[20] and by Peterson and co-workers.[21] However, the latter — irregular tubes or ribbons or long tubular structure (diameter approximately 9 nm) — were obtained from LPS D21 from *E. coli* K12 (Ra-type LPS) or LPS from *E. coli* O111:B4 (wild type), respectively, which contain considerably longer sugar chains. It should, furthermore, be considered that, as Coughlin et al.[20] pointed out, "electron micrographs are unfortunately misleading and can lead to erroneous interpretations with regard to molecular organization".

The X-ray diffraction measurements with Re LPS from *S. minnesota* R595 by Naumann et al.[23] as dried films and as fully hydrated samples (suspensions or centrifuged pellets) in the temperature range 10 to 60°C gave no evidence for nonlamellar structures. Moreover, neither pH variation between 4 and 10 nor cation bridging gave any significant signs of loss of overall lamellar organization of the lipid matrix. The authors state that, on the contrary, divalent cation bridging (e.g., by Ca^{2+}), which plays an important role for stabilization of the outer membrane under physiological conditions, caused highly ordered multibilayer stacking, overall rigidification of the matrix, increase of T_c, and reduction of cooperativity of chain melting.

Using more indirect techniques, evidence for the existence of nonlamellar structures for lipid A and Re LPS was obtained by our group and by the group of McGroarty.[22,30,42] From a pH titration study applying ESR (spin probe CAT_{12}) Coughlin et al.[30] proposed a complex structural polymorphism for Re LPS from *E. coli* D21f2: at high pH values, the LPS should assume

H_I or tubular micelle structures, at low pH inverted micelles or a hexagonal H_{II} structure, and at normal pH, lamellar structures. These ESR findings were backed by DSC measurements showing no detectable change in the enthalpy of the $\beta \leftrightarrow \alpha$ chain melting transition at high pH which is indicative of a lack of cooperative association between LPS molecules and, with that, of a loss of a lamellar structure. A similar decrease of the phase transition enthalpy was also observed at low pH and was assumed to result from a decrease in headgroup size, concomitant with a decrease in charge, and a maximum in hydrogen bonding. For Ra-type LPS from *E. coli* D21, the authors concluded that at high pH there was a behavior very similar to that of LPS from strain D21f2, i.e., the existence of a micellar structure. For the same compound at low pH, they presume that the longer polysaccharide and the higher charge density should hinder the formation of inverted micelles. From our experimental data with fluorescence spectroscopy and entrapment of hydrophilic solutes (e.g., 6-carboxyfluorescein), we proposed the existence of nonlamellar structures for lipid A and, at least partially, for Re LPS from *S. minnesota* R595 above their respective phase transition temperatures T_c.[22] These results were supported by orientation measurements applying infrared polarization which proved the existence of an "isotropic" phase for Re LPS at temperatures above 40°C and the absence of a pure lamellar phase for free lipid A over the investigated temperature range of 10 to 60°C.[43]

The results presented so far do not seem to convey a unique picture of the supramolecular structure of endotoxin. This is not only due to the different chemical structures investigated, but may also be explained by the different experimental conditions under which the samples were investigated. As emphasized in the previous section, the phase behavior of endotoxin expresses strong lyotropism and is, furthermore, dependent on the concentration of divalent cations. Considering the influence of these parameters, and in view of possible biological implications of endotoxin supramolecular structures, we have, for free lipid A and rough mutant LPS, established complete phase diagrams. For this we have measured the phase transition region and determined the supramolecular structure over a wide range of water content (20 to 95%), concentrations of Mg^{2+} ([lipid]:[Mg^{2+}] from 1:0 to 1:1 M), and in dependence on temperature.[44,45] These parameter ranges include near physiological conditions. The phase transition behavior was investigated with FT-IR and DSC and the supramolecular structure with small-angle X-ray diffraction utilizing the high brilliance of synchrotron radiation. The phase diagrams established for free lipid A of *S. minnesota* strain R595 in the absence and at equimolar concentrations of Mg^{2+} are depicted in Figure 5A and 5B, respectively. The main features include

- The $\beta \leftrightarrow \alpha$ acyl chain melting transition is, under all conditions, associated with a change in the supramolecular structure of the lipid assembly. Within the transition range, usually different phases coexist.

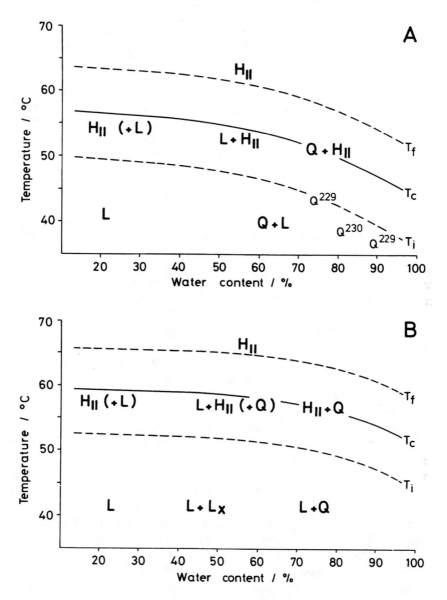

FIGURE 5. Phase diagrams of the (A) lipid A-water and the (B) lipid A-Mg^{2+}-water systems. In the latter, the [lipid A]:[Mg^{2+}] ratio is 1:1 M. The solid line represents the midtemperature T_c, the dashed lines the onset (T_i) and end (T_f) temperatures of the acyl chain-melting process, respectively. L: lamellar phase; Q: cubic phase; H$_{II}$: hexagonal phase; L$_x$: lamellar phase with yet unknown details. (From Brandenburg, K., Koch, M. H. J., and Seydel, U., *J. Struct. Biol.*, 105, 11, 1990. With permission.)

- In the gel (β) state cubic (Q) phases are predominant at higher water content (>60%) and at high [lipid A]:[Mg^{2+}] molar ratios. In the low water concentration range, the lamellar phase (L) is the exclusive structure at all Mg^{2+} contents.
- In the liquid crystalline (α) state the H$_{II}$ phase is predominant. Its contribution is weak at low Mg^{2+} concentrations and high water contents and becomes stronger at higher Mg^{2+} concentrations and lower water contents.
- Under near physiological conditions of water content, divalent cation concentration, and temperature the lipid A assemblies nearly exclusively adopt cubic structures.

These findings were, in principle, also found to be valid for free lipid A from *E. coli* F515; however, the temperatures above which H$_{II}$ structures are formed are higher.

The results so far available for LPS Re (from *S. minnesota* R595) indicate a similarly complex structural polymorphism as observed for free lipid A.[46] Briefly, under physiological conditions of temperature, divalent cation concentration, and water content, Re LPS aggregates preferentially assume nonlamellar (cubic) structures. Inverted H$_{II}$ structures are observed only at high, physiologically irrelevant temperatures (>70°C), and the transition from cubic to H$_{II}$ is decoupled from the β↔α chain melting transition. It seems reasonable to assume that the tendency to form nonlamellar three-dimensional configurations decreases with increasing length of the sugar moiety. However, evidence for the existence of cubic structures has been obtained also for LPS from an Rd mutant of *S. minnesota* under physiological conditions.[47]

In Section II.A we have outlined the correlation between the molecular shape of the individual amphiphilic molecules and the supramolecular structure of aggregates of a large number of identical molecules. Therefore, the knowledge of the molecular shape of an endotoxin molecule should render valuable information on the three-dimensional structure of endotoxin aggregates. Labischinski et al.[32] have published results from energy minimization calculations on the conformation of lipid A from *S. minnesota* which are reproduced in Figure 6. The most striking feature of this model is the orientation of the bisphosphoryl disaccharide backbone with respect to the fatty acid residues forming an angle of approximately 45°. The fatty acid chains occupy positions lying almost exactly on a hexagonal lattice and do not end in one plane. Similar results were obtained for Re LPS from *E. coli* with respect to the inclination of the backbone relative to the fatty acid chains and their arrangement.[23] The 2-keto-3-deoxyoctonate moiety of the Re LPS was found to be on top of the molecule and to fill the space provided by the tilted diglucosamine backbone and thus not to contribute to the overall length of the molecule. In Figure 7, the area functions, describing the area of the molecule in dependence on its long axis position as calculated by Naumann

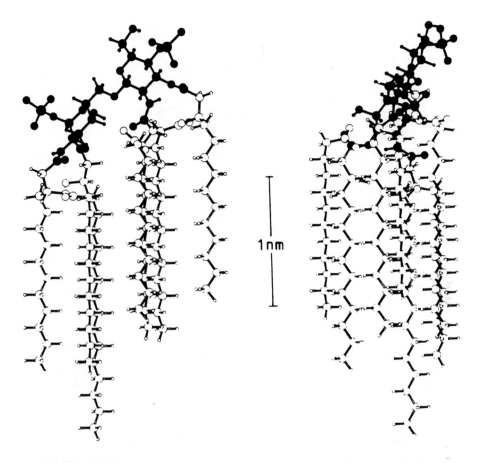

FIGURE 6. Models of *S. minnesota* lipid A. The model was calculated by energy minimization techniques. The two views are rotated 90° against each other. Backbone atoms are blackened. C, O, and N atoms are represented by spheres, H atoms by small spheres. (From Labischinski, H., Barnickel, G., Bradaczek, H., Naumann, D., Rietschel, E. Th., and Giesbrecht, P., *J. Bacteriol.*, 162, 9, 1985. With permission.)

et al.,[23] for the most representative but slightly different molecular conformations of Re LPS in the β-phase are shown. The authors found the situation described by their Re LPS data to be very similar to that calculated for phosphatidylethanolamines (PE) which are known to form inverted H_{II} structures at higher temperatures. The authors conclude from the fact that their model calculation was performed on an Re LPS in the β-phase that one might expect that these molecules could also form inverted phases at elevated temperatures. However, since they did not observe the H_{II} structure for Re LPS, they deduced that a purely geometric, static interpretation of such data has inherent limitations. They argue, furthermore, that the effective polar headgroup area of charged lipids is much larger than that of a noncharged lipid

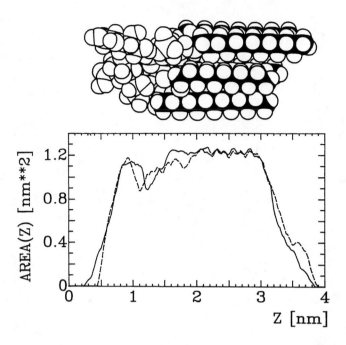

FIGURE 7.   Area function of Re LPS from *E. coli* calculated for two slightly different conformations obtained with energy minimization procedures. The space-filling model corresponds to the area function represented by the solid line. (Adapted from Naumann, D., Schultz, C., Sabisch, A., Kastowsky, M., and Labischinski, H., *J. Mol. Struct.*, 214, 213, 1989. With permission.)

and, in this respect, Re LPS should be much more comparable, for example, with phosphatidic acid which, as they state, adopts lamellar phases at physiologically relevant conditions, although their model calculations for this compound suggest the preference for the H_{II} phase. This phase, indeed, was observed for dilauroylphosphatidic acid under the influence of Mg^{2+} above 50°C,[44,48] for dioleyol phosphatidic acid in the presence of divalent cations at room temperature,[49] and for natural phosphatidic acid at 37°C.[50] Thus, these model calculations do not exclude the formation of inverted structures by Re LPS and, even less, that by free lipid A.

IV. ENDOTOXIC CONFORMATION

At present, the details of the mechanisms underlying the interaction of LPS with host cell membranes or with particular components of these, leading to the induction of the well-known biological effects, are not well understood. Thus, it cannot be definitely decided whether the stimulation of the host cells proceeds solely via "hydrophobic interaction" leading to the intercalation of the LPS molecules or whether the interaction is mediated by particular

LPS-binding proteins which may be either integral constituents of the host cell membrane[51,52] or serum components which are then, in a second step, bound to a receptor on the cell surface.[53,54] Very recently, evidence has been published that a highly disaggregated form of Re LPS (possibly the monomer) may function as the active unit.[55] In any case, the configuration and size of the LPS aggregates as well as the state of order of the acyl chains should strongly influence the biological activity, independent of the particular mechanism of cell activation.

From the foregoing, it should be evident that there are obviously considerable discrepancies in the experimental findings on the three-dimensional structures of LPS and lipid A. The results on the phase behavior are less controversial. Assuming that our own results, in particular those obtained under near physiological conditions, reliably describe the three-dimensional structure of lipid A and Re LPS, there are some striking correlations between the phase transition temperatures of various LPS and lipid A, as well as of the supramolecular structure and some biological effects. However, since the reversible chain melting transition frequently goes along with a transition between different structures, it is, in many cases, difficult to decide whether the phase states of the acyl chains or the structural behavior of the whole lipid assembly, or a combination of both, are the governing processes. Nevertheless, it is tempting to correlate, in the following examples, the observed biological effects with the value of T_c, i.e., with the state of order of the acyl chains at 37°C (see Figure 4):

- The induction of leukotriene C4 (LTC-4) release from mouse peritoneal macrophages after stimulation with wild-type, rough mutant LPS and lipid A, all from *S. minnesota*.[56] The release of LTC-4 was highest for Re LPS (lowest T_c), lowest for free lipid A (highest T_c), and decreased with increasing completeness of the sugar moiety (T_c increases with respect to Re LPS).
- The lipid A- and LPS-induced hemagglutination of erythrocytes which is low for S form, significantly higher for Ra, further increasing with increasing deficiency of the core oligosaccharide, i.e., highest for Re LPS, and lower again for lipid A.[57]
- The binding of complement factors C1 and C1q, determined as the amount of sample necessary for 50% consumption which was found to be highest for Re to Ra LPS, lowest for S-form LPS, while the effect of lipid A was intermediate.[58]
- The stimulation of polymorphonuclear neutrophil chemiluminescence and inhibition of chemotaxis which is highest for lipid A and decreases from Rc and Rd to Re LPS, and increases again over Rb and Ra to S-form LPS.[59]
- The effect of complement and anti-lipid A and homologous antisera on lecithin liposomes prepared with lipid A or different LPS expressed as

the release of trapped glucose. The release is again highest for Re LPS, decreases with increasing length of the sugar moiety, being lowest for wild-type LPS, and is intermediate for lipid A.[60]

- The spleen cell proliferative response to various R- and S-chemotype LPS and lipid A preparations which are high for all rough mutant LPS (highest for Rd LPS), while the responses for lipid A and S-form LPS are significantly lower.[61]

- The susceptibility of various mutant strains of *S. minnesota* to hydrophobic antibiotics which correlates directly with the state of order at 37°C of the hydrocarbon chains of the LPS.[62]

- The antimicrobial capacity of magainin 2 also correlates with the acyl chain mobility of LPS at 37°C, that is, the sensitivity of wild type is lowest, increases over Ra, Rb, and Rc mutants to deep rough mutant Re, all from *S. typhimurium*.[63] It should be noted that the different mutant LPS from *S. typhimurium* have a very similar structure as those from *S. minnesota*.

The following examples may serve as illustrations of the correlation of particular biological effects with the three-dimensional structure:

- The mitogenic activity of lipid A- and LPS-lysozyme complexes in comparison to the native LPS was strongly reduced in complexes with Re LPS and lipid A, but showed no reduction with Ra and S-form LPS.[64] These observations might be explained by the tendency of lipid A and Re LPS, but most likely not of Ra-mutant and S-form LPS (because of their larger headgroups) to adopt inverted structures under physiological conditions.

- In the passive hemolysis inhibition assay, free lipid A has comparably low activity when tested in a free form, but a considerably higher activity when incorporated into the lipid matrix of phospholipid liposomes.[65] This behavior can readily be explained by the existence of an inverted state in the former and of a lamellar phase in the latter case.

In the context of attempting to correlate the three-dimensional structure with the endotoxic activity, it might be interesting to refer the former to the primary chemical structure. The bisphosphorylated *E. coli* lipid A carrying six fatty acid residues, four at the nonreducing and two at the reducing end of the diglucosamine, adopts inverted structures under physiological conditions and is known to have a high biological activity in many test systems. The lipid A tetraacyl precursor Ia carrying only four 3-OH C 14:0 fatty acid residues, two on each glucosamine unit, and the lipid A from *Pseudomonas aeruginosa*, the main constituent of which is acylated with only five hydrocarbon chains with a maximal length of 12 carbon atoms, are known to be considerably less active than *E. coli* lipid A.[66-68] From geometrical considerations, the hydrocarbon area is reduced, while the hydrophilic area is unchanged as compared to *E. coli* lipid A; it may be assumed that these compounds adopt pure lamellar (*P. aeruginosa* lipid A) or even micellar (precursor Ia) structures.

A very interesting aspect concerning the induction of biological activity was raised by Tahri-Jouti and Chaby in a recent paper.[69] They found that LPS (from *Bordetella pertussis* carrying a polysaccharide consisting of 12 sugar residues, from *E. coli* J5, and from *S. minnesota* R595, respectively) bound to mouse peritoneal macrophages by both specific and nonspecific interactions. The latter interactions most probably occur as a result of the insertion of LPS into the lipid layer of the cellular membrane. These investigations suggest that LPS-LPS associations may also contribute to the nonspecific binding and that the nonspecific binding was less temperature dependent than specific binding. Above 22°C (and particularly at 37°C) specific binding was completely obscured. The authors concluded that the latter observation could be explained by increased nonspecific binding interactions with the lipid layer, resulting from a modification of the fluidity of the cellular membrane at this temperature. Of course, the fluidity of the LPS aggregates is also temperature dependent in a characteristic way (see Section III.A).

We have pointed out that the ability of lipids to adopt nonlamellar structures may have implications for various functions of the bacterial membrane in which they constitute particularly for membrane permeability and the regulation of protein activity. In recent experiments with the reconstituted lipid matrix of the outer membrane of *E. coli* deep rough mutant as a planar asymmetric bilayer, in which one leaflet was composed from the natural bacterial phospholipid mixture and the other exclusively from Re LPS, we could demonstrate that α-toxin from *Staphylococcus aureus* does not form pores, whereas in symmetric phospholipid bilayers it does form pores.[70]

A completely different aspect of the possible implications put forth by the ability of lipid A and rough mutant LPS to adopt inverted structures relates to the biosynthesis of the outer leaflet of the outer membrane. Osborn[71] (see also Reference 72) has discussed different models which require either flip-flop processes of O-antigen, core/lipid A, and/or mature LPS across one or two lipid bilayers or involve budding of synthesized LPS from the inner membrane and subsequent fusion with the outer membrane. Alternatively, bilayer continuity between the cytoplasmic and the outer membrane may be required for LPS translocation to the outer membrane. We have proposed a model based upon physicochemical considerations, involving the following steps:[22] (1) lipid A or lipid A/2-keto-3-deoxyoctonate "precursor" molecules are biosynthesized in the inner leaflet of the cytoplasmic membrane, (2) the precursor-rich areas are budded off, forming inverted micelles which (3) move across the periplasmic space through the Bayer's junction, and (4) are intercalated into the outer leaflet of the outer membrane.

A budding process as postulated in our model, which is induced by the ability of the involved molecules to adopt inverted structures, might also explain the observed instability of the outer membrane of deep rough bacterial organisms. The LPS leaflet disrupts from the bacterial cell under formation of blebs, particularly in the stationary growth phase.[73,74]

V. CONCLUDING REMARKS

The data reviewed above reveal that physicochemical parameters such as phase state and supramolecular structure of endotoxin aggregates, in addition to ambient conditions, depend strongly on the primary chemical structure of the LPS. From the examples for correlations between these parameters and particular biological effects, it is tempting to assume that certain requirements of endotoxin molecules concerning these parameters must be fulfilled for the expression of biological activity. Thus, endotoxic activity should be favored by a lower state of order of the lipid A portion of LPS under physiological conditions and, most likely, by the disposition of LPS and lipid A to adopt nonlamellar structures. However, we are far from being able to propose a unique endotoxic conformation or to provide any unequivocal lines which would enable the decision on the actual activation mechanism. The long-term objective should be the establishment of structure-activity relationships which would allow the prediction of the biological activity of any given endotoxin (see also Chapter 1). This requires the determination of the phase transition temperature (and with that, the state of order at the physiological temperature), either predictive from the acylation patterns or from measurements, and the prediction of the most probable supramolecular structure from the molecular shape which should be obtained from theoretical energy minimization calculations.

ACKNOWLEDGMENTS

We are indebted to Mrs. M. Lohs and Mrs. B. Köhler for preparing the drawings and the photographs, respectively.

REFERENCES

1. **Rietschel, E. Th., Brade, H., Brade, L., Brandenburg, K., Schade, U., Seydel, U., Zähringer, U., Galanos, C., Lüderitz, O., Westphal, O., Labischinski, H., Kusumoto, S., and Shiba, T.**, Lipid A, the endotoxic center of bacterial lipopolysaccharides: relation of chemical structure to biological activity, in *Detection of Bacterial Endotoxins with the LAL Test*, (*Progr. Clin. Biol. Res.* Vol. 231) Watson, S. W., Levin, J., and Novitsky, Th. J., Eds., Alan R. Liss, New York, 1987, 25.
2. **Luzzati, V.**, X-ray diffraction studies of lipid-water systems, in *Biological Membranes, Physical Fact and Function*, Chapman, D., Ed., Academic Press, London, 1968, 71.
3. **Luzzati, V. and Tardieu, A.**, Lipid phases: structure and structural transitions, *Annu. Rev. Phys.*, 25, 79, 1974.

4. **Sackmann, E.**, Dynamic molecular organization in vesicles and membranes, *Ber. Bunsenges. Phys. Chem.*, 82, 891, 1978.

5. **Silvius, J. R.**, Thermotropic phase transitions of pure lipid in model membranes and their modification by membrane proteins, *Lipid-Protein Interactions*, 2, 239, 1982.

6. **Brandenburg, K. and Seydel, U.**, Investigations of the order-disorder behaviour of various phospholipids of natural and synthetic origin by optical and calorimetric techniques, *Thermochim. Acta*, 69, 71, 1983.

7. **Jacobsen, K. and Papahadjopoulos, D.**, Phase transitions and phase separations in phospholipid membranes induced by changes in temperature, pH, and concentration of bivalent cations, *Biochemistry*, 14, 152, 1975.

8. **Chapman, D., Williams, R. M., and Ladbrooke, B. D.**, Physical studies of phospholipids. VI. Thermotropic and lyotropic mesomorphism of some 1,2-diacylphosphatidylcholines (lecithins), *Chem. Phys. Lipids*, 1, 445, 1967.

9. **Israelachvili, J., Marcelja, S., and Horn, R. G.**, Physical principles of membrane organization, *Q. Rev. Biophys.*, 13, 121, 1980.

10. **Seddon, J. M.**, Structure of the inverted hexagonal (H_{II}) phase, and non-lamellar phase transitions of lipids, *Biochim. Biophys. Acta*, 1031, 1, 1990.

11. **Luzzati, V., Mariani, P., and Gulik-Krzywicki, T.**, The cubic phases of liquid-containing systems: physical structure and biological implications, in *Physics of Amphiphilic Layers*, (Springer-Proceedings in Physics, Vol. 21), Mennier, J., Langevin, D., and Boccara, N., Eds., Springer-Verlag, Berlin, 1987, 131.

12. **Lindblom, G. and Rilfors, L.**, Cubic phases and isotropic structures formed by membrane lipids — possible biological relevance, *Biochim. Biophys. Acta*, 988, 221, 1989.

13. **Cullis, P. R., De Kruijff, B., Hope, M. J., Nayar, R., and Schmid, S. L.**, Phospholipids and membrane transport, *Can. J. Biochem.*, 58, 1091, 1980.

14. **Cullis, P. R., de Kruijff, B., Hope, M. J., Verkleij, A. J., Nayar, R., Farren, S. B., Tilcock, C., Madden, T. D., and Bally, M. B.**, Structural properties of lipids and their functional roles in biological membranes, in *Membrane Fluidity in Biology*, Vol. 1, *Concepts of Membrane Structure*, Aloia, R. C., Ed., Academic Press, New York, 1983, 39.

15. **Shands, J. W.**, The physical structure of bacterial lipopolysaccharides, in *Microbial Toxins*, Vol. 4, Weinbaum, G., Kadis, S., and Ajl, S. J., Eds., Academic Press, New York, 1971, 127.

16. **Cevc, G.**, Isothermal phase transitions, *Chem. Phys. Lipids*, 57, 293, 1991.

17. **Brandenburg, K. and Seydel, U.**, Investigations into the fluidity of lipopolysaccharide and free lipid A membrane systems by Fourier-transform infrared spectroscopy and differential scanning calorimetry, *Eur. J. Biochem.*, 191, 229, 1990.

18. **Emmerling, G., Henning, U., and Gulik-Krzywicki, T.**, Order-disorder conformational transition of hydrocarbon chains in lipopolysaccharides, *Eur. J. Biochem.*, 78, 503, 1977.

19. **Van Alphen, L., Lugtenberg, B., Rietschel, E. Th., and Mombers, C.**, Architecture of the outer membrane of *Escherichia coli* K12, *Eur. J. Biochem.*, 101, 571, 1979.

20. **Coughlin, R. T., Haug, A., and McGroarty, E. J.**, Physical properties of defined lipopolysaccharide salts, *Biochemistry*, 22, 2007, 1983.

21. **Peterson, A. A., Haug, A., and McGroarty, E. J.**, Physical properties of short- and long-O-antigen-containing fractions of lipopolysaccharide from O111:B4, *J. Bacteriol.*, 165, 116, 1986.

22. **Brandenburg, K. and Seydel, U.**, Physical aspects of structure and function of membranes made from lipopolysaccharides and free lipid A, *Biochim. Biophys. Acta*, 775, 225, 1984.

23. **Naumann, D., Schultz, C., Sabisch, A., Kastowsky, M., and Labischinski, H.**, New insights into the phase behaviour of a complex anionic amphiphile: architecture and dynamics of bacterial deep rough lipopolysaccharide membranes as seen by FTIR, X-ray, and molecular modelling techniques, *J. Mol. Struct.*, 214, 213, 1989.

24. **Jiao, B., Freudenberg, M., and Galanos, C.,** Characterization of the lipid A component of genuine smooth-form lipopolysaccharide, *Eur. J. Biochem.,* 180, 515, 1989.

25. **Naumann, D., Schultz, C., Born, J., Labischinski, H., Brandenburg, K., von Busse, G., Brade, H., and Seydel, U.,** Investigations into the polymorphism of lipid A from lipopolysaccharides of *Escherichia coli* and *Salmonella minnesota* by Fourier-transform infrared spectroscopy, *Eur. J. Biochem.,* 164, 159, 1987.

26. **Wollenweber, H.-W., Seydel, U., Lindner, B., Lüderitz, O., and Rietschel, E. Th.,** Nature and location of amide-bound *(R)*-3-acyloxyacyl groups in lipid A of lipopolysaccharides from various gram-negative bacteria, *Eur. J. Biochem.,* 145, 265, 1984.

27. **Mayer, H. and Weckesser, J.,** 'Unusual' lipid A's: structures, taxonomical relevance and potential value for endotoxin research, in *Handbook of Endotoxin,* Vol. 1, Rietschel, E. Th., Ed., Elsevier, Amsterdam, 1984, 221.

28. **Krauss, J. H., Seydel, U., Weckesser, J., and Mayer, H.,** Structural analysis of the nontoxic lipid A of *Rhodobacter capsulatus* 37b4, *Eur. J. Biochem.,* 180, 519, 1989.

29. **Brandenburg, K. and Blume, A.,** Investigations into the thermotropic phase behaviour of natural membranes extracted from Gram-negative bacteria and artificial membrane systems made from lipopolysaccharides and free lipid A, *Thermochim. Acta,* 119, 127, 1987.

30. **Coughlin, R. T., Peterson, A. A., Haug, A., Pownall, H. J., and McGroarty, E. J.,** A pH titration study on the ion binding within lipopolysaccharide aggregates, *Biochim. Biophys. Acta,* 821, 404, 1985.

31. **Brandenburg, K. and Seydel, U.,** Thermodynamic investigations on mono- and bilayer membrane systems made from lipid components of Gram-negative bacteria, *Thermochim. Acta,* 85, 473, 1985.

32. **Labischinski, H., Barnickel, G., Bradaczek, H., Naumann, D., Rietschel, E. Th., and Giesbrecht, P.,** High state of order of isolated bacterial lipopolysaccharide and its possible contribution to the permeation barrier property of the outer membrane, *J. Bacteriol.,* 162, 9, 1985.

33. **Nikaido, H., Takeuchi, Y., Ohnishi, S.-I., and Nakae, T.,** Outer membrane of *Salmonella typhimurium* electron spin resonance studies, *Biochim. Biophys. Acta,* 465, 152, 1977.

34. **Vaara, M., Plachy, W. Z., and Nikaido, H.,** Partitioning of hydrophobic probes into lipopolysaccharide bilayers, *Biochim. Biophys. Acta,* 1024, 152, 1990.

35. **Milner, K. C., Anacker, R. L., Fukushi, K., Haskins, W. T., Landy, M., Malmgren, B., and Ribi, E.,** Structure and biological properties of surface antigens from Gram-negative bacteria, *Bacteriol. Rev.,* 27, 352, 1963.

36. **Rothfield, L., Takeshita, M., Pearlman, M., and Horne, R. W.,** Role of phospholipids in the enzymatic synthesis of the bacterial cell envelope, *Fed. Proc.,* 25, 1495, 1966.

37. **Burge, R. E. and Draper, J. C.,** The structure of the cell wall of the Gram-negative bacterium *Proteus vulgaris.* III. A lipopolysaccharide "unit membrane", *J. Mol. Biol.,* 28, 205, 1967.

38. **Shands, J. W., Graham, J. A., and Nath, K.,** The morphologic structure of isolated bacterial lipopolysaccharide, *J. Mol. Biol.,* 25, 15, 1967.

39. **Wawra, H., Buschmann, H., Formanek, H., and Formanek, S.,** Strukturuntersuchung mit Röntgenbeugungsmethoden an Lipopolysacchariden von *Salmonella minnesota* Mutanten S SF 1111 und R 595 SF1167, *Z. Naturforsch.,* 34c, 171, 1979.

40. **Labischinski, H., Naumann, D., Schultz, C., Kusumoto, S., Shiba, T., Rietschel, E. Th., and Giesbrecht, P.,** Comparative X-ray and Fourier-transform-infrared investigations of conformational properties of bacterial and synthetic lipid A of *Escherichia coli* and *Salmonella minnesota* as well as partial structures and analogues thereof, *Eur. J. Biochem.,* 179, 659, 1989.

41. **Labischinski, H., Vorgel, E., Uebach, W., May, R. P., and Bradaczek, H.,** Architecture of bacterial lipid A in solution, A neutron small-angle scattering study, *Eur. J. Biochem.,* 190, 359, 1990.

42. **Hayter, J. B., Rivera, M., and McGroarty, E. J.**, Neutron scattering analysis of bacterial lipopolysaccharide phase structure, *J. Biol. Chem.*, 262, 5100, 1987.
43. **Brandenburg, K. and Seydel, U.**, Orientation measurements on membrane systems made from lipopolysaccharides and free lipid A by FT-IR spectroscopy, *Eur. Biophys. J.*, 16, 83, 1988.
44. **Brandenburg, K., Koch, M. H. J., and Seydel, U.**, Phase diagram of lipid A from *Salmonella minnesota* and *Escherichia coli* rough mutant lipopolysaccharide, *J. Struct. Biol.*, 105, 11, 1990.
45. **Seydel, U. and Brandenburg, K.**, Conformations of endotoxin and their relationship to biological activity, in *Cellular and Molecular Aspects of Endotoxin Reactions*, (Endotoxin Research Series, Vol. 1), Nowotny, A., Spitzer, J. J., and Ziegler, E. J., Eds., Excerpta Medica, Amsterdam, 1990, 61.
46. **Brandenburg, K., Koch, M. H. J., and Seydel, U.**, Structural Polymorphism of Deep Rough Mutant Lipopolysaccharide, Annu. Rep. HASYLAB (Hamburger Synchtron strahlungslabor am Deutschen Elektronen-Synchrotron DESY, Hamburg, Germany), 1990, 499.
47. **Seydel, U., Brandenburg, K., Koch, M. H. J., and Rietschel, E. Th.**, Supramolecular structure of lipopolysaccharide and free lipid A under physiological conditions as determined by synchrotron small-angle X-ray diffraction, *Eur. J. Biochem.*, 186, 325, 1989.
48. **Miner, V. W. and Prestegard, J. H.**, Structure of divalent cation-phosphatidic acid complexes as determined by ^{31}P-NMR, *Biochim. Biophys. Acta*, 774, 227, 1984.
49. **Verkleij, J., De Maagd, R., Leunissen-Bijvelt, J., and De Kruiff, B.**, Divalent cations and chlorpromazine can induce non-bilayer structures in phosphatidic acid-containing model membranes, *Biochim. Biophys. Acta*, 684, 255, 1982.
50. **Brandenburg, K. and Seydel, U.**, unpublished results.
51. **Lei, M.-G. and Morrison, D. C.**, Specific endotoxic lipopolysaccharide-binding proteins on murine splenocytes. I. Detection of lipopolysaccharide-binding sites on splenocytes and splenocyte subpopulations, *J. Immunol.*, 141, 996, 1988.
52. **Lei, M.-G. and Morrison, D. C.**, Specific endotoxic lipopolysaccharide-binding proteins on murine splenocytes. II. Membrane localization and binding characteristics, *J. Immunol.*, 141, 1006, 1988.
53. **Tobias, P. S., Soldau, K., and Ulevitch, R. J.**, Identification of a lipid A binding site in the acute phase reactant lipopolysaccharide binding protein, *J. Biol. Chem.*, 264, 10867, 1989.
54. **Wright, S. D., Tobias, P. S., Ulevitch, R. J., and Ramos, R. A.**, Lipopolysaccharide (LPS) binding protein opsonizes LPS-bearing particles for recognition by a novel receptor on macrophages, *J. Exp. Med.*, 170, 1231, 1989.
55. **Takayama, K., Din, Z. Z., Mukerjee, P., Cooke, P. H., and Kirkland, T. N.**, Physicochemical properties of the lipopolysaccharide unit that activates B lymphocytes, *J. Biol. Chem.*, 265, 14023, 1990.
56. **Lüderitz, Th., Brandenburg, K., Seydel, U., Roth, A., Galanos, C., and Rietschel, E. Th.**, Structural and physico-chemical requirements of endotoxins for the activation of arachidonic acid metabolism in mouse peritoneal macrophages *in vitro*, *Eur. J. Biochem.*, 179, 11, 1989.
57. **Kirikae, T., Inada, K., Hirata, M., Yoshida, M., Galanos, C., and Lüderitz, O.**, Hemagglutination induced by lipopolysaccharides, *Microbiol. Immunol.*, 30, 269, 1986.
58. **Clas, F., Schmidt, G., and Loos, R.**, The role of the classical pathway for the bactericidal effect of normal sera against Gram-negative bacteria, *Curr. Top. Microbiol. Immunol.*, 121, 261, 1988.
59. **Pugliese, C., LaSalle, M., and DeBari, V. A.**, Relationships between the structure and function of lipopolysaccharide chemotypes with regard to their effects on the human polymorphonuclear neutrophil, *Mol. Immunol.*, 25, 631, 1988.

60. **Kataoka, T., Inoue, K., Galanos, C., and Kinsky, S. C.,** Detection and specificity of lipid A antibodies using liposomes sensitized with lipid A and bacterial lipopolysaccharide, *Eur. J. Biochem.,* 24, 123, 1971.

61. **Flebbe, L., Vukajlovitch, and Morrison, D. C.,** Immunostimulation of C3H/HeJ lymphoid cells by R-chemotype lipopolysaccharide preparations, *J. Immunol.,* 142, 642, 1989.

62. **Schlecht, G. and Schmidt, G.,** Möglichkeiten zur Differenzierung von *Salmonella*-R-Formen mittels Antibiotica und antibakterieller Farbstoffe, *Zentralbl. Bakterol. Mikrobiol. Hyg. A,* 212, 505, 1969.

63. **Macias, E. A., Rana, F., Blazyk, J., and Modrzakowski, M. C.,** Bactericidal activity of magainin 2: use of lipopolysaccharide mutants, *Can. J. Microbiol.,* 36, 582, 1990.

64. **Ohno, N. and Morrison, D. C.,** Effects of lipopolysaccharide chemotype structure on binding and inactivation of hen egg lysozyme, *Eur. J. Biochem.,* 186, 621, 1989.

65. **Brade, L., Brandenburg, K., Kuhn, H.-M., Kusumoto, S., Macher, I., Rietschel, E. Th., and Brade, H.,** The immunogenicity and antigenicity of lipid A are influenced by its physicochemical state and environment, *Infect. Immunol.,* 55, 2636, 1987.

66. **Loppnow, H., Brade, H., Duerrbaum, I., Dinarello, C. A., Kusumoto, S., Rietschel, E. Th., and Flad, H.-D.,** Interleukin 1 induction capacity of defined lipopolysaccharide and partial structures, *J. Immunol.,* 142, 3229, 1989.

67. **Wang, M.-H., Feist, W., Herzbeck, H., Brade, H., Kusumoto, S., Rietschel, E. Th., Flad, H.-D., and Ulmer, A. J.,** Suppressive effect of lipid A partial structures on lipopolysaccharide or lipid A-induced release of interleukin 1 by human monocytes, *FEMS Microbiol. Immunol.,* 64, 179, 1990.

68. **Kropinski, A. M. B., Jewell, B., Kuzio, J., Milazzo, F., and Berry, D.,** Structure and function of *Pseudomonas aeruginosa* lipopolysaccharide, *Antibiot. Chemother.,* 36, 58, 1985.

69. **Tahri-Jouti, M.-A. and Chaby, R.,** Specific binding of lipopolysaccharides to mouse macrophages. I. Characteristics of the interaction and inefficiency of the polysaccharide region, *Mol. Immunol.,* 27, 751, 1990.

70. **Seydel, U., Schröder, G., and Brandenburg, K.,** Reconstitution of the lipid matrix of the outer membrane of Gram-negative bacteria as asymmetric planar bilayer, *J. Membr. Biol.,* 109, 95, 1989.

71. **Osborn, M. J.,** Biosynthesis and assembly of the lipopolysaccharide of the outer membrane, in *Bacterial Outer Membranes,* Inouye, M., Ed., John Wiley & Sons, New York, 1979, 15.

72. **Raetz, C. R. H.,** Biochemistry of endotoxins, *Annu. Rev. Biochem.,* 59, 129, 1990.

73. **Nikaido, H. and Vaara, M.,** Molecular basis of bacterial outer membrane permeability, *Microbiol. Rev.,* 49, 1, 1985.

74. **Lugtenberg, B. and Van Alphen, L.,** Molecular architecture and functioning of the outer membrane of *Escherichia coli* and other Gram-negative bacteria, *Biochim. Biophys. Acta,* 737, 51, 1983.

Section B: Cell Biology of Endotoxins

Chapter 10

CELLULAR MEMBRANE RECEPTORS FOR LIPOPOLYSACCHARIDE

Mei-Guey Lei and Tai-Ying Chen

TABLE OF CONTENTS

I. INTRODUCTION

Although the effects of lipopolysaccharide (LPS) on host cells have been clearly elucidated, the molecular mechanisms by which LPS interacts with cellular membranes and the subsequent signal transduction pathways are still not well defined. Many investigators have shown that the binding of LPS to the cell surface may involve nonspecific interactions with the membrane lipid bilayer (reviewed in Reference 1) as well as interactions with specific membrane gangliosides.[2-4] However, in the past several decades, extensive efforts to demonstrate specific LPS binding to membrane receptor glycoproteins have been pursued by many investigators (reviewed in References 5 to 8).

In 1974, Springer et al.[9] were the first to identify an LPS receptor, a 265-kDa lipoglycoprotein, on human erythrocytes. However, these investigators could not identify specific binding sites for LPS on human granulocytes, mononuclear leukocytes, and platelets.[10] Because of nonspecific interactions of LPS with the cellular membrane, the actual identification of specific LPS receptors has been difficult. Nevertheless, with the development of more advanced and improved techniques, a variety of membrane receptors for LPS has been reported by different laboratories. Since several reviews for LPS receptors have been published recently,[5-8] and the topics of LPS-nonspecific interactions, interactions with gangliosides, and signal transductions are reviewed separately by others in these volumes, we shall focus this review on the recent studies demonstrating specific binding of LPS to the cell surfaces, thus providing evidence of LPS receptors. We shall discuss those physically identified and characterized specific LPS-binding proteins and their functional roles as candidate cellular membrane receptors for LPS.

In general, LPS isolated from smooth-type Gram-negative bacteria consists of three distinct, covalently linked regions: the polysaccharide (O-antigen), the core oligosaccharide (outer core and inner core), and the lipid A region. The specific binding of LPS to the cellular membrane may be mediated through any of these regions. For example and as will be discussed in detail below, polysaccharide region was shown to bind specifically to the cellular mannose receptor,[11] core oligosaccharide region was shown to bind to the KDO receptor,[12] and lipid A region was shown to bind to the 73-kDa receptor[13] or CD11/CD18 molecules.[14] Structurally, the O-antigens of LPS from different gram-negative bacteria are heterogeneous, whereas the core oligosaccharides (especially inner core) and the lipid A regions are relatively conserved.[15] Functionally, lipid A has been defined as the primary active principle of LPS,[16] although more recent studies have identified inner core KDO determinants as relevant factors in interleukin-1 (IL-1) induction.[12,17] Based on these structural-functional relationships, we shall focus this review on the cellular membrane receptors for LPS inner core oligosaccharide and lipid A.

II. CELLULAR MEMBRANE REQUIREMENT FOR LPS BINDING

One of the earliest significant studies showing the importance of the cell membrane in LPS-triggering events was carried out by Jakobivitz and colleagues.[18] These investigators demonstrated that the LPS hyporesponsiveness of C3H/He3J mouse B-lymphocytes could be corrected with a purified membrane fraction from LPS-responsive C3HeB/FeJ mouse B-lymphocytes. In these studies the membrane fraction isolated from C3HeB/FeJ B-lymphocyte was fused to intact C3H/HeJ B-lymphocytes, thus rendering these cells mitogenically responsive to LPS. This experiment not only indicated that the defect of C3H/HeJ B-lymphocytes is in the cell membrane, but also demonstrated that the LPS-triggered events are initiated from the cell membrane.

Using tritium-labeled LPS (³H-LPS) from *Salmonella typhi*, Larsen and Sullivan[19] showed that the binding of LPS to human peripheral blood monocytes was specific and saturable. The binding of LPS to the cells was initially a rapid ($t_{1/2} < 5$ min), reversible, and temperature-independent surface adsorption followed by a slower ($t_{1/2} > 20$ min), irreversible, temperature-dependent interaction. In addition, these investigators used ⁵¹Cr-labeled lipid A to study the binding of lipid A to human monocytes. Although ⁵¹Cr-lipid A only showed a rapid ($t_{1/2} < 5$ min) binding to the cells, it exhibited a similar binding affinity to that of ³H-LPS. Both ³H-LPS and ⁵¹Cr-lipid A binding sites decreased when the cells were pretreated with unlabeled LPS, indicating a competitive binding. The binding characteristics of LPS to the monocytes provided some evidence supporting the concept of membrane receptors for LPS. However, the binding of ³H-LPS and ⁵¹Cr-lipid A to the isolated human monocyte membranes turned out to be less specific and showed a lower apparent affinity than that observed with intact cells.[20] Based on these results, Larsen and Sullivan suggested that the recognition and attachment of LPS to monocyte membranes may not be a simple classical ligand-receptor type of binding. Rather, structural relationships of membrane proteins (receptors) and phospholipids, and factors affecting hydrophobic and ionic associations between LPS and cellular membrane molecules, all contribute to an effective binding.[20] This interesting concept has yet to be totally refuted.

III. LECTIN-LIKE RECEPTORS FOR THE POLYSACCHARIDE STRUCTURE OF LPS

Raichvarg et al.[11] prepared a polysaccharide fraction by mild acid hydrolysis of LPS from *Haemophilus infuenzae*, which contained glucosyl, galactosyl, rhamnosyl, glucosaminyl, and mannosyl residues at 4:1:1:2:2 molar ratio. They used this polysaccharide to stimulate mouse peritoneal macrophages *in vitro* and found that the macrophages showed an increased capacity for the phagocytosis. The induced phagocytosis could be inhibited specifically by mannosyl polymers, but not by other sugar polymers. Based on these results, Raichvarg et al. suggested that there is a mannosyl-poly-saccharide binding site on mouse peritoneal macrophages. Recently, Ste-phenson and Shepherd[21] isolated a mannose receptor (175-kDa protein) from human alveolar macrophages which may have a relationship to the mannose receptor defined by Raichvarg et al.[11] However, Akagawa and Tokunaga[22] failed to find specific binding site for LPS on mouse alveolar macrophages. Therefore, the relationship of the 175-kDa mannose receptor to the mannose binding sites suggested by Raichvarg et al.[11] for the LPS on mouse peritoneal macrophages remains to be determined.

Haeffner-Cavaillon et al.[23] prepared a tritium-labeled LPS (^{3}H-LPS) from *Bordetalla pertussis* and demonstrated that the binding of LPS to both resident and elicited rabbit peritoneal macrophages was specific, saturable, and re-versible. They further showed that the binding of ^{3}H-LPS to rabbit macro-phages was significantly inhibited by one of the two polysaccharides they prepared (PS-1), but not by another polysaccharide (PS-2) or lipid A (PS-1, PS-2, and lipid A were prepared from the same LPS). Therefore, these in-vestigators suggested a lectin-like receptor for LPS on the membrane of rabbit peritoneal macrophages. Later, these investigators extended their studies[24] and showed specific binding of LPS to mouse-elicited peritoneal macrophages and human blood monocytes. Again the binding of LPS was implicated to be mediated by polysaccharide region. They also reported that the specific binding of LPS to mouse macrophages and human monocytes was serum dependent. Since the specific binding of LPS to human monocytes could not be detected in complement component C3-depleted human serum, these in-vestigators suggested that specific binding of LPS to human monocytes may be dependent on the C3-cleavage products utilizing C3 receptors.

The functional role of inner core oligosaccharide in the IL-1 production by human monocytes has been studied by Cavaillon's group.[25] Initially an isolated polysaccharide component of LPS, (PS-1), which contains inner core oligosaccharide, was shown to be as potent as intact LPS and lipid A to induce human monocytes to produce IL-1 at a concentration of about 10 μg/ml. Later, the minimal structure required for the induction of IL-1 was determined to be the heptosyl-2-keto-3-deoxyoctulosonate [hep(1-5)-KDO] disaccharide of the inner core oligosaccharide.[17] Furthermore, within the inner core, KDO was shown to play the major role to induce IL-1 production; however, mon-omeric KDO and KDO (2-4)-KDO disaccharide have no such activity.[12,17] Cavaillon and colleagues later reported that inner core KDO residues mediate both IL-1 production and IL-1 release; in contrast, lipid A induces IL-1 production but not IL-1 release.[26,27] On the basis of these studies, Cavaillon and colleagues have proposed a specific membrane receptor for the KDO region of LPS in which ionic interactions are likely to be involved in the ligand-receptor interactions.[12,26]

The report that the isolated polysaccharide chain (PS-1) was as potent as intact LPS and lipid A to induce IL-1 production by human monocytes was further studied by Loppnow et al.,[28] who reported that isolated inner core hep-hep-KDO trisaccharide at relatively lower concentration (<0.2 μg/ml) did not induce IL-1 production. Rather, the minimal concentration of inner core oligosaccharide required to induce IL-1 production was 10^5- to 10^6-fold higher than that of LPS and lipid A. Recently, Flebbe et al.[29] reported that Re LPS and Rd LPS (From *Salmonella minnesota*) have similar ability to induce IL-1 production and release by C3HeB/FeJ mouse peritoneal elicited macrophages. If the inner core were to be responsible for IL-1 production and release, and lipid A were only to induce IL-1 production but not release,[26,27] then the induction and release of IL-1 by Re LPS and Rd LPS must be mediated by inner core LPS structural determinants. Since the core oligosaccharide of Re LPS does not contain heptose, whereas Rd LPS does, the minimal structure required for IL-1 induction reported by Cavaillon's laboratory[17] might need to be reevaluated. If, however, the minimal structure required for IL-1 induction is as reported, namely, the hep-KDO disaccharide, Re LPS-induced IL-1 production and release[29] must be mediated by the lipid A moiety of the LPS. Thus, the notion that lipid A can only induce IL-1 production but not IL-1 release[27] needs to be clarified. However, it is also possible that the IL-1 induction mechanisms are totally different between human monocytes and mouse macrophages, a possibility which would also explain these results.

IV. MEMBRANE RECEPTORS FOR LIPID A

In contrast to the conclusion of Cavaillon and colleagues[24] that the specific binding of LPS to mouse peritoneal elicited macrophages, rabbit peritoneal macrophages, and human monocytes are mediated by the inner core oligosaccharide region, other researchers have reported that lipid A is the region responsible for the specific LPS binding to mouse macrophages.

Akagawa and Tokunaga[22] showed that the induction of mouse macrophages to tumor cytotoxicity by LPS was tissue dependent and corresponded with the capacity of the macrophages for specific LPS binding. These investigators found that LPS-activated mouse peritoneal macrophages (both resident and exudate) were tumor cytotoxic. However, they were unable to stimulate mouse alveolar macrophages by LPS to induce cytotoxicity under many experimental conditions. While evaluating the binding of FTIC-LPS on the surface of macrophages, they found that greater than 90% of peritoneal macrophages were stained with FITC-LPS, whereas less than 5% of alveolar macrophages were stained. Of interest, by treating lung macrophages with IFN-γ, the binding of LPS to the cell surface was restored, and the macrophages were induced for cytotoxicity in response to LPS. The results suggested that IFN-γ promotes the expression of LPS-binding sites on mouse lung macrophages. Further, these authors demonstrated that the binding of LPS to both peritoneal macrophages and IFN-γ-treated lung macrophages were inhibitable by either polymyxin B, which forms complex with lipid A and neutralizes the activity of lipid A,[30] or unlabeled LPS. Thus, Akagawa and Tokunaga concluded that the binding of LPS to the macrophages may be mediated by the lipid A region of LPS and is not a simple nonspecific interaction.

Akagawa and colleagues[31] later reported that FITC-LPS negative mouse alveolar macrophages stimulated with colony-stimulating factor (CSF-1) or L929-cell-conditioned medium were stained positively with FITC-LPS. The stimulated cells significantly increased the expression of surface Mac-1 antigen and manifest peritoneal macrophage-like morphology. These results suggested to the authors that LPS binding to the murine macrophages may be developmentally regulated.

Evidence that LPS binding to macrophages may be dependent on microenvironment was further provided by the work of Erroi and colleagues.[32] These investigators found that macrophages associated with certain poorly immunogenic murine tumors were unresponsive to LPS, as assessed by the induction of procoagulant activity. Only 6% of these tumor-associated macrophages were stained with FITC-LPS, whereas more than 80% of resident peritoneal macrophages were positive for FITC-LPS. Treating these tumor-associated macrophages with IFN-γ cound neither restore the binding of FITC-LPS to the cells or induce these macrophages to produce procoagulant activity in response to LPS. These results suggested that these tumor-associated macrophages either lack totally or express very low levels of binding sites for LPS.

Using tritium-labeled LPS from *B. pertussis,* Tahri-Jouti and Chaby[33] have more recently demonstrated a dose-dependent, time-dependent, and saturable LPS binding to the mouse peritoneal macrophages and macrophage-like cell lines (J774 and P388D1). The kinetics of the specific binding was temperature dependent. They further showed that neither the polysaccharide region (PS-1) of the LPS nor bovine serum albumin (BSA)-coupled KDO bound to these macrophages, wherease BSA-conjugated lipid A clearly manifests a specific binding. The binding was not inhibited by ligands of scavenger receptors (methylated BSA) or complement receptors (zymosan) but was inhibited by dexamethasone (modulate the expression of other macrophage surface markers). In an accompanying paper, Tahri-Jouti et al.[34] reported that monoclonal antibodies to the lipid A region of LPS inhibited the specific binding of LPS to the macrophages and suggested that ester-bound fatty acids of the lipid A region may be involved in the binding of LPS to the macrophages. Considerably reduced LPS binding to the macrophages was observed with the fatty acid esters of LPS were removed by either hydroxylamine or carboxylic ester hydrolase treatment. Using various synthetic lipids, the substructures of lipid A which may interact with cellular membranes were determined. It was suggested that there may be three classes of LPS binding sites (i.e., multiple receptors) present on the macrophage surface.[34]

All of the above studies provided good evidence for the concept of the existence of cell surface LPS receptors; however, none of the above-cited studies established molecular identity of specific cell surface receptors for LPS.

V. CANDIDATES FOR LPS RECEPTOR(S)

Wright and Jong[14] showed that LPS-coated erythrocytes bind to human macrophages. The binding was inhibited with monoclonal antibodies directed to the adhesion molecules (a family of cell surface receptors: CR3, CD11b/CD18; LFA-1, CD11a/CD18; and P150,95, CD11c/CD18). Using individual monoclonal anti-CD11 antibodies they did not find significant inhibition of LPS binding to the macrophages. Nevertheless, a profound inhibition of binding of the LPS-coated erythrocytes was observed by treating the cells with either anti-CD18 antibody or a combination of anti-CD11 antibodies. These results suggested that all three members of CD11/CD18 family can bind LPS-coated erythrocytes. Furthermore, polymyxin B sulfate could block the binding of LPS-coated erythrocytes to macrophages, suggesting the lipid A region is responsible for the recognition of LPS-coated erythrocytes to macrophages.

Using monoclonal antibodies to different epitopes on CD11b, Wright and Levin[35] found that the binding site of CD11b/CD18 for LPS is different from the site recognized by Arg-Gly-Asp region of C3bi or synthetic peptides. In addition, phagocytes from CD18-deficient patients showed no binding activity for LPS-coated erythrocytes. However, Wright et al. reported later[36] that LPS could, nevertheless, induce monocytes and macrophages isolated from CD18-deficient patients to produce normal amount of IL-1 and tumor necrosis factor alpha (TNFα). In addition, LPS could prime monocytes and PMN from CD18-deficient patients normally for enhanced release of O_2^-. Thus, although Wright and colleagues' earlier studies have clearly documented a role for CD11/CD18 molecules on phagocytes binding of LPS-coated particles to these cells. CD11/CD18 molecules may not be essential for cellular responses to LPS.[36]

More recently, Wright et al.[37] reported that a monocyte/macrophage surface antigen, CD14, may serve as a receptor for the complex of LPS and LBP (an acute-phase LPS-binding protein). LBP promoted LPS-coated erythrocytes to bind to mononuclear phagocytes, but not to polymorphonuclear or other cell types. This enhanced binding was also observed with monocytes isolated from CD18-deficient patients. Monoclonal antiboides to CD14 down-modulated the binding of the complex of LBP and LPS-coated erythrocytes to the macrophages and prevented the induction of TNF-α by whole blood incubated with LPS (1 ng/ml). These authors suggested that, at very low concentrations of LPS, CD14 may serve to bind LBP-LPS complexes to the cell surface, thus potentiating the monocyte response to LPS. However, at higher concentrations of LPS (e.g., 10 ng/ml), LPS could induce macrophages to produce TNF-α without either LBP or CD14.[37] It is also noteworthy that CD14-deficient patients have been reported to produce IL-1 upon LPS stimulation *in vitro*.[27] Thus, while CD14 may serve as a receptor for LPS-LBP complexes, it may not be crucial for LPS-induced responses in macrophages.

Hampton et al.[38] have employed radiolabeled lipid A precursor, 4'-^{32}P-lipid IV$_A$, as a probe in a ligand-blotting assay to identify specific lipid A binding sites on the RAW 264.7 macrophage-like cell line. The binding of 4'-^{32}P-lipid IV$_A$ to RAW 264.7 cells was shown to be saturable, inhibitable with excess unlabeled lipid IV$_A$, and sensitive to proteinase K. The binding could be inhibited by both Re LPS and lipid A. Of interest, however, binding was not inhibitable by S LPS. To detect lipid IV$_A$-binding proteins, the proteins of whole cell lysate, membrane, and nuclear fractions were resolved by sodium dodecyl sulfate-polyacrylamide gel electrophoresis (SDS-PAGE), immobilized on nitrocellulose paper, and then probed with 4'-^{32}P-lipid IV$_A$ in the absence or presence of excess unlabeled lipid IV$_A$. Under these experimental conditions, Hampton et al. detected several lipid IV$_A$-binding proteins, in which proteins of 31 and 95 kDa were predominant. The 31-kDa protein was detected in the nuclear fraction and was suggested by the authors to be a histone, whereas the 95-kDa protein was detected in the membrane fraction. The binding of 4'-^{32}P-lipid IV$_A$ to the 95-kDa protein was inhibitable with excess unlabeled lipid IV$_A$. Based on the above results, these investigators suggested that the 95-kDa protein is a binding site for lipid A and may be a candidate receptor for lipid A on the cell surface. Using a similar approach, Kirikae et al.[39] have detected a 96-kDa Re LPS binding protein on mouse erythrocytes. However, the relationship of this protein to the 95-kDa lipid A binding protein on macrophage cell line reported by Hampton et al. remains to be defined.

Employing a different experimental approach, Wollenweber and Morrison developed a radioiodinated, disulfide reducible, photoactivatable LPS derivative, ^{125}I-ASD-LPS, for detecting LPS-binding proteins.[40] Using this LPS probe allowed Lei and Morrison to detect specific LPS-binding proteins on mouse splenocytes.[13] Splenocytes were photoaffinity labeled and the reduced and solubilized cell lysates analyzed by two-dimensional polyacrylamide gel electrophoresis and autoradiography to detect the radiolabeled LPS-binding proteins. An 80-kDa protein with pI of about 6.5 was thus identified as a major LPS-binding protein. The binding of LPS to the 80-kDa protein was saturable and inhibitable by both homologous and heterologous underivatized LPS as well as by purified lipid A.[41] As determined by sucrose gradient centrifugation, the 80-kDa LPS-binding protein was found to be associated with membrane fraction of the splenocytes. This protein could be noncytotoxically released by a nonionic detergent, octylglucoside.

Lei and Morrison further demonstrated that the 80-kDa LPS-binding protein was present in mouse splenic B-lymphocytes, T-lymphocytes, and macrophages. The protein was detected on mouse 70Z/3 pre B cell line. YAC-1 and EL4 T cell lines but not on undifferentiated mouse Sp2/0 myeloma cell line. The 80-kDa protein was also detected on splenocytes of LPS-unresponsive C3H/HeJ (*lps^d*) mice, which appeared to be indistinguishable to that on splenocytes of C3HeB/FeJ (*lps^n*) mice.[13] The data indicated that the defect of C3H/HeJ may not be necessarily the result of the absence of specific LPS-binding proteins or receptors. Possibly, a component required for effective transmembrane triggering subsequent to LPS-receptor interaction may be defective or missing in C3H/HeJ mice. Alternatively, a mutation in the receptor which does not affect LPS binding and/or could not detected by two-dimensional gel electrophoresis may be responsible for the defect.

In addition to the mouse splenocytes and cell lines, proteins similar to the mouse 80-kDa-specific LPS-binding protein were detected on lymphoid cells of various mammalian specific including man, but not on those of chicken and frog.[42] The presence of the 80-kDa protein on the lymphoid cells may reflect the sensitivity of these species to LPS. The data also implicated the 80-kDa-specific LPS-binding proteins are relatively conserved among mammalian species. Recently, Halling et al.[43] have extensively analyzed specific LPS-binding proteins on human peripheral blood cell subpopulations. Again, these investigators found the 80-kDa protein as a predominant LPS-binding site on monocytes, B- and T-lymphocytes, neutrophils, and platelets. However, the specific LPS-binding proteins were not detected on human erythrocytes. Of interest, these authors further demonstrated that the 80-kDa protein detected on human blood cells may be antigenically related to that of mouse splenocytes.

Using photoaffinity labeling, butanol extraction, preparative SDS-PAGE, and electro-elution, Chen et al.[44] partially purified the 80-kDa protein from C3HeB/FeJ splenocytes. Immunizing Armenian hamsters with the partially purified 80-kDa protein, Bright et al.[45] developed a panel of hamster-mouse monoclonal antibodies with specificity for the 80-kDa LPS-binding protein. One monoclonal antibody, MAb5D3, which has been extensively characterized, exhibited binding specificity for highly purified 80-kDa protein as determined by enzyme-linked immunosorbent assay (ELISA). This antibody competitively inhibited the binding of LPS to the 80-kDa protein and vice versa.

To evaluate the functional properties of MAb5D3, Morrison et al. demonstrated that this monoclonal antibody functions as an agonist *in vitro* and *in vivo*.[46,47] MAb5D3 was found to activate bone marrow culture-derived C3H/HeN macrophages for tumor cell killing. This antibody failed to stimulate C3H/HeJ macrophages, although quantitative ELISA showed the bindings of this monoclonal antibody to these two macrophage populations to be almost identical.[46] The antibody also showed a full capacity to protect mice against lethal effects of endotoxin.[47] Significantly, after treating the antibody at 100°C for 1 h, the biological activity of the antibody was totally abrogated, indicating that the antibody was not due to the contamination of LPS. These results provided strong evidence that the 80-kDa LPS-binding protein may serve as a functional receptor for LPS.

Using the procedures developed by Wollenweber and Morrison,[40] Kirkland and co-workers also synthesized a ^{125}I-ASD-Re595 LPS probe to identify LPS-binding proteins on 70Z/3 pre B cell line. They found a major LPS-binding protein with apparent molecular weight of 18 kDa and two minor proteins of 25 and 28 kDa by SDS-PAGE.[48] However, the functional aspects of these proteins are yet to be established.

Hara-Kuge et al.[49] applied the ^{125}I-ASD-LPS photoaffinity labeling method described by Lei and Morrison[13] and identified two proteins with molecular weight of 65 and 55 kDa on mouse J774.1 macrophage-like cell line by SDS-PAGE. The binding of ^{125}I-ASD-LPS to 65- and 55-kDa proteins was inhibited by unlabeled LPS or lipid X. Significantly, neither of these proteins were detected in LPS-resistant mutant LR-9 cells isolated from the J774.1 cell line. The authors thus concluded that one or both of the two LPS-binding proteins might be related to the specific membrane receptors for LPS. It would be of considerable interest to compare the 80-kDa LPS-binding protein characterized by Lei and Morrison and the 65-kDa protein described by Hara-Kuge and colleagues. It is possible that the proteins reported by these two groups are identifcal, since the same LPS probes were used in both studies. The disagreement in apparent molecular weights may be due to different electrophoresis conditions. If this is to be the case, the results reported by Hara-Kuge and colleagues have further provided genetic evidence to support that the 80-kDa (or 65-kDa) protein is a functional receptor for LPS.

More recently, Dziarski[50] has identified peptidoglycan (PGN, isolated from *Staphylococcus aureus*) and LPS-binding proteins on mouse B-lymphocytes by photoaffinity-labeling procedure with ^{125}I-ASD-PGN and ^{125}I-ASD-LPS as probes. Both PGN- and LPS-binding proteins have similar apparent molecular weight of 70 kDa and pI of 6.5. Binding of PGN probe to the 70-kDa protein was inhibited by LPS, Re LPS, and lipid A. Binding of LPS probe to the 70-kDa protein was inhibited by PGN. Bindings of both LPS probe and PGN probe to the 70-kDa protein were inhibitable by dextran sulfate, a (GlcNAc)$_2$-specific lectin. The proteins labeled with LPS and PGN probes were inseparable and had identical peptide maps when they were digested with chymotrypsin, subtilisin, staphylococcal protease V8, or papain. Dziarski has concluded that the 70-kDa protein functions as a binding protein for both LPS and PGN and suggested that the binding was specific for the (GlcNAc)$_2$ moiety of lipid A and (GlcNAc-MurNAc)$_n$ backbone of PGN. In an accompanying paper, Dziarski[51] established that the 70-kDa LPS- and/or PGN-binding protein is membrane localized. Since the LPS probe used by Dziarski was made according to the method of Wollenweber and Morrison,[40] the 70-kDa protein reported by Dziarski may be identical to the 80-kDa protein reported by Lei and Morrison. (These two proteins also showed identical pI value.) Dziarski has confirmed[50] this assumption by using LPS probe made in Morrison's laboratory. The difference in the molecular weight was again probably due to the difference in the gel electrophoresis. Recently, in Morrison's laboratory, the apparent molecular weight of the 80-kDa LPS-binding protein has been reevaluated in detail by one-dimensional SDS-PAGE and shown to have an apparent molecular mass of 73 kDa.[52] Since bacterial PGN and LPS are both potent macrophage activators and polyclonal B-lymphocyte mitogens, the results reported by Dziarski provided further support that the 73- (or 70-) kDa LPS/PGN-binding protein may be a functional receptor.

Although Dziarski showed that both LPS and PGN bind to the same 70- (or 73-) kDa protein, Lei et al. found that the 73-kDa protein is specific for LPS but not for PGN. A variety of peptidoglycan-polysaccharide (PGN-PS) and PGN from *Streptococcus pyrogen, S. faecium,* and *Lactobacillus casei,* pseudomurein-polysaccharide from *Methanobacterium formicium,* and synthetic *N*-acetylmuramyl-L-alanyl-D-isoglucamine (muramyl dipeptide) showed no inhibition to the binding of LPS probe to the 73-kDa protein.[52] The reason for the discrepancy found between these two laboratories is not known, it may be due to the different sources of PGNs used and the procedures applied for the PGN preparation.

Other than lymphoreticular cells, Parent[53,54] found high-affinity LPS-binding sites on rat hepatocytes, which recognized the heptose region of the core oligosaccharide. Using photoactivable, iodinated LPS probes,[40] he identified a 47-kDa integral membrane protein on hepatocytes by SDS-PAGE. The author suggested that the 47-kDa protein may be a subunit of the LPS receptor on hepatocytes.

Following the identification of a 73-kDa specific LPS-binding protein (candidate LPS receptor for lipid A region), more recently, Lei and Morrison using the ^{125}I-ASD-LPS probe have identified a 38-kDa KDO-specific LPS-binding protein on mouse lymphocytes and macrophages. The binding was found to be inhibited by unlabeled Re595-LPS, but not by purified lipid A.[55] It would be of interest to speculate that the 38-kDa LPS-binding protein may be a candidate receptor for the KDO region of LPS. Further investigation of this protein is in progress.

VI. SUMMARY AND CONCLUSIONS

To date, several laboratories have provided strong evidence supporting the concept of LPS receptors on the membranes of mammalian cells. It appears that there may exist lectin-like receptors for core oligosaccharide region of LPS as well as specific receptors for lipid A. The functional roles of 96-, 95-, 47-, 38-, 28-, 25-, and 18-kDa LPS-binding proteins are not clear yet and further research is needed. There exist some experimental evidence that adhesion molecules, CD11/CD18, and monocyte/macrophage surface antigen, CD14, may serve as LPS receptors. However, experimental results obtained from patients deficient in CD18 or CD14 indicate that these molecules may not be essential for LPS-triggered cellular responses. In contract, accumulating experimental evidence has implicated the 73- (or 70-), 65-, and 55-kDa proteins to be possible candidate receptors for LPS. Alternative approaches are however clearly warranted in order to prove and define unequivocally the functional receptor for LPS. Undoubtedly, molecular biology approaches will provide a definitive answer for the question of membrane receptors for LPS.

ACKNOWLEDGMENTS

This work was supported by NIH grants AI 23447-05 and IA 22948-06.

REFERENCES

1. **Morrison, D. C.**, Non-specific interactions of bacterial lipopolysaccharides with membranes and membrane components. in *Handbook of Endotoxin*, Vol. 3, Berry, L. J., Ed., Elsevier Science, New York, 1985, 25.
2. **Coleman, D. L., Morrison, D. C., and Ryan, J. L.**, Gangliosides block the inhibition of macrophage Fc-dependent phagocytosis by lipopolysaccharide, *Cell. Immunol.*, 100, 288, 1986.
3. **Cavaillon, J. M., Fitting, C., Hauttecoeur, B., and Haeffner-Cavaillon, N.**, Inhibition by gangliosides of the specific binding of lipopolysaccharide (LPS) to human monocytes prevents LPS-induced interleukin-1 production, *Cell. Immunol.*, 106, 293, 1987.
4. **Yohe, H. C., Berenson, C. S., Cuny, C. L., and Ryan, J. L.**, Altered B-lymphocyte membrane architecture indicated by ganglioside accessibility in C3H/HeJ mice, *Infect. Immun.*, 58, 2888, 1990.
5. **Haeffner-Cavaillon, N., Cavaillon, J. M., and Szabo, L.**, Cellular receptors for endotoxin, in *Handbook of Endotoxin*, Vol. 3, Berry, L. J., Ed., Elsevier Science, New York, 1985, 1.
6. **Morrison, D. C.**, Minireview — the case for specific lipopolysaccharide receptors expressed on mammalian cells, *Microb. Pathogenesis*, 7, 389, 1989.
7. **Lei, M. G., Chen, T. Y., and Morrison, D. C.**, Lipopolysaccharide receptors on mammalian lymphoreticular cells, *Int. Rev. Immunol.*, 6, 223, 1990.
8. **Chen, T. Y., Lei, M. G., Suzuki, T., and Morrison, D. C.**, The macrophage receptor for bacterial lipopolysaccharide, in *Macrophages and Macrophage Activation*, Gordon, S. and Russell, S. W., Ed., Springer-Verlag, Heidelberg, in press.
9. **Springer, G. F., Adye, J. C., Bezkorovainy, A., and Jirgensons, B.**, Properties and activity of the lipopolysaccharide-receptor from human erythrocytes, *Biochemistry*, 13, 1379, 1974.
10. **Springer, G. F. and Adye, J. C.**, Endotoxin binding substances from human leukocytes and platelets, *Infect. Immun.*, 12, 978, 1975.
11. **Raichvarg, D., Hatat, D., Sarfati, G., and Agneray, J.**, Action of a polysaccharide fraction of *Haemophilus influenzae* lipopolysaccharide on macrophage: implication of receptor for mannosyl-polysaccharides, *Med. Microb. Immol.*, 171, 91, 1982.
12. **Haeffner-Cavaillon, N., Caroff, M., and Cavaillon, J. M.**, Interleukin-1 induction by lipopolysaccharide: structure requirements of the 3-deoxy-D-manno-2-octulosonic acid (KDO), *Mol. Immunol.*, 26, 485, 1989.

13. **Lei, M. G. and Morrison, D. C.**, Specific endotoxic lipopolysaccharide binding proteins on murine splenocytes. I. Detection of LPS binding sites on splenocytes and splenocyte subpopulation, *J. Immunol.*, 141, 996, 1988.

14. **Wright, S. D. and Jong, M. T. C.**, Adhesion-promoting receptor on human macrophages recognize *Escherichia coli* by binding to lipopolysaccharide, *J. Exp. Med.*, 164, 1876, 1986.

15. **Rietschel, E. Th., Brade, L., Holst, O., Kulshin, V. A., Lindner, B., Moran, A. P., Schade, U. F., Zahringer, U., and Brade, H.**, Molecular structure of bacterial endotoxin in relation to bioactivity, in *Cellular and Molecular Aspects of Endotoxin Reactons*, Nowotny, A., Spitzer, J. J., and Ziegler, E. J., Ed., Elsevier Science, New York, 1990, 15.

16. **Rietschel, E. Th., Brade, H., Brade, L., Brandenburg, K., Schade, U., Seydel, U., Zahringer, U., Galanos, C., Luderitz, O., Westphal, O., Labischinski, H., Kusumoto, S., and Shiba, T.**, Lipid A, the endotoxic center of bacterial lipopolysaccharide: relation of chemical structure to biological activity, *Prog. Clin. Biol. Res.*, 231, 25, 1987.

17. **Lebbar, S., Cavaillon, J. M., Caroff, M., Ledur, A., Brade, H., Sarfati, R., and Haeffner-Cavaillon, N.**, Molecular requirement for interleukin 1 induction by lipopolysaccharide-stimulated human monocytes: involvement of the heptosyl-2-keto-3-dexoyoctulosonate region, *Eur. J. Immunol.*, 16, 87, 1986.

18. **Jakobivitz, A., Sharon, N., and Zan-Bar, I.**, Acquisition of mitogenic responsiveness by non-responding lymphocytes upon insertion of appropriate membrane components, *J. Exp. Med.*, 156, 1274, 1982.

19. **Larsen, N. E. and Sullivan, R.**, Interaction between endotoxin and human monocytes: characteristics of the binding of ^{3}H-labeled lipopolysaccharides and ^{51}Cr-labeled lipid A before and after the induction of endotoxin tolerance, *Proc. Natl. Acad. Sci. U.S.A.*, 81, 3491, 1984.

20. **Larsen, N. E. and Sullivan, R.**, Interaction of radiolabeled endotoxin molecules with human monocyte membranes, *Biochim. Biophys. Acta*, 774, 261, 1984.

21. **Stephenson, J. D. and Shepherd, V. L.**, Purification of the human alveolar macrophage mannose receptor, *Biochem. Biophys. Res. Commun.*, 148, 883, 1987.

22. **Akagawa, K. S. and Tokunaga, T.**, Lack of binding of bacterial lipopolysaccharide to mouse lung macrophages and restoration of binding by γ-interferon, *J. Exp. Med.*, .162, 1444, 1985.

23. **Haeffner-Cavaillon, N., Chaby, R., Cavaillon, J. M., and Szabo, L.**, Lipopolysaccharide receptor on rabbit peritoneal macrophages. I. Binding characteristics, *J. Immunol.*, 128, 1950, 1982.

24. **Haeffner-Cavaillon, N., Cavaillon, J. M., Etievant, M., Lebbar, S., and Szabo, L.**, Specific binding of endotoxin to human monocytes and mouse macrophages: serum requirement, *Cell. Immunol.*, 91, 119, 1985.

25. **Haeffner-Cavaillon, N., Cavaillon, J. M., Moreau, M., and Szabo, L.**, Interleukin 1 secretion by human monocytes stimulated by the isolated polysaccharide region of the *Bordetella pertussis* endotoxin, *Mol. Immunol.*, 21, 389, 1984.

26. **Haeffner-Cavaillon, N. and Cavaillon, J. M.**, Involvement of the LPS receptor in the induction of interleukin-1 in human monocytes stimulated with endotoxins, *Ann. Inst. Pasteur/Immunol.*, 138, 461, 1987.

27. **Cavaillon, J. M., Munoz, C., Fitting, C., Couturier, C., and Haeffner-Cavaillon, N.**, Signals involved in interleukin-1 production induced by endotoxins, in *Cellular and Molecular Aspects of Endotoxin Reactions*, Nowotny, A., Spitzer, J. J., and Ziegler, E. J., Ed., Elsevier Science, New York, 1990, 257.

28. **Loppnow, H., Brade, H., Durrbaum, I., Dinarello, C. A., Kusumoto, S., Rietschel, E. Th., and Flad, H. D.**, IL-1 induction-capacity of defined lipopolysaccharide partial structures, *J. Immunol.*, 142, 3229, 1989.

29. **Flebbe, L. M., Chapes, S. K., and Morrison, D. C.,** Activation of C3H/HeJ macrophage tumoricidal activity and cytokine release by R-chemotype lipopolysaccharide preparations. Differential effects of IFN-γ, *J. Immunol.*, 145, 1505, 1990.

30. **Morrison, D. C. and Jacobs, D. M.,** Binding of polymyxin B to the lipid A portion of bacterial lipopolysacchardies, *Immunochemistry*, 13, 813, 1976.

31. **Akagawa, K. S., Kamoshita, K., and Tokunaga, T.,** Effects of granulocyte-macrophage colony-stimulating factor and colony-stimulating factor-1 on the proliferation and differentiation of murine alveolar macrophages, *J. Immunol.*, 141, 3383, 1988.

32. **Erroi, A., Casali, B., Donati, M. B., Mantovani, A., and Semeraro, N.,** Mouse tumor-associated macrophages do not generate procoagulant activity in response to different stimuli, *Int. J. Cancer*, 41, 65, 1988.

33. **Tahri-Jouti, M. and Chaby, R.,** Specific binding of lipopolysaccharides to mouse macrophages. I. Characteristics of the interaction and inefficiency of the polysaccharide region, *Mol. Immunol.*, 27, 751, 1990.

34. **Tahri-Jouti, M., Mondange, M., Dur, A. L., Auzanneau, F., Charon, D., Girard, R., and Chaby, R.,** Specific binding of lipopolysaccharides to mouse macrophages. II.Involvement of distinct lipid A substructures, *Mol. Immunol.*, 27, 763, 1990.

35. **Wright, S. D. and Levin, S. D.,** CR3 (CD11b/CD18) expresses one binding site for Arg-Gly-Asp-containing peptides and a second site for bacterial lipopolysaccharide, *J. Exp. Med.*, 169, 175, 1989.

36. **Wright, S. D., Detmers, P. A., Aida, Y., Adamowski, R., Anderson, D. C., Chad, Z., Kabbash, L. G., and Pabst, M. J.,** CD18-deficient cells respond to lipopolysaccharide *in vitro*, *J. Immunol.*, 144, 2566, 1990.

37. **Wright, S. D., Ramos, R. A., Tobias, P. S., Ulevitch, R. J., and Mathison, J. C.,** CD14, a receptor for complexes of lipopolysaccharide (LPS) and LPS binding protein, *Science*, 249, 1431, 1990.

38. **Hampton, R. Y., Golenbock, D. T., and Raetz, R. H. C.,** Lipid A binding sites in membranes of macrophage tumor cells, *J. Biol. Chem.*, 263, 14802, 1988.

39. **Kirikae, T., Inada, K., Hirata, M., Yoshida, M., Kondo, S., and Hisatsune, K.,** Identification of Re lipopolysaccharide-binding protein on murine erythrocyte membrane, *Microb. Immunol.*, 32, 33, 1988.

40. **Wollenweber, H. W. and Morrison, D. C.,** Synthesis and biochemical characterization of a photoactivatable, iodinatable, cleavable bacterial lipopolysaccharide derivative, *J. Biol. Chem.*, 260, 15068, 1985.

41. **Lei, M. G. and Morrison, D. C.,** Specific endotoxic lipopolysaccharide binding proteins on murine splenocytes. II. Membrane localization and binding characteristics, *J. Immunol.*, 141, 1006, 1988.

42. **Roeder, D., Lei, M. G., and Morrison, D. C.,** Specific endotoxic lipopolysaccharide (LPS) binding proteins on lymphoid cells of various animal species-correlation with endotoxin susceptibility, *Infect. Immun.*, 57, 1054, 1989.

43. **Halling, J. L., Hamill, D. R., Lei, M. G., and Morrison, D. C.,** Identification and characterization of LPS binding proteins on human peripheral blood cell subpopulations, *Infect. Immun.*, 60, 845, 1992.

44. **Chen, T. Y., Lei, M. G. and Morrison, D. C.,** Partial purification of specific LPS binding proteins from murine splenocytes, *FASEB J.*, 3, A1082, 1989.

45. **Bright, S. W., Chen, T. Y., Flebbe, L. M., Lei, M. G., and Morrison, D. C.,** Generation and characterization of hamster-mouse hybridomas secreting monoclonal antibodies with specificity for lipopolysaccharide receptor, *J. Immunol.*, 145, 1, 1990.

46. **Chen, T. Y., Bright, S. W., Pace, J. L., Russell, S. W., and Morrison, D. C.,** Induction of macrophage mediated tumor cytotoxicity by a hamster monoclonal antibody with specificity for LPS receptor, *J. Immunol.*, 145, 8, 1990.

47. **Morrison, D. C., Silverstein, R., Bright, S. W., Chen, T. Y., Flebbe, L. M., and Lei, M. G.,** Monoclonal antibody to mouse lipopolysaccharide receptor protects mice against the lethal effects of endotoxin, *J. Infect. Dis.*, 162, 1063, 1990.

48. **Kirkland, T. N., Virca, G. D., Kuus-Reichel, T., Multer, F. K., Kim, S. Y., Ulevitch, R. J., and Tobias, P. S.,** Identification of lipopolysaccharide-binding proteins in 70Z/3 cells by photoaffinity cross-linking, *J. Biol. Chem.*, 265, 9520, 1990.
49. **Kara-kuge, S., Amano, F., Nishijima, M., and Akamatsu, Y.,** Isolation of a lipopolysaccharide (LPS)-resistant mutant, with defective LPS binding, of cultured macrophage-like cells, *J. Biol. Chem.*, 265, 6606, 1990.
50. **Dziarski, R.,** Peptidoglycan and lipopolysaccharide bind to the same binding site on lymphocytes, *J. Biol. Chem.*, 266, 4719, 1991.
51. **Dziarski, R.,** Demonstration of peptidoglycan binding sites on lymphocytes and macrophages by photoaffinity crosslinking, *J. Biol. Chem.*, 266, 4713, 1991.
52. **Lei, M. G., Stimpson, S. A., and Morrison, D. C.,** Specific endotoxic lipopolysaccharide-binding receptors on murine splenocytes. III. Binding specificity and characterization, *J. Immunol.*, 147, 1925, 1991.
53. **Parent, J. B.,** Core specific receptor for lipopolysaccharide on hepatocytes, *Circ. Shock*, 27, 341, 1989.
54. **Parent, J. B.,** Membrane receptors on rat hepatocytes for the inner core region of bacterial lipopolysaccharides, *J. Biol. Chem.*, 265, 3455, 1990.
55. **Lei, M. G. and Morrison, D. C.,** Identification of an LPS binding protein with specificity for inner core region (KDO) determinant, *FASEB J.*, 5, A1363, 1991.

Chapter 11

PLASMA MEMBRANE GANGLIOSIDES AS POTENTIAL BINDING SITES FOR BACTERIAL ENDOTOXINS

Herbert C. Yohe, Leora Suprun Brown, and John L. Ryan

TABLE OF CONTENTS

I. INTRODUCTION

Gangliosides are sialic acid-containing glycosphingolipids found in the plasma membrane of eucaryotic cells. By virtue of their negatively charged carbohydrate moiety, these membrane components are amphipathic molecules with detergent-like properties. Their lipophilic portion, ceramide, is comprised of a long-chain amino alcohol, sphingosine, with an N-acyl-linked fatty acid moiety. The hydrophilic portion of the ganglioside molecule is comprised of an oligosaccharide moiety to which is attached one or more sialic acid residues.

Sialic acid is a generic term for the naturally occurring modifications of neuraminic acid, a nine-carbon acidic amino sugar. The major modification is acetyl or glycolyl acylation of the amino nitrogen. As a result, the two major sialic acids are *N*-acetylneuraminic acid (NeuAc) and *N*-glycolylneuraminic acid (NeuGc). Minor sialic acids occur in which various hydroxyl groups are acylated in addition to or independently of the amino nitrogen.

Gangliosides are generally classified according to their oligosaccharide "family". Except for sialosylgalactosyl ceramide, gangliosides derive from lactosyl ceramide. Structures from which the family names are obtained are shown in Table 1. To date, the vast majority of gangliosides which contain three or more sugars in the oligosaccharide backbone are in the ganglio or neolacto families.

Gangliosides were originally discovered by Klenk[1] as components of neuronal tissue and a vast majority of data concerning their structures and metabolism comes from that source.[2] Gangliosides were eventually found in virtually all extraneuronal vertebrate tissue examined but peripheral tissues were generally considered to have much lower concentrations than that found in brain.[2] However, recent evidence from this laboratory and from others indicate that certain cells from the immune system contain ganglioside levels approaching that found in brain.[3-5]

Ganglioside patterns found in brain are generally conserved within a species, whereas the patterns of many peripheral tissues vary from strain to strain.[6] However, increasing evidence indicates that, in the absence of a genetic defect, ganglioside patterns of immune tissues are also conserved.[6,7]

Attempts to determine the function of gangliosides in the eukaryotic cell have resulted in a wide array of data and a variety of theoretical functions as well.[8] Changes in ganglioside patterns occur in differentiation and oncogenesis.[9] Furthermore, addition of exogenous gangliosides has been found to either induce differentiation or operate in a bimodal fashion dependent on experimental conditions.[10-12] Gangliosides have been found to be receptors or components of receptors of agents known to affect the immune system.[13] The ability of gangliosides to act as receptors for certain protein bacterial toxins is well established.[14] However, the association of gangliosides with bacterial endotoxin has not been well studied.

TABLE 1
Naming of the Ganglioside Families

Hemato series	cer-glc-gal-SA-
Lacto series	cer-glc-gal-glcNAc-
Ganglio series	cer-glc-gal-galNAc-
Globo series	cer-glc-gal-gal-

II. GANGLIOSIDE ANALYSES

Bacterial lipopolysaccharides (LPS) share many of the properties of gangliosides. LPS are amphipathic molecules with a lipophilic fatty acid-containing portion and a polysaccharide structure which contains a 2-ketoacid, ketodeoxyoctulosonic acid. Early investigations indicated LPS could readily form aggregates with gangliosides resulting in modification of LPS effects on the immune system.[15] However, there has been no evidence supporting an association of LPS with *in situ* gangliosides in the plasma membrane.

A. GANGLIOSIDE PATTERNS OF MURINE LYMPHOCYTES AND MACROPHAGES

Prior to any investigations of membrane ganglioside-LPS interactions, the ganglioside patterns of murine immune cells were determined. These investigations were made on gangliosides isolated from murine thymus, T-lymphocytes, B-lymphocytes, and macrophages. Several murine strains exist which have genetic defects in their immune system; the one of major importance to endotoxin research is the defect found by Sulzer,[16] in which the C3H/HeJ strain was determined to be hyporesponsive to certain preparations of LPS. Therefore, a comparison of gangliosides obtained from this hyporesponsive strain with the normally responsive parent strain (C3H/HeN) formed the basis of our approach to examine the potential significance of LPS-ganglioside interactions.

Initial studies on murine peritoneal macrophages immediately demonstrated profound differences in the ganglioside pattern of resident and stimulated cells.[3] The pattern of resident cells appeared rather simple being comprised of 15 or fewer spots on two-dimensional thin-layer chromatograms; there were only 3 major gangliosides running as doublets. The pattern of stimulated cells was quite complex with over 40 spots seen. A major cluster of gangliosides running in the center of the plates contained eight gangliosides which in thioglycolate-elicited cells accounted for half of the total ganglioside sialic acid content (Figure 1). Neither careful visual examination nor densitometric analysis indicated any ganglioside distributional differences between C3H/HeJ or C3H/HeN resident or thioglycolate-stimulated macrophages.[7] However, when macrophages were obtained from the two strains following i.p. injection of isolated LPS preparations or killed Gram-negative organisms, readily apparent different ganglioside patterns were evident.[17,18] When phenol-

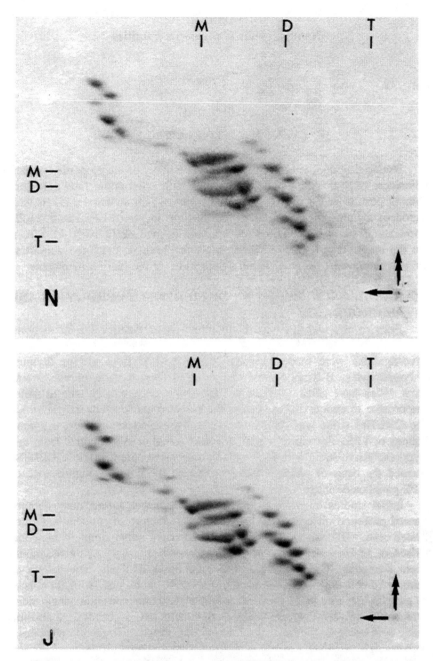

FIGURE 1.   Two-dimensional thin-layer chromatography patterns of the total ganglioside fraction obtained from thioglycolate-elicited murine macrophages from C3H/HeN (top) and C3H/HeJ (bottom) strains. Chromatograms were generated as described in Reference 3. M, D, and T show the relative mobility of standard gangliosides GM1, GD1a, and GT1b, respectively. Directions of first and second solvents are indicated by single- and double-beaded arrows of the origin.

extracted LPS was used as the activating agent, the ganglioside pattern of the hyporesponsive C3H/HeJ macrophages barely differed from the resident pattern.[17] This is consistent with the observation that this strain is nearly nonresponsive to such LPS preparations. The use of butanol-extracted LPS as the activating agent resulted in a minor change in the macrophage ganglioside pattern, consistent with the partial response of the HeJ strain to this LPS preparation.[17] When killed *Escherichia coli* K-235 was injected as the activating agent, a complex ganglioside pattern was produced; however, the ganglioside pattern produced was still not the same as that produced by the normal C3H/HeN strain (Figure 2). While the same components were present in the pattern, distinct shifts in the distribution of major gangliosides were seen between the normal HeN and the hyporesponsive HeJ strain.[18]

B. SIALIC ACID COMPOSITION

Analysis of macrophage ganglioside sialic acid indicated the presence of both NeuAc and NeuGc. Resident cells contained predominantly NeuGc (>80%). Cells elicited by thioglycolate had nearly equal quantities of NeuAc- and NeuGc-containing gangliosides As expected, when cells were activated with LPS preparations or killed *E. coli,* the ratio of NeuAc- to NeuGc-containing gangliosides differed between the hyporesponsive and normal strains; the normal strain consistently contained a higher proportion of NeuAc-containing gangliosides (Table 2).[17,18]

C. GANGLIOSIDE STRUCTURES FROM MACROPHAGES

Structural analyses were performed on the 11 major gangliosides from stimulated murine macrophages, 8 of which were determined to be monosialylated. The complex thin-layer chromatographic patterns were determined to be due to both sialic acid and ceramide fatty acid heterogeneity. For each ganglioside characterized, NeuGc- and NeuAc-containing species were present. Ceramide fatty acid ranged from C16 to C24 moieties but C18 sphingosine was the only long-chain base present.[7] By using a battery of analytical techniques, it was determined that all the major tetraosyl structures present were of the ganglio family series. GM1 gangliosides were the major monosialo species comprising 43 to 60% of those from thioglycolate-stimulated cells and 60 to 70% of those from *E. coli*-activated cells. Four species of GM1 were present: sialidase-resistant NeuAc-GM1a and NeuGc-GM1a and sialidase-sensitive NeuAc-GM1b and NeuGc-GM1b. The dramatic pattern differences between *E. coli*-activated macrophage gangliosides from C3H/HeJ and C3H/HeN mice was a shift in the ratio of GM1a to GM1b. In the endotoxin-hyporesponsive C3H/HeJ strain, the sialidase-sensitive GM1b was markedly decreased.[7]

The monosialoganglioside GM1b has been previously reported to be a major component of other murine immunocytes[19] but an extremely minor component of brain.[20,21] Hirabayashi and others[22] found GM1b in hepatoma

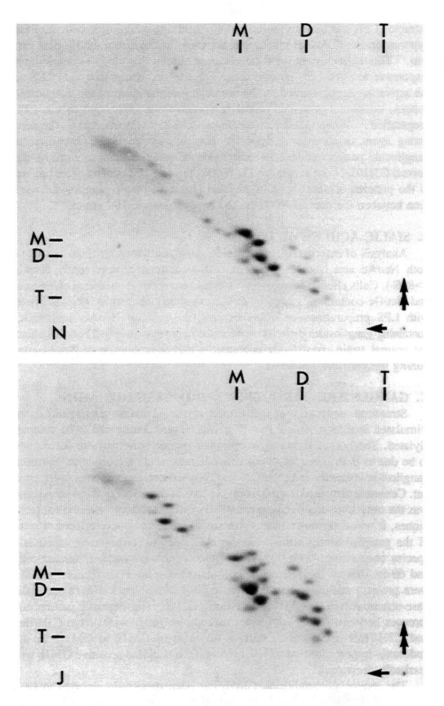

FIGURE 2. Two-dimensional thin-layer chromatography patterns of the total ganglioside fraction obtained from *E. coli* K-235-elicited macrophages from C3H/HeN (top) and C3H/HeJ (bottom) strains. Conditions are the same as described for Figure 1.

TABLE 2
Distribution and Ratios of Ganglioside Sialic Acid

	Thioglycolate-elicited macrophages		E. coli-activated macrophages	
	HeN	HeJ	HeN	HeJ
Polysialo	20 ± 5	18 ± 1	17 ± 3	20 ± 6
GM1	58 ± 7	58 ± 4	72 ± 13	64 ± 16
GM2	2 ± 2	5 ± 0	2 ± 2	1 ± 1
GM3	16 ± 4	20 ± 1	3 ± 2	8 ± 8
Ratio GM1b/GM1a	1.27	0.91	4.89	0.17
Ratio NeuAc/NeuGc	0.95	0.95	1.27	0.79

cells and based on their data predicted GM1b would be found in cells containing asialoGM1. Therefore, the finding of GM1b in stimulated murine macrophages is not surprising as Mercurio et al.[23] found high concentrations of asialoGM1 in these cells. Furthermore, the deficit of GM1b in the *E. coli*-activated macrophage from the C3H/HeJ mouse is not the first report of a deficit in a sialidase-sensitive GM1 from an immunologically mutant mouse. A genetic mutant with natural killer cell deficiency, beige mice, has reduced lymphocyte levels of asialoGM1 and sialidase-sensitive GM1.[24] The defect in C3H/HeJ mice is not, however, in ganglioside metabolism per se as these mice produce a normal thioglycolate-stimulated macrophage ganglioside pattern. The altered ganglioside patterns of their *E. coli*-activated macrophages is an additional manifestation of their defect in the processing of an endotoxin stimulus.

D. T-LYMPHOCYTE AND THYMOCYTE PATTERNS

Although the data on the macrophage gangliosides was the first demonstrated molecular difference found between the hyporesponsive and normal mouse strains, it was a post-LPS-exposure event. We, therefore, began to examine the ganglioside patterns of other cells of the murine immune system: thymocytes, T-lymphocytes and B-lymphocytes with the view that one or more of these cell types may show an altered ganglioside pattern in the hyporesponsive mouse.

Ganglioside patterns derived from resting murine thymocytes and splenic T-lymphocytes displayed dramatically different patterns.[25] Thymocytes contained a complex pattern of 28 to 30 gangliosides with 8 major gangliosides. The T-lymphocyte pattern was much simpler, containing 6 or 7 major components and 12 or 13 minor ones. A densitometric comparison of C3H/HeJ and C3H/HeN thymocyte ganglioside patterns revealed no major differences. In contrast, examination of the T-lymphocyte pattern indicated a marked difference in the relative proportions of three homologous gangliosides between the two strains. Consistent with our observations on endotoxin-activated macrophages, the ratio of NeuAc-containing gangliosides to NeuGc-containing gangliosides was higher in both thymocytes and T-lymphocytes from the normal LPS-responsive strain.[25]

E. B-LYMPHOCYTE PATTERNS

Initial studies on murine B lymphocyte gangliosides from LPS-responsive and LPS-hyporesponsive strains suggested the ganglioside patterns were similar with only minor differences in the relative proportions of four gangliosides. However in the course of study, Chaby et al.[26] reported one of five galactose oxidase-sensitive gangliosides found in the B cells of responsive mice was missing in the hyporesponsive strain. The total ganglioside content or resorcinol-positive thin-layer chromatographic patterns was not reported. The question immediately arose as to whether there was a true alteration in ganglioside content or whether the surface accessibility of the B cell gangliosides of the HeJ strain was altered. Densitometric analyses indicated complete homology in the resorcinol patterns between strains and verified only the minor differences mentioned above. Of 17 resorcinol-positive moieties, only 12 were accessible (or susceptible) to surface labeling following galactose oxidase treatment in both strains. However, labeling of the more polar components was greatly reduced in the hyporesponsive strain indicating an alteration exists in the B cell membrane structure that may be associated with the LPS hyporesponsiveness seen in the C3H/HeJ strain.[27]

The total B lymphocyte results of Chaby et al.[26] combined with our own[27] suggested that similarity of B lymphocyte ganglioside patterns did not indicate a similarity in surface exposure between the HeN and HeJ strains. Despite our observations that the thioglycolate-stimulated macrophage ganglioside patterns of C3H/HeN and C3H/HeJ were similar, the distinct possibility of differences in surface accessibility existed and, therefore, a possible difference in *in situ* ganglioside-LPS interactions. Work by Lei and Morrison,[28,29] who, despite a careful and extensive effort, were unable to demonstrate a difference in LPS-membrane protein interactions between C3H/HeJ and normal mice, left open the possibility that LPS-lipid interactions might vary between the two strains. Thus we have endeavored to investigate the interaction of LPS with membrane gangliosides in murine macrophages from the C3H/HeN and C3H/HeJ strains.

III. GANGLIOSIDE-LPS INTERACTIONS

A. METHODS

Several experiments were done using the ^{125}I-labeled photoaffinity labeling reagent, SASD, to investigate whether gangliosides on the cell surface interact directly with LPS. SASD has been used successfully to identify LPS-specific binding proteins in lymphoid cells.[28-29] The technique is most useful for investigating receptors for large or aggregated ligands because the resulting cross-linked complexes can be dissociated in the presence of sulfhydryl reducing reagents. The ^{125}I-labeled probe remains attached to the receptor (Figure 3) and facilitates subsequent identification by two-dimensional thin-layer chromatography and radioautography.

LPS (smooth, 0111:B4 and rough, Rd) was covalently coupled to the cross-linking reagent, sulfosuccinimidyl-2-(*p*-azidosalicylamido)-1,3'-dithio-proprionate (SASD) to yield a photoactivatable, cleavable derivative (LPS-ASD) based on the procedure by Wollenweber and Morrison.[30]

This procedure was modified as follows: a stock solution of SASD dissolved in dimethyl sulfoxide at 250 μg/μl was diluted 1:200 (by a two-step dilution) with 0.1 M sodium phosphate buffer (pH 7.4). This SASD working solution containing 231 nmol of ligand was added to 42 nmol of LPS (0.5 mg) dissolved in 0.1 M borate buffer (pH 8.5). The reaction mixture was ultrasonicated and kept at room temperature for 30 min. LPS-ASD was freed from excess ligand by dialysis against PBS (pH 7.2) overnight at 4°C.

[125]I-iodination of LPS-ASD (smooth, 0111:B4 and rough, Rd) was carried out according to Ulevitch[31] with minor modifications. Na[125]I (1 mCi in 10 μl of NaOH), KI (10 μl, 10^{-4} M), and chloramine-T (30 μl, 0.1 mg/ml) were added to 200 μg of LPS dissolved in 200 μl of PBS (pH 7.2), mixed well, and incubated for 10 min at room temperature with occasional shaking. Iodination was stopped by adding 30 μl of a 10-mM tyrosine stock solution. The reaction mixture was directly applied onto a Sephadex G-50 glass column (10-ml bed volume), pretreated with PBS/0.2% gelatin and PBS/2.0 M NaCl. Fractions were collected and eluting radioactivity was monitored by measuring aliquots. Pooled peak fractions were dialyzed against PBS (ph 7.2) at 4°C until counts in the dialysate stabilized.

The iodination reaction of SASD was carried out in an IODO-GEN-coated culture tube containing 9 μg IODO-GEN, 72 μl of SASD (166 nmol) in 0.1 M sodium phosphate buffer (pH 7.4), 120 μCi of Na[125]I and 55.5 nmol KI in 10 μl of 0.1 M sodium phosphate buffer pH 7.4. The reaction was allowed to proceed for 30 s and stopped by removing the ^{125}I-SASD from the IODO-GEN tube. All reactions with the photosensitive compound were carried out under reduced lighting using a 25-W red light source.

B. LPS INTERACTIONS WITH MACROPHAGE GANGLIOSIDES

Peritoneal macrophages for the photoaffinity labeling experiments were first washed twice with 30-ml volumes of PBS (pH 7.2) to remove unbound medium components. Approximately 3 μg of LPS (smooth, 0111:B4 and rough, Rd) was added to 20 ml of PBS containing 3 to 4 × 10^8 macrophages. The ^{125}I-ASD-LPS was allowed to react with the cells for 20 min in the dark at 37°C in the presence of 5% CO_2. The reaction mixture was then exposed to a long-wave UV light source (emission between 300 and 400 nm) for 20 min at a fixed distance of approximately 7 cm. Following the cross-linking of the compound to the macrophage surface, cell suspension was washed three times with 30-ml volumes of PBS. Cell pellet was resuspended in an

Step 1.

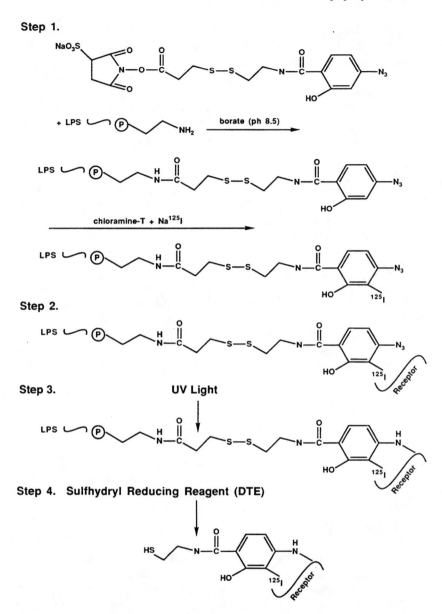

Step 2.

Step 3.   **UV Light**

Step 4. Sulfhydryl Reducing Reagent (DTE)

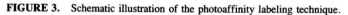

FIGURE 3.   Schematic illustration of the photoaffinity labeling technique.

Step 1.   Covalent attachment of SASD to primary amino groups (phosphoethanolamine) of LPS
          at alkaline pH and subsequent iodination by chloramine-T.
Step 2.   Noncovalent interaction of ^{125}I-ASD-LPS with receptors on the macrophage surface in
          the dark.
Step 3.   Covalent attachment of the photoactivated azido group of ^{125}I-ASD-LPS to a receptor
          complex for LPS.
Step 4.   Reduction of the disulfide bond of ^{125}I-ASD-LPS leaving the ^{125}I-labeled probe attached
          to the receptor.

isotonic solution of 0.05 M dithioerythritol and incubated for 30 min at 37°C in order to reduce the disulfide bond of ^{125}I-ASD-LPS followed by removal of the ligand moiety. Once the reduction was complete cells were spun down and the dithioerythritol was removed.

In reactions without cells, an equivalent amount of ^{125}I-ASD-LPS in PBS was exposed to UV light and reduced. The reducing agent was removed by dialysis overnight against pyrogen-free H_2O at 4°C, followed by lyophilization of the dialyzed material.

In control reactions when ^{125}I-SASD rather than ^{125}I-ASD-LPS was used, 3 μg of the labeled ligand was added to cells in order to be equivalent to the amount of ^{125}I-ASD-LPS added to cells.

Gangliosides were displayed for visualization (orcinol spray) and autoradiography using two-dimensional thin-layer chromatography.[18] Thin-layer chromatography plates with ^{125}I-labeled macrophage gangliosides were visualized using Kodak XAR X-ray film which had been presensitized by exposure to 92% N_2, 8% H_2 at 48°C for 16 h. The film was developed using standard methods.

A thin-layer chromatography display of gangliosides isolated from responder, C3H/HeN mice and hyporesponder, C3H/HeJ-stimulated macrophages showed a similar pattern between the two groups of gangliosides. More than 40 individual gangliosides were identified by orcinol spray in each group (Figure 4).

When ^{125}I-derivatized *E. coli* 0111:B4 LPS reacted with macrophages derived from either C3H/HeN mice or C3H/HeJ mice, five individual gangliosides were labeled on the macrophage surface from each group of mice. These gangliosides were not detected on the control two-dimensional thin-layer chromatography autoradiogram when ^{125}I-ASD-0111:B4 was irradiated with UV light in the absence of macrophages (Figure 5). Three of the five labeled gangliosides (3, 4, and 5) from both HeN and HeJ macrophages appear to be identical based on their migration on the chromatogram. These five individually labeled gangliosides were not detected when ^{125}I-SASD (without LPS) reacted with macrophages derived from C3H/HeN and C3H/HeJ mice (results not shown).

When the radiolabeled derivatized rough LPS (Rd) reacted with macrophages derived from C3H/HeN and C3H/HeJ mice, fewer gangliosides appeared to be labeled in comparison to labeling by 0111:B4. ^{125}I-ASD-Rd bound to two individual macrophage gangliosides derived from HeN mice and to only one macrophage ganglioside derived from HeJ mice (Figure 5). One of the labeled gangliosides from HeN macrophages (3) appears to be identical to a corresponding ganglioside (3) labeled by 0111:B4 in both HeN and HeJ macrophage gangliosides. However, (6) appears to be a newly labeled ganglioside specific to HeN macrophage labeling by Rd LPS. The single ganglioside (2) labeled by Rd in HeJ macrophage gangliosides appears to be identical to a corresponding ganglioside (2) labeled by 0111:B4 in HeN macrophage gangliosides.

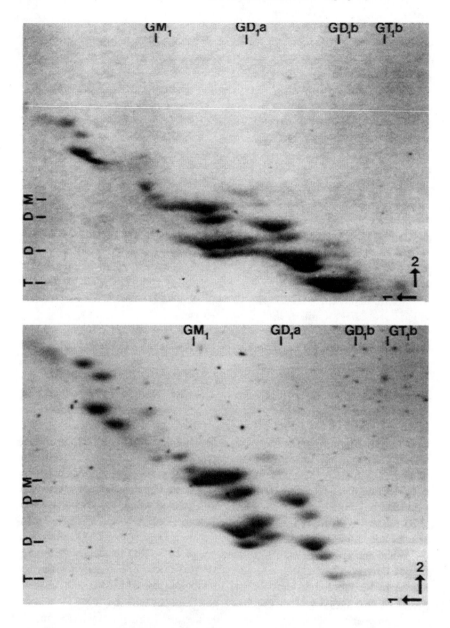

FIGURE 4. Two-dimensional thin-layer chromatography pattern of the total ganglioside fraction obtained from thioglycolate-elicited murine macrophages from C3H/HeN (top) and C3H/HeJ (bottom) strains labeled by ^{125}I-ASD-LPS. Conditions are the same as described for Figure 1.

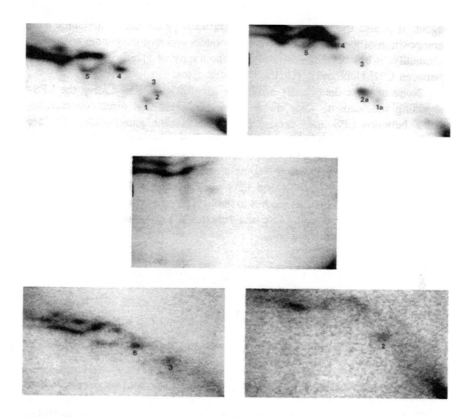

FIGURE 5. Two-dimensional thin-layer chromatography autoradiograms of thioglycolate-elicited murine macrophage gangliosides from C3H/HeN and C3H/HeJ strains labeled by ^{125}I-ASD-0111:B4 and ^{125}I-ASD-Rd. Gangliosides derived from HeN murine macrophages (top left) and from HeJ macrophages (top right) both labeled by ^{125}I-ASD-0111:B4 are shown. ^{125}I-ASD-0111:B4 irradiated with UV light and displayed on a two-dimensional thin-layer chromatogram is shown in center autoradiogram. Gangliosides derived from HeN macrophages (bottom left) and HeJ macrophages (bottom right) both labeled by ^{125}I-ASD-Rd are shown. Conditions are the same as described for Figure 1.

The rapidly migrating irregular dark areas in the upper left corner of the autoradiograms are most likely degradation products of the labeled ligand, which are hydrophobic by nature such that they migrate rapidly in this thin-layer solvent system.

C. DISCUSSION

These data show that LPS becomes closely associated with specific gangliosides when it interacts with murine macrophages. Whether this LPS-ganglioside interaction is critical for LPS signaling remains to be proven. It is clear, however, that significant differences exist in the resting and stimulated macrophage ganglioside patterns when LPS is used as the activating agent. It is also evident that there is significant difference in the sialic acid composition of gangliosides between responder and hyporesponder cells. Additionally, it has been shown that the topography of gangliosides is distinct between C3H/HeN and C3H/HeJ B-lymphocytes.

None of these data clarify a role for gangliosides in mediating the LPS-signaling mechanism. They do indicate, however, that direct interactions occur between LPS and specific plasma membrane gangliosides. Further studies are needed to define the relationship, if any, between the LPS-binding gangliosides and the proteins in the cell membrane which have been shown to bind LPS. It is possible that a complex receptor composed of both protein and glycolipid exists to transduce LPS signals to the cell interior. The development of monoclonal antibodies to the LPS receptor protein and to specific gangliosides may help elucidate the functional importance of these LPS-ganglioside interactions.

REFERENCES

1. **Klenk, E.,** Über die Natur der Phosphatide und onderer Lipoide des Gehirns und der Leber in Niemann-Pickscher Krankheit, *Hoppe-Seyler's Z. Physiol. Chem.,* 235, 24, 1935.
2. **Ledeen, R. W. and Yu, R. K.,** Gangliosides: structure, isolation and analysis, *Methods Enzymol.,* 83, 139, 1982.
3. **Yohe, H. E., Coleman, D. L., and Ryan, J. L.,** Ganglioside alterations in stimulated murine macrophages, *Biocheim. Biophys. Acta,* 818, 81, 1985.
4. **Mercurio, A. M., Schwarting, G. A., and Robbins, P. W.,** Glycolipids of the mouse peritoneal macrophage, *J. Exp. Med.,* 160, 1114, 1984.
5. **Ritter, K., Hartl, R., Bandlow, G., and Thomssen, R.,** Cytostatic effect of gangliosides present in the membrane of macrophages, *Cell. Immunol.,* 97, 248, 1986.
6. **Nakamura, K., Hashimoto, Y., Yamakawa, T., and Suzuki, A.,** Genetic polymorphism of ganglioside expression in mouse organs, *J. Biochem.,* 103, 201, 1988.
7. **Yohe, H. C., Cuny, G. L., Macala, L. J., Saito, M., McMurray, W. J., and Ryan, J. L.,** The presence of sialidase sensitive sialosylgangliotetraosyl ceramide (GM1b) in stimulated murine macrophages; deficiency of GM1b in *E. coli*-activated macrophages from the C3H/HeJ mouse, *J. Immunol.,* 146, 1900, 1991.
8. **Hakomori, S. -I.,** Bifunctional role of glycosphinoglipids, *J. Biol. Chem.,* 256, 18713, 1990.
9. **Hakomori, S. -I.,** Glycosphinoglipids in cellular interaction, differentiation, and oncogenesis, *Annu. Rev. Biochem.,* 50, 733, 1981.
10. **Nojiri, H., Takaku, F., Terui, Y., Miura, Y., and Saito, M.,** Ganglioside GM3: an acidic membrane component that increases during macrophage-like cell differentiation can induce monocytic differentiation of human myeloid and monocytoid leukemic cell lines HL-60 and U937, *Proc. Natl. Acad. Sci. U.S.A.,* 83, 782, 1986.
11. **Nojiri, H., Kitagawa, S., Nakamura, M., Kirito, K., Enomoto, Y., and Saito, M.,** Neolacto-series gangliosides induce granulocyte differentiation of human promyelocytic leukemia cell line HL-60, *J. Biol. Chem.,* 263, 7443, 1988.

12. **Spiegel, S. and Fishman, P. H.**, Gangliosides as bimodal regulators of cell growth, *Proc. Natl. Acad. Sci. U.S.A.*, 84, 141, 1987.
13. **Symington, F. W. and Hakomori, S. -I.**, Glycosphingolipids of immune cells, *Lymphokines*, 12, 201, 1985.
14. **Fishman, P. H.**, Role of membrane gangliosides in the binding and action of bacterial toxins, *J. Membr. Giol.*, 69, 85, 1982.
15. **Ryan, J. L., Inouye, L. N., Gobran, L., Yohe, W. B., and Yohe, H. C.**, Lack of specificity of brain gangliosides in the modulation of lymphocyte activation, *Yale J. Biol. Med.*, 58, 459, 1985.
16. **Sulzer, B.**, Genetic control of leukocyte response to endotoxin, *Nature*, 219, 1253, 1968.
17. **Ryan, J. L., Yohe, H. C., and Malech, H. L.**, Changes in membrane gangliosides: differentiation of human and murine monocytic cells, *Yale J. Biol. Med.*, 58, 125, 1985.
18. **Yohe, H. C. and Ryan, J. L.**, Ganglioside expression in macrophages from endotoxin responder and nonresponder mice, *J. Immunol.*, 137, 3921, 1986.
19. **Müthing, J., Egge, H., Kniep, B., and Muhbradt, P. F.**, Structural characterization of gangliosides from murine T lymphocytes, *Eur. J. Biochem.*, 163, 407, 1987.
20. **Ariga, T. and Yu, R. K.**, Isolation and characterization of ganglioside GM1b from normal human brain, *J. Lipid Res.*, 28, 285, 1987.
21. **Hirabayashi, Y., Hyogo, A., Nakao, T., Tsuchiya, K., Suzuki, Y., Matsumoto, M., Kon, K., and Ando, S.**, Isolation and characterization of extremely minor gangliosides, GM1B and GD12, in adult bovine brains as developmentally regulated antigens, *J. Biol. Chem.*, 265, 8144, 1990.
22. **Hirabayashi, Y., Taki, Y., and Matsumoto, M.**, Tumor ganglioside-natural occurrence of GM1b, *FEBS Lett.*, 100, 253, 1979.
23. **Mercurio, A. M., Schwarting, G. A., and Robbins, P. W.**, Glycolipids of the mouse peritoneal macrophage, *J. Exp. Med.*, 160, 1114, 1984.
24. **Schwarting, G. A. and Gajewski, A.**, Glycolipids of murine lymphocyte subpopulations: a defect in the levels of sialidase-sensitive sialosylated asialo GM1 in beige mouse lymphocytes, *J. Immunol.*, 126, 2403, 1981.
25. **Yohe, H. C., Cuny, C. L., Berenson, C. S., and Ryan, J. L.**, Comparison of thymocyte and T lymphocyte gangliosides from C3H/HeN and C3H/HeJ mice, *J. Leukocyte Biol.*, 44, 521, 1988.
26. **Chaby, R., Morelec, M. -J., Enserqueix, D., and Girard, R.**, Membrane glycolipid and phospholipid composition of lipopolysacchairde-responsive and nonresponsive murine B lymphocytes, *Infect. Immun.*, 52, 777, 1986.
27. **Yohe, H. C., Berenson, C. S., Cuny, C. L., and Ryan, J. L.**, Altered B-lymphocyte membrane architecture indicated by ganglioside accessibility in C3H/HeJ mice, *Infect. Immun.*, 58, 2888, 1990.
28. **Lei, M. -G. and Morrison, D. C.**, Specific endotoxic lipopolysaccharide-binding proteins on murine splenocytes. I. detection of lipopolysaccharide-binding sites on splenocytes and splenocyte populations, *J. Immunol.*, 141, 996, 1988.
29. **Lei, M. -G. and Morrison, D. C.**, Specific endotoxic lipopolysaccharide-binding proteins on murine splenocytes. II. membrane localization and binding characteristics, *J. Immunol.*, 141, 1006, 1988.
30. **Wollenweber, H. -W. and Morrison, D. C.**, Synthesis and biochemical characterization of a photo-activatable, iodinatable, cleavable bacterial lipopolysaccharide derivative, *J. Biol. Chem.*, 260(28), 15068, 1985.
31. **Ulevitch, R. J.**, The preparation and characterization of a radioiodinated bacterial lipopolysaccharide, *Immunochemistry*, 15, 157, 1978.

Chapter 12

LPS-INITIATED SIGNAL TRANSDUCTION PATHWAYS IN MACROPHAGES

Dolph O. Adams

TABLE OF CONTENTS

I. INTRODUCTION

Bacterial endotoxin or lipopolysaccharide (LPS) is a potent stimulant of a wide variety of host cells.[1-3] One of the principal targets of LPS is the macrophage, which undergoes a strikingly broad array of changes after interaction with LPS.[4] These changes, which are frequently long lasting, can be observed after minutes or hours and comprise changes both in the fundamental metabolism of macrophages and in the expression of many genes. Of interest, numerous LPS-initiated responses in macrophages can be potentiated or primed by prior treatment of the macrophages with cytokines such as interferon-γ (IFN-γ).

The specific membranous molecule(s) by which LPS interacts with cells in general and macrophages in particular remain the subject of intense scrutiny and interest[2,3] (see Chapters 10, 11). The active chemical portion of LPS has recently been reviewed and, for most biolgoical effects, is thought to be the lipid A component.[3] Recently, a membranous protein of ~80 kDa to which LPS and lipid A bind has been reported.[5,6] Though the functional consequences of such binding remain to be established,[3] an antibody to the ~80-kDa protein has been reported to trigger functional effects in macrophages, including induction of activation of cytolysis.[7] Other LPS-binding proteins of ~95, 65, 55, 28, 25, and 18 kDa have also been reported (for reviews, see References 3 and 8). A binding site on macrophages, T-lymphocytes, and B-lymphocytes for LPS has recently been found to be an ~70-kDa protein of 6.5 PI which binds bacterial cell wall peptidoglycan (PGN) as well as LPS;[8,9] it has been suggested that binding of PGN and lPS are mediated via the same membranous protein, which has specificity for the $(GlcNac-MurNac)_n$ in backbone of PGN and the $(GlcNac)_2$ of lipid A.[8,9] LPS has also been reported to bind to mononuclear phagocytes via the CD11/CD18 complex of leukocyte-adhesion molecules.[10] Another mechanism binding of a complex formed between LPS and a serum LPS-binding protein to the CD14 molecule formed in leukocyte membranes has recently been suggested to initiate functional responses.[11]

LPS may also exert some of its effects via one or more of the family of scavenger receptors (see Reference 3 for details). Macrophages bear a unique receptor which recognizes maleylated or acetylated proteins/lipoproteins, in good part by alterations in the net negative charge of the ligand; this receptor has been termed the scavenger receptor.[12,13] Recent evidence indicates that the receptor actually is one of a family of distinct receptors. Two receptors, both binding acetylated low-density lipoproteins (LDL) but having distinct structures, have now been identified.[14,15] Furthermore, other workers have shown that other modified lipoproteins such as oxidized LDL are recognized and endocytosed via different receptors.[16] From these data, it is likely that there exists a family of two or more (and probably at least three) receptors which recognizes variously altered polyanionic molecules.

These observations on scavenger receptors are pertinent to LPS in that a disaccharide precursor of lipid A which is not fully acetylated (e.g., lipid$_{IVA}$) can produce some of the functional activities of lipid A (see Reference 3). Specific and saturable binding of lipid$_{IVA}$ to macrophages and macrophage-like tumor cells has been observed, and acetylated lipoproteins can compete with the lipid A for such binding.[17] Lipid A may in fact bind to multiple sites on phagocytic cells, since the binding of lipid$_{IVA}$ is to a ~95-kDa protein rather than to C18, which recognizes LPS by a different mechanism.[18]

LPS and lipid A can apparently bind to a number of membranous proteins and receptors on mononuclear phagocytes. It remains to be determined fully which of the broad array of metabolic effects of LPS on macrophages can be attributed to any of these particular LPS/lipid A-membrane protein interactions.

The mechanism(s) by which these various membranous proteins initiate signaling in macrophages after binding of LPS also remains the subject of intensive investigation. Several reports have indicated that pertussis-toxin-sensitive, guanidine, nucleotide-binding protein(s) (G proteins) can mediate certain LPS-initiated responses in B cells, B like cell lines, and macrophage-like cell lines.[19-21] cDNAs for several members of the inhibitory G protein family (G$_i$ family) including Gi$_2$ and Gi$_3$ have been identified in monocytes and macrophages (for reviews, see Reference 20). Whether these or other G proteins can be linked to specific LPS-initiated responses, transductional or biological, in macrophages remains to be established.

Current views of the regulation of macrophages indicate that these cells are regulated at two distinct temporal levels.[4,22] First, certain surface-active agents rapidly initiate functional responses in macrophages (i.e., within minutes). A classic example is the initiation of chemotaxis, respiratory burst, and secretion of enzymes and products of the arachidonic acid cascade by chemotactic stimuli. In this regard, LPS can stimulate macrophages to release products of arachidonic acid, albeit not at a high level.[4] Second, the fundamental potential of macrophages to react to such stimuli can be slowly altered (i.e., within hours). This process has been termed *macrophage activation,* which can be formally defined as alterations in the ability of macrophages to execute or complete a complex function (e.g., the ability to present antigen to T-lymphocytes or to destroy neoplastic cells). Activation for various functions in macrophages is a highly diverse, pleiotropic process, and it is currently estimated that there may be as many as several hundred different states of macrophage activation.[22] Many, though certainly not all, of the biochemical changes in macrophages constituting any one form of activation can be attributed to precise alterations, both increases and decreases, in the expression of various macrophage gene products.[23] These alterations, in turn, are frequently, though not invariably or exclusively, linked to corresponding alterations in transcription of the genes encoding the various proteins.[23] Here, too, LPS is a potent signal for macrophages, altering expression of a number of major gene products, e.g., increased secretion of tumor necrosis factor α (TNF-α) and interleukin-1 (IL-1) and decreased membrane expression of class II MHC molecules.

The changes in macrophages of both types are doubtless attributable to the initiation of intracellular mechanisms of signal transduction by LPS and other potent macrophage-activating agents such as IFNγ and by suppressive agents such as α₂-macroglobulin-protease complexes. The precise mechanisms of signal transduction initiated in macrophages by these agents have been investigated in a number of laboratories over the past 5 years (see References 23 to 27 for reviews). At present, four distinct intracellular cascades of transductional events initiated by extracellular stimuli have now been identified (Figure 1).[24,25,27] The first cascade (cascade I), an inductive one, is initiated by IFNγ. The second two cascades (cascades II and III) are initiated by LPS. Cascade II, at present, appears to be a purely inductive cascade, whereas cascade III can be either inductive or suppressive, depending on the particular function or gene studied. Lastly, we have identified a fourth cascade which is suppressive; recent evidence indicates that this cascade is modified but not initiated by LPS. These cascades interact with one another (Figure 2).

The purpose of this chapter is to describe the two cascades initiated by LPS in macrophages.

II. LPS-INITIATED SIGNALS RELATED TO THE BREAKDOWN OF POLYPHOSPHOINOSITIDES

LPS and lipid A initiate the rapid hydrolysis of phosphatidyl inositol-4,5-bisphosphate (PIP₂) in macrophages (Figure 3).[28] The breakdown of this product is complex and involves generation of inositol-1,4,5-trisphosphate and the subsequent phosphorylation of this compound to inositol-1,3,4,5-tetrakis phosphate, the breakdown of the tetrakis compound into inositol-1,3,4-triphosphate, and the breakdown of all of the preceding into various inositol bisphosphates and ultimately into inositol itself. The formation of inositol-1,4,5-triphosphate and of inositol-1,3,4,5-tetrakis phosphate have been linked in multiple cells to elevations in intracellular levels of Ca^{2+} ($[Ca^{2+}]_i$) from intracellular and extracellular stores, respectively. Direct elevations in intracellular levels of $[Ca^{2+}]_i$ have been observed in macrophages exposed to either lipid A or LPS, though the magnitude of elevation in $[Ca^{2+}]_i$ induced by lipid A is significantly greater than that induced by LPS.[28] LPS and lipid A also lead to rises in intracellular levels of diacylglycerol (DAG).[29,30] The principal source of DAG production in macrophages, as observed in other leukocytes is not, however, the breakdown of PIP₂, which characteristically supplies ~10 to 20% of the DAG formed in these cells.[31,32] Rather, initiation of PIP₂ hydrolysis leads to the subsequent degradation of phosphatidylcholine (PC) into DAG, which provides a major and sustained source of this metabolite. A precursor of lipid A has been shown to lead to significant increases

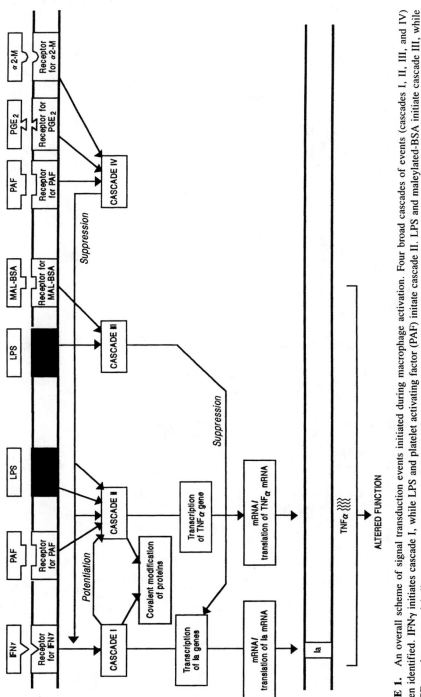

FIGURE 1. An overall scheme of signal transduction events initiated during macrophage activation. Four broad cascades of events (cascades I, II, III, and IV) have been identified. IFNγ initiates cascade I, while LPS and platelet activating factor (PAF) initiate cascade II. LPS and maleylated-BSA initiate cascade III, while PAF, PGE$_2$, and α$_2$-macroglobulin protease complexes (α$_2$M) initiate cascade IV. Cascades I and II have proven to be purely inductive to date, while cascade IV appears to be purely suppressive at present. Cascade III can be either inductive or suppressive. The net result of these cascades is alteration of membrane and secreted proteins, which in turn lead to altered functions. (Adapted from Reference 27.)

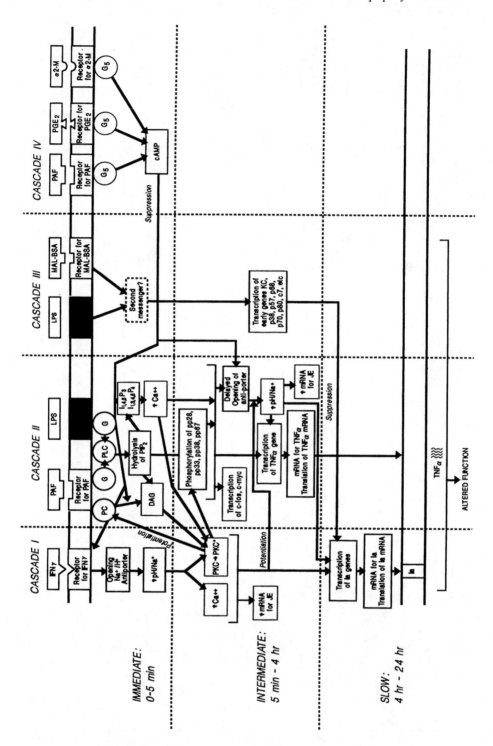

FIGURE 2. Three cascades leading to two forms of macrophage activation — activation for tumor cytolysis and activation of presentation of antigen (see Figure 1). The first two cascades (cascades I and II) are inductive. Cascade III is amphipathic, while cascade IV is suppressive. The cascades begin at the top of the figure, with application of extracellular ligands which include IFNγ in cascade I; platelet-activating factor (or PAF) and bacterial lipopolysaccharide (or LPS) in cascade II; LPS and maleylated bovine serum albumin (or mal-BSA) in cascade III; and PAF, prostaglandin E_2 (or PGE_2), and the fast form of α_2-macroglobulin (or α_2M) in cascade IV. These respectively act upon membranous receptors or binding sites to initiate a variety of intracellular events. In the immediate time frame of 0 to 5 min, rapid second messengers are induced. A second messenger, in this time frame, for cascade III has not been identified. PC stands for phosphatidylcholine; PLC stands for phospholipase C; GC stands for guanidine binding protein C; DAG stands for diacylglycerol; cAMP stands for cyclic AMP; and GS stands for guanidine-binding nuclear proteins. In the intermediate time frame of 5 min to 4h, delayed second messenger effects and certain gene effects occur, which include transcription and stabilization of mRNA. PKC stands for protein kinase C; PCK* stands for the enhanced functional form of PKC; TNFα stands for tumor necrosis factor α. In the slow time frame of 4 to 24 h, other genomic events including transcription of Ia genes is observed. The end result of these cascades is either surface expression of proteins such as Ia or secretion of proteins such as TNFα. The combined alterations in expression of membranous and secreted proteins lead to altered functions. Note that several arrows show potential interactions between several of the cascades. (Adapted from Reference 27.)

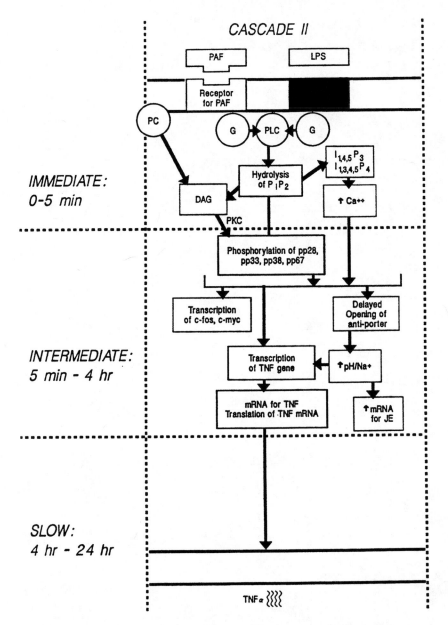

FIGURE 3. A closeup of the events initiated in cascade II by LPS and PAF.

(i.e., rises of four- to eightfold) of lysophosphatidylinositol in the RAW264.7 macrophage-like cell lines, though the functional significance of elevations in this phospholipid remains to be determined.[33]

Subsequently, LPS or lipid A initiate a characteristic pattern of heightened protein phosphorylation.[34] Specifically, increased phosphorylation of five distinct phosphoproteins (pp28, pp33, pp38, pp67, and pp103) is induced or enhanced in macrophages treated with either of these two agonists. The pattern of phosphorylation so induced closely resembles that induced phorbol myristilic acetate (PMA), a specific stimulant of protein kinase C (PKC). Partial proteolysis of pp28, pp38, and pp67 by either staphylococcal V8 protease or chymotrypsin indicates that the phosphopeptide breakdown products of these proteins, whether induced by LPS or PMA, are extremely similar.[34] The ~68-kDa substrate protein phosphorylated in response to LPS has been of particular interest because it appears to be quite similar to an ~87-kDa protein that is a major substrate for PKC in a variety of cells.[35,36] Myristoylation of the ~68-kDa protein enhances the rate of phosphorylation via PKC. This 68-kDa substrate has been recently cloned and identified as homologous to the ~87-kDa protein, which has been termed myristolyated alanine-rich, C-kinase substrate or MARCKS.[37] The ability of LPS to prime for release or metabolites of arachidonate has been linked to the altered phosphorylation of this protein.[38]

The induction of electrically silent Na^+/H^+ exchange via the sodium/hydrogen exchanger or antiporter is a widely observed phenomenon in numerous cells pulsed with mitogenic agents.[39,40] Current evidence indicates that sodium/hydrogen exchange can be initiated by receptors which have tyrosine kinase activity, such as the receptor for CSF-1 or by stimulation of PKC through PMA. Despite the almost ubiquitous observation of Na^+/H^+ exchange in cells undergoing mitogenosis, a critical role for such in the induction of mitogenosis itself or in any specific gene regulation has not been forthcoming.[40] Recently, attention has been drawn to the fact that IFN-γ activates macrophages, at least in part, via induction of Na^+/H^+ exchange and that such induction leads to enhanced expression of message of two distinct genes.[41] LPS can also initiate Na^+/H^+ exchange in macrophages and B-cells.[42,43] Unlike IFN-γ, which initiates such exchange within seconds after its addition to macrophages *in vitro,* the LPS-mediated initiation is observed after an interval of ~10 to 15 min as monitored by influx of radioactive sodium. The initial effect of LPS on intracellular pH (pH_i) is acidification of the cytosol; at about 10 to 15 min, pH_i returns to basal values and thereafter steadily climbs into alkaline range. The initial acidification has been observed in many leukocytes in response to stimulants of the respiratory burst and has been attributed to a variety of causes, including protons spun off during the activation of the membrane-bound oxidiase complex.[44] The influx of Na^+ and the elevation in pH_1 induced by LPS, once established at 10 to 15 min, persists for at least 30 to 60 min; similar effects can be induced by PMA in macrophages and B-cells. LPS and PMA have recently been reported to initiate tyrosine phosphorylation of a number of proteins ranging from ~41 to ~142 kDa 5 to 10 min after application to macrophages; herbimycin A was found to inhibit both LPS-stimulated phosphorylation of tyrosine residues and LPS-stimulated release of arachidonic acid metabolites.[45]

III. LPS-MEDIATED TRANSDUCTIONAL EVENTS RELATED TO THE INDUCTION OF IMMEDIATE EARLY GENES

Stimulation of a wide variety of cell types is accompanied by the very rapid expression (i.e., within a few minutes) of proteins, which are frequently expressed only transiently (see Reference 27 for review). The genes encoding these proteins, by analogy to phenomenologically similar genes in viral replication, have been termed early or immediate early genes. Similar genes have been observed in the induction of competence for division in fibroblasts by platelet-derived growth factor (PDGF) and have been termed competence genes. These diverse genes, now often termed early genes and including many protooncogenes, encode several families of important proteins. At present, early genes comprise at least two large families: (1) nuclear regulatory proteins such as AP-1, NFκ-B, c-*fos*, c-*myc*, and c-*jun*, and (2) proteins that serve autocrine or paracrine functions such as JE, KC, and *gro* (for review, see Reference 27).

LPS and lipid A initiate the expression of a variety of early genes in mononuclear phagocytes. (See Figure 4.) Early gene expression was initially described in experiments using one- or two-dimensional, gel-electrophoretic analysis of macrophage proteins labeled metabolically with ^{35}S-methionine.[46-49] Subsequently, early expression of numerous genes has been demonstrated using Northern hybridization analysis with cDNA probes coding for known gene products. Such studies have included the proto-oncogenes c-*fos* and c-*myc* and cytokines such as JE or monocyte chemotaxis protein 1 (MCP-1), KC or *gro,* TNF-α, IL-1α and IL-1β, IL-6, and IL-8 (see References 22 and 27). In addition to these products which have been identified and characterized functionally in a number of experimental settings, additional genes have been identified through differential hybridization analysis of cDNA libraries made from LPS-activated macrophages.[50] For example, a gene originally termed C7 by Hamilton and colleagues has been shown to be structurally related to human IP-10, a gene first identified in IFN-γ-treated U937 cells also using the differential hybridization strategy.[51,52] IP-10 is a member of the supergene family encoding small chemotactic and/or mitogenic proteins, which includes IL-8, MCP-1, and *gro* though a specific function for IP-10 per se has not been reported. Interestingly, murine macrophages are more sensitive to LPS than IFN-γ for induction of IP-10, while human monocytes do not express IP-10 in response to LPS. In addition to known products, several genes with relatively novel structure have also been identified in LPS-treated macrophage cDNA libraries (e.g., D2, D3, D7, D8).[50] While the functional significance of many of these products remains obscure, their expression is stringently controlled by inflammatory stimulation. All are induced by LPS, but some are also sensitive to induction by other agents, including IFN-γ; IFN-β, PDGF, and TNF-α (see Reference 22 for details). In this regard, the

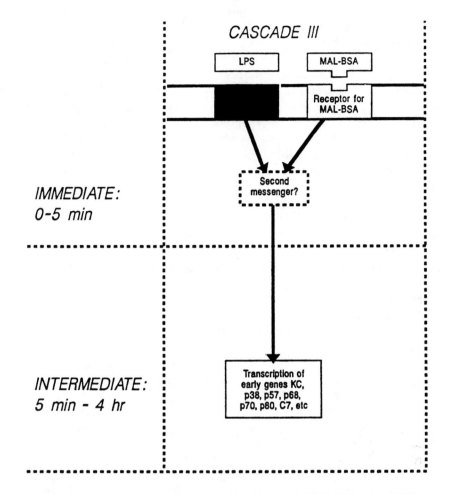

FIGURE 4. A closeup of the events initiated in cascade III by LPS and mal-BSA.

molecular mechanisms through which different stimuli induce expression of these genes may be distinct.

By virtue of the very rapid transcription of these genes as well as the fact that most inductive gene events require a second messenger, we have postulated for several years that a rapid second messenger links surface interaction of LPS or mal-BSA on macrophages with transcription of the genes.[23,24,27] To date, this putative, rapid, second messenger has yet to be identified.

IV. INTERRELATIONSHIPS BETWEEN THE TWO LPS-INITIATED SIGNAL CASCADES

Several lines of evidence indicate that the two cascades of transductional events initiated by LPS (i.e., signal cascade II involving the breakdown of polyphosphoinositides and delayed initiation of Na^+/H^+ exchange and cascade III involving the initiation of early genes) are distinct.

Events in cascade II can be selectively mimicked by platelet-activating factor (PAF). Recent evidence indicates that murine mononuclear phagocytes bear high-affinity surface receptors for PAF, ligation of which recapitulates very closely, in *qualitative* terms, the events described above, that is, PAF initiates the breakdown of polyphosphoinositides, elevations in $[Ca^{2+}]_i$, elevations in DAG, characteristic phosphorylation via PKC, and delayed activation of Na^+/H^+ exchange.[29,42] It is important to note, however, that the effects of LPS and PAF are *quantitatively* distinct. The breakdown of polyphosphoinositides initiated by PAF rises to greater levels but is initiated and terminated much more rapidly than that induced by LPS, which rises more slowly to lower levels and persists for longer periods of time.[28,34,42] Concomitantly, the characteristic pattern of protein phosphorylation and activation of the antiport by PAF, as compared to the same events initiated by LPS, are initiated and terminated more rapidly. Mal-BSA has not been found to initiate these events.[30]

The effects of LPS on early gene expression can be recapitulated by application of maleylated or acetylated proteins, which bind to one of the family of scavenger receptors.[47] Current evidence indicates that these effects are initiated by binding to low-affinity scavenger receptor which recognizes maleylated-BSA and α-casein but not by the high-affinity scavenger receptor which recognizes malondialdehyde-LDL.[53] These effects initiated by mal-BSA are not reproduced by PAF.[30]

Another line of evidence for distinguishing between the two transductional cascades has been provided by analysis of the requirements for inducing elevated levels of mRNA specific for JE and KC in macrophages.[54] JE is initiated in macrophages by PAF, which does not induce KC. Pharmacologically, induction of JE can be mimicked by addition of PMA plus calcium ions in the presence of the ionophore A23187 but induction of KC cannot. Last, the inductive effects of LPS on JE can be inhibited by provision of extracellular dibutyryl cyclic AMP (cAMP), while those for KC are not[55] (*vide infra*).

V. INTERRELATIONSHIPS BETWEEN SIGNAL CASCADES INDUCED BY LPS AND OTHER CASCADES

Over the past several years, at least one of the immediate transductional events initiated by IFNγ has been identified, namely rapid activation of sodium/hydrogen exchange via the antiport[41] (Figure 5). Following this, two additional changes are observed. First is delayed efflux or pumping of Ca^{2+} from macrophages stimulated with IFN-γ, though alterations in $[Ca^{2+}]_i$ are not seen.[56,57] Second, there is a change in the potential ability of PKC to

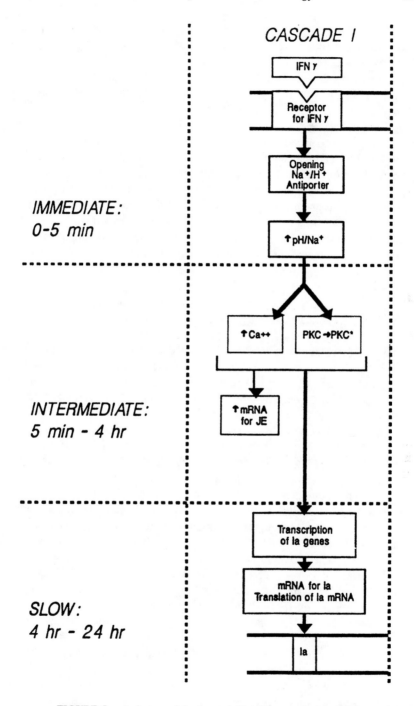

FIGURE 5. A closeup of the events initiated in cascade I by IFNγ.

effect substrate phosphorylation. This change is manifest by enhanced phosphorylation (by four- to fivefold) of substrates in isolated preparations when the enzyme is supplied with the requisite co-factors of calcium and phosphatidylserine as well as a stimulant of the regulatory domain such as PMA.[58,59] Correspondingly, phosphorylation mediated by agonist-induced stimulation of PKC produced by enhanced levels of DAG is proportionately enhanced.[29,34] Specifically, LPS and PAF induce enhanced phosphorylation of several characteristic substrate proteins when applied to macrophages previously prepared or primed by treatment with IFN-γ. As noted above, the production of DAG after stimulation with LPS in macrophages comes predominantly from the breakdown of phosphatidylcholine rather than the breakdown of phosphatidylinositol-4,5-biphosphate (PIP$_2$). The breakdown of PC into DAG is potentiated severalfold by prior treatment of macrophages with IFN-γ.[60] Thus, at least two mechanisms are established whereby prior treatment of macrophages can enhance phosphorylation due to stimulation of PKC (Figure 6).

We have further observed that the activation of sodium/hydrogen exchange induced by either LPS or PAF may be altered by prior treatment with IFN-γ.[41] Specifically, the rise of pH$_i$ into the alkaline range from the acid can be observed earlier. Greater alterations in cellular alkalinity can be observed; these can persist for longer periods of time. A reasonable hypothesis to explain these findings is that potentiated activation of the antiporter is attributable to enhanced phosphorylation of the antiport molecule itself via PKC, and studies to test this possibility are currently underway. In sum, several mechanisms exist whereby prior treatment of macrophages can potentiate the events initiated in cascade II by LPS or by PAF.

Prior treatment with IFN-γ also modulates the events observed in cascade III. Specifically, IFN-γ significantly lowers the dose of LPS (i.e., by two orders of magnitude) required to initiate synthesis of the characteristic new protein products identified by gel electrophoresis in macrophages treated with LPS.[46,47] The molecular basis of this has not been defined.

Elevating intracellular levels of cAMP represents a potent suppressor mechanism for most leukocytes, although the molecular basis of such suppression has not been well established to date.[61,62] Surface-active agonists such as PGE$_2$ can raise intracellualr levels of cAMP, which in turn can suppress a wide variety of leukocyte functions (Figure 6). One target of such elevations in cAMP levels has recently been established to be a reduction in Na$^+$/H$^+$ exchange induced by IFNγ (Figure 7).[63] Subsequent studies have shown that activation of Na$^+$/H$^+$ exchange by other agonists such as PAF is also suppressed by elevating intracellular levels of cAMP.[64] Whether or not cAMP is able to suppress other transductional mechanisms in macrophages, such as the breakdown of PIP$_2$ or PC, remains to be established.

These observations are pertinent to transductional effects initiated by LPS. In addition to the obvious possibility that LPS-mediated transductional events, such as induction of Na$^+$/H$^+$ exchange, are suppressed by elevating cAMP,

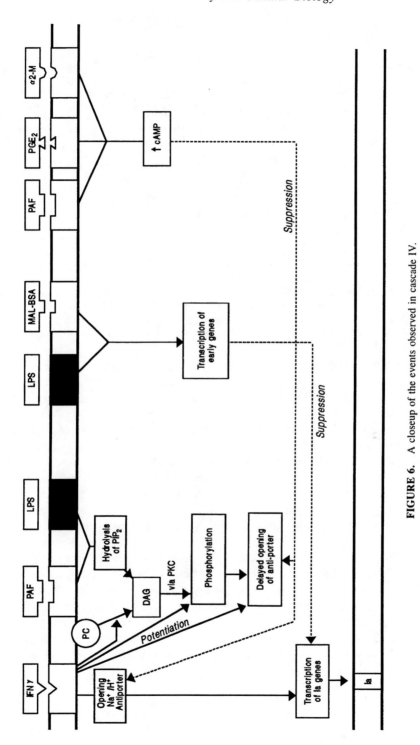

FIGURE 6. A closeup of the events observed in cascade IV.

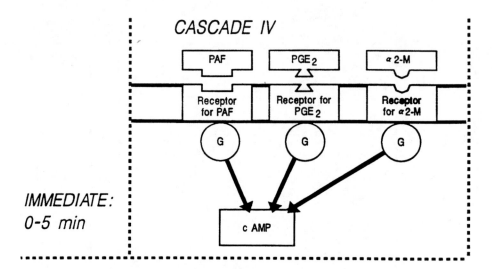

FIGURE 7. A schematic highlighting interrelationships between the various cascades. Induction of cascade I can potentiate events in cascade III at at least three levels. Events in cascade III can suppress transcription of Ia genes, whereas events in cascade IV suppress activation of sodium/hydrogen exchange whether rapidly induced by IFNγ or slowly induced by LPS or PAF.

recent evidence indicates that LPS may have countersuppressive effects.[65] We have observed that pretreatment of macrophages with LPS for 60 min markedly decreases PGE-mediated elevations in cAMP, by decreasing the overall maximum level of cAMP elevation, accelerating the decrease of cAMP to basal levels, and by abolishing a subsequent and sustained phase of cAMP elevation.[65] LPS can selectively activate phosphodiesterase(s) and thereby accelerate the degradation of recently formed cAMP. It thus becomes of interest to ask whether the LPS-mediated effects on levels of cAMP are carried out through interactions with G proteins. Although the major effects of G_i proteins (those that inhibit cAMP) established to date in modulating levels of cAMP are exerted through other mechanisms,[66] recent reports have suggested coupling of phosphodiesterase(s) to G proteins in the retina (see Reference 67). Whether these reports are pertinent to other cells remains to be determined. The basic observation is, however, of interest because it suggests that LPS, a potent bacterial product, may prolong its cellular activating effects by inhibiting natural mechanisms suppressing the transductional mechanisms originally initiated by LPS (i.e., countersuppression). In this regard, it is interesting to compare the temporal course of phosphoinositide breakdown and subsequent activation of Na^+ exchange between LPS and PAF (*vide supra*).

VI. FUNCTIONAL EFFECTS OF LPS-INITIATED
TRANSDUCTIONAL CASCADES

The criteria originally described by Sutherland for determining whether cAMP is implicated in a given response remain very useful.[68] These criteria are (1) the physiologic ligand should cause alterations in the level of the second messenger in question; (2) inhibition of the second messenger in question should block induction of the function; and (3) pharmacologic mimicry of the second messenger in question should produce the physiologic response.

One of the most widely studied functional effects of LPS on macrophages is triggering cytolysis of tumor cells during macrophage-mediated tumor cytotoxicity (MTC). Although the biological basis for this potent effect remains to be fully established, it is known that LPS can trigger or stimulate the release of toxic molecules, such as a novel cytolytic serine proteinase and TNF-α from macrophages previously prepared or primed by treatment with IFN-γ.[4] The dose and time requirements for LPS in initiating both signal cascade II and III are consistent with the requirements for triggering of cytolysis (for reviews, see References 23 and 24). As noted above, IFN-γ can potentiate the response of LPS and lower the dose requirement in these two signal cascades, just as is observed with the effects of IFN-γ on lowering the dose requirements and magnifying the effects of LPS in triggering cytolysis. Macrophages from mice of the C3H/HeJ strain are well known to respond ineffectively to LPS in terms of triggering cytolysis. Macrophages from these mice also do not initiate either characteristic protein phosphorylation or synthesis of the characteristic new proteins when treated with LPS. Application of an alternative second or triggering signal to such macrophages (such as heat-killed *Listeria monocytogenes*) induces cytolysis and, concomitantly, the characteristic patterns of both protein phosphorylation and synthesis of new proteins. The triggering effects of LPS on cytolysis are blocked by the inhibitor of protein synthesis, cycloheximide, at concentrations which inhibit production of new proteins. Lastly, when one correlates ability to induce cytolysis and induce these two characteristic biochemical responses by analyzing a group of diverse stimuli, one consistently observes induction of cytolysis only by stimuli which also induce the two characteristic biochemical changes. Taken together, these lines of evidence suggest that both LPS-initiated transductional pathways may be necessary for induction of cytolytic competence.

An interesting functional consequence of initiation of cascade III is suppression of transcription of class II major histocompatibility complex (MHC) genes in macrophages (Figure 6). It has been well established for some time that intermediate doses of LPS (e.g., 100 ng/ml) are capable of suppressing surface expression of class II MHC molecules *in vitro*,[69] and this effect has subsequently been shown to be dependent upon suppression of IFN-γ-induced transcription.[70] The suppressive effects of LPS upon such gene expression can be mimicked by mal-BSA.[71] Furthermore, administration of either LPS or mal-BSA with cycloheximide for several hours blocks the ability of these agents to suppress surface expression almost completely.[71] It is tempting to speculate that one of the early genes induced by macrophages in response to LPS/mal-BSA is a nuclear binding protein that may have a suppressive function, but this interesting possibility has yet to be verified. Acetyl-LDL, presumably acting through the scavenger receptor, has recently been found to suppress LPS-induced production of IL-1.[72]

The regulation of certain genes initiated in macrophages by LPS can be ascribed to one of these cascades. Alterations in level of specific message for c-*fos*, c-*myc*, and JE can be reasonably ascribed to initiation of cascade II.[54,73] These effects can be pharmacologically mimicked by stimulating the macrophages with PMA in the presence of extracellular calcium plus an ionophore such as A23187. Furthermore, PAF can initiate heightened expression of message specific for JE.[41] By contrast, KC is initiated in macrophages by mal-BSA but not by the combination of PMA plus A23187 in the presence of extracellular calcium.[54] At present, it remains unclear as to whether the LPS-mediated initiation of transcription of TNF-α can clearly be ascribed to a single cascade. Prior treatment of macrophages with IFN-γ potentiates the magnitude of such transcription.[3,30,73] Inhibitors of PKC can block LPS-mediated initiation of transcription and PMA and/or A23187 can produce enhanced mRNA and secretion of TNF-α.[74,75] Enhancement of intracellular cAMP partially suppresses accumulation of mRNA as does amiloride — an inhibitor of Na^+/H^+ exchange.[76-79] Key structural elements of the promoter of this gene have recently been elucidated and found to include several sequences which resemble the recognition site for NF$\kappa\beta$.[80,81] Although the precise second signals required to initiate TNF-α transcription by LPS remain undefined, it seems reasonable at present to assign this effect provisionally — at least in part — to cascade II. Inhibitors of PKC can also block LPS-mediated stimulation of message specific for IL-1β.[82]

Although LPS itself is not a powerful stimulant of release of metabolites of arachidonic acid, LPS does potently prime or prepare macrophages for such release, in that application of a second signal such as PAF ieads to markedly enhanced release of these metabolites after pretreatment of the macrophages with LPS (see Reference 83). LPS-mediated priming has been ascribed to phosphorylation of a specific protein substrate described above as pp67 or the murine equivalent of MARCKS.[37,38] Concomitantly, LPS also induces the synthesis of MARCKS as well as potentiating its phosphorylation and myristoylation. Priming by LPS is sensitive to inhibition by actinomycin-D, while PAF-stimulated PGE$_2$ production in primed macrophages is blocked by cyclohexamide, which may implicate a protein that turns over rapidly.[83] Events in cascades II and III, albeit still not fully defined, may thus participate in the priming of macrophages for enhanced release of metabolites of arachidonate.

VII. CODA

Over the past several years, two major cascades of transductional events initiated in macrophages by LPS have been identified. The first of these, termed overall signal cascade II, involves the breakdown of polyphosphoinositides and a subsequent delayed activation of Na^+/H^+ exchange. The second of these, termed overall signal cascade III, involves the induction of certain early genes. All evidence to date strongly indicates that these two large, complex cascades of events are quite distinct.

Progress is now being made on ascribing various functional effects of LPS on macrophages to one or more of these cascades. LPS-mediated priming for enhanced release of arachidonate has been ascribed to phosphorylation of a specific protein substrate (i.e., pp67).[37,38] Priming for increased release of arachidonate has also been related in neutrophils to increases in $[Ca^{2+}]_i$.[84] Finally, such priming apparently requires the presence of rapidly synthesized proteins since such priming can be blocked by inhibitors of protein synthesis.[38,83] Certain early genes, which include c-*fos*, c-*myc*, and JE, can be ascribed to cascade II; enhanced transcription of TNF-α can be provisionally so ascribed — at least in some part (*vide supra*). Other early genes are ascribed to the second cascade (cascade III); these include KC and C7, which has been subsequently identified as a homolog of IP10. Initiation of this latter broad set of genes has been indirectly linked to suppression and transcription of at least one family of genes of considerable importance in macrophages, namely, genes encoding for class II MHC molecules. Lastly, considerable evidence indicates that both transductional cascades are necessary for LPS-mediated triggering of cytolysis. A major question of importance will now be to assess systematically and critically the other functional and metabolic effects of LPS on macrophages and assign these initially to one or more of the two large cascades and subsequently to specific molecular events within them.

Another question of key importance is to define the relationship between the binding of LPS or lipid A to particular membranous proteins and the initiation of signal cascades. At present, information on this topic is quite limited. Because a precursor of lipid A (e.g., lipid$_{IVA}$) may bind to the scavenger receptor, it is possible that LPS exerts some of the effects observed in cascade III by ligation of one or more of the family of scavenger receptors, but no direct evidence as to whether this possibility is likely or unlikely currently exists.

It will be interesting to observe the further evolution of our understanding, in specific molecular terms, of the transductional events initiated in macrophages and indeed in other cells by LPS. One can reasonably anticipate that the two cascades now defined will be amplified in considerable depth and detail, including determining the rapid second messenger which putatively initiates signal cascade III. Given the rapid expansion of information on the transductional effects of LPS over the past several years as well as the pleiotrophic effects of this molecule, it would not be surprising if an additional large cascade(s) is observed, in addition to filling out details of the two cascades already identified.

Over a number of years, the interrelationships between IFN-γ and LPS have proven to be both complex and perplexing.[4] Overall, these two powerful stimulants of macrophages may produce similar, disparate, or cooperative effects. We are now beginning to appreciate in specific molecular terms how such a disparate array of effects can be created (compare Figure 5 with Figures 3 and 4). Both agonists, however, initiate sodium/hydrogen exchange in macrophages, so at least one potential mechanism of induced similar effects has been identified. Although LPS initiates elevated $[Ca^{2+}]_i$ while IFN-γ does not, IFN-γ does initiate enhanced efflux of Ca^{2+} from macrophages, so it is possible that both agonists could also induce localized membrane pumping and cycling of calcium.[85] Lastly, at least two mechanisms have been observed (see Figure 6) by which IFN-γ can potentiate or magnify the effects of LPS. It has also been observed, without defining the molecular basis of the observation, that IFN-γ can lower the dose of LPS required for certain other effects.

A recent but very intriguing observation relates to how LPS might exert its potent and long-lasting effects. Certain transductional events in signal cascades I and III can be suppressed by elevating intracellular levels of cAMP. It has now been found that LPS reduces PGE_2-mediated increases in levels of cAMP. Although the biological significance of this observation remains to be established, it is possible that this countersuppressive effect of LPS could prolong transductional effects initiated by LPS and thus explain, at least in part, why LPS-mediated effects are both potent and long lasting.

Over the past several years, considerable progress has been made in identifying some of the major transductional events initiated by LPS. Over the coming years, it will be reasonable to anticipate amplification of our knowledge of these transductional events and define the critical relationships between these transductional events on the one hand to membrane binding sites for LPS/lipid A and, on the other hand, to genomic and metabolic effects initiated by LPS.

ACKNOWLEDGMENTS

Thanks to Tom Hamilton for unpublished data and critical review of the manuscript. Research was supported in part by U.S. Public Health Service grants CA 29586, CA 16784, and CA 02922.

REFERENCES

1. **Takada, H. and Kotani, S.**, Structural requirements of lipid A for endocytoxicity and other biological activities, *Crit. Rev. Microbiol.*, 16, 477, 1989.
2. **Morrison, D. C.**, The case for specific lipopolysaccharide receptors expressed on mammalian cells, *Microb. Pathogenesis*, 7, 389, 1989.
3. **Raetz, C. R. H.**, Biochemistry of endotoxins, *Annu. Rev. Biochem.*, 59, 129, 1990.
4. **Adams, D. O. and Hamilton, T. A.**, The cell biology of macrophage activation, *Annu. Rev. Immunol.*, 2, 283, 1984.
5. **Lei, M. -G. and Morrison, D. C.**, Specific endotoxic lipopolysaccharide binding proteins on murine splenocytes. I. Detection of lipopolysaccharide-binding sites on splenocytes and splenocyte subpopulations, *J. Immunol.*, 141, 996, 1988.
6. **Lei, M. -G. and Morrison, D. C.**, Specific endotoxic lipopolysaccharide binding proteins on murine splenocytes. II. Membrane localization and binding characteristics, *J. Immunol.*, 141, 1006, 1988.
7. **Chen, T. -Y., Bright, S. W., Pace, J. L., Russell, S. W., and Morrison, D. C.**, Induction of macrophage-mediated tumoricidal toxicity by a hamster monoclonal antibody with specificity for lipopolysaccharide receptor, *J. Immunol.*, 145, 8, 1990.
8. **Dziarski, R.**, Peptidoglycan and lipopolysaccharide bind to the same binding site on lymphocytes, *J. Biol. Chem.*, 266, 4719, 1991.
9. **Dziarski, R.**, Demonstration of peptidoglycan binding sites on lymphocytes and macrophages by photoaffinity crosslabelling, *J. Biol. Chem.*, 266, 4713, 1991.
10. **Wright, S. D. and Jong, M. T-C.**, Adhesion-promoting receptors on human macrophages recognize *E. coli* by binding to lipopolysaccharide, *J. Exp. Med.*, 164, 1876, 1986.
11. **Wright, S. D., Ramos, R. A., Tobias, P. S., Uhlevich, R. J., and Mathison, J. C.**, C14, a receptor for complexes of lipopolysaccharide (LPS) and LPS-binding protein, *Science*, 249, 1431, 1990.
12. **Brown, M. S. and Goldstein, J. L.**, Lipoprotein metabolism in the macrophage: implications for cholesterol deposition in atherosclerosis, *Annu. Rev. Biochem.*, 52, 233, 1983.
13. **Fogelman, A. M., Van Lenten, B. J., Wardin, C., Haberland, M. E., and Edwards, P. A.**, Macrophage lipoprotein receptors, *J. Cell Sci.*, 9(Suppl.), 135, 988.
14. **Kodama, T., Freeman, M., Rohrer, L., Zabrecky, J., Matsudaira, P., and Krieger, M.**, Type I macrophage scavenger receptor contains α-helical and collagen-like coiled coils, *Nature*, 343, 531, 1990.
15. **Rohrer, L., Freeman, M., Kodama, T., Pinman, M., and Kriger, M.**, Coiled-coil fibrous domains mediate ligand binding by macrophage scavenger receptor type II, *Nature*, 343, 570, 1990.
16. **Sparrow, C. P., Parthasarathy, F. S., and Steinberg, D.**, A macrophage receptor that recognizes oxidized low density lipoprotein but not acetylated low density lipoprotein, *J. Biol. Chem.*, 264, 2599, 1989.
17. **Hampton, R. Y., Golenbock, D. T., and Rates, C. R. H.**, Lipid A binding sites in membranes of macrophage tumor cells, *J. Biol. Chem.*, 263, 14802, 1988.
18. **Golenbock, D. T., Hampton, R. Y., Raetz, C., and Wright, S. D.**, Human phagocytes have multiple lipid A-binding sites, *Infect. Immun.*, 58, 469, 1990.
19. **Dziarski, R.**, Correlation between ribosylation of pertussis toxin substrates and inhibition of peptidoglycan-, murimyldipeptide-, and lipopolysaccharide-induced mitogenic stimulation in B-lymphocytes, *Eur. J. Immunol.*, 19, 125, 1989.
20. **Issakani, S. D., Spiegel, A. M., and Strulovic, P.**, Lipopolysaccharide responses linked to the GTP binding protein, Gi_2 in the promonocyte cell line U937, *J. Biol. Chem.*, 264, 240, 1990.

21. **Jakway, J. P. and DeFranko, A. L.**, Pertussis toxin inhibition of B-cell and macrophage responses to bacterial lipopolysaccharides, *Science,* 234, 743, 1986.

22. **Adams, D. O. and Hamilton, T. A.**, Molecular bases of macrophage activation: diversity and its origins, in *The Natural Immune System,* Vol. II, Lewis, C., Ed., University of Oxford Press, Oxford, England, 1992, 77.

23. **Adams, D. O. and Koerner, T. J.**, Gene regulation in macrophage development and activation, *Year Immunol.,* 4, 159, 1988.

24. **Adams, D. O. and Hamilton, T. A.**, Molecular bases of signal transduction in macrophage activation induced by IFNγ and by second signals, *Immunol. Rev.,* 97, 5, 1987.

25. **Hamilton, T. A. and Adams, D. O.**, Molecular mechanisms of signal transduction in macrophage activation, invited review, *Immunol. Today,* 8, 151, 1987.

26. **Riches, D. W. H., Channon, J. Y., Leslie, C. C., and Hinson, P. M.**, Receptor-mediated signal transduction in mononuclear phagocytes, *Prog. Analogy,* 42, 65, 1988.

27. **Adams, D. O., Johnson, S. P., and Uhing, R. J.**, Early gene expression in the activation of mononuclear phagocytes, in *Mechanisms of Leukocyte Activation,* Grinstein, S. and Rothstein, O. D., Eds., *Current Topics in Membranes and Transport.,* 35, 587, 1990.

28. **Prpic, V., Weiel, J. E., Somers, S. D., Gonias, S. L., Pizzo, S. V., DiGusieppi, J., Hamilton, T. A., Herman, B., and Adams, D. O.**, The effects of bacterial endotoxin on the hydrolysis of phosphatidylinositol-4,5-bisphosphate in murine peritoneal macrophages, *J. Immunol.,* 139, 526, 1987.

29. **Prpic, V., Uhing, R. J., Weiel, J. E., Jakoi, L., Gawdi, G., Herman, B., and Adams, D. O.**, Biochemical and functional responses stimulated by platelet activating factor in murine peritoneal macrophages, *J. Cell. Biol.,* 107, 363, 1988.

30. **Adams, D. O.**, unpublished data, 1988.

31. **Uhing, R. J., Prpic, V., Hollenbach, P. W., and Adams, D. O.**, Involvement of protein kinase C in platelet activating factor-stimulated diacylglycerol accumulation in murine peritoneal macrophages, *J. Biol. Chem.,* 264, 9224, 1989.

32. **Sebaldt, R. J., Prpic, V., Hollenbach, P. W., Adams, D. O., and Uhing, R. J.**, IFN-gamma potentiates the accumulation of diacylglycerol in murine macrophages, *J. Immunol.,* 145, 684, 1990.

33. **Zoeller, R. A., Wightman, P. D., Anderson, M. S., and Raetz, C. R. H.**, Accumulation of lysophosphatidylinositol and RAW264.7 macrophage tumor cells stimulated by lipid A precursors, *J. Biol. Chem.,* 262, 17212, 1987.

34. **Weiel, J., Hamilton, T., and Adams, D. O.**, LPS induces altered phosphate labeling of proteins in murine peritoneal macrophages, *J. Immunol.,* 136, 3012, 1986.

35. **Aderem, A. A., Albert, K. A., Keum, M. M., Wang, J. K. T., Greengard, P., and Cohn, Z. A.**, Stimulus-dependent myristoylation of a major substrate for protein kinase C, *Nature,* 332, 362, 1988.

36. **Rosen, A., Nairn, A. C., Greengard, P., Cohn, Z. A., and Aderem, A. A.**, Bacterial lipopolysaccharide regulates the phosphorylation of the 68K protein kinase C substrate in macrophages, *J. Biol. Chem.,* 264, 9118, 1989.

37. **Seykora, J., Ravetch, J. V., and Aderem, A.**, Cloning and molecular characterization of the murine macrophage "68 kDa" protein kinase C substrate and its regulation by bacterial lipopolysaccharide, *Proc. Natl. Acad. Sci. U.S.A.,* 88, 2502, 1991.

38. **Aderem, A. A.**, Murine myristoylation as an intermediate step during signal transduction in macrophages: its role in arachidonic acid metabolism and in response to IFNγ, *J. Cell Sci.,* 9(Suppl.), 151, 1988.

39. **Moolenaar, W.**, Effects of growth factors on intracellular pH regulation, *Annu. Rev. Physiol.,* 48, 363, 1986.

40. **Grinstein, S., Rotin, D., and Mason, M. J.**, Na^+/H^+ exchange and growth factor-induced cytosolic pH changes. Role in cellular proliferation, *Biochim. Biophys. Acta,* 988, 73, 1989.

41. **Prpic, V., Yu, S. -F., Figueiredo, F., Hollenbach, P. W., Gawdi, G., Herman, B., Uhing, R. J., and Adams, D. O.**, Role of Na$^+$/H$^+$ exchange by interferon-γ in enhanced expression of JE and I-Aβ genes, *Science*, 244, 469, 1989.
42. **Prpic, V., Yu, S. F., and Adams, D. O.**, LPS and PAF activate the Na$^+$/H$^+$ antiporter and induce the immediate early gene JE in murine peritoneal macrophages, in preparation.
43. **Gaidano, G., Ghigo, D., Schena, M., Bergui, L., Treves, S., Turrini, F., Cappio, F. C., and Bosia, A.**, Na$^+$/H$^+$ exchange, activation mediates the lipopolysaccharide-induced proliferation of human B-lymphocytes and is impaired in malignant B-chronic lymphocytic leukemia lymphocytes, *J. Immunol.*, 142, 913, 1989.
44. **Weismann, S. J., Punzo, A., Ford, C., and Sha-Afi, R. I.**, Intracellular pH changes during neutrophil activation: Na$^+$/H$^+$ antiport, *J. Leukocyte Biol.*, 41, 25, 1987.
45. **Weinstein, S. L., Gold, M. R., and DeFranco, A. L.**, Bacterial lipopolysaccharide stimulates protein tyrosine phosphorylation in macrophages, *Proc. Natl. Acad. Sci. U.S.A.*, 88, 4148, 1991.
46. **Hamilton, T. A., Somers, S. D., Jansen, M. M., and Adams, D. O.**, Effects of bacterial lipopolysaccharide on protein synthesis in murine peritoneal macrophages: relationship to activation for macrophage tumoricidal function, *J. Cell. Physiol.*, 128, 9, 1986.
47. **Johnston, P. A., Jansen, M. M., Somers, S. D. Adams, D. O., and Hamilton, T. A.**, Maleyl-BSA and fucoidan induce expression of a set of early proteins in murine mononuclear phagocytes, *J. Immunol.*, 138, 1551, 1987.
48. **Largen, M. T. and Tannenbaum, C. S.**, LPS-regulation of specific protein synthesis in murine peritoneal macrophages, *J. Immunol.*, 136, 988, 1986.
49. **Mackay, R. J. and Russell, S. W.**, Protein changes associated with stages of activation of mouse macrophages for tumor cell killing, *J. Immunol.*, 137, 1392, 1986.
50. **Tannenbaum, C. S., Koerner, T. J., Jansen, M. M., and Hamilton, T. A.**, Characterization of lipopolysaccharide-induced macrophage gene expression, *J. Immunol.*, 140, 3640, 1988.
51. **Ohmori, Y. and Hamilton, T. A.**, A macrophage LPS-inducible macrophage gene encodes the murine homologue of human IP-10, *Biochem. Biophys. Res. Comm.*, 168, 1261, 1990.
52. **Luster, A. D., Unkeless, J. C., and Ravetch, J. V.**, γ-Interferon transcriptionally regulates an early-response gene containing homology to platelet protein, *Nature*, 315, 672, 1985.
53. **Haberland, M. E., Tannenbaum, C. S., Williams, R. E., Adams, D. O., and Hamilton, T. A.**, Role of maleyl-albumin receptor in activation of murine peritoneal macrophages *in vitro*, *J. Immunol.*, 142, 855, 1989.
54. **Introna, M., Bast, R. C., Hamilton, T. A., Tannenbaum, C. S., and Adams, D. O.**, The effect of LPS on expression of the early competence genes JE and KC in murine peritoneal macrophages, *J. Immunol.*, 138, 3891, 1987.
55. **Tannenbaum, C. S. and Hamilton, T. A.**, Lipopolysaccharide-induced gene expression in murine peritoneal macrophages is selectively suppressed by agents which elevate intracellular cAMP, *J. Immunol.*, 142, 1274, 1989.
56. **Somers, S. D., Weiel, J., Hamilton, T., and Adams, D. O.**, Phorbol esters and calcium ionophore can prime murine peritoneal macrophages for tumor cell destruction, *J. Immunol.*, 136, 4199, 1986.
57. **Prpic, V. and Adams, D. O.**, unpublished observations, 1989.
58. **Hamilton, T. A., Becton, D. A., Somers, S. D., and Adams, D. O.**, Interferon gamma modulates protein kinase C activity in murine peritoneal macrophages, *J. Biol. Chem.*, 260, 1378, 1985.
59. **Becton, D. L., Hamilton, T. A., and Adams, D. O.**, Characterization of protein kinase C activity in interferon gamma treated murine peritoneal macrophages, *J. Cell. Physiol.*, 125, 485, 1985.

60. **Sebaldt, R. J., Prpic, V., Hollenbach, P. W., Adams, D. O., and Uhing, R. J.,** Interferon-gamma potentiates the accumulation of diacylglycerol in murine macrophages, *J. Immunol.,* 145, 684, 1990.

61. **Schade, U. F., Burmeister, I., Elekes, E., Engel, R., and Wolter, D. T.,** Mononuclear phagocytes and eicosanoids: aspects of their synthesis and biological activities, *Blut,* 59, 475, 1989.

62. **Uhing, R. J., Cowlen, M., and Adams, D. O.,** Mechanisms regulating the release of arachidonate metabolites in mononuclear phagocytes, in *Current Topics in Membranes and Transport,* Grinstein, S. and Rotstein, O. D., Eds., Academic Press, 35, 349, 1990.

63. **Figueiredo, F., Uhing, R. J., Okonogi, K., Gettys, T., Johnson, S., Adams, D. O., and Prpic, V.,** Activation of the cAMP cascade inhibits an early event involved in murine macrophage Ia expression, *J. Biol. Chem.,* 265, 12317, 1990.

64. **Cowlen, M. S., Prpic, V., Figueiredo, F., Okonogi, K., Yu, S. F., Uhing, R. J., and Adams, D. O.,** Regulation of interferon-γ- and platelet-activating factor-stimulated Na$^+$/H$^+$ antiport activity and gene expression by cAMP in murine peritoneal macrophages, *FASEB Proc.,* 4(4), PA5050(Abstr.), 1990.

65. **Okonogi, K., Gettys, T. W., Uhing, R. J., Tarry, W., Adams, D. O., and Prpic, V.,** Inhibition of PGE$_2$-stimulated cAMP accumulation by lipopolysaccharide (LPS) in macrophages, *J. Biol. Chem.,* 266, 10305, 1991.

66. **Birnbaumer, L.,** Transduction of receptor signal into modulation of effector activity by G proteins: the first twenty years or so ..., *FASEB J.,* 4, 3068, 1990.

67. **Sadler, S. E. and Maller, J. L.,** A similar pool of cAMP phosphodiesterase in *Xenopus* oocytes is stimulated by insulin-like growth factor 1 and [Val-12, THR59] HA-*ras* protein, *J. Biol. Chem.,* 264, 856, 1989.

68. **Robinson, G. A., Buttcher, R. W., and Sutherland, E. W.,** in *Cyclic AMP,* Academic Press, New York, 1971, 36.

69. **Steeg, P. S., Johnson, H. M., and Oppenheim, J. J.,** Regulation of murine macrophage I-A antigen expression by an immune interferon-like lymphokine: inhibitory effect of endotoxins, *J. Immunol.,* 129, 2402, 1982.

70. **Figueiredo, F., Koerner, T. J., and Adams, D. O.,** Molecular mechanisms regulating the expression of class II histocompatibility molecules on macrophages: Effects of inductive and suppressive signals on gene transcription, *J. Immunol.,* 143, 3781, 1989.

71. **Hamilton, T. A., Gainey, P. V., and Adams, D. O.,** Maleylated-BSA suppresses IFNγ-mediated Ia expression in murine peritoneal macrophages, *J. Immunol.,* 138, 4063, 1987.

72. **Haga, Y., Takata, K., Araki, N., Sakamoto, K., Akagi, M., Morino, Y., and Horiuchi, S.,** Intracellular accumulation of cholesteryl esters suppresses production of lipopolysaccharide-induced interleukin-1 by rat peritoneal macrophages, *Biochem. Biophys. Res. Commun.,* 1, 6874, 1989.

73. **Koerner, T. J., Adams, D. O., and Hamilton, T. A.,** Regulation of tumor necrosis factor (TNF) expression: interferon gamma enhances the accumulation of mRNA for TNF induced by lipopolysaccharide in murine peritoneal macrophages, *Cell. Immunol.,* 109, 437, 1987.

74. **Dong, Z., Lew, S., and Zhang, Y.,** Effects of pretreatment with protein kinase C activities on macrophage activation for tumoricidal toxicity secretion of tumor necrosis factor and its mRNA expression, *Immunobiology,* 179, 382, 1989.

75. **Kovac, S., Radziouchd, E. J., Young, H. A., and Faresio, L.,** Differential inhibition of IL-1 and TNFα mRNA expression by agents which block second messenger pathways in murine macrophages, *J. Immunol.,* 141, 3101, 1988.

76. **Renz, H., Gong, J. H., Schmidt, A., Naine, M., and Gimpser, D.,** Release of tumor necrosis factor-α from macrophages. Enhancement and suppressment are dose-dependently regulated by prostaglandin E$_2$ and cyclic nucleotides, *J. Immunol.,* 141, 2388, 1988.

77. **Taffet, S. M., Singhei, K. J., Overholtzer, J. F., and Shirtleft, S. A.,** Regulation of tumor necrosis factor expression in macrophage-like cell line by lipopolysaccharide and cyclic AMP, *Cell. Immunol.,* 120, 291, 1989.
78. **Spengler, R. M., Spengler, M. L., Strieter, R. M., Remick, D. G., Larrick, J. W., and Conchel, S. L.,** Modulation of tumor necrosis factor-α gene expression. Desensitization of prostaglandin E₂-suppression, *J. Immunol.,* 142, 4346, 1989.
79. **Yu, S. F. and Adams, D. O.,** unpublished observations, 1989.
80. **Collart, M. A., Baeuerle, P., and Vassalli, P.,** Regulation of tumor necrosis factor alpha transcription in macrophages: involvement of four κB-like motifs and of constitutive and inducible forms of NF-κB, *Mol. Cell. Biol.,* 10, 1498, 1990.
81. **Shakkhof, A. N., Collart, M. A., Vassalli, P., Nedospasov, S. A., and Jongeneel, C. V.,** κB-type enhancers are involved in the lipopolysaccharide-mediated transcriptional activation of the tumor necrosis factor alpha gene in primary macrophages, *J. Exp. Med.,* 171, 35, 1990.
82. **Kovac, S., Brock, B., Faresio, L., and Young, H. A.,** IL-2 induction of IL-1β mRNA expression in monocytes. Regulation by agents that block second messenger pathways, *J. Immunol.,* 143, 3532, 1989.
83. **Galser, K. B., Asnis, R., and Dennis, E. A.,** Bacterial lipopolysaccharide priming of p388D1 macrophage-like cells for enhanced arachidonic acid metabolism, *J. Biol. Chem.,* 265, 8658, 1990.
84. **Forehand, T. R., Pabst, M. J., Phillips, W. A., and Johnston, R. B., Jr.,** Lipopolysaccharide priming of human neutrophils for an enhanced respiratory burst. Role of intracellular free calcium, *J. Clin. Invest.,* 83, 74, 1989.
85. **Alkon, D. L. and Rasmussen, H.,** A spatial-temporal model of cell activation, *Science,* 239, 998, 1988.

Chapter 13

THE *Lps* MUTATIONAL DEFECT IN C3H/HeJ MICE

Masayasu Nakano and Hiroto Shinomiya

TABLE OF CONTENTS

I. INTRODUCTION

The C3H/HeJ strain of mice is a unique mutant strain which was derived in 1947 from the C3H/He strain. C3H/HeJ mice possess a profound defect in their ability to respond to the lipid A component of endotoxin or lipopolysaccharide (LPS) derived from the cell walls of Gram-negative bacteria.[1] The mutation responsible for the inability of these mice to respond to LPS occurred some time between 1960 to 1965 and has been mapped to a single gene (*Lps*) located on the fourth chromosome.[2] The mutational defect of this strain to LPS is manifested not only in *in vivo* responses but also in cellular responses *in vitro*. The diminished capacity of C3H/HeJ mice to respond to LPS is expressed as decreased LPS responsiveness of their B lymphocytes, T lymphocytes, fibroblasts, and macrophages.[2] Though a vast amount of investigation has helped unravel the genetic basis of non- or low responsiveness to LPS,[2,3] the product of the *Lps* locus and the exact molecular mechanisms of the hyporesponsiveness of C3H/HeJ cells to LPS have remained undefined.

Another LPS-hyporesponsive mouse strain, C57BL/10ScCr, was described by Coutinho et al.[4] The progenitor of the strain C57BL/10ScN was also been found to be LPS hyporesponsive.[5] Phenotypically, these C57BL/10ScCr and C57BL/10ScN mice behave like the C3H/HeJ, since they do not respond to LPS and lipid A, but respond to other agents that stimulate C3H/HeJ mice or cells. There is no evidence to suggest that the mutation leading to the LPS defect in these strains is different from that in the C3H/HeJ strain, and the defect in these mice has also been shown to be in the *Lps* gene locus in chromosome 4 by F1 hybrid cross-analysis between C57BL/10ScCr and C3H/HeJ mice.[2,3,5]

The hyporesponsiveness of C3H/HeJ cells to LPS has generally been attributed to a defect in the ability of these cells to recognize the LPS molecule, and specifically the lipid A moiety. To date, studies directly demonstrating reduced binding of LPS to the cells from C3H/HeJ mice have not been presented. Non- or low responsiveness of C3H/HeJ cells is presumed not to be due to the inaccessibility of lipid A to its receptors on the surfaces of cells in the mice, but to refractoriness of receptors for lipid A and/or the dysfunction of intracellular signal transmission after the binding of lipid A to the receptors.

This chapter will review and discuss recent publications on the cellular and molecular basis of endotoxin-refractory cells.

II. RELATION BETWEEN LPS PREPARATION AND REFRACTORY STATE

A. ENDOTOXIN-ASSOCIATED PROTEINS

Though C3H/HeJ cells show a profound abnormality in endotoxin responsiveness when highly purified LPS from certain Gram-negative bacteria is used, the cells can still respond to Boivin-type and other types of endotoxin. In the past, there has been much confusion about whether C3H/HeJ hyporesponsiveness can be associated with endotoxin other than that which resides in purified LPS. Part of the problem has been the extraction method employed, since contaminating protein does indeed affect cells from nonresponder strains.[6] A highly purified endotoxin-associated protein was separated by subjecting a classic Boivin extract to the usual hot phenol extraction procedure. It was shown to be active in both LPS-responder mice and LPS-nonresponder C3H/HeJ mice.[7] Brade et al.[8] examined mitogenic activities of synthetic *Escherichia coli* lipid A and synthetic *S*-[2,3-bis-(palmitoyloxy) propyl]-*N*-palmitoylpentapeptide (tripalmitoyl pentapeptide [TPP]: the mitogenic principle of lipoprotein) on splenocytes of LPS-responder BALB/c (*Lps*ⁿ) and LPS-low-responder C3H/HeJ (*Lps*ᵈ) mice. These synthetic substances represent the mitogenically active principles of LPS and lipoprotein, respectively. They found that TPP resulted in mitogenic activation of B lymphocytes from both strains of mice, whereas lipid A was active only in LPS-responder mice.

B. PROTEIN-FREE LPS

Recently, a number of studies have appeared suggesting that certain forms of LPS and lipid A can stimulate DNA synthesis in C3H/HeJ spleen cells to a limited extent. Vukajlovich and Morrison[9] demonstrated that a protein-free and lipid A-enriched homogeneous monomeric component of *E. coli* LPS induced a limited amount of [³H]thymidine incorporation by C3H/HeJ spleen cells. Kumazawa et al.[10] reported that a synthetic lipid A analog (no. 314) which was nonphosphorylated and contained ester-bound (*R*)-3-hydroxytetradecanoic acid and an amine-bound (*R*)-3-tetradecanoyloxytetradecanoic acid stimulated DNA synthesis in C3H/HeJ spleen cells to a limited but statistically significant degree. Tanamoto et al.[11] also examined the ability of synthetic lipid A analogs to induce prostaglandin E_2 release by C3H/HeJ macrophages. The synthetic preparations comprised monomeric or dimeric derivatives of D-glucosamine with different patterns of substitution by phosphate and tetradecanoic, (*R*)-3-hydroxytetradecanoic, and (*R*)-3-tetradecanoyloxytetradecanoic acid. Some of these nontoxic lipid A analogs (nos. 308 and 313; both monomeric preparations) stimulated C3H/HeJ macrophages to the same extent in NMRI (*Lps*ⁿ) mice, while others (no. 304, 305, 307, 316, and 317; all dimeric preparations, each with a strong ability to activate normal cells) hardly affected C3H/HeJ macrophages. They suggested, however, that the response of C3H/HeJ macrophages to the monomeric preparations is presumably endotoxin nonspecific.

Sultzer et al.[12] found that certain protein-free preparations of pertussis LPS extracted by the phenol-chloroform-petroleum ether method used for R mutants activated C3H/HeJ B cells in a fashion resembling cells from normal responder mice. They[13,14] also reported that splenic B lymphocytes and macrophages from C3H/HeJ mice do respond to the pertussis LPS and *Salmonella minnesota* R595 LPS as measured by DNA synthesis, polyclonal antibody synthesis, and interleukin-1 (IL-1) activity. The activity could not be ascribed to endotoxin-associated protein, since mitogenesis was inhibited by polymyxin B which is a potent lipid A inhibitor. Wild-type, smooth LPS inhibited activation of the C3H/HeJ B cells not only by the LPS but also by mitogenic proteins, including purified protein derivative of tuberculin, suggesting that the C3H/HeJ cells that are turned on by protein mitogens or R-type LPS can be turned off by the wild-type smooth LPS. The results reflect the complexity of the response of C3H/HeJ B cells to endotoxin.

Vogel et al.[15] reported that a lipid A precursor obtained from a temperature-sensitive mutant of *S. typhimurium* activated C3H/HeJ B cells and macrophages, but their preparation contained a trace of protein. Tomai et al.[16] showed a monophosphoryl lipid A (MPL) from *S. minnesota* and the Re glycolipid from one strain of *E. coli* activated C3H/HeJ splenocytes in culture. Similarly, Hiernaux et al.[17] reported that C3H/HeJ spleen cells produced a weak mitogenic response to both preparations of nontoxic MPL from polysaccharides, heptoseless Re mutants of *S. typhimurium* and *S. minnesota in vitro*. Furthermore, C3H/HeJ mice showed a significant increase in serum IgM levels without an increase in numbers of splenic IgM-secreting plaque-forming cells (antibody-secreting cells) after *in vivo* treatment with MPL.

Flebbe et al.[18,19] also suggested that the refractory state of B cells from LPS-low-responder strains to mitogenic stimulation does not extend to R-type LPS preparations from a variety of rough strains of *Salmonella* or *E. coli*. They confirmed that the activation was not due to contaminating proteins. The activity depended on a structural requirement of the LPS wherein the 2-keto-3-deoxyoctulosonate linkage of lipid A with core oligosaccharides was left intact. The addition of either polymyxin B or S-chemotype LPS to R-LPS-stimulated C3H/HeJ splenocytes had only minimal effects on the mitogenesis. They also demonstrated the relative immunostimulatory activities of R-chemotype LPS preparations on secretion of tumor necrosis factor-α (TNF-α) and IL-1 and tumor cell destruction by C3H/HeJ peritoneal macrophages *in vitro*.

These investigations indicate that C3H/HeJ cells can respond to some protein-free LPS preparations, especially R-chemotype ones. However, the responses are generally weaker than those of cells from normal LPS-responder mice. The responses of C3H/HeJ cells to some of these LPS preparations might be due to LPS-nonspecific effects.[11] The triggering mechanisms of C3H/HeJ cells by the R-chemotype and MPL preparations remain unclear.

C. DOSES OF LPS USED AND ASSAY METHODS TO DETECT LPS RESPONSE

C3H/HeJ peritoneal exudate cells were reported to be deficient in LPS-induced TNF production by Beutler et al.[20] They found that C3H/HeJ macrophages did not produce detectable TNF-α protein by immunoprecipitation with specific antibody, but C3H/He macrophages did, in response to LPS at a concentration of 10^1 or 10^3 ng/ml. However, high levels of TNF-α mRNA were expressed by macrophages obtained from both C3H/HeJ and C3H/He strains when the cells were exposed to 10^3 ng/ml LPS (no detectable TNF-α mRNA in C3H/HeJ macrophages exposed to 10^1 ng LPS). They suggested that the inability of C3H/HeJ macrophages to produce TNF-α in response to LPS was attributable to defects in both transcription (no TNF-α mRNA expression and no TNF-α secretion to low dose of LPS) and translation (TNF-α mRNA expression but no TNF-α secretion to high dose of LPS). Lattime et al.[21] also found the production of TNF-α by spleen cells from C3H/HeJ mice after *in vitro* stimualtion with LPS, although the levels were significantly less than the LPS-responsive C3H/HeSn cells. Using a limiting dilution assay, northern blot analysis, and *in situ* hybridization with a cDNA probe for murine TNF-α, they detected LPS-induced TNF-α-producing cells and the cells expressing TNF-α mRNA in the C3H/HeJ spleen cells. The frequency of LPS-induced TNF-α-producing cells in C3H/HeJ spleen cells was less than half in comparison with the normal C3H/HeSn spleen cell population. These results indicate that the dose of LPS used for experiments and techniques for detection of LPS responses must be considered to evaluate the refractoriness of C3H/HeJ cells.

D. DIFFERENT SENSITIVITY TO LPS AMONG C3H/HeJ CELLS

Hiernaux et al.[17] studied adjuvant activity on IgM anti-type III pneumococcal polysaccharide antibody response and polyclonal or mitogenic B cell activation of *Salmonella*-derived MPL in C3H/HeJ and C3H/HeSnJ (*Lps*^{n}) mice. They found that the enhanced antibody response and the B cell activation were observed in C3H/HeSnJ mice, while C3H/HeJ mice did not show any enhancement of antibody response, but showed enhanced B cell activation to some extent. Since the adjuvant action was attributed to the inactivation of suppressor T cells by MPL and since the suppresser T cell activity was demonstrable in both C3H/HeSnJ and C3H/HeJ mice, they concluded that the suppressor T cells of C3H/HeJ mice are refractory to inactivation by MPL, while the polyclonal and mitogenic effects produced in C3H/HeJ mice are due to the direct activity of MPL on B lymphocytes.

III. ALTERATION OF ENDOTOXIN SENSITIVITY

A. BACCILLUS CALMETTE-GUÉRIN (BCG)

The *Lps* gene locus regulates the response in C3H/HeJ mice to LPS, but the phenotypic expression of the *Lps* gene locus can be altered experimentally.[2] Many years ago, Suter et al.[22] showed that infection of mice with BCG increases their sensitivity to endotoxin. The increase of the sensitivity to LPS was seen in not only normally LPS-sensitive mice but also in LPS-resistant mice. Vogel et al.[23] reported that the enhancement of LPS sensitivity in

C3H/HeJ mice induced by BCG infection is presumed to be mediated primarily through an effect on T cells and/or macrophages rather than B cells. They also showed that BCG infection produced a population of activated T cells which in turn activated macrophages and rendered them endotoxin sensitive, but the isolated macrophages from BCG-infected C3H/HeJ mice remained endotoxin unresponsive unless they were cultured together with T cells from the same animals or with T cells from *Lps*[n] mice.[24]

BCG infection can enable C3H/HeJ mice to produce interferon (IFN) after the stimulation with LPS.[24] LPS is considered to be able to induce INF-α/β, but not IFN-γ except under certain conditions.[25] Matsumura and Nakano[26] demonstrated that the IFN produced by cultured cells prepared from spleens and peritoneal exudate cells of BCG-infected C3H/HeJ mice is mainly IFN-γ, which is produced by both L3T4[+] T cells and asialo-GM1[+] natural killer (NK) cells in the presence of LPS. The IFN production requires BCG-primed macrophages as accessory cells, since neither the cultures reconstituted with the BCG-primed nonadherent cell population (T and NK cell populations) plus unprimed adherent cells (macrophages) nor those with unprimed non-adherent cells plus BCG-primed adherent cells could produce IFN.

B. IFN-γ

Chronic infections with BCG and other intracellular parasites cause lymphocytes to produce "macrophage activating factor(s)" (MAF). IFN-γ is a lymphokine known to have potent MAF activity. Vogel and Fertsch[27] suggested that the functional deficiency of C3H/HeJ macrophages is not related to an intrinsic defect but rather to a lack of maturation which can be partially corrected by exogenous interferon. Indeed, activation of C3H/HeJ macrophages by lipid A-associated proteins to a fully tumoricidal state has been reported to require priming with IFN-γ.[28,29] Beutler et al.[30] reported that the deficiency of C3H/HeJ cells was reversed by simultaneous treatment with IFN-γ and LPS as evidenced by the production of both TNF mRNA and immunoreactive TNF. The restoration of LPS-mediated cytotoxic C3H/HeJ macrophage function by IFN-γ was also demonstrated by Akagawa et al.[31] They reported that, not only the LPS activation of tumor cytotoxicity of the proteose peptone-induced peritoneal exudate macrophages, but also the sensitivity of the macrophages to the lethal toxicity of LPS was restored by IFN-γ. Furthermore, IFN-γ action in restoring the LPS responsiveness of macrophages was mimicked by a calcium ionophore, A23187, but not by phorbol 12-myristate 13-acetate. They suggested that the restoration of C3H/HeJ macrophage function by IFN-γ involves the elevation of intracellular calcium levels.

However, different results were reported by Hogan and Vogel.[32] They found that IFN-γ-primed C3H/HeJ macrophages failed to become cytolytic against tumor cells or to produce TNF in response to any concentration of protein-free, phenol-water-extracted LPS preparation, although lipid

A-associated protein could induce responses. They also demonstrated that IFN-γ-primed peritoneal macrophages from C3H/HeJ and C3H/OuJ (*Lps^n*) mice can be triggered in a dose-dependent manner by C5a to effect high levels of tumoricidal activity.[33] LPS has the ability to induce macrophage C5 production and enzyme release. The macrophage-derived C5 may be locally converted to C5a (or other biologically active C5 cleavage fragments). They could function as autocrine trigger signals for the induction of tumoricidal activity. Thus, in some circumstances, protein-free LPS as well as lipid A-associated protein provoke C3H/HeJ macrophages.

C. TRYPSIN

Kuus-Reichel and Ulevitch[34] showed partial restoration of endotoxin responsiveness in C3H/HeJ B lymphocytes by treating the cells with the proteolytic enzyme trypsin. Trypsin had no significant effect when added alone or when added together with LPS to cells from LPS-responder strains. It was concluded that the effect of trypsin was LPS specific, allowing transmission of an LPS-induced signal normally lacking in B cells from C3H/HeJ mice. However, Cardell and Möller[35] have reported that trypsin has a moderate mitogenic effect on B cells of both LPS-responder and -nonresponder strains of mice with enhanced B cell-proliferation and IgM secretion in spleen cell cultures stimulated by LPS. They concluded that the increased LPS-induced proliferation observed when trypsin was added to cells from the C3H/HeJ strain was therefore not due to a reconstitution of the LPS response.

D. CELL FUSION

Ritchie and Zuckerman[36] reported that the fusion of peritoneal macrophages from C3H/HeJ mice to an HPRT-negative variant of the murine macrophage cell line P388D1 has resulted in the derivation of a series of macrophage hybrids which produce the cytokine hepatocyte-stimulating factor in the presence of LPS. Cell fusion techniques and/or transfection techniques will be useful tools to analyze the defective *Lps* gene of C3H/HeJ cells.

IV. RECEPTORS FOR ENDOTOXIN

It is generally considered that the recognition of LPS in *Lps^n* cells must precede subsequent biological activity. LPS is a macromolecular substance, which consists of repeated O polysaccharides, R-core oligosaccharides, and a lipid A region. Each molecular part of LPS may be recognized by receptors with different specificity on or in *Lps^n* cells. Lipid A is the essential structure defining *Lps^n* cell responsiveness, but other portions of LPS must influence the cognition by the receptors. For example, when the cultured human mononuclear cells are stimulated with different structures of natural and synthetic lipid A as well as oligosaccharide partial structures of LPS, the induction capacity of IL-1 is very different among them.[37]

Lei and Morrison[38,39] succeeded in detecting a specific LPS-binding 80-kDa protein on the membranes of splenic B lymphocytes, T lymphocytes, and macrophages of C3H/HeB/FeJ (*Lps*[n]) mice, using a unique method with radio-iodinated, disulfide-reducible, photoactivatable LPS deivative. The 80-kDa protein was also detected in C3H/HeJ (*Lps*[d]) splenocytes. This protein was expressed on splenocytes from both phenotypically normal and LPS-nonresponder mice and appeared to be indistinguishable as assessed by migration on two-dimensional gel electrophoresis. Roeder et al.[40] detected the LPS-binding 80-kDa protein in peripheral blood mononuclear cells from human, rabbit, pig, dog, and sheep, but could not demonstrate the presence of a similar LPS-binding protein in the leukocytes of frog or chicken which are endotoxin-resistant species. Bright et al.[41] raised monoclonal antibodies against this LPS-binding 80-kDa protein. One of the antibodies (mAb5D3) as well as LPS could activate murine bone marrow-derived C3H/HeN (*Lps*[n]) macrophages to become tumoricidal for mastocytoma cells *in vitro*. Although it bound equivalently to macrophages from both C3H/HeN and C3H/HeJ mice, it was unable to activate macrophages from the C3H/HeJ mice.[42] These results suggest that the LPS-binding 80-kDa protein must represent one of the receptors for LPS in the cells, but the functional defect of C3H/HeJ is not due to inability of this receptor-protein to bind LPS.

V. POSSIBLE REFRACTORY SITES IN C3H/HeJ CELLS

A. ALTERATION OF MEMBRANE STRUCTURE

Although the difference between LPS receptors on the cell membranes of *Lps*[n] and *Lps*[d] cells is not defined, an alteration in the membrane structure is presumed to be associated with endotoxin hyporesponsiveness in C3H/HeJ cells. Gangliosides are sialic acid-containing glycosphingolipids which are located predominantly in the plasma membrane of eukaryotic cells, and have been shown to represent receptors or components of receptor complexes in a variety of ligand-membrane interactions.[43] Chaby et al.[44] examined the ability of B-cell gangliosides from both normal and C3H/HeJ mice to react with galactose oxidase, and reported that one of five gangliosides present in the LPS-responsive B-cell population was not detectable in the C3H/HeJ strain, but there was a concomitant increase in another ganglioside in the C3H/HeJ strain. Yohe et al.[45,46] reported a difference in sialic acid composition between gangliosides of C3H/HeJ and C3H/He (*Lps*[n]) cells. Furthermore, they compared the total ganglioside patterns and galactose oxidase-accessible ganglioside patterns of B cells from both strains to investigate whether an alteration in the amounts of surface accessibility of specific gangliosides may be involved in the expression of the *Lps* gene.[47] They found that the total ganglioside compositions of the two strains were nearly equivalent, but there were dramatic differences in ganglioside surface accessibility. They suggested that this alteration in membrane structure may be associated with the endotoxin hyporesponsiveness observed in C3H/HeJ B lymphocytes.

B. BLOCKADE OF INTRACELLULAR SIGNAL TRANSMISSION

Since the binding of LPS to the cell surfaces of C3H/HeJ cells is equivalent to that of normal cells, it has been speculated that intracellular signal transduction after LPS binds to cell surface may be hampered. However, little is known about the signal transmissive pathways that are related to the initiation of physiologic processes in cells stimulated by LPS. LPS-induced elevation of intracellular cyclic AMP is presumed to be necessary for cytotoxic activation of macrophages by LPS.[48] Vogel et al.[49] showed that cyclic AMP-elevating agents partially restored the Fc-receptor function of C3H/HeJ macrophages. However, the cyclic AMP cannot account for all LPS-mediated intracellular pathways, since the refractoriness of LPS-induced IL-1 production by CeH/HeJ macrophages does not recover in the presence of cyclic AMP or cyclic GMP.[50]

It is well known that activation of protein kinase C (PKC) and an increase in intracellular Ca^{2+} ($[Ca^{2+}]_i$) initiate signal transduction.[51] LPS pretreatment of mouse peritoneal macrophages results in the myristoylation of a 68-kDa protein of unknown function, rendering this protein much more susceptible to phosphorylation by PKC.[52-54] LPS activates PKC partially purified from a macrophage cell line in a cell-free system.[55] LPS also enhances the hydrolysis of phosphatidylinotytol-4,5-bisphosphate in macrophages, to which PKC activation is presumably related,[56] and stimulates IL-1 production through an interaction with PKC and another unknown pathway.[57] However, the phenotypic expression encoded by Lps^d seems to be irrelevant to activation of PKC, since C3H/HeJ macrophages as well as C3H/HeN macrophages possess an equal number of receptors for phorbol diesters on their cell membrane.[58] These receptors are now known to be PKC, and extracts from C3H/HeJ macrophages show similar PKC activity to that of C3H/He macrophages.[59] C3H/HeJ macrophages displayed the same ability to produce H_2O_2[58] or IL-1[60] in response to phorbol myristate acetate (PMA), a potent PKC activator, as did C3H/He macrophages. It was also shown that PMA-pretreated macrophages, in which PKC is depleted, can still respond to LPS to produce IL-1, while they can no longer be stimulated with PMA to produce IL-1.[57,61] Weiel et al.[62] reported that PMA induces protein phosphorylation in both C3H/HeN and C3H/HeJ macrophages, whereas LPS induces protein phosphorylation only in the macrophages of C3H/HeN but not in C3H/HeJ macrophages. Moreover, Mohri et al.[63] have recently shown that LPS stimulates a macrophage-like cell line to produce TNF without activation of PKC. This observation suggests that LPS-initated intracellular processes which are deficient in C3H/HeJ mice have little relation to PMA-initiated intracellular processes including PKC activation.

The calcium ionophore A23187 has the ability to increase $[Ca^{2+}]_i$ by binding and moving Ca^{2+} across the cell membranes. We found that A23187 induces IL-1 production by C3H/He macrophages, but it does not induce the production by C3H/HeJ macrophages.[60] Thus, the Lps^d defect shares the inability to respond to LPS and the calcium ionophore. However, the mode of action of A23187 is different from that of LPS in that it increases intracellular-free Ca^{2+} in C3H/HeJ macrophages as well as in C3H/He macrophages, while LPS does not elevate $[Ca^{2+}]_i$ in either strain. Furthermore, calmodulin antagonists such as W-5 and W-7 inhibit A23187-induced IL-1 production by C3H/He macrophages at reasonable concentrations, but minimally inhibit LPS-induced IL-1 production. These findings suggest that the Lps^d defect of LPS-induced IL-1 production in C3H/HeJ macrophages is related to A23187-initiated intracellular processes, although the actions of LPS and A23187 are distinct. Lps^d dysfunction seems to affect either directly or indirectly a common pathway in both Ca^{2+}-calmodulin-initiated processes and the LPS-initiated processes. We found that the activity of calmodulin itself in C3H/HeJ macrophages was not different from C3H/He macrophages. However, one species of calmodulin-binding proteins was minimally detected using a cross-linking assay with ^{125}I-calmodulin in C3H/HeJ macrophages stimulated by LPS.[59] These results suggest that the LPS-refractory site in C3H/HeJ macrophages is related to the lack of this calmodulin-binding protein.

Enzyme-mediated phosphorylation-dephosphorylation of cellular proteins is a major regulatory mechanism by which signals from the extracellular environment alter intracellular functions and coordinate control of cellular events. Weiel et al.[62] demonstrated that treatment of Lps^n murine peritoneal macrophages with nanogram quantities of LPS consistently resulted in altered ^{32}Pi labeling of a specific set of proteins measured by autoradiography after sodium dodecyl sulfate-polyacrylamide gel electrophoresis, while C3H/HeJ macrophages showed no alteration in the pattern of phosphorylation after treatment with LPS. The results suggest that the LPS signal fails to trigger phosphorylation in C3H/HeJ cells. We confirmed this phenomenon and found that LPS-induced phosphorylated proteins (pp) which were characteristically located in cytosol and/or membrane of C3H/He macrophages were detectable after 15 to 30 min of incubation, but no noticeable change in protein phosphorylation was observed in C3H/HeJ macrophages when stimulated with LPS under various conditions.[50] Among these LPS-induced partial peptides in C3H/He macrophages, pp65 was the most heavily phosphorylated cytosol protein. Although LPS-induced phosphorylation of pp65 was not observed in C3H/HeJ macrophages, C3H/HeJ macrophages had about the same amounts of unphosphorylated protein of 65 kDa (p65) as C3H/He macrophages. After LPS-plasma membrane interaction, the pp65 was found to be phosphorylated at serine residues, indicating the activation of serine kinase(s). Cyclic AMP-dependent protein kinase, PKC, and Ca^{2+}-calmodulin-dependent protein kinase are major serine kinases. Analysis of partial peptide sequences of pp65 revealed that pp65 is a novel protein whose sequence has not been reported yet, and that it is different from the 68-kDa protein reported by Aderem et al.[53] (Shinomiya et al., submitted for publication).

VI. ABNORMALITY RELATED TO *Lps*[d] GENE LOCUS

A. NATURAL RESISTANCE TO INFECTION

The *Lps*[d] gene locus exerts a profound influence on the natural resistance of mice to a number of different pathogenic organisms, such as *S. typhimurium, Klebsiella pneumoniae, Neisseria gonorrhoeae, Rickettsia akari, Chlamydia psittaci,* herpes simplex virus-1, *Babesia microti,*[2] *E. coli,*[64-67] *Haemophilus pleuropneumoniae,*[68] *N. meningitidis,*[69] and vesicular stomatitis virus.[70] C3H/HeJ mice are more resistant than other C3H subline mice to *N. gonorrhoeae,* herpes simplex virus-1, and *H. pleuropneumoniae,* while the mice are more susceptible to *S. typhimurium, K. pneumoniae, R. akari, C. psittaci, B. microti, E. coli, N. meningitidis,* and vesicular stomatitis virus.

Most of the work in this area has involved the study of *S. typhimurium.* O'Brien et al.[71] demonstrated the linkage between *Lps* gene and susceptibility to infection. They noted that C3H/HeJ mice were significantly more susceptible to this strain than all other C3H substrains tested. This susceptibility was due to a gene that was identical or very closely linked to the *Lps* locus. However, dissociation of innate susceptibility to infection and LPS responsiveness in one C3H substrain was reported; the LPS-responsive C3HeB/FeJ (*Lps*[n]) mice were more susceptible than C3H/HeJ mice to *Salmonella* or *E. coli* infection.[72-74] The susceptibility of mice to *Salmonella* infection is also controlled by the *salmonella* susceptibility locus (*Ity*). C3H/HeJ mice carry a *Salmonella*-resistant gene (*Ity*[r]). Weinstein et al.[75] demonstrated that the C3H/HeJ (*Lps*[d]/*Ity*[r]) mice were more resistant than *Lps*[n]/*Ity*[s] mice to strains of *S. typhimurium* of reduced virulence but less resistant than *Lps*[n]/*Ity*[r] mice. Furthermore, Svanborg-Edén et al.[65,66] reported the difference between the genetic determinants of host resistance to urinary tract infection with *E. coli* or *S. typhimurium* at mucosal and systemic sites in C57BL/10ScCr (*Lps*[d], *Ity*[s]) and C3H/HeJ (*Lps*[d], *Ity*[r]) mice. They demonstrated that additional unknown determinants conferred increased resistance to mucosal infection in the C57BL mice compared to the C3H mouse. The susceptibility or resistance to *Salmonella* infection in mice is regulated under control of several or many genes such as *Lps,* the *salmonella* susceptibility locus (*Ity*), the *Xid* antibody production locus, major histocompatibility (*H-2*) complex, and other unknown loci.[76-78] The susceptibility expressed by the *Lps* gene may be influenced by the other regulatory genes and/or factors.

The *Lps* locus also influences susceptibility to *N. meningitidis* infection. Woods et al.[69] demonstrated that there were significant differences in levels of the bacteremia between C3H/HeJ mice and each of the other strains (all *Lps*[n]). The association of *Lps*[d] genotype with susceptibility was confirmed by using coisogenic strains from two widely separated mouse lineages. Since there was no difference in meningococcal growth *in vitro* in serum from C3H/HeJ and C3H/HeN (*Lps^n*) mice, they suggested that the *Lps*-related difference in susceptibility might involve a cellular response.

Immediate inflammatory responses triggered by LPS may be a key to the elimination of Gram-negative bacteria from local infection. Shahin et al.[64] reported that, 24 h after ascending *E. coli* urinary tract infection, no significant neutrophil influx into the urine and phagocytosis were observed in C3H/HeJ mice, while the obvious influx and phagocytosis occurred in C3H/HeN mice. Therefore, they suggested that the inflammatory responses were inversely related to bacterial clearance. Similarily, Cross et al.[67] suggested involvement of TNF-α in host protection, since the C3H/HeJ mice were unable to produce this inflammatory protein directly in response to LPS, whereas C3H/HeN mice did, and pretreatment of C3H/HeJ mice with a combination of recombinant murine TNF-α and IL-1-α protected the mice against *E. coli* infection. The increased susceptibility of C3H/HeJ mice to *S. typhimurium, R. akari,* and *C. psittaci* is probably secondary to a defect in recognition of lipid A and a subsequent failure to develop the "activated" macrophages that are required to eliminate these organisms.[2]

Macrophages of C3H/HeJ mice fail to develop microbicidal and tumoricidal activity after treatment *in vivo* or *in vitro* with a variety of activating agents, and the susceptibility of these mice strain to *S. typhmurium* is presumed to be due to a failure of activation and recruitment of macrophages.[79] C3H/HeJ macrophages fail to produce normal levels of endogenous IFN.[27,70] IFN is known to be one of the regulatory cytokines of the immune system and to be capable of enhancing defense mechanisms against infection. Endogenously poor production of IFN in C3H/HeJ mice may hinder differentiation and activation of macrophages. Vogel and Fertsch[80] reported that C3H/HeJ macrophages were significantly more susceptibile to vesicular stomatitis virus infection *in vitro* than their LPS-responsive counterparts of C3H/OuJ strain. The results were reversed by treating C3H/HeJ macrophages with exogenous IFN or by allowing the permissive C3H/HeJ macrophages to mature in the presence of the IFN-inducing cytokine, CSF-1. Gessani et al.[70] showed that the capacity of resident peritoneal macrophages to transfer an anti-viral state (to vesicular stomatitis virus) *in vitro* correlated directly with LPS response phenotype, and addition of anti-IFN-α/β antibodies to cultures blocked the anti-viral protection afforded by LPS-responsive macrophages. This was confirmed in genetic studies by them with F1 hybrids between *Lps^n* and *Lps^d* mice and with *Lps^n* and *Lps^d* recombinant inbred strains.

Although the correlation of phenotypic defects of LPS-nonresponsive C3H/HeJ mice to decreased IFN production by their macrophages suggests the physical proximity on chromosome 4 of the gene which encodes sensitivity to LPS (*Lps*) and the structural gene cluster which encodes IFN-γ (*Ifa*), the physical separation of these gene loci has been observed in C57BL/6J × C3H/HeJ recombinant inbred mice.[81]

B. OTHER INTRINSIC ABNORMALITIES

A variety of defects that do not relate directly to lipid A recognition have been ascribed to C3H/HeJ cells. They show, for example, low expression of certain Fc receptors, poor phagocytosis mediated by these receptors, unresponsiveness to an activating stimulus by formation of cytotoxic macrophages, reduced tumoricidal activity, inability to support LPS- or Fc fragment-mediated polyclonal B-cell activation, etc.[2,82] C3H/HeJ B cells have a characteristically low cloning efficiency in semisolid agar cultures,[83] and C3H/HeJ macrophages are low producers of the soluble mediator(s) to support the proliferation of lymphocytes.[83,84] It is not clear whether these abnormalities are closely associated with the control of Lps^d gene locus and/or arise secondarily to the inability of the cells to recognize lipid A.

VII. LPS-NONRESPONSIVE CELL LINES

Another approach to elucidate the mechanisms of LPS-specific refractoriness has been conducted by using cultured tumor cell lines. A mutation that has occurred on any point in genes encoding the cascade of transduction pathways for LPS signal may result in refractoriness of the cells to LPS. LPS-resistant mutants which did not respond to LPS or lipid A were isolated from murine 70Z/3 pre-B-lymphocyte tumor line.[85,86] The mutants were unable to respond to LPS and lipid A when monitored by measuring the synthesis of the κ-immunoglobulin chain, though some LPS-nonresponsive strains were still able to synthesize κ-chains in response to IFN-γ. The defective site of these unresponsive mutants to LPS has not been determined. Another LPS-nonresponsive mutant was isolated by Hara-Kuge et al.[87] from the murine J774.1 macrophage-like tumor cell line. LPS enhanced O_2^- generation and the release of arachidonic acid in J774.1 cells but not in the mutant, named LR-9, cells. LR-9 was found to be defective in LPS/lipid A-binding activity and lacked two distinct LPS-binding proteins with molecular masses of about 65 and 55 kDa, which were present on the surface of the parental J774.1 cells. They suggested that the 65- and 55-kDa proteins are the physiological LPS receptors responsible for activation of J774.1 cells by LPS, and the unresponsiveness of LR-9 mutant is due to the missing of the receptor proteins. There is no evidence that these LPS-resistant cells lines are relevant to mutation in the Lps gene locus.

REFERENCES

1. **Sultzer, B. M.,** Genetic control of host responses to endotoxin, *Infect. Immun.*, 5, 107, 1972.
2. **Rosenstreich, D. L.,** Genetic control of endotoxin response: C3H/HeJ mice, in *Handbook of Endotoxin*, Vol. 3, Berry, L. J., Ed., Elsevier Science, Amsterdam, 1985, chap. 4.

3. **Scibienski, R. J.**, Defects in murine responsiveness to bacterial lipopolysaccharide. The C3H/HeJ and C57BL/10ScCr strains, in *Immunologic Defects in Laboratory Animals 2*, Gershwin, M. E. and Merchant, B., Eds., Plenum Press, New York, 1981, chap. 10.

4. **Coutinho, A., Forni, L., Melchers, F., and Watanabe, T.**, Genetic defect in responsiveness to the B cell mitogen lipopolysaccharide, *Eur. J. Immunol.*, 7, 324, 1977.

5. **Vogel, S. N., Weinblatt, A. C., and Rosenstreich, D. L.**, Inherent macrophage defects in mice, in *Immunologic Defects in Laboratory Animals 1*, Gershwin, M. E. and Merchant, B., Eds., Plenum Press, New York, 1981, chap. 15.

6. **Burrell, R.**, Immunomodulation by bacterial endotoxin, *Crit. Rev. Microbiol.*, 17, 189, 1990.

7. **Johns, M. A., Sipe, J. D., Melton, L. B., Strom, T. B., and McCabe, W. R.**, Endotoxin-associated protein: interleukin-1-like activity on serum amyloid A synthesis and T-lymphocyte activation, *Infect. Immun.*, 56, 1593, 1988.

8. **Brade, L., Bessler, W. G., and Brade, H.**, Mitogenic activities of synthetic *Escherichia coli* lipid A and a synthetic partial structure (tripalmitoyl pentapeptide) of *E. coli* lipoprotein, *Infect. Immun.*, 56, 1382, 1988.

9. **Vukajlovich, S. W. and Morrison, D. C.**, Conversion of lipopolysaccharides to molecular aggregates with uniform subunit composition: demonstration of LPS-responsiveness in "endotoxin-unresponsive" C3H/HeJ B-lymphocytes, *J. Immunol.*, 130, 2804, 1983.

10. **Kumazawa, Y., Matsumura, Y., Nakatsuru-Watanabe, Y., Fukumoto, M., Nishimura, C., Homma, J. Y., Inage, M., Kusumoto, S., and Shiba, T.**, Mitogenic and polyclonal B cell activation activities of synthetic lipid A analogues, *Eur. J. Immunol.*, 14, 109, 1984.

11. **Tanamoto, K., Shade, U., Rietschel, E. T., Kusumoto, S., and Shiba, T.**, Endotoxic induction of prostaglandin release from macrophages by nontoxic lipid A analogs synthesized chemically, *Infect. Immun.*, 58, 217, 1990.

12. **Sultzer, B. M., Craig, J. P., and Castagna, R.**, Immunomodulation by outer membrane protein associated with the endotoxins of Gram-negative bacteria. *Prog. Leukocyte Biol.*, 6, 113, 1987.

13. **Sultzer, B. M. and Castagna, R.**, Inhibition of activated nonresponder C3H/HeJ lymphocytes by lipopolysaccharide endotoxin, *Infect. Immun.*, 56, 3040, 1988.

14. **Sultzer, B. M. and Castagna, R.**, The activation of C3H/HeJ cells by certain types of lipopolysaccharides, *Adv. Exp. Med. Biol.*, 256, 149, 1990.

15. **Vogel, S. N., Madonna, G. S., Wahl, L. M., and Rick, P. D.**, Stimulation of spleen cells and macrophages of C3H/HeJ mice by a lipid A precursor derived from *Salmonella typhymurium*, *Rev. Infect. Dis.*, 6, 535, 1985.

16. **Tomai, M. A., Johnson, A. J., and Ribi, E.**, Glycolipid induced proliferation of lipopolysaccharide hyporesponsive C3H/HeJ splenocytes, *J. Leuk. Biol.*, 43, 11, 1988.

17. **Hiernaux, J. R., Stashak, P. W., Cantrell, J. L., Rudbach, J. A., and Baker, P. J.**, Immunomodulatory activity of monophosphoryl lipid A in C3H/HeJ and C3H/HeSnJ mice, *Infect. Immun.*, 57, 1483, 1989.

18. **Flebbe, L., Vukajlovich, S. W., and Morrison, D. C.**, Immunostimulation of C3H/HeJ cells by chemotype lipopolysaccharide preparations, *J. Immunol.*, 142, 642, 1989.

19. **Flebbe, L. M., Chapes, S. K., and Morrison, D. C.**, Activation of C3H/HeJ macrophage tumoricidal activity and cytokine release by R-chemotype lipopolysaccharide preparations, *J. Immunol.*, 145, 1505, 1990.

20. **Buetler, B., Krochin, N., Milsark, I. W., Leudke, C., and Cerami, A.**, Control of cachectin (tumor necrosis factor) synthesis: mechanisms of endotoxin resistance, *Science*, 232, 977, 1986.

21. **Lattime, E. C., Stoppacciaro, A., and Stutman, O.**, Limiting dilution analysis of TNF producing cells in C3H/HeJ mice, *J. Immunol.*, 141, 3422, 1988.

22. **Suter, E., Ullman, G. E., and Hoffman, R. G.**, Sensitivity of mice to endotoxin after vaccination with BCG (Bacillus Calmette-Guérin), *Proc. Soc. Exp. Biol. Med.*, 99, 167, 1958.

23. **Vogel, S. N., Moore, R. N., Sipe, J. D., and Rosenstreich, D. L.**, BCG-induced enhancement of endotoxin sensitivity in C3H/HeJ mice. I. In vivo studies, *J. Immunol.*, 124, 2004, 1980.

24. **Vogel, S. N., Weedon, L. L., Wahl, L. M., and Rosenstreich, D. L.**, BCG-induced enhancement of endotoxin sensitivity in C3H/HeJ mice. II. T cell modulation of macrophage sensitivity to LPS *in vitro*, *Immunobiology*, 160, 479, 1982.

25. **Blanchard, K., Djeu, J. Y., Klein, T. W., Friedman, H., and Stewart, W. E., II**, Interferon-γ induction by lipopolysaccharide: dependence on interleukin 2 and macrophages, *J. Immunol.*, 136, 963, 1986.

26. **Matsumura H. and Nakano, M.**, Endotoxin-induced interferon-γ in culture cells derived from BCG-infected C3H/HeJ mice, *J. Immunol.*, 140, 494, 1988.

27. **Vogel, S. N. and Fertsch, D.**, Endogenous interferon production by endotoxin-responsive macrophages provides an autostimulatory differentiation signal, *Infect. Immun.*, 45, 417, 1984.

28. **Chapes, S. K., Killion, J. W., and Morrison, D. C.**, Tumor cell killing and cytostasis by C3H/HeJ macrophages activated *in vitro* by lipid A-associated protein and interferon-gamma, *J. Leukocyte Biol.*, 43, 232, 1988.

29. **Hogan, M. M. and Vogel, S. N.**, Lipid A-associated proteins provide an alternate "second signal" in the activation of recombinant interferon-γ-primed C3H/HeJ macrophages to a fully tumoricidal state, *J. Immunol.*, 139, 3697, 1987.

30. **Beutler, B., Tkacenko, V., Milsark, I., Krochin, N., and Cerami, A.**, Effect of γ-interferon on cachectin expression by mononuclear phagocytes: reversal of the Lps^d (endotoxin resistance) phenotype, *J. Exp. Med.*, 164, 1791, 1986.

31. **Akagawa, K. S., Kamoshita, K., Onodera, S., and Tokunaga, T.**, Restoration of lipopolysaccharide-mediated cytotoxic macrophage induction C3H/HeJ mice by interferon-γ or a calcium ionophore, *Jpn. J. Cancer Res.*, 78, 279, 1987.

32. **Hogan, M. M. and Vogel, S. N.**, Production of tumor necrosis factor by rIFN-γ-primed C3H/HeJ (Lps^d) macrophages requires the presence of lipid A-associated proteins, *J. Immunol.*, 141, 4196, 1988.

33. **Hogan, M. M. and Vogel, S. N.**, Role of C5a in the induction of tumoricidal activity in C3H/HeJ (Lps^d) and C3H/OuJ (Lps^n) macrophages, *J. Leukocyte Biol.*, 46, 565, 1990.

34. **Kuus-Reichel, K. and Ulevitch, R. J.**, Partial restoration of the lipopolysaccharide-induced proliferative response in splenic B cells from C3H/HeJ mice, *J. Immunol.*, 137, 472, 1986.

35. **Cardell, S. and Möller, G.**, Trypsin does not reconstitute responsiveness to lipopolysaccharide in the strain C3H/HeJ, but is a B-cell mitogen-like lipopolysaccharide, stimulating a different subpopulation, *Scand. J. Immunol.*, 29, 143, 1989.

36. **Ritchie, D. G. and Zuckerman, S. H.**, Restoration of the LPS responsive phenotype in C3H/HeJ macrophage hybrid: LPS regulation of hepatocyte-stimulating factor production, *Immunology*, 61, 429, 1987.

37. **Loppnow, H., Brade, H., Dürrbaum, I., Dinarello, C. A., Kusumoto, S., Rietschel, E. T., and Flad, H. D.**, IL-1 induction-capacity of defined lipopolysaccharide partial structures, *J. Immunol.*, 142, 3229, 1989.

38. **Lei, M. -G. and Morrison, D. C.**, Specific endotoxic lipopolysaccharide-binding proteins on murine splenocytes. I. Detection of lipopolysaccharide-binding sites on splenocytes and splenocyte subpopulations, *J. Immunol.*, 141, 996, 1988.

39. **Lei, M. -G. and Morrison, D. C.**, Specific endotoxic lipopolysaccharide-binding proteins on murine splenocytes. II. Membrane localization and binding characteristics, *J. Immunol.*, 141, 1006, 1988.

40. **Roeder, D. J., Lei, M. -G., and Morrison, D. C.,** Endotoxin-lipopolysaccharide-specific binding proteins on lymphoid cells of various animal species: association with endotoxin susceptibility, *Infect. Immun.,* 57, 1054, 1989.

41. **Bright, S. W., Chen, T., Flebbe, L. M., Lei, M. -G., and Morrison, D. C.,** Generation and characterization of hamster-mouse hybridomas secreting monoclonal antibodies with specificity for lipopolysaccharide receptor, *J. Immunol.,* 145, 1, 1990.

42. **Chen, T., Bright, S. W., Pace, J. L., Russell, S. W., and Morrison, D. C.,** Induction of macrophage-mediated tumor cytotoxicity by a hamster monoclonal antibody with specificity for lipopolysaccharide receptor, *J. Immunol.,* 145, 8, 1990.

43. **Symington, F. W. and Hakomori, S. -I.,** Glycosphingolipids of immune cells, *Lymphokines,* 12, 201, 1985.

44. **Chaby, R., Morelec, M. -J., Enserqueix, D., and Girad, R.,** Membrane glycolipid and phospholipid composition of lipopolysaccharide-responsive and -nonresponsive murine B lymphocytes, *Infect. Immun.,* 52, 777, 1986.

45. **Yohe, H. C. and Ryan, J. L.,** Ganglioside expression in macrophages from endotoxin responder and nonresponder mice, *J. Immunol.,* 137, 3921, 1986.

46. **Yohe, H. C., Cuny, C. L., Berenson, C. S., and Ryan, J. L.,** Comparison of thymocyte and T lymphocyte gangliosides from C3H/HeN and C3H/HeJ mice, *J. Leukocyte Biol.,* 44, 521, 1988.

47. **Yohe, H. C., Berenson, C. S., Cuny, C. L., and Ryan, J. L.,** Altered B-lymphocyte membrane architecture indicated by ganglioside accessibility in C3H/HeJ mice, *Infect. Immun.,* 58, 2888, 1990.

48. **Akagawa, K. S., Kamoshita, K., Tomita, T., Yasuda, T., and Tokunaga, T.,** Regulatory mechanisms of expression of LPS binding site(s) and signaling events by LPS in macrophages, *Adv. Exp. Med. Biol.,* 256, 467, 1990.

49. **Vogel, S. N., Weedon, L. L., Oppenheim, J. J., and Rosenstreich, D. L.,** Defective Fc-mediated phagocytosis in C3H/HeJ macrophages. II. Correction by cAMP agonists, *J. Immunol.,* 126, 441, 981.

50. **Nakano, M. and Shinomiya, H.,** Molecular mechanisms of macrophage activation by LPS, in *Cellular and Molecular Aspects of Endotoxin Reactions,* Nowotny, A. H., Spitzer, J. J., and Ziegler, E. J., Eds., Elsevier Science, Amsterdam, 1990, 205.

51. **Nishizuka, Y.,** The role of protein kinase C in cell surface signal transduction and tumor promotion, *Nature,* 308, 693, 1984.

52. **Aderem, A. A., Keum, M. M., Pure, E., and Cohn, Z. A.,** Bacterial lipopolysaccharides, phorbol myristate acetate, and zymosan induce the myristoylation of specific macrophage proteins, *Proc. Natl. Acad. Sci. U.S.A.,* 83, 5817, 1986.

53. **Aderem, A. A., Albert, K. A., Keum, M. M., Wang, J. K. T., Greengard, P. and Cohn, Z. A.,** Stimulus-dependent myristoylation of major substrate for protein kinase C, *Nature,* 332, 362, 1988.

54. **Rosen, A., Nairn, A. C., Greengard, P., Cohen, Z. A., and Aderem, A.,** Bacterial lipopolysaccharide regulates the phosphorylation of the 68K protein kinase C substrate in macrophages, *J. Biol. Chem.,* 264, 9118, 1989.

55. **Wightman, P. D. and Raetz, C. R. H.,** The activation of protein kinase C by biologically active lipid moieties of lipopolysaccharide, *J. Biol. Chem.,* 259, 10048, 1984.

56. **Pritc, V., Weiel, J. E., Somers, S. D., DiGuiseppi, J., Gonias, S. D., Pizzo, S. V., Hamilton, T. A., Herman, B., and Adams, O.,** Effects of bacterial lipopolysaccharide on the hydrolysis of phosphatidylinositol-4,5-bisphosphate in murine peritoneal macrophages, *J. Immunol.,* 139, 526, 1987.

57. **Katakami, Y., Nakao, Y., Matsui, T., Koizumi, T., Kaibuchi, K., Takai, Y., and Fujita, T.** Possible involvement of protein kinase C in interleukin 1 production by mouse peritoneal macrophages, *Biochem. Biophys. Res. Commun.,* 135, 355, 1986.

58. **Weinberg, J. B. and Misukonis, M. A.,** Phorbol diester-induced H_2O_2 production by peritoneal macrophages. Different H_2O_2 production by macrophages from normal and BCG-infected mice despite comparable phorbol diester receptors, *Cell. Immunol.,* 80, 405, 1983.

59. **Terada, Y., Shinomiya, H., and Nakano, M.,** Defect of calmodulin-binding protein in expression of interleukin-1β gene by LPS-nonresponder C3H/HeJ mouse macrophages, *Biochem. Biophys. Res. Commun.,* 158, 723, 1989.

60. **Shinomiya, H. and Nakano, M.,** Calcium ionophore A23187 does not stimulate lipopolysaccharide nonresponsive C3H/HeJ peritoneal macrophages to produce interleukin 1, *J. Immunol.,* 139, 2730, 1987.

61. **Fenton, M. J., Vermeulen, M. W., Clark, B. D., Webb, A. C., and Auron, P. E.,** Human pro-IL-1β gene expression in monocytic cells is regulated by two distinct pathways, *J. Immunol.,* 140, 2267, 1988.

62. **Weiel, J. E., Hamilton, T. A., and Adams, D. O.,** LPS induces altered phosphate labeling of proteins in murine peritoneal macrophages, *J. Immunol.,* 136, 3012, 1986.

63. **Mohri, M., Spriggs, D. R., and Kufe, D.,** Effects of lipopolysaccharide on phospholipase A_2 activity and tumor necrosis factor expression in HL-60 cells, *J. Immunol.,* 144, 2678, 1990.

64. **Shahin, R. D., Engberg, I., Hagberg, L., and Svanborg-Edén, C.,** Neutrophil recruitment and bacterial clearance correlated with LPS responsiveness in local Gramnegative infection, *J. Immunol.,* 138, 3475, 1987.

65. **Svanborg-Edén, C., Hagberg, L., Hull, R., Hull, S., Magnusson, K. -E., and Öhman, L.,** Bacterial virulence versus host resistance in the urinary tracts of mice, *Infect. Immun.,* 55, 1224, 1987.

66. **Svanborg-Edén, C., Shahin, R., and Briles, D.,** Host resistance to mucosal gramnegative infection. Susceptibility of lipopolysaccharide nonresponder mice, *J. Immunol.,* 140, 3180, 1988.

67. **Cross, A. S., Sadoff, J. C., Kelly, N., Bernton, E., and Gemski, P.,** Pretreatment with recombinant murine tumor necrosis factor α/cachectin and murine interleukin 1 α protects mice from lethal bacterial infection, *J. Exp. Med.,* 169, 2021, 1989.

68. **Fenwick, B. W., Osburn, B., and Olander, H. J.,** Resistance of C3H/HeJ mice to the effect of *Hemophilus pleuropneumoniae, Infect. Immun.,* 53, 474, 1986.

69. **Woods, J. P., Frelinger, J. A., Warrack, G., and Cannon, J. G.,** Mouse genetic locus *Lps* influences susceptibility to *Neisseria meningitidis* infection, *Infect. Immun.,* 56, 1950, 1988.

70. **Gessani, S., Belardelli, F., Borghi, P., and Gresser, I.,** Correlation between the lipopolysaccharide response of mice and the capacity of mouse peritoneal cells to transfer an antiviral state. Role of endogenous interferon, *J. Immunol.,* 139, 1991, 1987.

71. **O'Brien, A. D., Rosenstreich, D. L., Scher, I., Campbell, G. H., MacKermott, R. P., and Formal, S. B.,** Genetic control of susceptibiltiy to *Salmonella typhimurium* in mice. Role of the *Lps* gene, *J. Immunol.,* 124, 20, 1980.

72. **Eisenstein, T., Deakins, W. L., Killar, L., Saluk, P. H., and Sultzer, B. M.,** Dissociation of innate susceptibility to *Salmonella* infection and endotoxin responsiveness in C3HeB/FeJ mice and other strains in C3H lineage, *Infect. Immun.,* 36, 696, 1982.

73. **O'Brien, A. D. and Rosenstreich, D. L.,** Genetic control of the susceptibility of C3HeB/FeJ mice to *Salmonella typhimurium* is regulated by a locus distinct from known salmonella response genes, *J. Immunol.,* 131, 2613, 1983.

74. **Hagberg, L., Briles, D. E., and Eden, C. S.,** Evidence for separate genetic defects in C3H/HeJ and C3HeB/FeJ mice, that affect susceptibiltiy to gram-negative infections, *J. Immunol.,* 134, 4118, 1985.

75. **Weinstein, D. L., Lissner, C. R., Swanson, R. N., and O'Brien, A. D.,** Macrophage defect and inflammatory cell recruitment dysfunction in *Salmonella* susceptible C3H/HeJ mice, *Cell. Immunol.,* 102, 68, 1986.

76. **O'Brien, A. D., Rosenstreich D. L., Metcalfe, E. S., and Scher, I.,** Differential sensitivity of inbred mice to *Salmonella typhimurium:* a model for genetic regulation of innate resistance to bacterial infection, in *Genetic Control of Natural Resistance to Infection and Malignancy,* Skamene, E., Kongshavn, P. A. L., and Landy, M., Eds., Academic Press, New York, 1980, 101.

77. **Plant, J. E. and Glynn, A. A.,** Control of resistance to *Salmonella typhimurium* in hybrid generations of inbred mice and Biozzi mice, in *Genetic Control of Natural Resistance to Infection and Malignancy,* Skamene, E., Kongshavn, P. A. L., and Landy, M., Eds., Academic Press, New York, 1980, 133.

78. **Nauciel, C., Ronco, E., Guenet, J. -L., and Pla, M.,** Role of *H-2* and non-*H-2* genes in control of bacterial clearance from the spleen in *Salmonella typhimurium*-infected mice, *Infect. Immun.,* 56, 2407, 1988.

79. **Schafer, R., Nacy, C. A., and Eisenstein, T. K.,** Induction of activated macrophages in C3H/HeJ mice by avirulent *Salmonella, J. Immunol.,* 140, 1638, 1988.

80. **Vogel, S. N. and Fertsch, D.,** Macrophages from endotoxin-hyporesponsive (*Lps*^{d}) C3H/HeJ mice are permissive for vesicular stomatitis virus because of reduced levels of endogenous interferon: possible mechanisms for natural resistance to virus infection, *J. Virol.,* 61, 812, 1987.

81. **Fultz, M. and Vogel, S. N.,** The physical separation of *Lps* and *Ifa* loci in BXH recombinant inbred mice, *J. Immunol.,* 143, 3001, 1989.

82. **Schultz, L. D. and Sidman, C. L.,** Genetically determined murine models of immunodeficiency, *Annu. Rev. Immunol.,* 5, 367, 1989.

83. **Vetvicka, V., Lee, G., and Kincade, P. W.,** Intrinsic B lymphocyte and macrophage defects in C3H/HeJ mice, *J. Immunol.,* 136, 2370, 1986.

84. **Souvannavong, V. and Adam, A.,** Macrophages from C3H/HeJ mice require an additional step to produce monokines: synergistic effects of silica and poly(I:C) in the release of interleukin 1, *J. Leukocyte Biol.,* 48, 183, 1990.

85. **Mains, P. E. and Sibley, C. H.,** LPS-nonresponsive variants of mouse B cell lymphoma, 702/3Z: isolation and characterization, *Somatic Cell Genet.,* 9, 699, 1983.

86. **Sibley, C. H., Terry, A., and Raetz, C. R. H.,** Induction of k light chain synthesis in 70Z/3 B lymphoma cells by chemically defined lipid A precursors, *J. Biol. Chem.,* 263, 5098, 1988.

87. **Hara-Kuge, S., Amano, F., Nishijima, M., and Akamatsu, Y.,** Isolation of a lipopolysaccharide (LPS)-resistant mutant, with defective LPS binding, of cultured macrophage-like cells, *J. Biol. Chem.,* 256, 6606, 1990.

Chapter 14

INTERACTION OF LIPOSOME-INCORPORATED LIPOPOLYSACCHARIDE WITH RESPONSIVE CELLS

Jan Dijkstra and John L. Ryan

TABLE OF CONTENTS

I. INTRODUCTION

Phospholipids in an aqueous environment are able to form closed membrane-vesicles or liposomes. Bacterial lipopolysaccharide (LPS) and its lipid A moiety have been incorporated into liposomal vesicles for a variety of reasons: first, to investigate the physical effects of LPS and lipid A incorporation on membrane systems and, conversely, on the endotoxin molecule itself. In these studies, the vesicles were prepared from either bacterial outer-membrane components or well-defined lipids and lipid mixtures. The second reason is to study the complement-mediated lysis of membranes treated with antibodies specific for LPS and lipid A. Third, liposomes were employed *in vivo* as carriers for LPS and lipid A to obtain an antibody response specific for these molecules. Alternatively, the vesicle-incorporated LPS and lipid A served as an adjuvant for other liposome-associated antigens. Fourth, the effect of liposomal incorporation on LPS in both responsive cells *in vitro* and pathophysiologic responses *in vivo* was studied.

The present chapter will focus on the manner by which liposome incorporation modifies biological responses triggered by the LPS molecule. Although LPS initially may trigger cells by nonspecific insertion into cellular membranes or binding to cell-surface receptors, the precise mechanism of LPS action on responsive cells is still unclear. For this reason, special attention will be paid to liposomal LPS studies and the initial interaction of LPS with responsive cells. Specific sections include a general introduction to liposomes, liposomal LPS, and virosomes. The latter structures are reconstituted viral membrane vesicles, which may be considered as fusogenic liposomes.

A. LIPOSOMES

Artificial lipid vesicles or liposomes are microscopic structures consisting of one or more concentric lipid bilayers enclosing an equal number of aqueous compartments. In 1965 Bangham and co-workers for the first time described arrangements of a large number of phospholipid bilayers as closed, vesicular structures.[1] These multilamellar vesicles (MLV) were simply prepared by evaporating chloroform from a phospholipid solution, followed by the dispersion of the dried lipid in water. The size distribution of such MLV is rather broad with dimensions up to several microns. Ultrasonic irradiation (sonication), vortex mixing, and membrane extrusion are methods to reduce the size of these vesicles. Prolonged sonication of MLV or repeated extrusion through small, defined pore-size polycarbonate membranes ultimately results in the formation of small unilamellar vesicles (SUV). The latter vesicles have, depending on the lipid composition, a minimum diameter of 22 to 50 nm. In addition, a number of procedures have been developed for the preparation of large unilamellar vesicles (LUV).

Both the lipid bilayers and the aqueous compartments of liposomes can be used to associate additional, nonstructural compounds with the vesicles.

Hydrophilic solutes can be entrapped in the aqueous compartments and lipophilic or amphiphatic molecules may be incorporated in the lipid membranes.

The incorporation of acidic phospholipids or other negatively charged amphipathic compounds in the bilayer provides the liposomes with a net negative charge, whereas positively charged vesicles are obtained by employing anionic amphipathic molecules. Cholesterol is often included in liposomal membranes to increase the stability of the vesicles under physiological conditions.

In theory, liposomes can intereact with cells via five major mechanisms:

1. Lipid exchange. Individual lipid molecules may be transferred from liposomes to the cell surface and vice versa, without concomitant association of the aqueous contents with the cells. The rate of exchange depends on the nature of the individual lipid molecules.
2. Cell-surface-induced leakage. Upon contact with the cell surface, part of the aqueous contents may leak out of the vesicles due to increased permeability of the liposomal membranes. The liposomes may subsequently interact by one of the following mechanisms.
3. Stable adsorption. The liposomes bind in a relatively stable manner to the cell surface and this interaction is not followed by fusion or endocytosis. Lipid exchange may also occur after stable attachment of liposomes to the cells.
4. Fusion. The outermost liposomal membrane fuses with the plasma membrane and the vesicle contents are released in the cytoplasm of the cell.
5. Endocytosis. The liposomes are internalized in endocytic vesicles which are derived from the plasma membrane of the cells. The endocytosed material often ends up in the lysosomal system of the cell.

In practice, endocytosis is the major mechanism by which professional phagocytic cells interact with liposomes. The interaction with other cell types is primarily restricted to stable adsorption. Endocytosis, however, may also contribute to the uptake of small liposomes by nonphagocytic cells. Fusion does not appear to play a significant role in the interaction of phospholipid vesicles with animal cells.

For a more detailed description of the various types of liposomes, their preparation, and interaction with cells, the reader is referred to a number of recent reviews and books.[2-7]

B. LIPOSOMAL LPS AND LIPID A

LPS, a major constituent of the outer membrane of Gram-negative bacteria, is exclusively located in the outermost leaflet of this membrane. Since LPS consists of a hydrophilic polysaccharide chain linked to a hydrophobic lipid A moiety, the molecule has an amphipathic character and is in fact a complex glycolipid.[8] Thus LPS can be incorporated in the phospholipid bilayers of liposomes.

Kataoka and co-workers were the first to report the incorporation of smooth and rough forms of *Salmonella minnesota* LPS and its lipid A moiety in liposomes.[9,10] In these studies, the LPS and lipid A were suspended in the aqueous buffer used to hydrate the dried liposomal lipids. Using this simple approach, a portion of the LPS and lipid A became either inserted into or stably associated with the liposomal bilayers, as indicated by complement-induced release of an encapsulated hydrophilic marker.

Later studies demonstrated that only small amounts of LPS and lipid A associate with liposomes when the phospholipids are hydrated with the bacterial products suspended in the aqueous phase.[11-15] This low incorporation efficiency resembles the general encapsulation behavior of other hydrophilic substances when MLV are prepared in this manner. In contrast to the efficient incorporation of lipophilic compounds in the bilayers, only a few percent of a hydrophilic solute will become entrapped in the aqueous compartments of such MLV.[3,5,7]

Since the preparation of classical MLV employs LPS in an aqueous form, it is relevant to consider its properties as a solute. Wild-type or smooth LPS is relatively rich in carbohydrates and moderately soluble in aqueous media. The aqueous solubility of rough-type LPS decreases in proportion to the core saccharide content. Lipid A is completely insoluble in water.[8] Solutions of LPS, however, are suspensions of aggregates of the monomeric form. The three-dimensional, supramolecular structure of these aggregates depends on both the chemical structure of the LPS and the physical conditions.[16] Thus, depending on the size of the aggregates, it is reasonable that a portion of the dissolved LPS will be entrapped in the aqueous space of liposomes, as a number of studies suggest.[12,14,17]

Thus, LPS-containing vesicles prepared by the classical procedure result in MLV with LPS entrapped in the aqueous space as well as inserted into or stably attached to the lipid bilayers. Aggregates of LPS attached to the liposomal surface but not inserted in the lipid bilayers may in biological systems behave similar to free aqueous endotoxin. Moreover, the majority of the added LPS does not become associated with the vesicles. This makes thorough separation of MLV-associated LPS from free LPS necessary, particularly when the activity of these LPS forms are to be compared.

A number of alternative approaches have resulted in considerably improved incorporation of lipid A and LPS in liposomes. First, the addition of chloroform-solubilized lipid A to the liposomal lipids in chloroform before drying and hydration of the lipids showed almost quantitative association of this relatively hydrophobic molecule with the vesicles.[11,13,18-22] This procedure was also effective for the preparation of Re LPS-containing liposomes.[13] Second, both smooth and rough-type LPS can be effectively incorporated in MLV by brief sonication of aqueous LPS-phospholipid mixtures, followed by a dehydration-rehydration step.[13,15,23,24] Prolonged sonication of LPS-lipid suspensions or the sonication of the MLV obtained after the dehydration-rehydration step will produce LPS-containing SUV.[15,24]

LPS and lipid A incorporated into such liposomes have been shown to have an at least 100-fold reduced potency to trigger *Limulus* amoebocyte lysate (LAL) coagulation.[13,15,21,22] LAL gelation is dependent on the fatty acids of the lipid A moiety and it is conceivable that membrane insertion hampers the interaction of the lysate enzymes with the hydrophobic region of this part of the endotoxin molecule. Thus, the observed decrease in LAL activity, together with the nearly quantitative association of the LPS and lipid A with the vesicles, strongly indicates that when using the latter procedures the endotoxin becomes inserted in the liposomal phospholipid bilayers.

LPS and lipid A inserted in phospholipid vesicles may form mixed bilayers, i.e., the LPS and phospholipid molecules are homogenously dispersed.[19,25] Other studies, however, indicate that segregated domains of LPS exist in such liposomal membranes.[24,26] These different observations may be explained by the occurrence of lipid domains that are dependent on many factors such as the nature of the LPS and phospholipid, the relative amounts of LPS in the bilayer,[24] the presence of ions as divalent cations, and other physical parameters. It is noteworthy that closed, vesicular structures may be prepared from pure LPS. Brandenburg and Seydel reported the preparation of closed vesicles from different types of *S. minnesota* LPS, with the exception of Re LPS and lipid A.[27] Furthermore, Van Alphen et al. observed in the electron microscope closed unilamellar vesicles when *Escherichia coli* K12 (R type) was hydrated at 37°C or higher temperatures.[28] Vaara and co-workers, however, were not able to make liposomes from pure LPS.[29]

C. VIROSOMES

Enveloped animal viruses enter cells via fusion with either the plasma membrane at neutral extracellular pH or intracellualr membranes after acidification of endocytic vesicles.[30] In general, the port of entry depends on the family to which a virus belongs.

In recent years, the envelopes of a number of viruses have been functionally reconstituted using detergent solubilization and removal methods.[31-34] The viral membrane vesicles, also known as virosomes,[35] are unilamellar and have dimensions corresponding to the intact virus particles. Similar to liposomes, hydrophilic, hydrophobic, or amphipathic molecules can be associated with the virosomes during their preparation. Moreover, all of the possible major mechanisms of liposome-cell interaction also apply to the interaction of virosomes with animal cells. In contrast to liposomes, however, functionally reconstituted viral envelopes are fusogenic. Thus, fusion will significantly contribute to the interaction of virosomes with target cells under appropriate conditions.

Influenza virus, for example, binds with high specificity to cell-surface sialic acid and enters cells via the endocytic route. Most animal cells contain terminal sialic-acid groups on the cell surface and consequently influenza virus is able to enter a wide range of host cells.[30] After internalization, the viral envelope fuses with the surrounding membrane upon acidification of the endocytic vesicle contents.[36,37] Fusion of influenza virus with the plasma membrane can be accomplished *in vitro* by a temporary reduction of the extracellular pH after low-temperature binding of the virus to the cells.[38]

Thus, due to their fusogenic capacity, virosomes are able to deliver incorporated molecules to different sites in the cell than liposomal carriers. As will be discussed below, influenza virus-derived virosomes have been employed to study the potential mechanism of *in vitro* activation of lymphocytes by LPS.

II. INTERACTION OF LIPOSOMAL LPS WITH MACROPHAGES

The activation of mononuclear phagocytic cells by LPS results in the induction of the tumoricidal state and the secretion of a variety of factors including prostaglandins, interferons, colony-stimulating factors, tumor necrosis factor (TNF) or cachectin, and interleukins (IL).[39,40] Since macrophages avidly internalize liposomes by endocytosis, a number of investigators have studied the effect of liposome incorporation on LPS-induced activation.

A. LPS-INDUCED TUMORICIDAL ACTIVITY

Fogler and co-workers incorporated *E. coli* 055:B5 LPS in MLV using the classical method and concluded from the quantities of LPS which became liposome associated that the endotoxin was captured in the aqueous compartments of the vesicles.[12] At the single concentration studied, liposome-encapsulated LPS was found to render peritoneal and alveolar macrophages isolated from C₃H/HeN mice tumoricidal. A corresponding amount of free LPS mixed with control liposomes (prepared without added LPS) did not activate the cells. In addition, macrophages isolated from LPS low-responder, C₃H/HeJ mice did not become cytotoxic after incubation with either free or liposomal LPS. These investigators concluded that liposome-encapsulated LPS represented a more efficient stimulus for activation of responsive macrophages than the free LPS.

Dumont et al. encapsulated LPS obtained from *S. abortus equi* in LUV.[17] The surface of the liposomes was provided with covalently linked mannosyl residues to promote uptake by human monocyte-derived macrophages via the mannose receptor. Unencapsulated LPS was separated from the liposomes by gel filtration on Sephadex G75. Liposomes and aggregates of LPS, however, will co-elute in the void volume of such columns. This probably explains the only threefold reduced *Limulus* activity observed in these preparations. At a concentration of 10 μg/ml, both free and liposomal LPS were found to increase the already high cytostatic activity of the macrophages from 50 to 67%. This effect, however, was not observed when LPS-containing, but untargeted, liposomes were used.

The induction of the tumoricidal state in isolated rat liver macrophages by liposomal *E. coli* 0127:B8 LPS was studied by Stukart and collaborators.[41] At the low concentration tested, free nor MLV-associated LPS induced cytotoxicity. The cells became activated only when γ-interferon was associated with the vesicles.

Taken together, the results of these studies suggest an increased, equal, and decreased potency of liposomal LPS to stimulate macrophage-mediated tumor cytotoxicity. Although different kinds of macrophages were used as effector cells, the discrepancies in these studies are probably a result of the less well-characterized liposomal-LPS preparations employed. As mentioned before, LPS in classical liposome preparations may physically become associated with the vesicles in various ways. The specific localization of liposome-associated LPS will determine its activity in biological systems. In addition, since LPS also associates with lipid membranes, the calculation of the total amount of LPS in the liposomes using the theoretical entrapped aqueous volume may lead to erroneous estimations. Another shortcoming in these studies is that only one LPS concentration was used to compare the activity of the free with the liposomal form. This can be misleading when a relatively high or low LPS concentration is tested.

Using liposomes containing various amounts of *S. minnesota* Re LPS and lipid A inserted in the phospholipid bilayers, it was shown that the liposomal formulations were at least 100-fold less potent than the free compounds in inducing the tumoricidal state of murine peritoneal macrophages and the macrophage cell line, RAW 264.7.[42] Similar results were obtained by Daemen and co-workers when rat Kupffer cells were employed as effector cells.[43] The majority of liposomal LPS was found to be endocytosed by the macrophages and appeared to be delivered to the lysosomal compartments.[42] In the lysosomes, the phospholipid vesicles will disintegrate[44-46] and presumably release the incorporated LPS. Since liposomal LPS had dramatically reduced activity, it is unlikely that endotoxin activates macrophages from the lysosomal system.

Consequently, it was concluded that liposome incorporation prevented the necessary interaction of the hydrophobic part of the lipid A moiety with cell-surface structures involved in signal transduction in these cells. However, the possibility that free LPS is able to activate macrophages from the endocytic-vesicle compartment can not be excluded from these experiments, since liposomes most probably remain intact during their passage through these subcellular structures.[46]

B. LPS-INDUCED CYTOKINE SECRETION

The production of cytokines by monocytes and macrophages upon incubation with LPS-containing liposomes has been the subject of further investigation.[13,14,42,47]

The incorporation of *S. minnesota* wild-type LPS, Re LPS, and its lipid A moiety in the membranes of both MLV and SUV decreased the potency of these compounds to induce IL-1 secretion from RAW 264.7 cells, between 100- and 1000-fold.[13] Since in this system cell-associated IL-1 never exceeded the amounts secreted, liposome insertion of LPS apparently prevented the synthesis of this cytokine. Similarly, liposomal LPS was found to have decreased activity on TNF secretion by RAW 264.7 macrophages.[42] These results are in agreement with the observation that, at least when the molecule is inserted into the vesicle membrane, liposomal LPS is an inefficient stimulus for macrophage-mediated tumoricidal activity.

Bakouche and co-workers investigated IL-1 production by human monocytes treated with *E. coli* 026:B6 LPS incorporated in MLV prepared in the classical manner.[14] Evidence was obtained that the greater part of the MLF-associated LPS was encapsulated in the aqueous space of the vesicles. The entrapped LPS did not induce IL-1 secretion, but was still able to trigger the production of significant amounts of intracellular and membrane-associated IL-1. Using a modified procedure involving lyophilization liposomes with the LPS predominantly associated with the vesicle, membranes were prepared and tested. These latter vesicles were found to be as effective as free LPS in inducing both cell-associated and extracellular IL-1. From these results, it was concluded that the density of LPS associated with the liposomal membranes determined the fate of the synthesized IL-1, i.e., higher membrane densities would favor the release of synthesized IL-1. It would have been of interest to know whether the LPS associated with the liposomal membranes had a reduced activity in the *Limulus* assay. If this LPS was not inserted, but was to a large extent adsorbed to the phsopholipid bilayers, the potency of these preparations could be explained by assuming that the hydrophobic part of the lipid A moiety was still accessible for cell-associated structures involved in monocyte stimulation.

The observation that in the monocyte system the aqueous space entrapped LPS still induced cell-associated IL-1, but not its secretion, is puzzling. It is possible that partial release of LPS from the vesicles during or after uptake by the cells results in the triggering of the synthesis, but not the release of IL-1. Another possibility might be that encapsulated LPS induces IL-1α. In contrast, IL-1β is the main form of this cytokine secreted when monocytes are stimulated by LPS.[48]

III. INTERACTION OF LIPOSOMAL LPS WITH LYMPHOCYTES

In vitro, about one third of the murine splenic B-lymphocyte population responds to LPS or its lipid A moiety by proliferation and differentiation into antibody-secreting plasma cells.[39] Previous studies on the interaction of liposomes with lymphocytes indicated that the uptake by these cells is primarily restricted to stable adsorption. Significant uptake by endocytosis, however, may occur when the liposomes are provided with antibodies directed to certain major histocompatibility complex proteins of the cells.[49] The effects of liposomal LPS on B-lymphocyte activation have been investigated employing untargeted phospholipid vesicles.

Alving and Richardson studied the effect of liposome incorporation on the potency of *Shigella flexneri*-derived lipid A to stimulate splenocyte proliferation and to trigger *Limulus* coagulation.[21,22] Both activities of lipid A were found to be markedly reduced when this molecule was inserted in the liposomal bilayers. The remaining stimulating activity appeared to increase with higher lipid A densities in the liposomal membranes. Since the hydrophobic fatty acids of the lipid A are essential for its activity, the authors concluded that the degree of surface exposure of the hydrophobic part of the lipid A probably determines the activity of the liposomal form. Thus, a higher lipid A density in the liposomal membrane would result in an increased surface exposure of hydrophobic groups. Alternatively, it is possible that vesicle preparations made at higher lipid A densities have relatively more lipid A which is either free or which is adsorbed to but not inserted in the liposomal membranes.

Dijkstra and co-workers have observed that *Salmonella minnesota* Re LPS incorporated in the membranes of extruded (0.1-μm diameter) unilamellar vesicles was about 100-fold less active than free LPS to stimulate splenocyte proliferation.[50] (See Figure 1 for a representative experiment.) At corresponding concentrations, similar amounts of radiolabeled free and liposomal LPS were found to be associated with the cells.[50] Moreover, control liposomes (prepared without added LPS) did not affect the proliferation stimulated by free LPS (Figure 1). This excludes the possibility that the liposomes per se were inhibitory in these experiments.

In conclusion, these studies indicate that LPS and lipid A incorporated in the membranes of liposomes have a reduced potency to stimulate B-lymphocytes in murine splenocyte cultures. The decreased activity of liposomal LPS in this system may suggest, as with activation of macrophages, a less-efficient interaction of the hydrophobic lipid A moiety with cell-surface structures involved in signal transduction in B-lymphocytes. Unlike macrophages, however, B-lymphocytes only minimally endocytose liposomes.[49] Thus, the possibility that liposomal LPS is able to stimulate B-lymphocytes from an intracellular compartment cannot be excluded from these results.

IV. INTERACTION OF VIROSOMAL LPS WITH LYMPHOCYTES

In contrast with liposomes, reconstituted viral membrane-vesicles or virosomes have intrinsically fusogenic properties. Thus, LPS-containing virosomes can be employed as a tool to insert LPS via fusion into the membranes

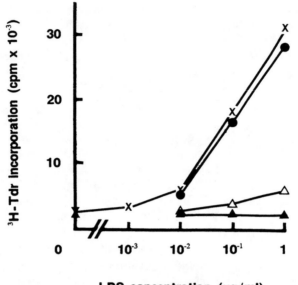

FIGURE 1.   Effect of liposome-incorporated *S. minnesota* Re LPS on lymphocyte proliferation. Murine splencytes (Balb/c; 0.1 × 10⁶ cells in 0.2 ml serum-free RPMI medium) were incubated at 37°C with the indicated amounts of free Re-LPS (X); liposomal Re-LPS (△); control liposomes (▲); and free Re-LPS mixed with control liposomes (●). The control liposomes were added in amounts corresponding to the LPS-containing vesicles. The liposomes consisted of egg phosphatiylcholine, egg phosphatidylglycerol, and cholesterol in a molar ratio of 4:1:4, and contained 100 μg Re LPS per micromole liposomal phospholipid. The vesicles were prepared as described[13] and subsequently extruded through 0.1-μm polycarbonate filters. Mitogenesis was assessed with 1 μCi ³H-Tdr per well during the last 4 h of the 46-h incubation period. The points shown are means of triplicates with the SEM all below 10%.

of responsive cells. Dijkstra and coworkers incorporated *S. minnesota* Re LPS in the membrane of virosomes derived from influenza virus and investigated the interaction of these vesicles with murine (Balb/c) splenocytes.[50] In these experiments two strains of influenza A virus were used, the X-31 and X-99 strains, with a pH threshold for fusion of 5.3 and 5.9, respectively. Free and virosomal LPS were prebound to the cell surface of splenocytes at low temperature. Subsequently, the cells were either treated at low extracellular pH for a few minutes to induce fusion of the virosomes with the plasma membrane or kept at neutral pH to allow endocytic uptake of the vesicles. The proliferation of the lymphocytes was assessed after continuous incubation at 37°C.

Virosomal LPS, fused at low extracellular pH into the plasma membranes, was found to be more-efficient stimulus for lymphocyte proliferation than free LPS, both when X-31- and X-99-derived virosomes were used as carriers.[50] Figure 2 shows the stimulation indices obtained at an LPS concen-

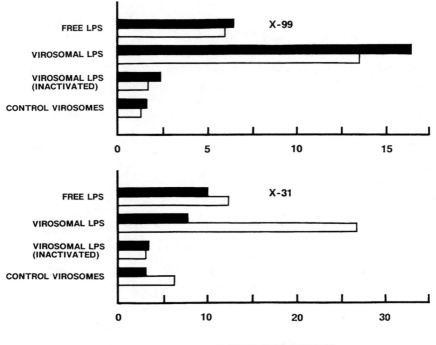

STIMULATION INDEX

FIGURE 2.   Effect of virosome-incorporated *S. minnesota* Re LPS on lymphocyte proliferation. Murine splenocytes (Balb/c; 35×10^6 cells per milliliter serum-free RPMI medium) were incubated for 90 min at 4°C with 1 µg/ml of either free LPS; virosomal LPS; acid-pretreated, fusion-inactivated virosomal LPS; or control virosomes, prepared without added LPS. The virosome concentration was 10 nmol phospholipid per milliliter. The virosomes were derived from the X-31 or X-99 strain of influenza A virus and prepared as described (the LPS was added with the detergent).[33] Subsequently, 3.5×10^6 cells were incubated for 2 min at 37°C in either 1 ml acid medium (final pH 4.9 and 5.6, for X-31 and X-99, respectively) (open bars) or 1 ml neutral medium (pH 7.4) (closed bars). After neutralization of the acid incubations, 0.6×10^6 cells in 0.2-ml medium were incubated in triplicate and labeled as described in the legend of Figure 1.

tration of 1 µg/ml. A similar increase in proliferation was observed with X-99 virus-derived virosomes after neutral pH treatment. At this pH, however, the X-31 virosomes appeared to be less efficient than the free LPS. Acid pretreatment of the LPS-containing virosomes. which irreversibly inactivates their fusogenic capacity,[33] abolished the stimulatory activity of the virosomal LPS at all incubation conditions. This suggests that a fusion step is necessary for optimal stimulation of the cells by virosomal LPS. Control virosomes as such were not (X-99) or minimally (X-31) stimulatory in these experiments (Figure 2). Control virosomes did not affect the proliferation stimulated by the free LPS (not shown).

Taken together, these results indicate that splenocytes, most likely the B-lymphocyte population, can be efficiently stimulated by LPS fused into both the plasma and endosomal membranes. The observed difference in potency of X-99 and X-31 incorporated LPS, when the vesicles are allowed to become internalized at neutral pH, is probably related to the difference in the pH threshold for fusion of these virosomes (pH 5.9 and 5.3, respectively). Since endocytosed material is gradually exposed to a decreasing pH environment during its transport to the lysosomal compartment, the results suggest that internalized virosomal LPS is optimally active when inserted into the membranes of early, less-acidified endosomes.

It is noteworthy that, in contrast to splenocytes derived from C_3H/HeN mice, the virosome-mediated fusion of Re LPS into the plasma or endosomal membranes of lymphocytes from C_3H/HeJ mice did not result in an activation different from the slight stimulatory effect that free Re LPS has in this mouse strain. This indicates that the low responsiveness of the C_3H/HeJ strain is not a result of impaired insertion of cell-associated LPS into the membranes of these mice (Dijkstra et al., unpublished results).

V. LIPOSOMAL LPS *IN VIVO* — PYROGENICITY AND LETHAL TOXICITY

Mononuclear phagocytes appear to be major effector cells for LPS-induced responses *in vivo*. Cytokines, including IL-1, IL-6, and TNF, are not only pyrogenic but also play an important role in the pathogenesis of endotoxic shock.[40] From the *in vitro* studies it appeared that both membrane-inserted and aqueous space-entrapped liposomal LPS have a reduced potency to induce cytokine secretion from mononuclear phagocytes. Liposomes of the type and lipid composition used in these studies are largely taken up by the macrophages of liver and spleen within a few hours after i.v. or i.p. administration.[6] Therefore, it was of interest to evaluate the *in vivo* effects of phospholipid-vesicle incorporated LPS and lipid A.

The pyrogenicity of three different types of lipid A inserted in the membranes of MLV was tested by Richards and colleagues (*S. minnesota* and *E. coli* derived di- and monophosphoryl lipid A).[51] The liposomal formulations were found to be 25- to 200-fold less pyrogenic in rabbits than the corresponding free suspensions of lipid A. This effect did not seem to depend on the lipid A density in the liposomal membranes.

The lethal toxicity of *S. minnesota* wild-type LPS and diphosphoryl lipid A in actinomycin D-sensitized mice appeared to be at least 10- to 60-fold reduced upon insertion in the phospholipid bilayers of both MLV and SUV.[52] Similar results were obtained with the smooth LPS in unsensitized animals. In addition, a sublethal amount of liposomal wild-type LPS was found to induce late-phase but not early phase tolerance to a subsequently administered lethal dose of free LPS. In this study, liposome incorporation also reduced the *in vivo* potency of lipid A to activate peritoneal macrophages, as indicated by the approximately 100-fold decreased capacity to prime these cells for the phorbol-myristate-acetate triggered release of H_2O_2.

Using actinomycin D-sensitized mice, Trubetskoy and co-workers observed a 200-fold reduction in the lethal toxicity of *Neisseria meningitidis* B125 (R-type) LPS upon incorporation in liposomal membranes.[24]

Furthermore, a 10- to 20-fold decrease in lethality was reported in sensitized, tumor-bearing mice when *S. abortus equi* LPS was encapsulated in LUV.[17]

So far, all the *in vivo* studies demonstrate that liposome incorporation considerably reduces the pyrogenic and toxic effects of LPS and lipid A. Thus, the *in vivo* results are consistent with the *in vitro* observation that liposomal LPS and lipid A, at least when inserted in the liposomal phospholipid bilayers, have a decreased potency to activate responsive cells.

VI. LPS-CELL INTERACTIONS: CONCLUDING REMARKS

The mechanism responsible for the initiation of the LPS-induced biological responses has still not been completely resolved. Previous studies suggested either a nonspecific or a receptor-mediated interaction of LPS with responsive cells (reviewed extensively in Reference 53 and 54, respectively). Nonspecific binding or adsorption to the cell surface, possibly mediated by ionic interactions, may result in hydrophobic insertion of the LPS into the plasma membrane and subsequently lead to the induction of cellular responses.[53,55] On the other hand, the activation of responsive cells might be triggered via cell-surface receptors specific for the carbohydrate or lipid moiety of LPS.[54]

Recently, a number of groups demonstrated the existence of membrane-associated binding proteins for LPS and lipid A on various cell types, including macrophages and lymphocytes.[56-62] Moreover, the monocyte differentiation antigen CD14 was reported to bind complexes of LPS with a serum LPS-binding protein.[63] The association of LPS with the binding proteins was found to be inhibited by lipid A, indicating specificity for this part of the LPS molecule.[58,60,62]

A monoclonal antibody with specificity for an 80-kDa LPS-binding protein appeared to induce macrophage-mediated tumor cytotoxicity *in vitro*.[64] In addition, this antibody protected mice from the lethal effects of LPS.[65] These latter results indicate that the antibody functions as an agonist for LPS-induced reponses and thus strongly support the receptor hypothesis.

The studies described in the previous sections clearly demonstrate that LPS and lipid A inserted into liposomal membranes have considerably reduced potency to stimulate both macrophages and splenocytes. This suggests that the interaction of cells with the hydrophilic part of the LPS molecule, which is accessible on the liposomal surface, is not a sufficient stimulus for

LPS-induced responses. Thus, liposome incorporation apparently prevents the interaction of the hydrophobic lipid A moiety of LPS with cellular structures involved in signal transduction.

Assuming that the LPS-binding proteins function as receptors for this molecule, the liposomal LPS studies suggest that the primary interaction between LPS and its receptor is hydrophobic in nature.

In agreement with this conclusion is the observation that LPS inserted into the plasma membrane of splenocytes via virosome-mediated fusion efficiently stimulated these cells. This indicates that cell membrane-inserted LPS is able to interact effectively with the LPS receptor. Taken together, it is conceivable that the association of LPS with its receptor coincides with the insertion of the lipid A moiety into the cellular membrane. The receptor molecule may facilitate the insertion or, vice versa, insertion may be a prerequisite for binding to the receptor.

The observation that virosomal LPS fused into the membranes of early endosomes also efficiently stimulated lymphocyte proliferation may indicate that LPS is able to trigger the cells from this compartment. Alternatively, the LPS molecule may have to recycle back to the cell surface to interact with the proposed LPS receptor. Within this context, it is of interest that both B-lymphocytes and mononulcear phaogcytes are able to endocytose free LPS.[42,66,67]

Future studies on the interaction of virosomal and liposomal LPS with macrophage and B-cell lines will contribute to the knowledge of the initial events of LPS-induced cellular responses.

ACKNOWLEDGMENTS

The authors wish to thank ImClone Systems, Inc., New York, for financial support to Jan Dijkstra at the University of Groningen and Rinske Kuperus for expert help in the preparation of the manuscript.

REFERENCES

1. **Bangham, A. D., Standish, M. M., and Watkins, J. C.,** Diffusion of univalent ions across the lamellae of swollen phospholipids, *J. Mol. Biol.,* 13, 238, 1965.
2. **Pagano, R. E. and Weinstein, J. N.,** Interactions of liposomes with mammalian cells, *Annu. Rev. Biophys. Bioeng.,* 7, 435, 1978.
3. **Szoka, F. and Papahadjopoulos, D.,** Comparative properties and methods of preparation of lipid vesicles (liposomes), *Annu. Rev. Biophys. Bioeng.,* 9, 467, 1980.
4. **Gregoriadis, G., Ed.,** *Liposome Technology,* Vol. I to III, CRC Press, Boca Raton, FL, 1984.
5. **Hope, M. J., Bally, M. B., Mayer, L. D., Janoff, A. S., and Cullis, P. R.,** Generation of multilamellar and unilamellar phospholipid vesicles, *Chem. Phys. Lipids,* 40, 89, 1986.

6. **Gregoriadis, G., Ed.,** *Liposomes as Drug Carriers,* John Wiley & Sons, Chichester, England, 1988.

7. **New, R. R. C., Ed.,** *Liposomes: A Practical Approach,* IRL Press, Oxford, England, 1990.

8. **Galanos, C. and Lüdertiz, O.,** Lipopolysaccharide: properties of an amphipathic molecule, in *Handbook of Endotoxin,* Vol. 1, Rietschel, E. Th., Ed., Elsevier, Amsterdam, 1984, 46.

9. **Kataoka, T., Inoue, K., Lüderitz, O., and Kinsky, S. C.,** Antibody- and complement-dependent damage to liposomes prepared with bacterial lipopolysaccharides, *Eur. J. Biochem.,* 21, 80, 1971.

10. **Kataoka, T., Inoue, G., Galanos, C., and Kinsky, S. C.,** Detection and specificity of lipid A antibodies using liposomes sensitized with lipid A and bacterial lipopolysaccharides, *Eur. J. Biochem.,* 24, 123, 1971.

11. **Schuster, B. G., Neidig, M., Alving, B. M., and Alving, C. R.,** Production of antibodies against phosphocholine, phosphatidylcholine, sphingomyelin, and lipid A by injection of liposomes containing lipid A, *J. Immunol.,* 122, 900, 1979.

12. **Fogler, W. E., Talmadge, J. E., and Fidler, I. J.,** The activation of tumoricidal properties in macrophages of endotoxin responder and nonresponder mice by liposome-encapsulated immunomodulators, *J. Reticuloendothel. Soc.,* 33, 165, 1983.

13. **Dijkstra, J., Mellors, J. W., Ryan, J. L., and Szoka, F. C.,** Modulation of the biological activity of bacterial endotoxin by incorporation into lipsomes, *J. Immunol.,* 138, 2663, 1987.

14. **Bakouche, O., Koff, W. C., Brown, D. C., and Lachman, L. B.,** Interleukin 1 release by human monocytes treated with liposome-encapsulated lipopolysaccharide, *J. Immunol.,* 139, 1120, 1987.

15. **Dijkstra, J., Ryan, J. L., and Szoka, F. C.,** A procedure for the efficient incorporation of wild-type lipopolysaccharide into liposomes for use in immunological studies, *J. Immunol. Methods,* 114, 197, 1988.

16. **Seydel, U., Brandenburg, K., Koch, M. H. J., and Rietchel, E. Th.,** Supramolecular structure of lipopolysaccharide and free lipid A under physiological conditions as determined by synchrotron small-angle X-ray diffraction, *Eur. J. Biochem.,* 186, 325, 1989.

17. **Dumont, S., Muller, C. D., Schuber, F., and Bartholeyns, J.,** Antitumoral properties and reduced toxicity of LPS targeted to macrophages via normal or mannosylated liposomes, *Anticancer Res.,* 10, 155, 1990.

18. **Dancey, G. F., Yasuda, T., and Kinsky, S. C.,** Enhancement of liposomal model membrane immunogenicity by incorporation of lipid A, *J. Immun.,* 119, 1868, 1977.

19. **Rottem, S.,** The effect of lipid A on the fluidity and permeability properties of phospholipid dispersions, *FEBS Lett.,* 95, 121, 1978.

20. **Alving, C. R., Shichijo, S., and Mattsby-Baltzer, I.,** Preparation and use of liposomes in immunological studies, in *Liposome Technology,* Vol. 2, Gregoriadis, G., Ed., CRC Press, Boca Raton, FL, 1984, 157.

21. **Alving, C. R. and Richardson, E. C.,** Mitogenic activities of lipid A and liposome-associated lipid A: effects of epitope density, *Rev. Infect. Dis.,* 6, 493, 1984.

22. **Richardson, E. C., Banerji, B., Seid, R. C., Levin, J., and Alving, C. R.,** Interactions of lipid A and liposome-associated lipid A with *Limulus polyphemus* amoebocytes, *Infect. Immun.,* 39, 1385, 1983.

23. **Nikaido, H., Takeuchi, Y., Ohnishi, S. -I., and Nakae, T.,** Outer membrane of *Salmonella typhimurium.* Electron spin resonance studies, *Biochem. Biophys. Act,* 465, 152, 1977.

24. **Trubetskoy, V. S., Koshkina, N. V., Omel'yanenko, V. G., L'vov, V. L., Dimitriev, B. A., Petrov, A. B., and Torchilin, V. P.,** FITC-labeled lipopolysaccharide: use as a probe for liposomal membrane incorporation studies, *FEBS Lett.,* 269, 79, 1990.

25. **MacKay, A. L., Nichol, C. P., Weeks, G., and Davis, J. H.**, A proton and deuterium nuclear magnetic resonance study of orientational order in aqueous dispersions of lipopolysaccharide and lipopolysaccharide/dipalmitoyl-phosphatidylcholine mixtures, *Biochim. Biophys. Acta*, 774, 181, 1984.

26. **Takeuchi, Y. and Nikaido, H.**, Persistance of segregated phospholipid domains in phospholipid-lipopolysaccharide mixed bilayers: studies with spin-labeled phospholipids, *Biochemistry*, 20, 523, 1981.

27. **Brandenburg, K. and Seydel, U.**, Physical aspects of structure and function of membranes made from lipopolysaccharides and free lipid A, *Biochim. Biophys. Acta*, 775, 225, 1984.

28. **Van Alphen, L., Verkleij, A., Burnell, E., and Lugtenberg, D.**, ^{31}P nuclear magnetic resonance and freeze-fracture electron microscopy studies on *Escherichia coli.* II. Lipopolysaccharide and lipopolysaccharide-phospholipid complexes, *Biochim. Biophys. Acta*, 597, 502, 1980.

29. **Vaara, M., Plachy, W. Z., and Nikaido, H.**, Partitioning of hydrophobic probes into lipopolysaccharide bilayers, *Biochim. Biophys. Acta*, 1024, 152, 1990.

30. **Marsh, M. and Helenius, A.**, Virus entry into animal cells, *Adv. Virus Res.*, 36, 107, 1989.

31. **Volsky, D. J. and Loyter, A.**, An efficient method for reassembly of fusogenic Sendai virus envelopes after solubilization of intact virions with Triton X-100, *FEBS Lett.*, 92, 190, 1978.

32. **Metsikkö, K., Van Meer, G., and Simons, K.**, Reconstitution of the fusogenic activity of vesicular stomatitis virus, *EMBO J.*, 5, 3429, 1986.

33. **Stegmann, T., Morselt, H. W. M., Booy, F. P., Van Breemen, J. F. L., Scherphof, G., and Wilschut, J.**, Functional reconstitution of influenza virus envelopes, *EMBO J.*, 6, 2651, 1987.

34. **Nussbaum, P., Lapidot, M., and Loyter, A.**, Reconstitution of functional influenza virus envelopes and fusion with membranes and liposomes lacking virus receptors, *J. Virol.*, 61, 2245, 1987.

35. **Almeida, J. D., Edwards, D. C., Brand, C. M., and Heath, T. D.**, Formation of virosomes from influenza subunits and liposomes, *Lancet*, ii, 899, 1975.

36. **Matlin, K. S., Reggio, H., Helenius, A., and Simons, K.**, The infective entry of influenza virus into MDCK cells, *J. Cell Biol.*, 19, 601, 1981.

37. **Yoshimura, A. and Ohnishi, S. -I.**, Uncoating of influenza virus in endosomes, *J. Virol.*, 51, 497, 1984.

38. **White, J., Matlin, K., and Helenius, A.**, Cell fusion by Semliki Forest, influenza, and vesicular stomatitis viruses, *J. Cell Biol.*, 89, 674, 1981.

39. **Morrison, D. C. and Ryan, J. L.**, Bacterial endotoxins and host immune responses, *Adv. Immunol.*, 28, 293, 1979.

40. **Morrison, D. C. and Ryan, J. L.**, Endotoxins and disease mechanisms, *Annu. Rev. Med.*, 38, 417, 1987.

41. **Stukart, M. J., Rijnsent, A., and Roos, E.**, Induction of tumoricidal activity in isolated rat liver macrophages by liposomes containing recombinant rat γ-interferon supplemented with lipopolysaccharide or muramyldipeptide, *Cancer Res.*, 47, 3880, 1987.

42. **Dijkstra, J., Larrick, J. W., Ryan, J. L., and Szoka, F. C.**, Incorporation of LPS in liposomes diminishes its ability to induce tumoricidal activity and tumor necrosis factor secretion in murine macrophages, *J. Leuk. Biol.*, 43, 436, 1988.

43. **Daemen, T., Veninga, A., Dijkstra, J., and Scherphof, G. L.**, Differential effects of liposome-incorporation on liver macrophage activating properties of rough lipopolysaccharide, lipid A, and muramyl dipeptide, *J. Immunol.*, 142, 2469, 1989.

44. **Dijkstra, J., Van Galen, W. J. M., Hulstaert, C. E., Kalicharan, D., Roerdink, F. H., and Scherphof, G. L.**, Interaction of liposomes with Kupffer cells *in vitro*, *Exp. Cell Res.*, 150, 161, 1984.

45. **Dijkstra, J., Van Galen, M., Regts, D., and Scherphof, G. L.**, Uptake and processing of liposomal phospholipids by Kupffer cells, *in vitro, Eur. J. Biochem.*, 148, 391, 1985.
46. **Harding, C. V., Collins, D. S., Slot, J. W., Geuze, H. J., and Unanue, E. R.**, Liposome-encapsulated antigens are processed in lysosomes, recycled, and presented to T cells, *Cell*, 64, 393k 1991.
47. **Koff, W. C., Cerny, E. H., Reeves, M. W., and Lachman, L. B.**, Interleukin 1 release by liposome-encapsulated macrophage activators, in *The Physiologic, Metabolic, and Immunologic Actions of Interleukin-1*, Kluger, M. H., Oppenheim, J. J., and Rowanda, M. C., Eds., Alan R. Liss, New York, 1985, 309.
48. **Cavaillon, J. -M., Munoz, C., Fitting, C., Couturier, C., and Haeffner-Cavaillon, N.**, Signals involved in interleukin-1 production induced by endotoxins, in *Cellular and Molecular Aspects of Endotoxin Reactions*, (Endotoxin Research Series, Vol. 1), Nowotny, A., Spitzer, J. J., and Ziegler, E. J., Eds., Excerpta Medica, Amsterdam, 1990, 257.
49. **Machy, P., Barbet, J., and Leserman, L. D.**, Differential endocytosis of T and B lymphocyte surface molecules evaluated with antibody-bearing fluorescent liposomes containing methrotrexate, *Proc. Natl. Acad. Sci. U.S.A.*, 79, 4148, 1982.
50. **Dijkstra, J., Wilschut J. C., and Ryan, J. L.**, Stimulation of murine lymphocyte proliferation by lipopolysaccharide incorporated in fusogenic, reconstituted influenza virus envelopes (virosomes), *J. Immunol.*, submitted.
51. **Richards, R. L., Swartz, G. M., Schultz, C., Hayre, M. D., Ward, G. S., Ballou, W. R., Chulay, J. D., Hockmeyer, W. T., Berman, S. L., and Alving, C. R.**, Immunogenicity of liposomal maleria sporozoite antigen in monkeys: adjuvant effects of aluminium hydroxide and non-pyrogenic liposomal lipid A, *Vaccine*, 7, 506, 1989.
52. **Dijkstra, J., Mellors, J. W., and Ryan, J. L.**, Altered *in vivo* activity of liposome-incorporated lipopolysaccharide and lipid A, *Infect. Immun.*, 57, 3357, 1989.
53. **Morrison, D. C.**, Nonspecific interactions of bacterial lipopolysaccharide with membranes and membrane components, in *Handbook of Endotoxin*, Vol. 3, Berry, L. J., Ed., Elsevier, Amsterdam, 1984, 25.
54. **Haeffner-Cavaillon, N., Cavaillon, J. -M., and Szabó, L.**, Cellular receptors for endotoxin, in *Handbook of Endotoxin*, Vol. 3, Berry, L. J., Ed., Elsevier, Amsterdam, 1984, 1.
55. **Jacobs, D. M. and Price, R. M.**, A model for lipopolysaccharide-membrane interaction, *Adv. Exp. Med. Biol.*, 216A, 691, 1987.
56. **Wright, S. D. and John, M. T. C.**, Adhesion promoting receptors on human macrophages recognize *Escherichia coli* by binding to lipopolysaccharide, *J. Exp. Med.*, 164, 1876, 1986.
57. **Lei, M. -G. and Morrison, D. C.**, Specific endotoxic lipopolysaccharide-binding proteins on murine splenocytes. I. Detection of lipopolysaccharide-binding sites on splenocytes and splenocyte subpopulations, *J. Immunol.*, 141, 996, 1988.
58. **Lei, M. -G. and Morrison, D. C.**, Specific endotoxic lipopolysaccharide-binding proteins on murine splenocytes. II. Membrane localization and binding characteristics, *J. Immunol.*, 141, 1006, 1988.
59. **Hampton, R. Y., Golenbock, D. T., and Raetz, C. R. H.**, Lipid A binding sites in membranes of macrophage tumor cells, *J. Biol. Chem.*, 263, 14802, 1988.
60. **Hara-Kuge, S., Amano, F., Nishijima, M., and Akamatsu, Y.**, Isolation of a lipopolysaccharide (LPS)-resistent mutant with defective LPS binding, of cultured macrophage-like cells, *J. Biol. Chem.*, 265, 6606, 1990.
61. **Golenbock, D. T., Hampton, R. Y., Raetz, C. R. H., and Wright, S. D.**, Human phagocytes have multiple lipid A-binding sites, *Infect. Immun.*, 58, 4069, 1990.
62. **Kirkland, T. N., Virca, G. D., Kuus-Reichel, T., Multer, F. K., Kim, S. Y., Ulevitch, R. J., and Tobias, P. S.**, Identification of lipopolysaccharide-binding proteins in 70Z/3 cells by photoaffinity labeling, *J. Biol. Chem.*, 265, 9520, 1990.

63. **Wright, S. D., Ramos, R. A., Tobias, P. S., Ulevitch, R. J., and Mathison, J. C.,** CD14, a receptor for complexes of lipopolysaccharide (LPS) and LPS binding protein, *Science,* 249, 1431, 1990.

64. **Chen, T. -Y., Bright, S. W., Pace, J. L., Russell, S. W., and Morrison, D. C.,** Induction of macrophage-mediated tumorcytotoxicity by a hamster monoclonal antibody with specificity for lipopolysaccharide receptor, *J. Immunol.,* 145, 8, 1990.

65. **Morrison, D. C., Silverstein, R., Bright, S. W., Chen, T. -Y., Flebbe, L. M., and Lei, M. -L.,** Monoclonal antibody to mouse lipopolysaccharide receptor protects mice against the lethal effects of endotoxin, *J. Infect. Dis.,* 162, 1063, 1990.

66. **Bona, C. A.,** Fate of endotoxin in macrophages: biological and ultrastructural aspects, *J. Infect. Dis.,* 128, S74, 1973.

67. **Bona, C., Juy, D., Truffa-Bachi, P., and Kaplan, G. J.,** Binding, capping and internalization of lipopolysaccharide in thymic and non-thymic lymphocytes of the mouse. Biological and autoradiographic study, *J. Microsc. Biol. Cell,* 25, 47, 1976.

Chapter 15

ENDOTOXIN-ENDOTHELIAL CELL INTERACTIONS

Timothy H. Pohlman and John M. Harlan

TABLE OF CONTENTS

I. INTRODUCTION

Until recently, interest in the endothelial lining of the vasculature focused primarily on the barrier function of this structure.[1] Endothelial damage caused by inflammatory reactions associated with endotoxemia produces increased permeability to fluid and macropmolecules and exposure of thrombogenic subendothelial surfaces, resulting in tissue and organ injury. It is now established, however, that cultured endothelial cells also acquire several functional properties after exposure to bacterial lipopolysaccharide (LPS). The expression of an altered phenotype *in vitro* following stimulation with LPS is referred to as endothelial cell "activation" and includes activities that promote inflammation and coagulation.[2-4] Furthermore, endothelial cells can be stimulated to release substances that profoundly affect vascular tone.[5] Endothelial cell activation also occurs following exposure to tumor necrosis factor-α (TNF)[2] and interleukin-1 (IL-1),[7] cytokines that are released by mononulcear phagocytes in response to microorganisms and microbial products such as endotoxin. Thus, LPS can activate endothelium directly, or indirectly, by inducing release of TNF and IL-1 from mononuclear phagocytes. Indeed, activation of endothelial cells by cytokines generated by LPS-stimulated mononuclear phagocytes may represent the major pathway by which LPS affects endothelial cell function *in vivo* (Figure 1). Therefore, as a target of destructive inflammatory reactions induced by Gram-negative infection, and as a potentiall important participant in host defense reactions, the interaction of LPS with endothelium has considerable significance.

Most of what is known about the interaction of LPS with endothelium is derived from the study of cultured endothelial cells isolated from a number of tissues.[8-10] Although the use of cultured cells vastly simplifies the study of endothelial biology, cell culture must be recognized as a highly artificial system. The phenotype of cultured cells may be significantly altered from that of endothelium *in vivo* by the use of proteases during the isolation of cells from vessels, growth on artificial matrix, and the use of exogenous growth factors. Also, endothelial cell cultures are most often established from large veins or arteries, and cells from these vessels may differ significantly from postcapillary venular endothelium where inflammatory responses in the circulation are usually initiated. It must be emphasized therefore that conclusions concerning endothelial activation that are based on observations *in vitro* should be considered as tentative until confirmed *in vivo*.

In this review we will summarize interactions of LPS with endothelial cells, focusing on those that result in endothelial cell injury or alterations in endothelial cell function. Since the effects of TNF and IL-1 on endothelial cell function often cannot be distinguished from alterations induced directly by LPS, the interactions of these two cytokines with endothelial cells will be considered together with LPS. We will also summarize recent studies from our laboratory that examine molecular mechanisms that mediate the interaction of LPS, TNF, and IL-1 with endothelium.

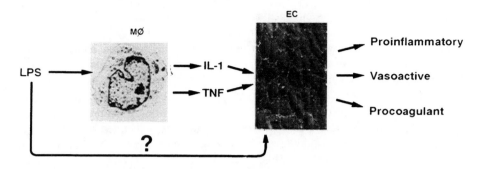

FIGURE 1. A paradigm of endothelial cell activation during endotoxemia. LPS stimulates mononuclear phagocytes (MO) to release the inflammatory cytokines, TNF, and IL-1. These mediators in turn activate endothelial cells (EC) to express properties that promote inflammation and coagulation, and influence vascular tone. When expressed at a local site of inflammation endothelial activation is beneficial to the host, serving to contain and eliminate infecting Gram-negative bacteria. However, generalized endothelial cell activation during endotoxemia may be detrimental to the host by promoting systemic inflammation and disseminated intravascular coagulation.

II. ENDOTHELIAL CELL INJURY

A. LPS-MEDIATED ENDOTHELIAL CELL INJURY *IN VIVO*

There is limited evidence demonstrating localization of LPS in endothelial cells during infection by Gram-negative organisms or following an intravenous injection of LPS in experimental models. Early studies identified LPS in endothelial cells following the intravenous injection of extremely large doses of a Boivin endotoxin preparation into mice or dogs.[11,12] Using indirect immunofluorescence microscopy, Tanaka et al. reported that conjugated LPS was observed in the capillary endothelium of the liver, lungs, intestine, and kidney within 3 h after injection into mice.[11] In a canine model of endotoxemia, Rubenstein et al. found endothelium with positive fluorescence in a variety of tissues including the liver and spleen, and, for the most part, only in capillary and venular endothelium.[12] Fluorescence was detected in endothelium within 10 min of the injection of endotoxin. In contrast, Mathison and Ulevitch, utilizing a more highly purified, radioiodinated preparation of LPS, did not detect localization of LPS in endothelial cells following intravenous bolus injection in rabbits.[13] Accumulation of LPS was observed in a number of other organs, including the liver, spleen, and lung, where it was primarily located in mononuclear phagocytes and other leukocytes. The apparent discrepancies between these studies on the localization of LPS to endothelium *in vivo* may reflect differences in the species examined, the methodologies used to detect LPS, or the amounts of LPS injected.

Increased endothelial cell turnover and morphologic evidence of endothelial cell injury following intravenous injection of LPS have been described in a number of experimental animal models. Using scanning electron microscopy, Reidy and Bowyer examined aortas from rabbits receiving a single large injection (400 mg) of *Serratia marcescens* LPS.[14] Within 1 h after the injection of LPS, aortic endothelial cells appeared spindle-like and were partially detached from arterial walls. Small denuded areas of aortic surface were noted to be covered with platelets. By 2 to 4 weeks, vascular injury had largely reversed and there was evidence of regenerating endothelium at the margins of the denuded areas. Other studies have demonstrated high mitotic indices in rabbit aortic endothelium following endotoxin injection, consistent with regeneration after injury.[15,16]

These investigations could not determine whether endothelial cell damage observed after the administration of LPS was a result of direct LPS-mediated cytotoxicity or a consequence of the intravascular activation of various inflammatory systems by LPS. Several studies, however, have examined the role of inflammatory mediators in LPS-induced vascular injury *in vivo*. Gaynor found no difference in the degree of endothelial injury induced by a bolus intravenous injection of endotoxin between control rabbits and those previously rendered neutropenic by nitrogen mustard treatment.[17] Desquamated endothelial cells circulating in blood were detected in both groups of animals. Histological examination of aortas from both neutropenic and control rabbits treated with LPS revealed similar abnormalities, including vacuolization, subendothelial edema, and desquamation. These abnormalities were not observed in neutropenic rabbits receiving saline alone. Similarly, Pingleton et al. found that Rhesus monkeys rendered leukopenic by total body irradiation developed significant endothelial swelling and perivascular edema in pulmonary capillaries following *Escherichia coli* endotoxemia.[18] The vascular abnormalities in leukopenic animals were equivalent to lesions noted in pulmonary capillaries of control animals. Light and electron microscopy revealed marked sequestration of polymorphonuclear leukocytes in specimens of lung from animals in the control group, whereas leukocytes were virtually absent in sections from lungs of the irradiated group. Further evidence in support of a direct endotoxin-endothelial cell interaction resulting in endothelial cell injury was found by using animals depleted of complement by cobra venom factor.[19] In this study, hypocomplementemic rats and normal rats developed an equivalent increase in the ^{3}H-thymidine labeling index of aortic endothelium following bolus injection of 500 mg of LPS. Treatment of animals with cobra venom factor by itself did not increase the labeling index.

Although these studies suggest that LPS-induced endothelial injury does not necessarily require the activation of leukocytes or complement, they do not exclude the possibility that activation of complement in the leukopenic animals or activation of leukocytes in the complement-depleted animals by LPS damaged endothelial cells. Moreover, it is likely that other inflammatory mediators including eicosanoids, the coagulation cascade, the kallikrein-kinin system, and the more recently described inflammatory cytokine network are involved in LPS-induced endothelial cell injury. Also, the metabolic and hemodynamic consequences of endotoxemia, such as acidosis and shock, may be sufficient to injure the endothelial lining of blood vessels. Thus, the question of whether LPS directly injures endothelial cells *in vivo* remains unresolved.

B. LPS-MEDIATED ENDOTHELIAL CELL INJURY *IN VITRO*

LPS has been demonstrated to injure endothelial cells directly *in vitro*.[20-24] Cultured bovine aortic and pulmonary artery endothelial cells were noted to first detach from monolayers and then lyse after exposure to LPS. The sensitivity of bovine endothelium to LPS was striking with detachment observed after incubation with as little as 100 pg/ml of LPS (*Salmonella minnesota* R595). A similar sensitivity of LPS-induced cytotoxicity was observed for endothelium derived from bovine mesenteric artery and vein, but not for bovine smooth muscle cells. In contrast, cultured endothelial cells obtained from goat aorta, canine vena cava, or human umbilical vein were not injured by LPS, suggesting that there are major species differences in susceptibility to direct injury.[20] LPS-mediated cytotoxicity of bovine endothelial cells was not dependent on the production of prostaglandins, protein synthesis, oxygen radical production, protease activity, or cytoskeletal function.[20] Subsequent studies showed that bovine endothelial cell injury by LPS required at a minimum the acyl-substituted disaccharide of lipid A, since a monosaccharide precursor of LPS (lipid X) did not cause detachment and/or lysis.[23,24]

Although LPS does not directly cause cultured endothelial cells from species other than bovine to detach or lyse, LPS does directly injure human umbilical vein endothelium (HUVE) in the presence of inhibitors of protein or RNA synthesis.[25,26] Like bovine aorta or pulmonary artery endothelium, individual HUVE cells were first observed to round up and detach followed shortly by lysis that correlated with visual disruption of the monolayer. LPS-induced HUVE cytotoxicity was not observed until the incorporation of ^{3}H-amino acids or ^{3}H-uridine into trichloroacetic acid-precipitable counts was reduced >75% by inhibitors of protein or RNA synthesis, respectively.[26] By itself, this degree of inhibition of protein synthesis or RNA synthesis was not cytotoxic to HUVE during the duration of assay. In contrast to HUVE, cultured human smooth muscle cells or human dermal fibroblasts were not injured after coincubation with LPS and cycloheximide.[26]

Tumor necrosis factor is cytostatic or cytotoxic for a number of tumor cell lines *in vitro*,[27] and toxicity can be potentiated with inhibitors of RNA synthesis such as actinomycin D. Many nontransformed cells in culture that express receptors for TNF are resistant to the cytostatic or cytotoxic effects of TNF even though these cells respond to TNF in a number of different ways.[28-34] HUVE cells respond to TNF with numerous alterations in cell function (see below), but, in addition, undergo cytolysis when coincubated with TNF and cycloheximide or actinomycin D.[26] The time course for cell detachment and lysis was similar to that observed with LPS. Furthermore, IL-1 which possesses little or no tumorcidal activity or toxicity *in vivo,* and, by itself is nontoxic to cells *in vitro,* was also markedly cytotoxic to HUVE in the presence of cycloheximide or actinomycin D.[26,35,36] Lipopolysaccharide-, TNF-, or IL-1-induced cytotoxicity in HUVE sensitized with a protein synthesis inhibitor was modulated by pretreatment of HUVE with LPS, TNF, or IL-1 alone for 4 h prior to the addition of cycloheximide or by pretreatment of HUVE with cycloheximide for 4 h prior to the addition of LPS, TNF, or IL-1. In both cases, cytotoxicity, determined after an additional period of incubation, was completely abolished.[26]

The mechanism by which LPS, TNF, or IL-1 produces a lethal injury to HUVE in the presence of RNA or protein synthesis inhibition is not known. For TNF it has been postulated that *de novo* RNA and protein synthesis is required for production of critical enzymes involved in detoxification or repair mechanisms.[37] Detoxification of LPS by deacylation has been described in leukocytes,[38-40] but a potential pathway for detoxification of LPS has not been identified in human endothelium. We postulate that the absence of cytotoxicity observed when HUVE were pretreated with these inflammatory mediators for several hours prior to the addition of cycloheximide is due to the induction of a repair process. Conversely, the protective effect of cycloheximide pretreatment on LPS-, TNF-, or IL-1-induced cytotoxicity may be due to the reduced synthesis of proteins involved in signal transduction following binding of LPS, TNF, or IL-1 to the cell. To satisfy this latter hypothesis, the critical protein targets in HUVE for LPS, TNF, and IL-1 would have to be constitutively synthesized and have a rapid turnover rate.

The observation that inhibition of protein synthesis sensitizes endothelial cells to the cytotoxic effects of LPS *in vitro* may have some relevance to previous *in vivo* studies.[41,42] Experimental animals treated with a protein synthesis inhibitor were found to be markedly more susceptible to the lethal effects of LPS with up to a 100- to 1000-fold reduction in the LD_{50}. Of interest, pretreatment of animals with a protein synthesis inhibitor 24 h prior to the injection of LPS did not increase lethality,[43] a finding consistent with the observation that pretreatment with cycloheximide prevents LPS-induced cytotoxicity to HUVE *in vitro.*

III. ENDOTHELIAL CELL ACTIVATION

It is now evident from several areas of investigation that endothelial cells can be activated by LPS to express a broad repertoire of functional activities categorized as proinflammatory, procoagulant (for review, see References 2 to 4), and vasoactive. Importantly, most of the functional properties that characterize LPS-activated endothelium are also induced by the inflammatory cytokines TNF and IL-1. In addition, endothelial cells respond to other cytokines such as the interferons[44] and colony-stimulating factors[45] that may circulate as part of the inflammatory milieu of endotoxemia. Furthermore, activated endothelial cells synthesize cytokines that may alter endothelial cell function in an autocrine fashion or affect the function of other cells that participate in the inflammatory response. Thus, endotheial cell function during Gram-negative infection is determined by the action of a complex cascade of inflammatory mediators initiated by LPS. Activation of endothelial cells in multiple vascular beds during endotoxemia could potentially have severe deleterious consequences by promoting systemic inflammation and intravascular coagulation.

A. COAGULATION

The interaction of LPS, TNF, and IL-1 with endothelium promotes coagulation by increasing the activity of the extrinsic pathway and by decreasing fibrinolytic activity. *In vitro*, LPS, TNF, and IL-1 induce surface expression of tissue factor[46-54] and down regulation of thrombomodulin (and thus protein C and protein S activities).[48,50,51] In concert with increased tissue factor expression, fibrinolytic activity on activated endothelium is reduced by increased expression of tissue plasminogen activator inhibitor-1.[55-58] The effects of TNF and IL-1 on coagulation have been examined *in vivo*. Infusion of IL-1 into rabbits increased tissue factor expression, decreased protein C activity, and resulted in deposition of fibrin strands on aortic endothelial surfaces.[48] Administration of TNF to normal subjects produces net procoagulant activity[59] with a rapid and sustained activation of the common pathway of coagulation, probably induced via the extrinsic pathway.[60] These experimental findings are potentially relevant to the pathogenesis of the intravascular coagulopathy that often complicates endotoxemia or Gram-negative sepsis. Activation of endothelium in multiple vascular beds by circulating inflammatory cytokines would create a large surface for initiation of coagulation within the vascular space and lead to uncontrolled coagulation and consumption of clotting factors.

B. INFLAMMATION

Inflammation is characterized by the accumulation of leukocytes at sites of tissue injury or microbial invasion. Leukocyte adherence to endothelium is a critical early event in the emigration of leukocytes to inflammation foci.[61,62] Several cell surface molecules involved in the adherence of leukocytes to endothelium have been identified functionally and immunochemically.[63-66] Most of the proteins involved in these adhesive interactions have been molecularly cloned permitting detailed analysis of their structure. Although leukocytes readily adhere to one or more endothelial cell surface proteins *in vitro*, the precise role of most endothelial adherence molecules in leukocyte emigration *in vivo* has not yet been defined.

Endothelial surface molecules that bind leukocytes include (1) immunoglobulin (Ig) superfamily proteins that interact with leukocyte integrins and

(2) LEC-CAM (or selectin) proteins that recognize specific carbohydrate counterstructures on glycoproteins or glycolipids of leukocytes.[63] The three Ig superfamily adherence molecules expressed on endothelial cell surfaces are intercellular adhesion molecule-1 (ICAM-1),[67,68] ICAM-2,[69] and vascular cell adhesion molecule-1 (VCAM-1).[70] ICAM-1 is expressed at low levels on HUVE *in vitro* and on endothelium *in vivo*.[71] Surface expression of ICAM-1 is upregulated on HUVE following exposure to LPS, TNF, IL-1, or interferon-γ.[72-75] ICAM-2 is constitutively expressed by HUVE *in vitro* and is not upregulated by LPS.

VCAM-1 (INCAM-110) is a 100-kDa glycoprotein expressed on HUVE following activation with LPS, TNF, or IL-1.[70,76,78] VCAM-1 is not expressed or is only minimally expressed on unstimulated HUVE,[70,76-78] and VCAM-1 is not significantly induced by cytokines other than TNF and IL-1. Maximal surface expression of VCAM-1 is observed after 4 h incubation with LPS, TNF, or IL-1 and expression persists with continued incubation. VCAM-1 binds peripheral blood lymphocytes, monocytes, eosinophils, and a variety of leukemic cell lines via the VLA-4 integrin receptor ($\alpha_4 \beta_1$, CD49d/CD29).[79] Neutrophils do not express VLA-4 and, consequently, do not bind to VCAM-1.

Endothelial selectins include granule membrane protein-140 (GMP-140, PADGEM, CD62)[80] and endothelial leukocyte adhesion molecule 1 (ELAM-1).[81-83] Both proteins consist of a cytoplasmic tail at the carboxy terminus, a hydrophobic transmembrane region, and three distinct extracellular domains.[84] The region adjacent to the cell membrane consists of tandem repetitive motifs that are homologus to complement regulatory proteins. Adjacent to this region is an epidermal growth factor motif and at the amino terminus there is a region homologous to several Ca^{2+}-dependent animal lectins.[84] Leukocyte binding involves the interaction of the lectin domain with specific carbohydrate structures on the leukocyte.[85-87] Adhesion mediated by selectins can be regulated by the degree of surface expression of the carbohydrate ligand on the leukocyte,[88,89] as well as by the number of selectin molecules expressed on the endothelial surface.

GMP-140 is contained within the Weibel-Palade bodies of endothelial cells and is mobilized to the surface of endothelial cells in response to thrombin or histamine.[80] This response is rapid, suggesting that GMP-140 may function as the initially deployed endothelial adherence molecule at sites of acute inflammation until other endothelial cell-dependent mechanisms of leukocyte adherence can be fully expressed. Since activation of endothelial cells by LPS, TNF, and IL-1 results in the surface expression of procoagulant activity and subsequent thrombin formation, GMP-140 may represent a point of linkage between the coagulation cascade and inflammation focused on the endothelial surface.

ELAM-1 is a 115-kDa single-chain glycoprotein that is transiently expressed on endothelial cells in response to LPS, TNF, or IL-1.[81,90-92] ELAM-1 expression requires only a brief exposure to any of the agonists, is maximal at 4 h after stimulation (even in the presence of continued stimulation), and declines somewhat by 24 h.[81] Monoclonal antibodies to ELAM-1 partially block the adherence of isolated, unstimulated neutrophils,[81] eosinophils,[93] monocytes,[94] and a subpopulation of T-lymphocytes[95] to LPS-, TNF-, or IL-1-treated HUVE monolayers. ELAM-1 binds to specific oligosaccharide structures on the surface of neutrophils.[85,86] This oligosaccharide is a lactosamine-repeating structure substituted with fucose and is also known as Le[x] antigen. This carbohydrate determinant exists in a neutral and a sialic acid-derivatized form, but it is only the sialic acid form of the fucose-substituted lactosamine, sialyated Le[x], that is recognized by ELAM-1.[85,86] In contrast, GMP-140 has been reported to bind to nonsialyated Le[x].[87]

ELAM-1 has been demonstrated by immunocytochemistry to be expressed at sites of inflammation[96] where it may participate in the coordinated interaction of a number of adhesion molecules that are necessary for leukocyte adherence to and migration across endothelial surfaces. In particular, ELAM-1 expression was observed in a number of microvascular beds in a baboon model of septic shock.[97] It is important to note, however, that, although expression of ELAM-1 can be demonstrated *in vivo,* the function of ELAM-1 in leukocyte emigration has not yet been established. Studies with blocking anti-ELAM-1 monoclonal antibodies in models of endotoxemia or bacteremia are necessary to determine the role of ELAM-1 in leukocyte-endothelial interactions in Gram-negative sepsis.

In addition to expressing adhesion molecules for leukocytes, endothelial cells may further promote inflammation by releasing inflammatory cytokines that activate leukocytes. LPS,[98] TNF,[99] and IL-1[100,101] have been reported to induce the release of hematopoietic colony-stimulating factors that not only promote the clonal proliferation of hematopoietic precursor cells, but also stimulate the functional activity of mature granulocytes, macrophages, and eosinophils.[102] For example, granulocyte-macrophage colony stimulating factor (GM-CSF) enhances the function of mature phagocytes by promoting antibody-dependent, cell-mediated, cytotoxicity, oxidative metabolism, and the phagocytosis of microorganisms.[100] GM-CSF also inhibits the migration of neutrophils. Therefore, release of GM-CSF by endothelium at local sites of inflammation would tend to retain granulocytes as well as enhance their function. GM-CSF may also augment the inflammatory response by promoting mononuclear phagocyte release of TNF.[103]

Activation of HUVE with LPS, TNF, or IL-1 induces the synthesis and release or surface expression of a number of other inflammatory mediators including IL-1, IL-6, IL-8, monocyte chemotactic protein-1 (MCP-1), and platelet-activating factor (PAF).[104-108] The capacity of endothelium to release IL-1 when stimulated *in vitro* suggests the possibility that an inflammatory response initiated *in vivo* by the local release of IL-1 could be amplified at the endothelial surface in a positive feedback loop.[104] IL-6 (IFN-β_2) subserves numerous host defense mechanisms including the stimulation of acute phase reactants.[109] IL-8 and the structurally related protein, MCP-1, are potent chemoattractants for neutrophils and monocytes, respectively.[106,107,110-112] These proteins may play an important role in the recruitment of phagocytes to sites of inflammation. Phospholipase products such as PAF and eicosanoids are also released from endothelial cells stimulated with LPS and these mediators may have multiple effects on inflammatory reactions.[113] Thus, in many respects, the endothelial cell response to LPS is similar to that of mononuclear phagocytes. Both cells may function as central coordinators of a complex series of reactions that characterize inflammation.

C. VASCULAR REACTIVITY

Endothelial cells synthesize and secrete a number of vasoactive mediators *in vitro* including prostacyclin, PAF, endothelin-1, and endothelium-derived relaxing factor (EDRF).[5,114] Recent studies suggest an important role for EDRF in endotoxin-mediated shock. EDRF is a labile vasodilator (half-life = 6 s) that is released from endothelial cells of both artery and vein in response to stimuli including bradykinin, histamine, thrombin, and LPS.[115-117] EDRF induces vessel dilatation by increasing cyclic GMP levels in vascular smooth muscle cells.[114] Nitric oxide (NO) has been identified as the primary active component of EDRF.[118,119] In endothelial cells, NO is released from the terminal guanidino nitrogen of L-arginine in a reaction that is catalyzed by a constitutive and inducible NO synthase.[120] Glucocorticoids inhibited the expression of LPS-induced NO synthase in porcine aortic endothelial cells.[120] Also, N^G-monomethyl-L-arginine, a specific inhibitor of NO synthase, blocked relaxation of rat aortic segments induced by LPS, an effect that required the presence of endothelium.[121] Notably, N^G-monomethyl-L-arginine has been reported to reverse systemic hypotension induced by TNF[122] or LPS.[123] Although it has not been established that NO released in response to TNF or LPS is derived from endothelium, it appears that NO may be an important contributor to shock in endotoxemia.

IV. MECHANISMS OF ENDOTHELIAL CELL ACTIVATION

Endothelial cell activation *in vitro* involves initially the interaction of LPS, TNF, or IL-1 with the endothelial cell membrane. High-affinity cell surface receptors for TNF and IL-1 have been identified. Two distinct TNF receptors have been cloned and characterized with approximate molecular weights of 55 and 75 kDa.[124-129] Both contain a single transmembrane-spanning domain and an extracellular cysteine-rich domain homologous to nerve growth factor receptor and B-cell activation protein, CD40. A monoclonal antibody that binds to the 55-kDa receptor mimics TNF, indicating that internalization of TNF is not necessary for cell activation. Many cells express both types of TNF receptor, although the ratio of the 55- and 75-kDa forms varies on different cell types. Of interest, HUVE were reported to express only the 55-kDa form and could not be induced to express the 75-kDa form.[130] The TNF receptor(s) has been reported to be down regulated by protein kinase C (PKC) agonists in a variety of cell types, but not in HUVE.[131]

A membrane receptor for IL-1 has been cloned[132] and is a member of the immunoglobulin supergene family. The human endothelial cell IL-1 receptor, isolated by chemical cross-linking and immunoprecipitation, is similar in molecular weight to the IL-1 receptor on T-lymphocytes.[133]

Membrane proteins involved in the activation of cells by LPS have been identified. Recently, Tobias et al identified an LPS-binding protein (LBP) in rabbit acute phase serum.[134-136] Wright et al. showed that CD14, a 55-kDa glycan phosphatidylinositol-linked membrane protein, was the mononuclear phagocyte receptor for LPS-LBP complexes, and that some monoclonal antibodies to CD14 inhibited LPS-stimulated release of TNF.[137,138] Using chemical cross-linking techniques, Lei and Morrison have shown that LPS interacts specifically with an 80-kDa surface protein on murine lymphocytes and macrophages.[139,140] Macrophage-mediated tumor cytotoxicity similar to that induced by LPS was triggered by a monoclonal antibody to this LPS receptor.[141] These studies suggest that the activation of endothelial cells with LPS could also be mediated by a specific membrane receptor. The possible existence of an endothelial cell receptor for LPS is supported by observations on the effect of deacylated LPS on endothelial cells. Following deacylation by neutrophil acyloxyacyl hydrolase, LPS lacks the nonhydroxylated fatty acids that are attached by acyloxyacyl linkage to the ester- and amide-linked long-chain fatty acid moieties of lipid A.[38,39,142] We have demonstrated that deacylation rendered LPS inactive for HUVE as assessed by the inability of deacylated LPS to stimulate endothelial adhesiveness for neutrophils.[143] Incubation of HUVE with monophosphoryl lipid A and lipid X also failed to induce adhesiveness for neutrophils. However, only pretreatment of HUVE with deacylated LPS, but not lipid X or monophosphoryl lipid A, inhibited neutrophil adherence to HUVE induced by incubation of HUVE with LPS. The inhibition of LPS-induced activation of HUVE by deacylated LPS was specific, since deacylated LPS did not prevent TNF- or IL-1-induced adhesiveness for neutrophils.[143] These observations are consistent with the hypothesis that deacylated LPS competively inhibits the binding of LPS to one or more cell surface or intracellular targets. Alternatively, the inhibition of endothelial activation by deacylated LPS may be due to the interference of LPS binding to LBP or another plasma constituent.

Endothelial cell activation by LPS, TNF, or IL-1 *in vitro* is associated with the accumulation of a number of species of mRNA transcribed from genes encoding information for the inflammatory and procoagulant properties of endothelial cells. Increased levels of mRNA following stimulation can result from increased transcription (transcriptional activation), stabilization of a constitutively transcribed, unstable mRNA, or a combination of both mechanisms (as occurs, for example, in an IL-1-mediated increase in GM-CSF mRNA levels in HUVE).[101] ELAM-1-specific transcripts are not detectable in quiescent HUVE, but within 2 h of exposure to LPS, TNF, or IL-1 ELAM-1 mRNA is readily detectable by Northern blot analysis.[82,144] The appearance of ELAM-1 mRNA is transient even in the continued presence of LPS, TNF, or IL-1 and begins to decay by 6 h after stimulation. Of note, the accumulation of ELAM-1 mRNA in response to LPS, TNF, and IL-1 was completely blocked by H7,[144] a relatively specific inhibitor of PKC.[145] Inhibition of ELAM-1 transcription by H7 did not appear to be due to a generalized inhibition of transcription, since transcription of α-tubulin, a short-lived "housekeeping" gene, was not inhibited by H7 (Figure 2). In contrast, H7 and other PKC inhibitors have been shown to augment procoagulant activity in HUVE induced by LPS, TNF, and IL-1.[146] These findings indicate that HUVE activation in response to LPS, TNF, and IL-1 likely involves several intracellular mechanisms. Although results with these kinase inhibitors suggested that LPS, TNF, and IL-1 activation of HUVE is mediated by a PKC-dependent pathway, direct measurement of the translocation of intracellular PKC to the plasma membrane failed to confirm activation of PKC in HUVE by LPS, TNF, or IL-1.[144] These results suggest that the inhibition of ELAM-1 gene transcription, or the augmentation of procoagulant activity, by H7 is not due to blockade of PKC. It is possible that H7, which blocks the ATP-binding site in PKC, inhibits a homologous ATP-binding site in an unidentified kinase that is necessary for activation of HUVE.

Nuclear runoff experiments suggested that the increase in ELAM-1-specific mRNA in response to LPS, TNF, or IL-1 was due primarily to transcriptional activation of the ELAM-1 gene.[144] However, sequence analysis of the 3'-untranslated region (UTR) of the ELAM-1 gene reveals an AT-rich sequence.[82] Numerous genes, including GM-CSF, TNF, other cytokines and c-*fos* and c-*jun* possess AT-rich sequences in the 3'-UTR.[147] Previous studies have shown that the presence of an AU-rich sequence in the 3'-UTR confers instability to associated mRNA. Whether the UA-rich region of the 3'-UTR of ELAM-1 mRNA contributes in some way to the final accumulation of ELAM-1 mRNA in HUVE activated with LPS, TNF, or IL-1 is not yet known.

Similar to the induction of ELAM-1 gene transcription by TNF, Scarpati and Sadler reported that the expression of procoagulant activity in HUVE treated with TNF resulted from increased transcription of the tissue factor (TF) gene.[148] In contrast, levels of TF mRNA in HUVE treated with LPS increased predominantly by mRNA stabilization, probably involving an AU-sequence in the 3'-UTR of TF mRNA.[149] Since inhibition of protein synthesis with cycloheximide rapidly induced TF mRNA without increasing transcription, destabilization of TF mRNA may have resulted from a protein-RNA interaction involving the AU-rich sites and a constitutively synthesized, short-lived protein (e.g., a ribonuclease). It is possible that LPS induced accumulation of TF mRNA in HUVE by blocking, in some manner, this protein-RNA interaction.

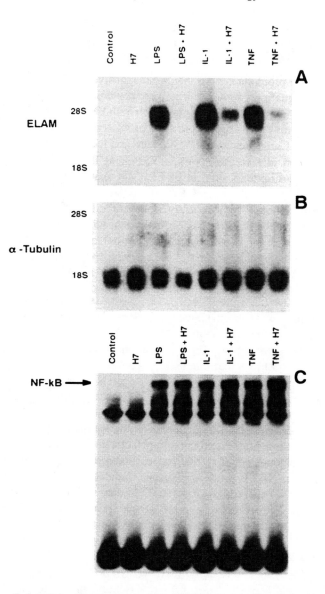

FIGURE 2. Endothelial cell activation by LPS, TNF, or IL-1 is associated with the accumulation of ELAM-1-specific transcripts and the appearance of a nuclear-binding protein, NF-κB in the nucleus. In A, HUVE were exposed to medium alone or medium containing LPS (100 ng/ml), TNF (100 μ/ml) or IL-1 (1 μ/ml) in the presence or absence of the protein kinase C inhibitor H7 (25 μM) for 4 h. Cells were then washed and total cellular RNA was isolated and electrophoresed. After transfer to nylon membranes, blots were probed with ^{32}P-labeled human ELAM-1 cDNA to identify ELAM-1-specific transcripts. In B, the same Northern blot as in A was probed with cDNA for human α-tubulin, a constitutive but relatively short-lived mRNA. In C, electrophoretic gel mobility shift assays were performed on cells under the same experimental conditions as in A, using crude nuclear protein extracts and a 20-bp ^{32}P-labeled oligonucleotide probe that contains the consensus sequence of the binding element for NF-κB.

A final common signaling pathway is one explanation for the virtually identical responses of HUVE to LPS, TNF, and IL-1. An alternative explanation is that HUVE genes activated by TNF, IL-1, or LPS may contain elements in the promoter region that are responsive to a number of different intracellular signaling pathways. This model would be analogous to the activation of the protooncogene c-*fos* in which both PKC- and cAMP-dependent signaling pathways converge on the c-*fos* promoter at SRF and CRE regions, respectively.[150] Consequently, we have examined factors that control ELAM-1 gene expression in response to LPS, TNF, and IL-1 in order to elucidate features of transcriptional control of ELAM-1 that may be common to the activation of other genes by these stimuli.

We[144] and others[151] have recently reported the DNA sequence for the 5'-flanking region of the ELAM-1 gene. Analysis of this region reveals consensus binding sequences for two known transcription factors, AP-1 and NF-κB. AP-1 includes the protein product of the c-*jun* protooncogene.[152-153] Activated AP-1 binds to a *cis*-acting element that mediates a transcriptional response to PKC agonists in genes such as collagenase,[154] stromelysin,[155] human metallothionein IIA,[156] and amyloid precursor protein.[157] NF-κB and a family of related proteins have been shown to increase transcription of a number of genes including immunoglobulin κ,[158] TNF,[159] and GM-CSF.[160] Although the mechanism(s) involved in NF-κB activation have not been completely identified, the observations that treatment of cultured cells with the PKC agonist, phorbol-12-myristate-13-acetate (PMA), suggests that PKC may play a key role in NF-κB activation, possibly via phosphorylation of its cytoplasmic inhibitor, IκB.[161] However, other studies have shown that TNF and IL-1 activate NF-κB through PKC-independent pathways.[162-164]

Using nuclear protein extracts from activated and control HUVE and ^{32}P-labeled oligonucleotide DNA probes incorporating the NF-κB binding sequence or an NF-κB binding sequence mutated at 2 base pairs, we have demonstrated NF-κB activation in HUVE treated with LPS, TNF, or IL-1.[144] Binding of the NF-κB probe to an NF-κB-like protein in HUVE was specific since binding to the mutant probe was not observed. Moreover, only LPS, TNF, or IL-1, but not IL-2, IL-4, IL-6, interferon-γ, histamine, or TGF-β, induced NF-κB-like activity in HUVE. In contrast, LPS, TNF, and IL-1 failed to induce AP-1-binding activity in HUVE. In preliminary experiments, deletion of the AP-1 site did not influence transcription when the 5'-flanking region of the ELAM-1 gene with this deletion was ligated to a reporter gene and transfected into TNF-activated HUVE (Montgomery and Pohlman, unpublished). However, induction of c-*jun* transcription in HUVE by TNF has been reported.[165] Of note, the PKC inhibitor H7, while inhibiting transcription of ELAM-1 gene, did not block activation of NF-κB in HUVE treated with LPS, TNF, or IL-1 (Figure 2). This observation together with the finding that LPS, TNF, or IL-1 do not directly activate PKC indicates that NF-κB activation in HUVE occurs by a PKC-independent mechanism. Furthermore,

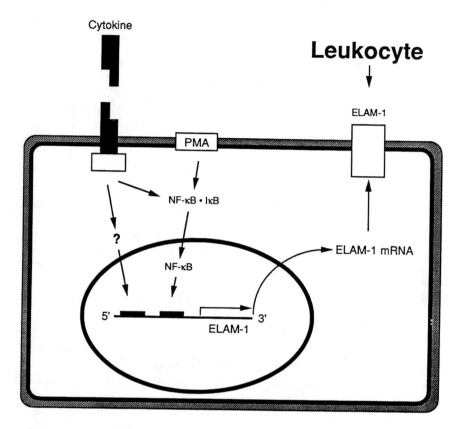

FIGURE 3. A model of ELAM-1 gene activation in human endothelial cells. The cytokines TNF and IL-1 bind to specific receptors at the endothelial surface. Signal transduction by a pathway not involving PKC results in the disassociation of NF-κB from IκB and migration of NF-κB to the nucleus where it binds to a specific sequence in the 5'-flanking region of ELAM-1 and perhaps other endothelial activation genes. Although NF-κB may be necessary for transcriptional activation of ELAM-1, it is not known what other pathways and nuclear-binding proteins are required for full promoter activity.

although induction of NF-κB activity may be necessary for LPS, TNF, or IL-1 activation of ELAM-1 gene transcription, the activity of other transcription factors may also be necessary for full promoter activity and ELAM-1 gene transcription. Based on available information, we propose a tentative model for transcriptional activation of ELAM-1 gene depicted in Figure 3.

V. SUMMARY

Endothelial cells are important participants in the host response to infection with Gram-negative organisms. Although direct interaction of LPS with endothelial cells can be demonstrated *in vitro*, endothelial responses during Gram-negative infection are likely mediated largely by the inflammatory cytokines, TNF, and IL-1, released by mononuclear phagocytes. Endothelial damage produced by LPS or the inflammatory response to LPS contributes significantly to organ dysfunction complicating Gram-negative sepsis. Activation of endothelial cells by LPS, TNF, or IL-1 stimulates leukocyte adherence, promotes procoagulant activity, and induces release of inflammatory mediators and vasoactive substances. Thus, as both critical targets of destructive inflammatory reactions and as potential effector cells in the inflammatory response, endothelial cells may profoundly influence the outcome and course of Gram-negative infection.

Available evidence suggests that the endothelial response to LPS, TNF, and IL-1 is highly coordinated at the level of gene transcription. However, the biochemical and molecular events responsible for gene activation by these inflammatory stimuli are only partially understood. Elucidation of these aspects of endothelial cell biology may suggest new therapeutic strategies to control deleterious host inflammatory responses induced by Gram-negative organisms.

ACKNOWLEDGMENTS

The authors are very grateful to Ms. Lenore D. Provan for her expert word processing skills and assistance in the preparation of the manuscript. This work was supported in part by USPHS Grant HL03174 and was performed during the tenure of an established investigator award of the American Heart Association.

REFERENCES

1. **Herwig, G., Esposito, C., and Stern, D. M.,** Modulation of endothelial hemostatic properties, an active role in the host response, *Annu. Rev. Med.,* 41, 15, 1990.
2. **Pober, J. S. and Cotran, R. S.,** The role of endothelial cells in inflammation, *Transplantation,* 50, 537, 1990.
3. **Pober, J. S. and Cotran, R. S.,** Cytokines and endothelial cell biology, *Physiol. Rev.,* 70, 427, 1990.
4. **Pober, J. S.,** Effects of tumour necrosis factor and related cytokines on vascular endothelial cells, in *Tumor Necrosis Factor and Related Cytotoxins,* (Ciba Foundation Symposium 131), Block, G. and Marsh, J., Eds., John Wiley & Sons, Chichester, England, 1987, 170.
5. **Vane, J. R., Anggard, E. E., and Botting, R. M.,** Regulatory functions of the vascular endothelium, *N. Engl. J. Med.,* 323, 27, 1990.
6. **Epstein, F. H., Beutler, B., and Cerami, A.,** Cachectin, more than a tumor necrosis factor, *N. Engl. J. Med.,* 316, 379, 1987.
7. **Dinarello, C. A. and Savage, N.,** Interleukin-1 and its receptor, *Crit. Rev. Immunol.,* 9, 1, 1989.

8. **Jaffe, E. A., Nachman, R. L., Becker, C. G., and Minick, C. R.,** Culture of human endothelial cells derived from umbilical veins, *J. Clin. Invest.,* 52, 2745, 1973.
9. **Gimbrone, M. A., Jr., Cotran, R. S., and Folkman, J.,** Human vascular endothelial cells in culture: growth and DNA synthesis, *J. Cell Biol.,* 60, 673, 1974.
10. **Striker, G. E., Harlan, J. M., and Schwartz, S. M.,** Human endothelial cells *in vitro, Methods Cell Biol.,* 21A, 135, 1980.
11. **Tanaka, N., Nishimura, T., and Yoshiyuki, T.,** Histochemical studies on the cellular distribution of endotoxin of *Salmonella enteritidis* in mouse tissue, *J. Microbiol.,* 3, 191, 1991.
12. **Rubenstein, H. S., Fine, J., and Coons, A. H.,** Localization of endotoxin in the walls of the peripheral vascular system during lethal endotoxemia, *Proc. Soc. Exp. Biol. Med.,* 111, 458, 1962.
13. **Mathison, J. C. and Ulevitch, R. J.,** The clearance, tissue distribution, and cellular localization of intravenously injected lipopolysaccharide in rabbits, *J. Immunol.,* 123, 2133, 1979.
14. **Reidy, M. A. and Bowyer, D.,** Scanning electron microscopy: morphology of aortic endothelium following injury by endotoxin and during subsequent repair, *Atherosclerosis,* 26, 319, 1977.
15. **Gaynor, E.,** Increased mitotic activity in rabbit endothelium after endotoxin, *Lab. Invest.,* 24, 318, 1971.
16. **Gerrity, R. G., Richardson, M., Caplan, B. A., Cade, J. F., Hirsh, J., and Schwartz, C. J.,** Endotoxin-induced vascular endothelial injury and repair, *Exp. Mol. Pathol.,* 24, 59, 1975.
17. **Gaynor, E.,** The role of granulocytes in endotoxin-induced vascular injury, *Blood,* 41, 797, 1973.
18. **Pingleton, W. W., Coalson, J. J., and Guenter, C. A.,** Significance of leukocytes in endotoxic shock, *Exp. Mol. Pathol.,* 22, 183, 1975.
19. **Evensen, S. A., Pickering, R. J., Bathouta, J., and Shepro, D.,** Endothelial injury induced by bacterial endotoxin: effect of complement depletion, *J. Clin. Invest.,* 5, 463, 1975.
20. **Harlan, J. M., Harker, L. A., Reidy, M. A., Gajdusek, C. M., Schwartz, S. M., and Striker, G. E.,** Lipopolysaccharide-mediated bovine endothelial cell injury *in vitro, Lab. Invest.,* 48, 269, 1983.
21. **Sage, H., Tupper, J., and Bramson, R.,** Endothelial cell injury *in vitro* is associated with increased secretion of an M$_r$ 43,000 glycoprotein ligand, *J. Cell. Physiol.,* 127, 373, 1986.
22. **Morel, D. W., DiCorleto, P. E., and Chisolm, G.,** Modulation of endotoxin-induced endothelial cell toxicity by low density lipoprotein, *Lab. Invest.,* 55, 419, 1986.
23. **Gartner, S. L., Sieckmann, D. G., Kang, Y. K., Watson, L. P., and Homer, L.,** Effects of lipopolysaccharide, lipid A, lipid X, and phorbol ester on cultured bovine endothelial cells, *Lab. Invest.,* 59, 181, 1988.
24. **Pohlman, T. H., Winn, R. K., Callahan, K. S., Maier, R. V., and Harlan, J. M.,** A glycolipid precursor of bacterial lipopolysaccharide (lipid X) lacks activity against endothelial cells *in vitro* and is not toxic *in vivo, J. Surg. Res.,* 45, 228, 1988.
25. **Harlan, J. M., Harker, L. A., Striker, G. E., and Weaver, L. J.,** Effects of lipopolysaccharide on human endothelial cells in culture, *Thromb. Res.,* 29, 15, 1983.
26. **Pohlman, T. H. and Harlan, J. M.,** Human endothelial cell response to lipopolysaccharide, interleukin-1, and tumor necrosis factor is regulated by protein synthesis, *Cell. Immunol.,* 119, 41, 1989.
27. **Pennica, D., Nedwin, G. E., Hayflick, J. S., Seeburg, P. H., Derynck, R., Palladino, M. A., Kohr, W. J., Aggarwal, B. B., and Goeddel, D. V.,** Human tumour necrosis factor: precursor structure, expression and homology to lymphotoxin, *Nature,* 312, 724, 1984.

28. **Reid, T. R., Torti, F. M., and Ringold, G. M.,** Evidence for two mechanisms by which tumor necrosis factor kills cells, *J. Biol. Chem.,* 264, 4583, 1989.

29. **Sugarman, B. J., Aggarwal, B. B., Hass, P. E., Figari, I. S., Palladina, M. A., Jr., and Shepard, H. M.,** Recombinant human tumor necrosis factor-alpha: effects on proliferation of normal and transformed cells *in vitro, Science,* 230, 943, 1985.

30. **Vilcek, J., Palombella, V. J., Henriksen-DeStefano, D., Swenson, C., Feinman, R., Hirai, M., and Tujimoto, M.,** Fibroblast growth enhancing activity of tumor necrosis factor and its relationship to other polypeptide growth factors, *J. Exp. Med.,* 163, 632, 1986.

31. **Shalaby, M. R., Aggarwal, B. B., Rinderkecht, E., Svedersky, L. P., Finkle, B. S., and Palladino, M. A., Jr.,** Activation of human polymorphonuclear neutrophil functions by interferon-gamma and tumor necrosis factors, *J. Immunol.,* 135, 2069, 1985.

32. **Takeda, K., Iwamoto, S., Sugimoto, H., Takuma, T., Kawatani, N., Noda, M., Masaki, A., Morise, H., Arimura, H., and Konno, K.,** Identity of differentiation inducing factor and tumour necrosis factor, *Nature,* 323, 338, 1986.

33. **Miller, S. C., Ito, H., Blau, H. M., and Torti, F. M.,** Tumor necrosis factor inhibits human myogenesis *in vitro, Mol. Cell. Biol.,* 8, 2295, 1988.

34. **Torti, F. M., Dieckman, H. B., Beutler, B., Cerami, A., and Ringold, G. M.,** A macrophage factor inhibits adipocyte gene expression, an *in vitro* model of cachexia, *Science,* 339, 867, 1985.

35. **Cavender, D. E., Haskard, D. O., Joseph, B., and Ziff, M.,** Interleukin 1 increases the binding of human B and T lymphocytes to endothelial cell monolayers, *J. Immunol.,* 136, 203, 1986.

36. **Cavender, D. E., Haskard, O., Foster, N., and Ziff, M.,** Superinduction of T lymphocyte-endothelial cell (EC) binding by treatment of EC with interleukin and protein synthesis inhibitors, *J. Immunol.,* 138, 2149, 1987.

37. **Wong, G. H. and Goeddel, D. V.,** Induction of manganous superoxide dismutase by tumor necrosis factor: possible protective mechanism, *Science,* 242, 941, 1988.

38. **Munford, R. and Hall, C. L.,** Detoxification of bacterial lipopolysaccharides (endotoxins) by a human neutrophil enzyme, *Science,* 234, 203, 1986.

39. **Hall, C. L. and Munford, R. S.,** Enzymatic deacylation of the lipid A moiety of *Salmonella typhimurium* lipopolysaccharides by human neutrophils, *Proc. Natl. Acad. Sci. U.S.A.,* 80, 6671, 1983.

40. **Munford, R. S. and Hall, C. L.,** Uptake and deacylation of bacterial lipopolysaccharides by macrophages from normal and endotoxin-hyporesponsive mice, *Infect. Immun.,* 48, 464, 1985.

41. **Rose, W. C. and Bradley, S. G.,** Enhanced toxicity for mice of combinations of antibiotics with *Escherichia coli* cells or *Salmonella typhosa* endotoxin, *Infect. Immun.,* 4, 550, 1971.

42. **Bradley, S. G.,** Interactions between endotoxin and protein synthesis, in *Handbook of Endotoxin,* Vol. 3, Berry, L. J., Ed., Elsevier, Amsterdam, 1985, 340.

43. **Bradley, S. G., Adams, A. C., and Smith, M. C.,** Potentiation of the toxicity of mithramycin by bacterial lipopolysaccharide, *Antimicrob. Agents Chemother.,* 7, 322, 1975.

44. **Billiau, A.,** Interferons and inflammation, *J. Interferon Res.,* 7, 559, 1987.

45. **Sieff, C. A.,** Hematopoietic growth factors, *J. Clin. Invest.,* 79, 1549, 1987.

46. **Bevilacqua, M. P., Pober, J. S., Majeau, G. R., Cotran, R. S., and Gimbrone, M. A.,** Interleukin-1 (IL-1) induces biosynthesis and cell surface expression of procoagulant activity in human vascular endothelial cells, *J. Exp. Med.,* 160, 618, 1984.

47. **Bevilacqua, M. P., Pober, J. S., Wheeler, M. E., Cotran, R. S., and Gimbrone, M. A., Jr.,** Interleukin 1 (IL-1) activation of vascular endothelium, effects on procoagulant activity and leukocyte adhesion, *Am. J. Pathol.,* 121, 393, 1985.

48. **Nawroth, P. P., Handley, D. A., Esmon, C. T., and Stern, D. M.,** Interleukin 1 induces endothelial cell procoagulant while suppressing cell-surface anticoagulant activity, *Proc. Natl. Acad. Sci. U.S.A.,* 48, 3460, 1986.

49. **Schorer, A. E., Kaplan, M. E., Raog, H. R., and Moldow, C. F.,** Interleukin-1 stimulates endothelial cell tissue factor production and expression by a prostaglandin-independent mechanism, *Thromb. Haemostasis,* 56, 256, 1986.

50. **Nawroth, P. P. and Stern, D. M.,** Modulation of endothelial cell hemostatic properties by tumor necrosis factor, *J. Exp. Med.,* 163, 740, 1986.

51. **Moore, K. L., Andreoli, S. P., Esmon, N. L., Esmon, C. T., and Bang, N. U.,** Endotoxin enhances tissue factor and suppresses thrombomodulin expression of human vascular endothelium *in vitro, J. Clin. Invest.,* 79, 124, 1987.

52. **Colucci, M., Balconi, G., Lorenzet, R., Pietra, A., Locati, D., Donati, M. B., and Semeraro, N.,** Cultured human endothelial cells generate tissue factor in response to endotoxin, *J. Clin. Invest.,* 71, 1893, 1983.

53. **Crossman, D. C., Carr, D. P., Tuddenham, E. G. D., Pearson, J. D., and McVey, J. H.,** The regulation of tissue factor mRNA in human endothelial cells in response to endotoxin or phorbol ester, *J. Biol. Chem.,* 265, 9782, 1990.

54. **Bevilacqua, M. P., Pober, J. S., Majeau, G. R., Fiers, W., Cotran, R. S., and Gimbrone, M. A., Jr.,** Recombinant tumor necrosis factor induces procoagulant activity in cultured human vascular endothelium, characterization and comparison with the actions of interleukin 1, *Proc. Natl. Acad. Sci. U.S.A.,* 83, 4533, 1986.

55. **Hanss, M. and Collen, D.,** Secretion of tissue-type plasminogen activator and plasminogen activator inhibitor by cultured human endothelial cells, modulation by thrombin, endotoxin, and histamine, *J. Lab. Clin. Med.,* 109, 97, 1987.

56. **Bevilacqua, M. P., Schleef, R. R., Gimbrone, M. A., Jr., and Loskutoff, D. J.,** Regulation of the fibrinolytic system of cultured human vascular endothelium by interleukin-1, *J. Clin. Invest.,* 78, 587, 1986.

57. **Emeis, J. J. and Kooistra, T.,** Interleukin and lipopolysaccharide induce an inhibitor of tissue-type plasminogen activator *in vivo* and in cultured endothelial cells, *J. Exp. Med.,* 163, 1260, 1986.

58. **Nachman, R. L., Hajjar, K. A., Silverstein, R. L., and Dinarello, C. A.,** Interleukin 1 induces endothelial cell synthesis of plasminogen activator inhibitor, *J. Exp. Med.,* 163, 1595, 1986.

59. **Bauer, K. A., Cate, H., Barzegar, S., Spriggs, D. R., Sherman, M. L., and Rosenberg, R. D.,** Tumor necrosis factor infusions have a procoagulant effect on the hemostatic mechanism of humans, *Blood,* 74, 165, 1989.

60. **Van Der Poll, T., Buller, H. R., Cate, H., Wortel, C. H., Bauer, K. A., Sander, J. H., Deventer, V., et al.,** Activation of coagulation after administration of tumor necrosis factor to normal subjects, *N. Engl. J. Med.,* 322, 1622, 1990.

61. **Harlan, J. M.,** Consequences of leukocyte-vessel wall interactions in inflammatory and immune reactions, *Semin. Thromb. Hemostasis,* 13, 434, 1987.

62. **Harlan, J. M.,** Leukocyte-endothelial interactions, *Blood,* 65, 513, 1985.

63. **Carlos, T. M. and Harlan, J. M.,** Membrane proteins involved in phagocyte adherence to endothelium, *Immunol. Rev.,* 114, 6, 1990.

64. **Larson, R. S. and Springer, T. A.,** Structure and function of leukocyte integrins, *Immunol. Rev.,* 114, 180, 1990.

65. **Wright, S. D. and Detmers, P. A.,** Adhesion-promoting receptors on phagocytes, *J. Cell. Sci.,* 9(Suppl.), 99, 1988.

66. **Springer, T. A.,** Adhesion receptors of the immune system, *Nature,* 346, 425, 1990.

67. **Rothlein, R., Dustin, M. L., Marlin, S. D., and Springer, T. A.,** A human intercellular adhesion molecule (ICAM-1) distinct from LFA-1, *J. Immunol.,* 137, 1270, 1986.

68. **Simmons, D., Makgoba, M. W., and Seed, B.,** ICAM, an adhesion ligand of LFA-1 is homologous to the neural cell adhesion molecule NCAM, *Nature,* 331, 624, 1988.

69. **Staunton, D. E., Dustin, M. L., and Springer, T. A.,** Functional cloning of ICAM-2, a cell adhesion ligand for LFA-1 homologous to ICAM-1, *Nature,* 339, 61, 1989.

70. **Osborn, L., Hession, C., Tizard, R., Vassallo, C., Luthowskyl, S., Chi-Rosso, G., and Lobb, R.,** Direct expression cloning of vascular cell adhesion molecule 1, a cytokine-induced endothelial protein that binds to lymphocytes, *Cell,* 59, 1203, 1989.

71. **Dustin, M. L., Rothlein, R., Bhan, A. K., Dinarello, C. A., and Springer, T. A.,** Induction by IL 1 and interferon-gamma: tissue distribution, biochemistry, and function of a natural adherence molecule (ICAM-1), *J. Immunol.,* 137, 245, 1986.

72. **Pober, J. S., Gimbrone, M. A., Jr., Lapierre, L. A., Mendrick, D. L., Fiers, W., Rothlein, R., and Springer, T. A.,** Overlapping patterns of activation of human endothelial cells by interleukin 1, tumor necrosis factor, and immune interferon, *J. Immunol.,* 137, 1893, 1986.

73. **Pober, J. S., Bevilacqua, M. P., Mendrick, D. L., Lapierre, L. A., Fiers, W., and Gimbrone, M. A., Jr.,** Two distinct monkines, interleukin 1 and tumor necrosis factor, each independently induce biosynthesis and transient expression of the same antigen on the surface of cultured human vascular endothelial cells, *J. Immunol.,* 136, 1680, 1986.

74. **Pober, J. S., Lapierre, L. A., Stolpen, A. H., Brock, T. A., Springer, T. A., Fiers, W., Bevilacqua, M. P., Mendrick, D. L., and Gimbrone, M. A., Jr.,** Activation of cultured human endothelial cells by recombinant lymphotoxin, comparison with tumor necrosis factor an interleukins 1 species, *J. Immunol.,* 138, 3319, 1987.

75. **Renkonen, R.,** Regulation of intercellular adhesion molecule-1 expression on endothelial cells with correlation to lymphocyte-endothelial binding, *Scand. J. Immunol.,* 29, 717, 1989.

76. **Rice, G. E. and Bevilacqua, M. P.,** An inducible endothelial cell surface glycoprotein mediates melanoma adhesion, *Science,* 246, 1303, 1989.

77. **Rice, G. E., Munro, J. M., and Bevilacqua, M. P.,** Inducible cell adhesion molecule 110 (INCAM-110) is an endothelial receptor for lymphocytes, *J. Exp. Med.,* 171, 1369, 1990/1991.

78. **Carlos, T. M., Schwartz, B. R., Kovach, N. L., Yee, E., Rosa, M., Osborn, L., Chi-Rosso, G., Newman, B., Lobb, R., and Harlan, J. M.,** Vascular cell adhesion molecule-1 mediates lymphocyte adherence to cytokine-activated cultured human endothelial cells, *Blood,* 76, 965, 1990.

79. **Elices, M. J., Osborn, L., Takada, Y., Crouse, C., Luhowskyj., S., Hemler, M. E., and Lobvb, R. R.,** VACAM-1 on activated endothelium interacts with the leukocyte integrin VLA-4 at a site distinct from the VLA-4/fibrinonectin binding site, *Cell,* 60, 577, 1990.

80. **Geng, J. G., Bevilacqua, M. P., Moore, K. L., McIntyre, T. M., Prescott, S. M., Kim, J. M., Bliss, G. A., Zimmerman, G. A., and McEver, R. P.,** Rapid neutrophil adhesion to activated endothelium mediated by GMP-140, *Nature,* 343, 757, 1990.

81. **Bevilacqua, M. P., Pober, J. S., Mendrick, D. L., Cotran, R. S., and Gimbrone, M. A., Jr.,** Identification of an inducible endothelial-leukocyte adhesion molecule, *Proc. Natl. Acad. Sci. U.S.A.,* 84, 9238, 1987.

82. **Bevilacqua, M. P., Stangelin, S., Gimbrone, M. A., Jr., and Seed, B.,** Endothelial leukocyte adhesion molecule 1, an inducible receptor for neutrophils related to complement regulatory proteins and lectins, *Science,* 243, 1160, 1989.

83. **Hession, C., Osborn, L., Goff, D., Chi-Rosso, G., Vassallo, C., Pasek, M., Pittack, C., Tizard, R., Goelz, S., McCarthy, K., Hopple, S., and Lobb, R.,** Endothelial leukocyte adhesion molecule 1, direct expression cloning and functional interactions, *Proc. Natl. Acad. Sci. U.S.A.,* 87, 1673, 1990.

84. **Springer, T. A.,** The sensation and regulation of interactions with the extracellular environment: the cell biology of lymphocyte adhesion receptors, *Annu. Rev. Cell Biol.,* 6, 359, 1990.

85. **Phillips, M. L., Nudelman, E., Gaeta, F. C. A., Perez, M., Singhal, A. K., Hakomori, S., and Paulson, J. C.,** ELAM-1 mediates cell adhesion by recognition of a carbohydrate ligand, sialyl-Le[x], *Science,* 250, 1130, 1990.
86. **Walz, G., Aruffo, A., Kolanus, W., Bevilacqua, M., and Seed, B.,** Recognition by ELAM-1 of the sialyl-Le[x] determinant on myeloid and tumor cells, *Science,* 250, 1132, 1990.
87. **Larsen, E., Palabrica, T., Sajer, S., Gilbert, G. E., Wagner, D. D., Furie, B. C., and Furie, B.,** PADGEM-dependent adhesion of platelets to monocytes and neutrophils is mediated by a lineage-specific carbohydrate, LNF III (CD15), *Cell,* 63, 467, 1990.
88. **Lowe, J. B., Stoolman, L. M., Nair, R. P., Larsen, R. D., Berhend, T. L., and Marks, R. M.,** ELAM-1-dependent cell adhesion to vascular endothelium determined by a transfected human fucosyltransferase cDNA, *Cell,* 63, 474, 1990.
89. **Goelz, S. E., Hession, C., Goff, D., Griffiths, B., Tizard, R., Newman, B., Chi-Rosso, G., and Lobb, R.,** A gene that directs the expression of an ELAM-1 ligand, *Cell,* 63, 1349, 1990.
90. **Gamble, J. R., Harlan, J. M., Klebanoff, S. J., and Vadas, M. A.,** Stimulation of the adherence of neutrophils to umbilical vein endothelium by human recombinant tumor necrosis factor, *Proc. Natl. Acad. Sci. U.S.A.,* 82, 8667, 1985.
91. **Schleimer, R. P. and Rutledge, B. K.,** Cultured human vascular endothelial cells acquire adhesiveness for neutrophils after stimulation with interleukin 1, endotoxin, and tumor-promoting phorbol diesters, *J. Immunol.,* 136, 649, 1986.
92. **Pohlman, T. H., Stanness, K. A., Beatty, P. G., Ochs, H. D., and Harlan, J. M.,** An endothelial cell surface factor(s) induced *in vitro* by lipopolysaccharide, interleukin 1, and tumor necrosis factor-alpha increases neutrophil adherence by a CDw18-dependent mechanism, *J. Immunol.,* 136, 4548, 1986.
93. **Kyan-Aung, U., Haskard, D. O., Poston, R. N., Thornhill, M. H., and Lee, T. H.,** Endothelial leukocyte adhesion molecule-1 and intercellualr adhesion molecule-1 mediate the adhesion of eosinophils to endothelial cells *in vitro* and are expressed by endothelium in allergic cutaneous inflammation *in vivo, J. Immunol.,* 146, 521, 1991.
94. **Carlos, T., Kovach, N., Schwartz, B., Osborn, L., Rosa, G., Newman, B., Wayner, E., Lobb, R., and Harlan, J. M.,** Human monocytes bind to two cytokine-induced adhesive ligands on cultured human endothelial cells, ELAM-1 and VCAM-1, *Blood,* 77, 2266, 1991.
95. **Schimizu, Y., Shaw, S., Graber, N., Vankat Gopal, T., Horgan, J. J., Van Seventer, G. A., and Newman, W.,** Activation-independent binding of human memory T cells to adhesion molecule ELAM-1, *Nature,* 349, 799, 1991.
96. **Cotran, R. S., Gimbrone, M. A., Jr., and Pober, J. S.,** Induction and detection of a human endothelial activation antigen *in vivo, J. Exp. Med.,* 164, 661, 1986.
97. **Redl, H., Schlag, G., Dinges, H. P., Buurman, W., Rothlein, R., Pober, J., and Cotran, R.,** Endothelial activation after polytrauma and sepsis in the baboon, *Circ. Shock,* 31, 79, 1990.
98. **Quesenbery, P. J. and Gimbrone, M. A., Jr.,** Vascular endothelium as a regulator of granulopoiesis: production of colony-stimulating activity by cultured human endothelial cells, *Blood,* 56, 1060, 1980.
99. **Broudy, V. C., Kaushansky, K., Segal, G. M., Harlan, J. M., and Adamson, J. W.,** Tumor necrosis factor type alpha stimulates human endothelial cells to produce granulocyte/macrophage colony-stimulating factor, *Natl. Acad. Sci. U.S.A.,* 83, 7467, 1986.
100. **Broudy, V. C., Kaushansky, K., and Adamson, J. W.,** Interleukin 1 stimulates human endothelial cells to produce granulocyte-macrophage colony-stimulating factor angd granulocyte colony-stimulating factor, *J. Immunol.,* 139, 464, 1987.
101. **Kaushansky, K.,** Control of granulocyte-macrophage colony-stimulating factor production in normal endothelial cells by positive and negative regulatory elements, *J. Immunol.,* 143, 2525, 1989.

102. **Gasson, J. C.,** Molecular physiology of granulocyte-macrophage colony-stimulating factor, *Blood,* 77, 1131, 1991.
103. **Cannistra, S. A., Rambaldi, A., Spriggs, D. R., Hermann, F., Klute, D., and Griffin, J. D.,** Human granulocyte-macrophage colony-stimulating factor induces expression of the tumor necrosis factor gene by the U937 cell line and by normal human monocytes, *J. Clin. Invest.,* 79, 1720, 1987.
104. **Warner, S. J. C., Auger, K. R., and Libby, P.,** Interleukin 1 induces interleukin 1. II. Recombinant human interleukin 1 induces interleukin 1 production by adult human vascular endothelial cells, *J. Immunol.,* 139, 1911, 1987.
105. **Shalaby, M. R., Waage, A., and Espevik, T.,** Cytokine regulation of interleukin 6 production by human endothelial cells, *Cell. Immunol.,* 121, 372, 1989.
106. **Strieter, R. M., Kunkel, S. L., Showell, H. J., Remick, D. G., Phan, S. H., Ward, P. A., and Marks, R.,** Endothelial cell gene expression of a neutrophil chemotactic factor by TNF, IL-1, and LPS, *Science,* 234, 1467, 1989.
107. **Rollins, B. J., Yoshimura, T., Leonard, E. J., and Pober, J. S.,** Cytokine-activated human endothelial cells synthesize and secrete a monocyte chemoattractant, MCP-1/JE, *Am. J. Pathol.,* 136, 1229, 1990.
108. **Bussolino, F., Camussi, G., and Baglioni, C.,** Synthesis and release of platelet-activating factor by human vascular endothelial cells treated with tumor necrosis factor or interleukin 1α, *J. Biol. Chem.,* 263, 11856, 1988.
109. **Heinrich, P. C., Castell, J. V., and Andus, T.,** Interleukin-6 and the acute phase response, *Biochem. J.,* 265, 621, 1990.
110. **Strieter, R. M., Wiggins, R., Phan, S. H., Wharram, B. L., Showell, H. J., Remick, D. G., Chensue, S. W., and Kunkel, S. L.,** Monocyte chemotactic protein gene expression by cytokine-treated human fibroblasts and endothelial cells, *Biochem. Biophys. Res. Commun.,* 162, 694, 1989.
111. **Leonard, E. J. and Yoshimura, T.,** Neutrophil attractant/activation protein-1 (NAP-1) (interleukin-8), *Am. J. Respir. Cell Mol. Biol.,* 2, 479, 1990.
112. **Djeu, J. Y.,** Functional activation of human neutrophils by recombinant monocyte-derived neutrophil chemotactic factor/IL-8, *J. Immunol.,* 144, 2205, 1990.
113. **Harlan, J. M. and Winn, R. K.,** The role of phospholipase products in the pathogenesis of vascular injury in sepsis, in *Contemporary Issues in Infectious Diseases,* Vol. 4, Root, R. K. and Sande, M. A., Eds., Churchill Livingstone, NY, 1985, 83.
114. **Furchgott, R. F. and Vanhouttee, P. M.,** Endothelium-derived relaxing and contracting factors, *FASEB J.,* 3, 2007, 1989.
115. **Gryglewski, R. J., Botting, R. G., and Vane, J. R.,** Mediators produced by the endothelial cell, *Hypertension,* 12, 530, 1988.
116. **Fleming, I., Gray, G. A., Julou-Schaeffer, G., Parratt, J. R., and Stoclet, J. C.,** Incubation with endotoxin activates the L-arginine pathway in vascular tissue, *Biochem. Biophys. Res. Commun.,* 171, 562, 1990.
117. **Salvemini, D., Korbut, R., Anggard, E., and Vane, J.,** Immediate release of a nitric oxide-like factor from bovine aortic endothelial cells by *Escherichia coli* lipopolysaccharide, *Proc. Natl. Acad. Sci. U.S.A.,* 87, 2593, 1990.
118. **Palmer, R. M. J., Ferrige, A. G., and Moncada, S.,** Nitric oxide release accounts for the biological activity of endothelium-derived relaxing factor, *Nature,* 327, 524, 1987.
119. **Ignarro, L. J., Buga, G. M., Wood, K. S., Byrns, R. E., and Chaudhuri, G.,** Endothelium-derived relaxing factor produced and released from artery and vein is nitric oxide, *Proc. Natl. Acad. Sci. U.S.A.,* 84, 9265, 1987.
120. **Radomski, M. W., Palmer, R. M. J., and Moncada, S.,** Glucocorticoids inhibit the expression of an inducible, but not the constitutive, nitric oxide synthase in vascular endothelial cells, *Proc. Natl. Acad. Sci. U.S.A.,* 87, 10043, 1990.
121. **Fleming, I., Gray, G. A., Julou-Schaeffer, G., Parratt, J. R., and Stoclet, J. C.,** Incubation with endotoxin activates the L-arginine pathway in vascular tissue, *Biochim. Biophys. Res. Commun.,* 171, 562, 1990.

122. **Kilbourn, R. G. and Belloni, P.,** Endothelial cell production of nitrogen oxides in response to interferon γ in combination with tumor necrosis factor, interleukin-1, or endotoxin, *J. Natl. Cancer Inst.,* 82, 772, 1990.

123. **Kilbourn, R. G., Jubran, A., Gross, S. S., Griffith, O. W., Levi, R., Adams, J., and Lodata, R. G.,** Reversal of endotoxin-mediated shock by N^G-methyl-L-arginine, an inhibitor of nitric oxide synthesis, *Biochem. Biophys. Res. Commun.,* 172, 1132, 1990.

124. **Loetscher, H., Schlaeger, E. J., Lahm, H. W., Pan, Y. C., Lesslauer, W., and Brockhaus, M.,** Purification and partial amino acid sequence analysis of two distinct tumor necrosis factor receptors from HL60 cells, *J. Biol. Chem.,* 265, 20131, 1990.

125. **Loetscher, H., Pan, Y. C. E., Lahm, H. W., Gentz, R., Brockhaus, M., Tabuchi, H., and Lesslauer, W.,** Molecular cloning and expression of the human 55 kd tumor necrosis factor receptor, *Cell,* 61, 351, 1990.

126. **Schall, T. J., Lewis, M., Koller, K. J., Lee, A., Rice, G. C., Wong, G. H. W., Gatanaga, T., Granger, G. A., Lentz, R., Raab, H., Kohr, W. J., and Goeddel, D. V.,** Molecular cloning and expression of a receptor for human tumor necrosis factor, *Cell,* 61, 361, 1990.

127. **Gray, P. W., Barrett, K., Chantry, D., Turner, M., and Feldmann, M.,** Cloning of human tumor necrosis factor (TNF) receptor cDNA and expression of recombinant soluble TNF-binding protein, *Proc. Natl. Acad. Sci. U.S.A.,* 87, 7380, 1990.

128. **Smith, C. A., Davis, T., Anderson, D., Solam, L., Beckmann, M. P., Jerzy, R., Dower, S. K., Cosman, D., and Goodwin, R. G.,** A receptor for tumor necrosis factor defines an unusual family of cellular and viral proteins, *Science,* 248, 1019, 1990.

129. **Nophar, Y., Kemper, O., Brakebusch, C., Engelmann, H., Zwang, R., Aderka, D., Holtmann, H., and Wallach, D.,** Soluble forms of tumor necrosis factor receptors (TNF-Rs). The cDNA for the type 1 TNF-R, cloned using amino acid sequence data of its soluble form, encodes both the cell surface and a soluble form of the receptor, *EMBO J.,* 9, 3269, 1990.

130. **Hohmann, H. P., Brockhaus, M., Baeuerle, P. A., Remy, R., Kolbeck, R., and van Loon, A. P.,** Expression of the types A and B tumor necrosis factor (TNF) receptors is independently regulated, and both receptors mediate activation of the transcription factor NF-κB, *J. Biol. Chem.,* 265, 22409, 1990.

131. **Aggarwal, B. B. and Eessalu, T. E.,** Effect of phorbol esters on down-regulation and redistribution of cell surface receptors for tumor necrosis factor-α, *J. Biol. Chem.,* 262, 16450, 1987.

132. **Sims, J. E., Acres, R. B., Grubin, C. E., McMahan, C. J., Wignall, J. M., March, C. J., and Dower, S. K.,** Cloning the interleukin 1 receptor from human T cells, *Proc. Natl. Acad. Sci. U.S.A.,* 22, 86, 1989.

133. **Thieme, T. R. and Wagner, C. R.,** The molecular weight of the endothelial cell IL-1 receptor is 78,000, *Mol. Immunol.,* 26, 249, 1989.

134. **Tobias, P. S., Soldau, K., and Ulevitch, R. J.,** Identification of a lipid A binding site in the acute phase reactant lipopolysaccharide binding protein, *J. Biol. Chem.,* 264, 10867, 1989.

135. **Tobias, P. S., Soldau, K., and Ulevitch, P. J.,** Isolation of a lipopolysaccharide-binding acute phase reactant from rabbit serum, *J. Exp. Med.,* 164, 777, 1986.

136. **Tobias, P. S., McAdam, K. P., Soldau, K., and Ulevitch, R. J.,** Control of lipopolysaccharide-high-density lipoprotein interactions by an acute-phase reactant in human serum, *Infect. Immun.,* 50, 73, 1985.

137. **Wright, S. D., Ramos, R. A., Tobias, P. S., Ulevitch, R. J., and Mathison, J C.,** CD14, a receptor for complexes of lipopolysaccharide (LPS) and LPS binding protein, *Science,* 249, 1431, 1990.

138. **Wright, S. D., Tobias, P. S., Ulevitch, R. J., and Ramos, R. A.**, Lipopolysaccharide (LPS) binding protein opsonizes LPS-bearing particles for recognition by a novel receptor on macrophages, *J. Exp. Med.*, 170, 1231, 1989.

139. **Lei, M. G. and Morrison, D. G.**, Specific endotoxic lipopolysaccharide-binding proteins on murine splenocytes. I. Detection of lipopolysaccharide-binding sites on splenocytes and splenoctye subpopulations, *J. Immunol.*, 141, 996, 1988.

140. **Lei, M. G. and Morrison, D. C.**, Specific endotoxic lipopolysaccharide-binding proteins on murine splenocytes. II. Membrane localization and binding characteristics, *J. Immunol.*, 141, 1006, 1988.

141. **Chen, T. Y., Bright, S. W., Pace, J. L., Russell, S. W., and Morrison, D. C.**, Induction of macrophage-mediated tumor cytotoxicity by a hamster monoclonal antibody with specificity for lipopolysaccharide receptor, *J. Immunol.*, 145, 8, 1990.

142. **Munford, R. S.**, Deacylation of bacterial endotoxins by neutrophils and macrophages: early observations and hypotheses, *Methods Exp. Clin. Pharmacol.*, 8, 63, 1986.

143. **Pohlman, T. H., Munford, R. S., and Harlan, J. M.**, Deacylated lipopolysaccharide inhibits neutrophil adherence to endothelium induced by lipopolysaccharide *in vitro*, *J. Exp. Med.*, 165, 1393, 1987.

144. **Montgomery, K. F., Osborn, L., Hession, C., Tizard, R., Goff, D., Vassalo, C., Tarr, P. I., Bomsztyk, K., Lobb, R., Harlan, J. M., and Pohlman, T. H.**, Endothelial leukocyte adhesion molecule 1 (ELAM-1) gene expression: roles of NF-kB, protein kinase C, and AP-1, *Proc. Natl. Acad. Sci. U.S.A.*, 88, 6523, 1991.

145. **Hidaka, H., Inagaki, M. K., and Sasaki, Y.**, Isoquinolinesulfonamides, novel and potent inhibitors of cyclic nucleotide dependent protein kinase and protein kinase C, *Biochemistry*, 23, 5036, 1984.

146. **Zuckerman, S. H. and Surprenant, Y. M.**, Augmentation of procoagulant activity in monokine stimulated human endothelial cells by calmodulin/protein kinase C inhibitors, *Thromb. Res.*, 40, 205, 1988.

147. **Shaw, G. and Kamen, R. A.**, A conserved AU sequence from the 3' untranslated region of GM-CSF mRNA mediates selective mRNA degradation, *Cell*, 46, 659, 1986.

148. **Scarpati, E. M. and Sadler, J. E.**, Regulation of endothelial cell coagulant properties: modulation of tissue factor, plasminogen activator inhibitors, and thrombomodulin by phorbol 12-myristate 13-acetate and tumor necrosis factor, *J. Biol. Chem.*, 264, 20705, 1989.

149. **Crossman, D. C., Carr, D. P., Tuddenham, G. D., Pearson, J. D., and McVey, J. H.**, The regulation of tissue factor mRNA in human endothelial cells in response to endotoxin or phorbol ester, *J. Biol. Chem.*, 265, 9782, 1990.

150. **Verman, M., Ransone, L. J., Visvader, J., Sassone-Corsi, P., and Lamph, V. W.**, *Fos-jun* conspiracy: implications for the cell, in *Proto-Oncogenes in Cell Development*, (Ciba Foundation Symp. No. 150), Block, G. and Marsh, J., Eds., John Wiley & Sons, Chichester, England, 1990, 128.

151. **Collins, T., Williams, A., Johnston, G. I., Kim, J., Eddy, R., Shows, T., Gimbrone, A., Jr., and Bevilacqua, M. P.**, Structure and chromosomal location of the gene for endothelial-leukocyte adhesion molecule 1, *J. Biol. Chem.*, 266, 2466, 1991.

152. **Bohmann, D., Bos, T. J., Admon, A., Nishimura, T., Vogt, P. K., and Tigan, R.**, Human proto-oncogene c-jun encodes a DNA binding protein with structural and functional properties of transcription factor AP-1, *Science*, 238, 1386, 1987.

153. **Angel, P., Allegretto, E. A., Okino, S. T., Hattori, K., Boyle, W. J., Hunter, T., and Karin, M.**, Oncogene jun encodes a sequence-specific trans-activator similar to AP-1, *Nature*, 332, 166, 1988.

154. **Whitham, S. E., Murphy, G., Angel, O. P., Rahmsdorf, H. J., Smith, B. J., Lyons, A., Harris, T. J. R., Herrlich, P., and Docherty, A. J. P.**, Comparison of human stromelysin and collagenase by cloning and sequence analysis, *Biochem. J.*, 240, 913, 1986.

155. **Matrisian, L. M., Leroy, P., Ruhlmann, C., Gesnel, C., and Breathnach, R.**, Isolation of the oncogene and epidermal growth factor-induced transin gene: complex control in rat fibroblasts, *Mol. Cell. Biol.*, 6, 1679, 1986.

156. **Angel, P., Imagawa, M., Chiu, R., Stein, B., Imbra, R. J., Rahmsdorf, H. J., Jonat, C., Herrlich, P., and Karin, M.**, Phorbol ester-inducible genes contain a common *cis* element recognized by a TPA-modulated *trans*-acting factor, *Cell*, 49, 729, 1987.

157. **Goldgaber, D., Harris, H. W., Hla, T., Maciag, T., Donnelly, R. J., Jacobsen, J. S., Vitek, M. P., and Gajdusek, D. C.**, Interleukin 1 regulates synthesis of amyloid beta-protein precursor mRNA in human endothelial cells, *Proc. Natl. Acad. Sci. U.S.A.*, 86, 7606, 1989.

158. **Sen, R. and Baltimore, D.**, Multiple nuclear factors interact with the immunoglobulin enhancer sequences, *Cell*, 46, 705, 1986.

159. **Shakhov, A. N., Collart, M. A., Vassalli, P., Nedospasov, S. A., and Jongeneel, C. V.**, Kappa B-type enhancers are involved in lipopolysaccharide-mediated transcriptional activation of the tumor necrosis factor alpha gene in primary macrophages, *J. Exp. Med.*, 171, 35, 1990.

160. **Schreck, R. and Baeuerle, P. A.**, NF-κB as inducible transcriptional activator of the granulocyte-macrophage colony-stimulating factor gene, *Mol. Cell. Biol.*, 10, 1281, 1990.

161. **Gosh, S. and Baltimore, D.**, Activation *in vitro* of NF-κB by phosphorylation of its inhibitor IκB, *Nature*, 344, 678, 1990.

162. **Osborn, L., Kunkel, S., and Nabel, G. J.**, Tumor necrosis factor alpha and interleukin 1 stimulate the human immunodeficiency virus enhancer by activation of the nuclear factor κB, *Proc. Natl. Acad. Sci. U.S.A.*, 86, 2336, 1989.

163. **Meichle, A., Schutze, S., Hensel, G., Brunsing, D., and Kronke, M.**, Protein kinase C-independent activation of nuclear factor κB by tumor necrosis factor, *J. Biol. Chem.*, 265, 8339, 1990.

164. **Bomsztyk, K., Rooney, J. E., Iwasaki, T., Rachie, N. A., Dower, S. K., and Sibley, C. H.**, Evidence that interleukin-1 and phrobol esters activate NF-κB by different pathways: role of protein kinase, *Cell Regul.*, 2, 329, 1991.

165. **Dixit, V. M., Marks, R. M., Sarma, V., and Prochownik, E. V.**, The antimitogenic action of tumor necrosis factor is associated with increased AP-1/c-*jun* proto-oncogene transcription, *J. Biol. Chem.*, 264, 16905, 1989.

Chapter 16

REGULATION OF CELL SURFACE RECEPTOR EXPRESSION BY LPS

Aihao Ding and Carl Nathan

TABLE OF CONTENTS

I. INTRODUCTION

Cell-cell communication by contact, as well as by soluble molecules such as cytokines, autacoids, most hormones, some cascade proteases, and certain plasma proteins, depends on receptors at the surface of responding cells. Among the myriad actions of LPS is its ability to affect cellular expression of receptors for signaling molecules from all these classes. This is one of the major if indirect ways in which lipopolysaccharide (LPS) affects cell function. Below we list many examples of this phenomenon that have emerged in recent years, and discuss several that illustrate our growing appreciation of the compexity of host responses to LPS.

II. *IN VIVO* STUDIES

Table 1 summarizes studies in which LPS has been administered to experimental animals, after which alterations were found in cell surface receptors for hormones, autacoids, or (poly)peptides. Discussion of two examples will underscore the point, that the obvious advantage of *in vivo* studies — increased potential for pathophysiologic relevance — tends to be offset by the difficulty of discerning primary from secondary or even tertiary effects.

A. ADRENERGIC HORMONE RECEPTORS

One of the metabolic hallmarks of septic shock is marked disturbance of glucose homeostasis, characterized by rapid depletion of liver glycogen, impaired glycogenesis, accelerated glycogenolysis, and depressed gluconeogenesis. Analysis of the underlying mechanisms led to the observation that administration of LPS to rats, guinea pigs, or dogs differentially affected the number of α- or β-adrenergic receptors in different tissues.

Alterations of binding of adrenergic ligands by the liver is of special interest, since this organ plays a critical role in glucose homeostasis, and contains the majority of tissue-bound radioactivity after an i.v. injection of radiolabeled LPS.[10] Liu and Ghosh[3] found a decrease in [^{3}H]-dihydroalprenolol binding to dog liver plasma membranes 2 h following endotoxin administration, reflecting a decrease both in the number of β-adrenergic receptors and their affinity. Impairment of β-adrenergic receptors and hyperreactivity of α$_1$-adrenergic receptors were also found in guinea pig heart after i.p. injection of endotoxin.[2] In a nonshock rat model of continuous endotoxin infusion via an implanted osmotic pump, Roth and Spitzer[1] found decreases in the number both of α$_1$-adrenergic receptors and vasopressin receptors, without changes in their affinity. Mechanisms considered for these effects include (1) ligand-induced down-regulation, as elevated catecholamine levels can be found in septic shock; (2) altered fluidity of membrane lipids; or (3) activation of protein kinase C, given that other protein kinase C activators (phorbol esters) cause a similar effect.[1,3]

TABLE 1
Effect of LPS on Heterologous Receptors *In Vivo*

Receptor	Species	Tissue/cell	Changes[a]	Assay	Ref.
α_1-Adrenergic R	Rat	Liver membrane	−	L-binding[b]	1
	Guinea pig	Heart vascular	+	Functional	2
β_1-Adrenergic R	Dog	Liver membrane	−	L-binding	3
	Guinea pig	Heart vascular	−	Functional	2
	Guinea pig	Spleen lymphocyte	−	L-binding	4, 5
Vasopressin R	Rat	Liver membrane	−	L-binding	1
Insulin R	Rat	Liver membrane	+	L-binding	6
Glucagon R	Rat	Liver membrane	+	L-binding	6
Tissue factor	Rabbit	Peritoneal leukocyte	+	Functional	7
CR1	Mouse	Peritoneal mϕ[c]	+	Functional	8
LTB$_4$ R	Rabbit	Liver membrane	−	L-binding	9
C5a R	Rabbit	Liver membrane	−	L-binding	9
fMLP R	Rabbit	Liver Membrane	+	L-binding	9

[a]   (+) indicates an increase, (−) a decrease.
[b]   By radioligand binding.
[c]   Mϕ, macrophage.

B. MACROPHAGE RECEPTORS FOR COMPLEMENT COMPONENT C3b

Griffin and Mullinax[8] detected LPS-induced alterations in the macrophage C3b receptor (CR1) by examining the cells' phagocytic ability. CR1 is normally immobilized in the plane of the plasma membrane and is unable to promote phagocytosis after binding erythrocytes coated with this ligand. Exposure to high concentrations of LPS *in vivo* and *in vitro* activated CR1, in the sense that CR1-dependent binding was followed by ingestion. However, the phenomenon was not reproduced during bacterial infection nor by exposure to lower concentrations of LPS, i.e., 100 ng/ml. *In vitro*, LPS-dependent activation of CR1 required the presence of T lymphocytes, and was mediated by a factor elaborated by the T cells in response to LPS.[8] This example of an indirect effect involving two different cell types illustrates the complexity of mechanisms involved after administration of LPS to whole animals.

III. *IN VITRO* STUDIES

Characteristics of responses by different cell types. Most studies on regulation of receptors by LPS have focused on relatively homogeneous populations of three kinds of cells *in vitro:* B cells, macrophages, or neutrophils.

LPS induces polyclonal proliferation and/or differentiation of resting mouse splenic B cells into immunoglobulin-producing cells. During these responses, LPS promotes the expression of Ia antigen, Fc receptors (FcR), IL-4 receptors, and transferrin receptors (Tf-R) (Table 2). The physiological role of B-cell

TABLE 2
Effects of LPS on Receptors of B Lymphocytes

Receptor	Dose (μg/ml)	Changes	Assay	Ref.
IL-4 R		+	L-binding	11
IL-4 R	10	+	L-binding	12
Ia	25	+	FACS	13
TfR	25	+	FACS	13
FcR	20–50	+	mAb binding	14, 15
FcR	10	+	FACS, Northern	16

Note: All the studies described here were carried out with mouse splenic B cells. See notes in Table 1.

responses to LPS has been challenged on the grounds that microgram per milliliter concentrations of LPS are generally required for these changes.

In contrast, both human and mouse macrophages respond to concentrations of LPS six orders of magnitude lower than that required to activate B cells, and display a greater diversity of responses, ranging from stimulatory to suppressive. On the one hand, LPS can activate macrophages by itself or in combination with IFN-γ, as evinced by ehanced tumoricidal activity and the ability to release the cytotoxic molecules tumor necrosis factor-α (TNF-α), interleukin-1 (IL-1), and nitric oxide (NO).[17] On the other hand, LPS can antagonize the action of IFN-γ, thereby suppressing other features of macrophage activation, such as the ability to generate reactive oxygen intermediates.[18] Some of these responses are thought to be central to the mediation of LPS-induced fever, hypotension, and coagulopathy.

Insofar as LPS effects on macrophage cell surface receptors are concerned (Table 3), most published data reveal suppressive effects, e.g., on class II (Ia) histocompability antigens, FcR, interleukin 6 receptors (IL-6R), receptors for colony-stimulating factor for macrophages (CSF-M-R), tumor necrosis factor receptors (TNF-R), mannose receptors (Man-R), and receptors for modified low-density lipoproteins (scavenger receptors). The most notable exception is tissue factor, which is undetectable in circulating blood cells, but is induced in monocytes by LPS.

Human neutrophils (PMN), like monocytes/macrophages, are exquisitely sensitive to LPS.[38] Exposure of PMNs to 1 to 10 ng/ml of LPS "primed" the cells to release more superoxide anion than control cells upon subsequent stimulation with fMLP. In part, this effect may reflect an LPS-induced increase of fMLP receptors on the cell surface of human PMNs.[39]

A. TISSUE FACTOR EXPRESSION

Tissue factor (TF) initiates the coagulation protease cascade by acting as the receptor and essential cofactor for coagulation factor VII and VIIa. TF

TABLE 3
Effects of LPS on Receptors of Macrophages

Receptor	Species/tissue	Dose	Changes	Assay	Ref.
Tissue factor	Human/monocytes	1–10 μg/ml	+	Functional	19, 20
	Rabbit/liver mφ	10 ng/ml	+	Functional	21
Ia	Mouse/mφ	100 ng/ml	−	Northern	22
		1 μg/ml	−	Functional	23
	Human/monocytes	20 μg/ml	−	Functional	24
FcR	Guinea pig/peritoneal	5 μg/ml	−	L-binding	25
	Human/monocytes	0.1 ng/ml	−	FACS	26
Scavenger R	Human/monocytes	1–200 ng/ml	−	L-degradation	27, 28
		100 μg/ml	−	L-degradation	29
TNF-R	Mouse/mφ	1–10 ng/ml	−	L-binding	30, 31
	Human/mφ	10 ng/ml	−	L-binding	30
IL-2 R	Human/monocytes	25 ng/ml	+	FACS	32
IL-6 R	Human/monocytes	10 ng/ml	−	Northern	33
CSF-M R	Mouse/mφ	10 ng/ml	−	L-binding	34, 35
		1 μg/ml	−	Northern N runoff	36
Man-R	rat/b.m. mφ	0.1 ng/ml	−	L-binding	37

Note: See notes in Table 1.

membrane glycoprotein is normally undetectable in circulating blood cells and endothelium, but is selectively expressed in certain tissues, particularly in vascular adventitial cells. Mononuclear leukocytes can be induced to express TF by a variety of stimuli, including LPS both *in vivo*[7] and *in vitro*.[40,41] This response has been implicated in the pathogenesis of disseminated intravascular coagulation associated with bacterial infection.

By fractionating human mononuclear cells and determining the relative procoagulant activity (PCA) in different cell subpopulations, Edwards et al.[19] demonstrated that monocytes are responsible for the bulk of mononuclear cell TF generation *in vitro*. The presence of nonadherent cells, however, greatly enhanced the capacity of adherent monocytes to generate TF in response to LPS. These results suggested that the induction of TF in monocytes is a complex process involving multiple cellular interactions.

Studies by Edgington and co-workers[20,42,43] added new insight into this collaborative cellular procoagulant response. They separated human peripheral blood lymphocytes into Fcγ-positive (Tγ) and Fcμ-positive (Tμ) cells by cytoaffinity, and showed that Tμ supported, but Tγ suppressed, the induction of monocyte PCA by LPS. More than 90% of TF activity of both intact and lysed cells could be blocked by adding anti-TF monoclonal antibodies to intact cells, and all surface membrane TF was capable of binding the ligand factor VII. These observations militated against the possibility that TF is stored intracellularly or is present in a cryptic form in the membrane before activation by LPS. Instead, the primary regulation of TF in monocytes appears to be at the transcriptional level, with additional control at the level of mRNA degradation.[43]

Endothelial cells can also respond to LPS by generating cell-associated TF. Exposure of freshly prepared cells from human umbilical veins to 10 to 100 ng/ml LPS for 4 to 6 h at 37°C resulted in 40- to 100-fold increase in PCA.[44-46] The presence of platelets, monocytes, granulocytes, or lymphocytes greatly enhanced lPS-induced PCA in endothelial cells. This may be one way in which endothelial cells contribute to thrombogenic events associated with bacterial infections.

B. MAJOR HISTOCOMPATIBILITY ANTIGENS

LPS enhances B-cell expression of class II (Ia) major histocompatibility complex (MHC) antigens,[13] but has the opposite effect in many studies involving mononuclear phagocytes.[23,24] For example, incubation of adherent mononuclear cells from human peripheral blood with 20 μg/ml of LPS for 40 h reduced both the Ia+-positive population and the presentation of antigen to T cells.[24] Likewise, treatment of mouse peritoneal macrophages with 1 μg/ml LPS for 2 d inhibited IFNγ-induced Ia antigen expression.[23] This effect could be mimicked by prostaglandin E_2 or dibutryl cyclic AMP, and was blocked by indomethacin, suggesting that LPS-induced prostagladin E_2 may be the mediator of this effect.

LPS can affect Ia expression at the mRNA level. I-Aβ mRNA was not detectable in explanted mouse macrophages obtained from sites of sterile inflammation, but was induced by IFN-γ, peaking at 24 h after stimulation. In the presence of 100 ng/ml LPS, the increase in levels of I-Aβ mRNA was reduced by 40 to 80%.[22] Levels of mRNA correlated closely with surface expression of Ia. Thus LPS could affect either the rate of Ia transcription or the stability of Ia mRNA.

C. Fc RECEPTORS

FcRs connect the humoral immune system to a wide spectrum of cell types whose physiological functions such as ingestion, secretion, cytotoxicity, and differentiation can be regulated by immune complexes. There are at least three distinct receptors for IgG in mouse and man.

Polyclonal activation of mouse spleen cells with LPS resulted in a five- to tenfold increase in cell surface Fcγ RII and the concomitant appearance in the culture medium of a soluble molecule antigenically related to Fcγ RII.[14,15] These changes can be attributed primarily to B cells, since depletion of B cells abolished the effects, while depletion of T cells or adherent cells did not, and results were similar in later studies with purified B cells.[16] Regulation of expression of FcR in B cells may be of physiological significance, since crosslinking these receptors triggers c-*myc* expression, phosphatidylinositol turnover, a transient of cytosolic-free calcium, and blastogenesis.[47-49]

In contrast to results with B cells, incubation of guinea pig peritoneal exudate macrophages with 0.5 to 5 ng/ml LPS for 3 d led to a 50% decrease in binding of radiolabeled immune complexes to the cell surface.[25] Binding studies suggested a reduction in the number of cell surface FcR rather than a change in their affinity. LPS can also inhibit IFN-γ-induced FcR expression on human monocytes,[26] and 30 to 70% decrease in Ig binding requiring 20-h exposure, to only 0.01 to 0.2 ng/ml LPS. Since LPS is a strong stimulus for synthesis of IL-1, and since purified IL-1 could also suppress IFNγ-induced enhancement of FcR, it was suggested that endogenous IL-1 may mediate the effects of LPS on macrophage expression of FcR. Diminished FcR expression may explain how LPS reduces the potential of macrophages to inflict antibody-dependent cell-mediated cytotoxicity and to release reactive oxygen intermediates in response to immune complexes.[50]

D. DOWN-REGULATION OF TNF RECEPTORS

TNF is a pivotal cytokine in inflammation. Among its many potential target cells are macrophages themselves, the chief source of TNF. Paradoxically, LPS, and major inducer of TNF production by macrophages, leads to rapid, profound, selective down-regulation of TNF-R on macrophages even before TNF is released.

Exposure of murine or human macrophages to trace levels of LPS resulted in a complete loss of cell surface TNFα binding sites.[30] Cells of the mouse macrophage-like line RAW 264.7 shared this property. The response was extremely sensitive: 50% decrease of TNF-R ensued after 1 h with 0.6 ng/ml LPS, or after 15 min with 10 ng/ml LPS. Binding studies with radiolabeled TNFα using digitonin-permeabilized macrophages indicated that TNF-R were rapidly internalized in response to LPS. Several lines of evidence argued that endogenous TNF-α cannot account for most of the down-regulation of TNF-R. About 50 ng/ml of exogenous TNF-α would have been required to mimic the effect of 10 ng/ml LPS. There is no evidence that macrophages can secrete such quantities of TNF-α rapidly enough in response to trace levels of LPS to account for this phenomenon. In addition, dexamethasone, which blocks TNF-α production by macrophages at both transcriptional and translational levels, had no effect on LPS-induced down-regulation of TNF-R.[51]

We recently discovered that the microtubule-active agent taxol mimics the effect of LPS on TNF-R. LPS-hyporesponsive C3H/HeJ mice were nonresponsive to this effect of taxol. Nonresponsiveness to the TNF-R down-regulating effect of taxol cosegregated with LPS-hyporesponsiveness in nine out of nine recombinant inbred strains tested.[51] This suggests that LPS directly or indirectly affects proteins associated with microtubules in order to transmodulate TNF-R.

E. DOWN-REGULATION OF IL-6 RECEPTORS

IL-6 is a cytokine that shares many actions with TNF-α, such as pyrogenicity, induction of hepatic synthesis of acute phase proteins, promotion of differentiation of hematopoietic precursor cells, and activation of lymphocytes.[52] Healthy individuals' peripheral blood monocytes, but not their lymphocytes, express IL-6 R. Exposure of monocytes to endotoxin caused a rapid down-regulation of IL-6 mRNA level within 4 h; a maximal effect required only 10 to 100 ng/ml LPS.[33] Suppression of IL-6 mRNA was specific, as shown by a concomitant increase in mRNA for IL-6 itself (as opposed to the receptor). Surprisingly, exogenous IL-6 and IL-1, both of which are LPS-induced monocyte products, mimicked the effect of LPS, down-regulating mRNA for IL-6-R with similar kinetics. In contast, hepatocyte IL-6-R mRNA was increased by IL-1 and IL-6. If these changes in receptor mRNA are reflected by parallel changes in the level of functional receptors at the cell surface, one can speculate that LPS-induced repression of monocyte IL-6-R and stimulation of hepatocyte IL-6-R may lead to a shift in the cell types upon which IL-6 predominantly acts during infection with Gram-negative bacteria.

F. DOWN-REGULATION OF CSF-M RECEPTORS

Macrophage receptors for CSF-M are also down-regulated by trace concentrations of LPS.[34,35] Treatment of mouse peritoneal macrophages for 1 h with as little as 0.1 ng/ml LPS was sufficient to cause a significant inhibitory effect, with complete loss of detectable CSF-M binding at 10 ng/ml LPS. Inhibition of binding of radiolabeled CSF-M to macrophages did not appear to be due to LPS-induced production of CSF-M, 100 U/ml exogenous CSF-M was required to mimic the action of LPS, but no CSF-M could be detected in medium conditioned by LPS-stimulated macrophages, using assays with a sensitivity of 2 to 5 U/ml. The CSF-M receptor is encoded by the proto-oncogene c-*fms*. Recently, Gusella et al.[36] showed that LPS induces down-regulation of c-*fms* mRNA in a murine macrophage cell line via decreased transcription.

G. DOWN-REGULATION OF MANNOSE RECEPTORS

Macrophages express on their surface a 175-kDa receptor that binds and internalizes ligands containing terminal mannose residues. Treatment of rat bone marrow-derived macrophages with 100 ng/ml LPS for 48 h resulted in a 40% loss of this receptor, as measured by the binding of ^{125}I-Man-BSA to the cells.[37] The protein kinase C activator PMA did not affect this process by itself, but potentiated the effect of LPS. Scatchard analysis showed that LPS/PMA caused a reduction in binding sites, but not in affinity for Man-BSA. An increase in endogenous ligands in response to LPS/PMA may be the cause of the observed down-regulation of Man-R. The suppression of this receptor might affect the ability of macrophages to clear pathogens or enzymes such as lysosomal hydrolases from the extracellular space during bacterial infections.

H. DOWN-REGULATION OF SCAVENGER RECEPTORS

Foam cells in atheromas are derived both from smooth muscle cells and from macrophages, the latter predominating at earlier stages in atherogenesis. *In vitro*, macrophages are most readily converted to foam cells by uptake of modified low-density lipoproteins (LDL) through the so-called scavenger receptor.[53] Recognition via this receptor depends on the increased negative charge conferred on LDL by various modifications, including oxidation, and oxidized LDL has been demonstrated within atheromas. The addition of low concentrations of LPS (1 ng/ml) to human monocyte/macrophages cultured in serum resulted in suppression of scavenger receptor activity without affecting the activity of the LDL receptor.[27,28] Different combinations of LPS and lipoproteins (LDL, HDL, reduced and methylated LDL) were compared for their ability to suppress the scavenger receptor in serum-free medium. Only LPS bound to LDL demonstrated this activity. The association of LPS with lipoproteins persisted after uptake of the complex by the cells. The LPS-LDL complex may therefore be transported into the artery wall and modify the evolution of the atheroma.

Recent work from Raetz's laboratory showed that uptake of ^{32}P-labeled lipid A into mouse macrophages and Chinese hamster ovary cells depends primarily on the native or transfected scavenger receptor, respectively.[53a] Binding of lipid A could be competed by several known ligands for the scavenger receptor. Thus, it appears that both native LPS and LPS-LDL complexes can bind to the scavenger receptor. It is not yet clear that binding of LPS to the scavenger receptor initiates any of the known effects of LPS on cell function. However, uptake via the scavenger receptor may be one mechanism for clearance of endotoxin from the circulation, indirectly modulating its effects.

I. UP-REGULATION OF CR3 IN PMNs

PMNs have large intracellular pools of C3bi receptors (CR3) in a compartment that comigrates with the secondary granules on sucrose density gradients. LPS is among the stimuli that induce rapid translocation of these pools to the plasma membrane; other effective agents include leukotrienes, cytokines, phorbol esters, and calcium ionophores.[54] LPS-induced up-regulation of CR3 is associated with, but apparently not responsible for, cell-cell aggregation.[55]

J. UP-REGULATION OF fMLP RECEPTORS IN PMNs

Likewise, PMN have an intracellular reservoir of fMLP receptors that is associated with specific granules on density gradients; LPS triggers the translocation of these intracellular receptors to the plasma membrane. Increased expression of fMLP receptors on the plasma membrane may account, in part, for priming, a functionally important effect of LPS on PMN.[56] Priming refers to the fact that LPS-treated cells are more chemotactically responsive and

TABLE 4
Mechanisms of Receptor Modulation by LPS

Possible mechanisms	Receptor	Ref.
Direct interaction with receptor	Scavenger R	53a
Change in membrane fluidity	β-Adrenergic R	1
Effect on transcription	CSF-M R	36
	TF	43
	Ia	22
	IL-6 R	33
Effect on mRNA stability	TF	43
Induction of the ligand	IL-6 R	33
	Man-R	37
Induction of effective cytokines	FcR	16, 26
	IL-6 R	33
	TF	20, 42
	CR1	8
Change in cAMP levels	Ia	23
Change in kinase activity	α_1-Adrenergic R	1
	Vasopressin R	1
Induction of transmodulation	TNF-R	30
Mobilization of internal pool	FMLP R	54

generate more superoxide than control PMN when subsequently challenged with fMLP. In the studies of Vosbeck et al. with human PMN,[39] priming for enhanced superoxide release was accompanied by a parallel two- to threefold increase in fMLP receptor number, as determined by flow cytometry and sodium dodecyl sulfate-polyacrylamide gel electrophoresis autoradiographic analysis of photoaffinity ligand binding. Results with human blood PMNs were similar, as assessed by binding of tritiated fMLP;[57] a threefold increase in fMLP binding required a 75-min exposure to as little as 50 ng/ml LPS. Intravenous injection of LPS led to a sevenfold increase in plasma membrane fMLP receptors on rabbit blood PMN.[9] The ability of LPS to prime PMN is augmented by complexation with LPS-binding protein, an acute phase reactant, and diminished by binding to high-density lipoprotein.[39]

IV. MECHANISMS FOR RECEPTOR MODULATION BY LPS

Effects of LPS on mammalian cells are exceedingly diverse, including changes of shape, induction of expression of specific genes, alteration of covalent co- or posttranslational modifications of specific proteins (myristoylation, phosphorylation), and promotion of release of autacoids and cytokines. There is a corresponding diversity in the mechanisms whereby LPS may alter the expression of cell surface receptors. Some of these mechanisms are illustrated in Table 4. It is possible that more than one mechanism may be involved in the regulation of a given receptor.

Nuclear factor κB (NF-κB) has been the focus of recent efforts to explain how LPS regulates some gene products at the transcriptional level. NF-κB is present as a nuclear DNA-binding protein in B lymphocytes and mature macrophages, but is found in the cytoplasm of many cells in a form unable to bind to DNA. NF-κB is involved in the induction of expression of IL-2, IL-2 R, IFN-γ, and TNF-α genes. A rapid increase of NF-κB activity observed in several cell types, including macrophages,[58] after treatment of the cells with LPS suggests that transcriptional control could be a primary mechanism for many LPS-mediated actions, including receptor modulation.

V. CONCLUSION

One major way in which LPS affects cell function is to regulate the plasma membrane expression of heterologous receptors. Both the types of the receptors regulated by LPS, and the mechanisms thought to underlie their regulation, are remarkably diverse. In this way, LPS can indirectly affect the host's response to hormones, cytokines, coagulation factors, lipoproteins, antibodies, complement components, and microbial peptides. Although these actions of LPS have been studied mostly by immunologists, and perhaps for this reason most extensively with B cells, macrophages, and PMN, it is clear that LPS also affects the expression of heterologous plasma membrane receptors on endothelial cells, hepatocytes, and possibly many other cells outside the hematopoietic lineage. These effects of LPS, along with its other effects on cell function, will be better understood when the various LPS receptors are identified and their mechanisms of signaling defined.

REFERENCES

1. **Roth, B. L. and Spitzer, J. A.,** Altered hepatic vasopressin and α_1-adrenergic receptors after chronic endotoxin infusion, *Am. J. Physiol.,* 252, E699, 1987.
2. **Van Heuven-Nolsen, D., de Wildt, D. J., and Nijkamp, F. P.,** Disturbed adrenergic regulation of coronary flow in the guinea heart after endotoxin, *Eur. J. Pharmacol.,* 118, 341, 1985.
3. **Liu, M. -S. and Ghosh, S.,** Changes in β-adrenergic receptors in dog livers during endotoxic shock, *A. J. Physiol.,* 244, R718, 1983.
4. **Van Oosterhout, A. J. M., van Heuven-Holsen, D., De Boer, S. F., Thijssen, J. H. H., and Nijkamp, F. P.,** Endotoxin-induced reduction of β-adrenergic binding sites on splenic lymphocytes *in vivo* and *in vitro, Life Sci.,* 44, 57, 1989.
5. **Van Oosterhout, A. J. M., Folkerts, G., Ten Have, G. A. M., and Nijkamp, F. P.,** Involvement of the spleen in the endotoxin-induced deterioration of the respiratory airway and lymphocyte β-adrenergic systems of the guinea pig, *Eur. J. Pharmacol.,* 147, 421, 1988.
6. **Abernathy, C. O., Bhathena, S. J., Recant, L., Zimmerman, H. J., and Utili, R.,** Effects of acute and chronic endotoxin treatment on glucagon and insulin receptors on rat liver plasma membranes, *Horm. Metab. Res.,* 14, 468, 1982.

7. **Niemetz, J.**, Coagulant activity of leukocytes: tissue factor activity, *J. Clin. Invest.*, 51, 307, 1972.

8. **Griffin, F. M., Jr. and Mullinax, P. J.**, High concentration of bacterial lipopolysaccharide, but not microbial infection-induced inflammation, activate macrophage C3 receptors for phagocytosis, *J. Immunol.*, 145, 697, 1990.

9. **Goldman, D. W., Enkel, H., Gifford, L. A., Chenoweth, D. E., and Rosenbaum, J. T.**, Lipopolysaccharide modulates receptors for leukotriene B$_4$, C5a, and formyl-methionyl-leucyl-phenylalanine on rabbit polymorphonuclear leukocytes, *J. Immunol.*, 137, 1971, 1986.

10. **Mathison, J. C. and Ulevitch, R. J.**, The clearance, distribution and cellular localization of lipopolysaccharide (LPS) in rabbits, *J. Immunol.*, 123, 2133, 1979.

11. **Ohara, J. and Paul, W. E.**, Receptors for B-cell stimulatory factor expressed on cells of haematopoietic lineage, *Nature*, 325, 537, 1987.

12. **Park, L. S., Friend, D., Grabstein, K., and Urdal, D. L.**, Characterization of the high-affinity cell-surface receptor for murine B-cell-stimulating factor 1, *Proc. Natl. Acad. Sci. U.S.A.*, 84, 1669, 1987.

13. **Ernst, D. N., McQuitty, D. N., Weigle, W. O., and Hobbs, M. V.**, Expression of membrane activation antigens on murine B lymphocytes stimulated with lipopolysaccharide, *Cell. Immunol.*, 114, 161, 1988.

14. **Pure, E., Durie, C. J., Summerill, C. K., and Unkeless, J. C.**, Identification of soluble Fc receptors in mouse serum and the conditioned medium of stimulated B cells, *J. Exp. Med.*, 160, 1836, 1984.

15. **Pure, E., Witmer, M. D., Lum, J. B., Mellman, I., and Unkeless, J. C.**, Properties of a second epitope of the murine Fc receptor for aggregated IgG, *J. Immunol.*, 139, 4152, 1987.

16. **Snapper, C. M., Hooley, J. J., Atasoy, U., Finkelman, F. D., Paul, W. E.**, Differential regulation of murine B cell FcγRII expression by CD4 T helper subsets, *J. Immunol.*, 143, 2133, 1989.

17. **Nathan, C. F.**, Perspectives: secretory products of macrophages, *J. Clin. Invest.*, 79, 319, 1987.

18. **Ding, A. H. and Nathan, C. F.**, Trace levels of bacterial lipopolysaccharide prevent interferon-γ or tumor necrosis factor-α from enhancing mouse peritoneal macrophage respiratory burst capacity, *J. Immunol.*, 139, 1971, 1987.

19. **Edward, R. L., Rickles, F. R., and Bobrove, A. M.**, Mononuclear cell tissue factor: cell of origin and requirements for activation, *Blood*, 54, 359, 1979.

20. **Levy, G. A., Schwartz, B. S., Curtiss, L. K., and Edgington, T. S.**, Regulatory roles of Tμ and Tγ cells in the collaborative cellular initiation of the extrinsic coagulation pathway by bacterial lipopolysaccharide, *J. Clin. Invest.*, 76, 548, 1985.

21. **Ulevitch, R. J., Tobias, P. S., and Mathison, J. C.**, Regulation of the host response to bacterial lipopolysaccharides, *Fed. Proc.*, 43, 2755, 1984.

22. **Koerner, T. J., Hamilton, T. A., and Adams, D. O.**, Suppressed expression of surface Ia on macrophage by lipopolysaccharide: evidence for regulation at the level of accumulation of mRNA, *J. Immunol.*, 139, 239, 1987.

23. **Steeg, P. S., Johnson, H. M., and Oppenheim, J. J.**, Regulation of murine macrophage Ia antigen expression by an immune interferon-like lymphokine: inhibitory effect of endotoxin, *J. Immunol.*, 129, 2402, 1982.

24. **Yam, A. W. and Parmely, M. J.**, Modulation of Ia-like antigen expression and antigen-presenting activity of human monocytes by endotoxin and zymosan A, *J. Immunol.*, 127, 2245, 1981.

25. **Yagawa, K., Kaku, M., Ichinose, Y., Nagao, S., Tanaka, A., Aida, Y., and Tomoda, A.**, Down-regulation of Fc receptor expression in guinea pig peritoneal exudate macrophages by muramyl dipeptide or lipopolysaccharide, *J. Immunol.*, 134, 3705, 1985.

26. **Arend, W. P., Ammons, J. T., and Kotzin, B. L.**, Lipopolysaccharide and interleukin 1 inhibit interferon-γ-induced Fc receptor expression on human monocytes, *J. Immunol.*, 139, 1873, 1987.

27. **Van Lenten, B. J., Fogelman, A. M., Seager, J., Ribi, E., Haberland, M. E., and Edwards, P. A.**, Bacterial endotoxin selectively prevents the expression of scavenger receptor activity on human monocyte-macrophages, *J. Immunol.*, 134, 3718, 1985.

28. **Van Lenten, B. J., Fogelman, A. M., Haberland, M. E., and Edwards, P. A.**, The role of lipoproteins and receptor-mediated endocytosis in the transport of bacterial lipopolysaccharide, *Proc. Natl. Acad. Sci. U.S.A.*, 83, 2704, 1986.

29. **Lopes-Virella, M. F., Klein, R. L., and Stevenson, H. C.**, Low density lipoprotein metabolism in human macrophages stimulated with microbial or microbial-related products, *Arteriosclerosis*, 7, 176, 1986.

30. **Ding, A. H., Sanchez, E., Srimal, S., and Nathan, C. F.**, Macrophages rapidly internalize their tumor necrosis factor receptors in response to bacterial lipopolysaccharide, *J. Biol. Chem.*, 264, 3924, 1989.

31. **Michishita, M., Yoshida, Y., Uchino, H., and Nagata, K.**, Induction of tumor necrosis factor-α and its receptors during differentiation in myeloid leukemic cells along the monocytic pathway: a possible regulatory mechanism for TNF-α production, *J. Biol. Chem.*, 265, 8751, 1990.

32. **Holter, W., Goldman, C. K., Casabo, L., Nelson, D. L., Greene, W. C., and Waldmann, T. A.**, Expression of functional IL-2 receptors by lipopolysaccharide and interferon-γ stimulated human monocytes, *J. Immunol.*, 138, 2917, 1987.

33. **Bauer, J., Bauer, T. M., Kalb, T., Taga, T., Lengyel, G., Hirano, T., Kishimoto, T., Acs, G., Mayer, L., and Gerok, W.**, Regulation of interleukin 6 receptor expression in human monocytes and monocyte-derived macrophages: comparison with the expression in human hepatocytes, *J. Exp. Med.*, 170, 1537, 1989.

34. **Chen, B. D. -M., Lin, H. -S., and Hsu, S.**, Lipopolysaccharide inhibits the binding of colony-stimulating factor (CSF-1) to murine peritoneal exudate macrophages, *J. Immunol.*, 130, 2256, 1983.

35. **Guilbert, L. J. and Stanley, E. R.**, Modulation of receptors for the colony-stimulating factor, CSF-1, by bacterial lipopolysaccharide and CSF-1, *J. Immunol. Methods*, 73, 17, 1984.

36. **Gusella, G. L., Ayroldi, E., Espinoza-Delgado, I., and Varesio, K.**, Lipopolysaccharide, but not IFN-γ, down-regulates c-fms mRNA proto-oncogene expression in murine macrophages, *J. Immunol.*, 144, 3574, 1990.

37. **Shepherd, V. L., Abdolrasulnia, R., Garrett, M., and Cowan, H. B.**, Down-regulation of mannose receptor activity in macrophage after treatment with lipopolysaccharide and phorbol esters, *J. Immunol.*, 145, 1530, 1990.

38. **Haslett, C., Guthrie, L. A., Kopaniak, M. M., Johnston, R. B., Jr., and Henson, P. M.**, Modulation of multiple neutrophil functions by preparative methods or trace concentrations of bacterial lipopolysaccharide, *Am. J. Pathol.*, 119, 101, 1985.

39. **Vosbeck, K., Tobias, P., Mueller, H., Allen, R. A., Arfors, K. -E., Ulevitch, R. J., and Sklar, L. A.**, Priming of polymorphonuclear granulocytes by lipopolysaccharides and its complexes with lipopolysaccharide binding protein and high density lipoprotein, *J. Leukocyte Biol.*, 47, 97, 1990.

40. **Rivers, R. P., Hathaway, W. E., and Weston, W. L.**, The endotoxin-induced coagulant activity of human monocytes, *Br. J. Haematol.*, 30, 311, 1975.

41. **Niemetz, J. and Morrison, D. C.**, Lipid A as the biologically active moiety in bacterial endotoxin (LPS)-initiated generation of procoagulant activity by peripheral blood leukocytes, *Blood*, 49, 947, 1977.

42. **Drake, T. A., Ruf, W., Morrissey, J. H., and Edgington, T. S.**, Functional tissue factor is entirely cell surface expressed on lipopolysaccharide-stimulated human blood monocytes and a constitutively tissue factor-producing neoplastic cell line, *J. Cell Biol.*, 109, 389, 1989.

43. **Gregory, S. A., Morrissey, H. J., and Edgington, T. S.,** Regulation of tissue factor gene expression in the monocyte procoagulant response to endotoxin, *Mol. Cell. Biol.,* 9, 2752, 1989.

44. **Colucci, M., Balconi, G., Lorenzet, R., Pietra, A., Locati, D., Donati, M. B., and Semeraro, N.,** Cultured human endothelial cells generate tissue factor in response to endotoxin, *J. Clin. Invest.,* 17, 1893, 1983.

45. **Lyberg, T., Galdal, K. S., Evensen, S. A., and Prydz, H.,** Cellular cooperation in endothelial cell thromboplastin synthesis, *Br. J. Haematol.,* 53, 85, 1983.

46. **Brox, J. H., Osterud, B., Bjorklid, E., and Fenton, J. W.,** Production and availability of thromboplastin in endothelial cells: the effects of thrombin, endotoxin and platelets, *Br. J. Haematol.,* 57, 239, 1984.

47. **Phillips, N. E. and Parker, D. C.,** Fcγ receptor effects induction of c-*myc* mRNA expression in mouse B lymphocytes anti-immunoglobulin, *Mol. Immunol.,* 24, 1199, 1987.

48. **Bijsterbosch, M. K. and Klaus, G. G. B.,** Crosslinking of surface immunoglobulin and Fc receptors on B lymphocytes inhibits stimulation of inositol phospholipid breakdown via the antigen-receptors, *J. Exp. Med.,* 162, 1825, 1985.

49. **Wilson, A. H., Greenblatt, D., Taylor, C. W., Putney, J. W., Tsien, R. Y., Finkelman, F. D., and Chused, T. M.,** The B lymphocyte calcium response to anti-Ig is diminished by membrane immunoglobulin cross-linkage to the Fcγ receptor, *J. Immunol.,* 138, 1712, 1987.

50. **Johnston, P. A., Adams, D. O., and Hamilton, T. A.,** Regulation of the Fc-receptor-mediated respiratory burst: treatment of primed murine peritoneal macrophages with lipopolysaccharide selectively inhibits H_2O_2 secretion stimulated by immune complexes, *J. Immunol.,* 135, 513, 1985.

51. **Ding, A. H., Porteu, F., Sanchez, E., and Nathan, C. F.,** Shared actions of endotoxin and taxol on TNF receptors and TNF release, *Science,* 248, 370, 1990.

52. **Muraguchi, A., Hirano, T., Tang, B., Matsuda, T., Horii, Y., Nakajima, K., and Kishimoto, T.,** The essential role of B cell stimulatory factor 2 (BSF-2/IL-6) for the terminal differentiation of B cells, *J. Exp. Med.,* 167, 332, 1988.

53. **Yla-Herttuala, S., Palinski, W., Rosenfeld, M. E., Parthasarathy, S., Carew, T. E., Butler, S., Witztum, J. L., and Steinberg, D.,** Evidence for the presence of oxidatively modified low density lipoprotein in atherosclerotic lesions of rabbit and man, *J. Clin. Invest.,* 84, 1086, 1989.

53a. **Hampton, R. V., Golenbock, D. T., Penman, M., Krieger, M., and Raetz, C. R. H.,** Recognition and plasma clearance of endotoxin by scavenger receptors, *Nature (London),* 352, 342, 1991.

54. **Gallin, J. I.,** Inflammation, in *Fundamental Immunology,* 2nd ed., Paul, W. E., Ed., Raven Press, New York, 1989, 721.

55. **Schleiffenbaum, B., Moser, R., Patarroyo, M., and Fehr, J.,** The cell surface glycoprotein Mac-1 (CD11/CD18) mediates neutrophil adhesion and modulates degranulation independently of its quantitative cell surface expression, *J. Immunol.,* 142, 3537, 1989.

56. **Guthrie, L. A., McPhail, L. C., Henson, P. M., and Johnston, R. G.,** The priming of neutrophils for enhanced release of oxygen metabolites by bacterial lipopolysaccharide: evidence for increased activity of the superoxide-producing enzymes, *J. Exp. Med.,* 160, 1656, 1984.

57. **Howard, T. H., Wang, D., Berkow, R. L.,** Lipopolysaccharide modulates chemotactic peptide-induced actin polymerization in neutrophils, *J. Leukocyte Biol.,* 47, 13, 1990.

58. **Alexander, N., Shakhov, N., Coliart, M. A., Vassalli, P., Nedospasov, S. A., and Jongeneel, C. V.,** κB-type enhancers are involved in lipopolysaccharide-mediated transcriptional activation of the tumor necrosis factor α gene in primary macrophages, *J. Exp. Med.,* 171, 35, 1990.

Chapter 17

LPS BINDING PROTEINS IN GRANULOCYTE LYSOSOMES

Naohito Ohno

TABLE OF CONTENTS

I. INTRODUCTION

Lipopolysaccharide (LPS) is well recognized as one of the major contributors to the development of the host defense systems, including immunity.[1-5] However, excess quantities of LPS generated in the course of Gram-negative infection can cause shock and multisystem organ failure. To date, appropriate therapy to effectively neutralize the toxicity has been limited. Many approaches, however, are currently being used to control endotoxin-mediated illness and death, as will be summarized by a number of contributors to these volumes. Host-derived LPS-binding proteins may well function as important candidates to modulate the biological activity of LPS. Such proteins are found in a variety of sources, including serum, host cell membranes, amoebocytes, and insects. In this chapter, I shall discuss the LPS-binding proteins found in granulocyte lysosomes and their potential relationship to the nonoxidative antimicrobial system of phagocytic cells.

Microbial constituents, as well as their products, are known to be potent inflammatory and immunomodulating substances. As a consequence, killing of the microorganisms is, in many cases, not necessarily sufficient to return the host to the preinfection status because of the remaining inflammatory and immunomodulating substances. Oxidative as well as nonoxidative degradation systems of the phagocyte may, therefore, be extremely important for these processes. Indeed, experimental evidence from both *in vivo* and *in vitro* studies suggests that LPS excreted from the macrophages which have engulfed intact *Escherichia coli* still showed significant biological activity (see Chapter 18; Chapter 11 in Volume II). Host self-defense mechanisms must, therefore, eventually destroy or inactivate these substances in some way. Multiple-host defense barriers with qualitatively as well as quantitatively different characteristics would provide optimal systems for such protection. The antimicrobial and/or LPS-binding proteins to be summarized below appear to act both coordinately and also independently in achieving the neutralization and detoxification of endotoxic LPS. The majority of this discussion will focus upon the former events, with particular emphasis upon the polymorphonuclear leukocyte.

In the mature neutrophil, at least two distinct classes of granules have been identified by classical cytochemical and immunocytochemical methods. The azurophile (primary) granules selectively contain myeloperoxidase, whereas the specific (secondary) granules lack myeloperoxidase but can be distinguished by the presence of lactoferrin and vitamin B12-binding protein. Simple calculations suggest that each of the major granule proteins is present in concentrations of the order of 1 to 5 mg/ml. The pH of the phagocytic vesicles is thought to be mildly alkaline initially following phagocytosis, but the pH grandually becomes acidic over a period of about 30 to 60 min. The ionic composition of the intraphagosomal environment in neutrophils has not been systematically studied.

A number of microbicidal and/or LPS-binding proteins have been purified and characterized from acid extracts of granules. Azurophil granules contain an array of nonoxidative antimicrobial proteins involved in the killing and degradation of microorganisms, including the defensins, lysozyme, cathepsin G, elastase, azurocidin, bactericidal/permeability-increasing protein (BPI), cationic antimicrobial proteins (CAP) 57 and 37, and bactericidal protein (BP). Different microbial species vary rather significantly in their relative susceptibility to individual proteins isolated from granule extracts. In addition, the conditions of the assay (ionic strength, serum and presence of other proteins, pH, duration) can influence markedly the differential microbicidal potency of various proteins. The most appropriate *in vitro* assay conditions, reflecting the environment of the phagosome, have not yet been fully defined, and it is still, therefore, not possible at this time to assess completely the relative role of the various granule proteins in microbicidal events. Because of the apparent specificity of some of these antimicrobial proteins for Gram-negative (LPS-containing) bacteria, it is likely that many of the granular proteins can interfere with LPS-inflammatory cell interactions and with resultant immunopharmacological activities.

II. SPECIFIC CHARACTERISTICS OF GRANULOCYTE LPS-BINDING PROTEINS

A. BACTERICIDAL/PERMEABILITY-INCREASING PROTEIN (BPI)

BPI is a cationic 50,000- to 60,000-MW protein first purified from human neutrophil granules by Weiss et al. in 1978.[6] BPI binds to the surface of susceptible Gram-negative bacteria *in vitro,* alters membrane permeability, and causes cell death.[7] A similar, if not identifcal 57,000-MW molecule has been isolated from neutrophils by Spitznagel et al. and designated CAP57.[8] A third antimicrobial protein, BP, has been described which damages *Pseudomonas aeruginosa in vitro.*[9] BPI, CAP57, and BP are highly similar proteins in size, amino acid composition, and antimicrobial activity.

The killing of Gram-negative bacteria by BPI of neutrophils requires surface binding, and is accompanied by a discrete increase in outer membrane permeability to small hydrophobic substances.[8,10,11] This outer membrane modification appears to be related to perturbation of outer membrane-localized LPS. BPI causes extracellular release of LPS, but only at suprasaturating doses. Neverthless, because the organization of LPS in the outer membrane is altered by pretreatment of bacteria with saturating doses of BPI, the amount of LPS released during tris-EDTA treatment is reduced by 80%. BPI markedly and selectively stimulates biosynthesis of LPS, suggesting an attempt by BPI-treated bacteria to repair outer membrane damage. The removal of surface-bound BPI by 40 mM Mg^{2+} initiates time- and temperature-dependent repair of the outer membrane permeability barrier and a further increase in LPS synthesis even though the bacteria are no longer viable. Inhibition of protein synthesis by chloramphenicol has little or no effect on repair. These findings indicate that the repair of the permeability barrier after the removal of BPI from the surface requires newly formed LPS, but apparently no biosynthesis of other outer membrane constituents, which strongly suggests that the effects of BPI on LPS are mainly responsible for the breakdown of the outer membrane permeability barrier.

Similar physiological changes have been seen in *P. aeruginosa* after exposure to a BP from the granules of human polymorphonulcear leukocytes.[9] Exposure of BP to *P. aeruginosa* markedly decreased the rate of cellular uptake and amino acids into trichloroacetic acid-insoluble material. The rate of O_2 consumption by *P. aeruginosa* was also decreased immediately after exposure to BP, and continued to decline exponentially until it ceased completely 30 min after exposure to BP. In the presence of 30 mM $CaCl_2$ or $MgCl_2$, bacteria were protected from death due to BP and respiration rates were unaffected. The cellular ATP pool of *P. aeruginosa* remained constant for up to 2 h after exposure to BP. Further, the cytoplasmic membrane of *P. aeruginosa* was partially depolarized after exposure to BP. Purified BP killed 95% of 5×10^6 colony-forming units of *P. aeruginosa* at a concentration of 60 to 100 ng of protein per milliliter. It is proposed that cytoplasmic membrane depolarization is the biochemical lesion responsible for the other physiological changes seen and ultimately for the death of *P. aeruginosa* induced by BP.

The sensitivity of *E. coli* to BPI depends mainly on the polysaccharide chain length of outer membrane LPS.[12] Thus, rough strains of *E. coli* producing only short-chain LPS are more sensitive to BPI than smooth strains that produce LPS with varied chain lengths. Changes in BPI sensitivity paralleled differences in binding affinity of *E. coli* for BPI and were closely correlated with changes in the chain length of LPS produced under different growth conditions. Rough strains showed little or no growth-dependent variation in BPI sensitivity. Thus, the BPI sensitivity of *E. coli* can be modulated not only by the genotypic conversion of the LPS phenotype, but also by environmental effects on LPS-polysaccharide formation in wild-type strains.

Marra et al. examined the neutralization capacity of BPI on the biological activity of LPS.[13] Human neutrophils can be activated both *in vitro* and *in vivo* by LPS. On stimulation, surface expression of CR1 and CR3 increases markedly. Using flow microfluorimetry, they analyzed surface expression of CR1 and CR3 as a measure of neutrophil stimulation in response to LPS. Purified BPI completely inhibited CR up-regulation on neutrophils stimulated with both rough and smooth LPS chemotypes at 1.8 to 3.6 nM (100 to 200 ng/ml). By comparison, the polypeptide antibiotic polymyxin B, which is also known to bind LPS, completely inhibited the same dose of LPS at 0.4 nM. The specificity of BPI for LPS was further demonstrated by inhibition of LPS activity in the limulus amebocyte lysate assay. Considering these facts, the role of BPI in infection may not be limited to its microbicidal activity, but it may also regulate the neutrophil response to LPS. Ooi et al. suggested that processing of the BPI molecule by certain proteases can result in the generation of a 25-kDa active fragment. Of interest, there is evidence to suggest that the 25-kDa fragment is essential for neutralizing biological activities of LPS. In their respect, following limited proteolysis of BPI of human neutrophils, the NH_2 terminal 25-kDa fragment was shown to possess the bactericidal and envelope-altering activities of the 60-kDa parent protein.[14] On a molar basis, the fragment was as potent as holo-human BPI against rough *E. coli,* was more potent than holo-BPI against more resistant smooth *E. coli,* and retained the specificity of BPI toward Gram-negative bacteria. These findings suggested that all of the molecular determinants of the antimicrobial properties of BPI reside within the NH_2-terminal 25-kDa segment, implying a novel structural/functional organization for a cytotoxic protein.

From an analysis of a full-length cDNA clone encoding human BPI, it was indicated that BPI may be organized into two domains. The amino-terminal half, previously shown to contain all known antimicrobial activity, contains a large fraction of basic and hydrophilic residues.[15] In contrast, the carboxyl-terminal half contains more acidic than basic residues and includes several potential transmembrane regions that may anchor the holoprotein in the granule membrane. As a result of a homology search, BPI was found to share significant homology with an LPS-binding protein of comparable size isolated from acute-phase rabbit serum (LPB).[16,17] Both of these proteins bind to LPS with high affinity. A computer search identified one other protein with significant homology to BPI: the cholesteryl ester transfer protein (22% identify and another 30% conservative substitutions). Both proteins are of similar size and the similar residues are consistently spread throughout the two sequences. They also indicated the significance of this similarity from the limited sequence analysis of the 5' end of each gene, which demonstrated that both genes contain an intervening sequences at homologous positions. Considering this evolutionary relationship, as well as the functional similarity, they suggested a common ancestry for the genes of these proteins. An analogous gene family has already been proposed as the 2-microglobulin superfamily. Similar to other bacterial and plant toxins, BPI also contains at least two distinct structural domains. A mechanism of antibacterial action in relation to the two domains has been proposed whereby the strongly cationic amino-terminal region is exposed to the interior of the granule, while the very hydrophobic carboxyl-terminal portion is embedded in the membrane. During phagocytosis and degranulation, proteases may be released and activated and cleave BPI at the junction of the amino and carboxyl portion of the molecule, providing controlled release of the amino-terminal antibacterial fragment when target bacteria are ingested.

B. CATIONIC ANTIMICROBIAL PROTEINS (CAP57 AND CAP37)

Spitznagel and co-workers have identified and characterized two granular proteins that exhibit *in vitro* antimicrobial action against *Salmonella typhimurium*. Using the serine proteinase inhibitor diisopropyl fluorophosphate minimized the otherwise significant and problematic proteolytic destruction of these proteins during purification.[8] Both of these proteins exhibited diminished activity against a polymyxin B resistant (PBr) mutant of *S. typhimurium* as discussed below, suggesting a similarity in the mechanism of action of these proteins and polymyxin B. These proteins differed substantially in molecular weight (57,000 and 37,000), amino acid composition, and antimicrobial activity. The characteristics of the larger protein are similar to those of human BPI prepared in 1978 by Weiss et al. as described above.

The action mechanism of CAP57 was investigated using *S. typhimurium*.[18-23] Antimicrobial action against three strains, i.e., LT-2, a smooth wild-type strain, SH9178, its rough (Rb LPS) strain, and SH7426, a polymyxin B-resistant pmr A mutant of SH9178 which contains lipid A phosphoryl groups and is 100% substituted with amino pentose, were compared. LT-2 and SH7426 are far more resistant than SH9178, suggesting a similarity with the mechanism of polymyxin B. Binding experiments using radioiodinated CAP57 also suggested specific binding to these bacterial cell surfaces compatible with polymyxin B. The surface of SH7426 was less anionic because of substitution of the lipid A phosphoryl group with amino pentose, suggesting that both hydrophobic and anionic properties of the microbial surface are important for resistance.

They produced two monoclonal antibodies to CAP57, termed P1G8 and P2A5. Binding of these monoclonals was shown to occur to closely related epitopes on CAP57. Of interest, however, only P1G8 inhibited the antimicrobial action of CAP57.[24] Immunocytochemical studies confirmed that CAP57 was localized in granules of PMN.

Spitznagel et al. have recently shown that a major granule-associated cationic protein CAP37 (Mr = 37 kDa) derived from human PMN has, in addition to its antimicrobial properties, monocyte-specific chemoattractant activity which mediates the second wave of inflammation.[25] Stimulation of PMN with opsonized *Staphylococcus aureus* resulted in a rapid release of 90% of CAP37 into the extracellular milieu. Since distribution of CAP37 is limited to PMN, and is not in mononuclear cells, eosinophils, and red blood cells, the role of this protein has considerable potential physiological relevance.

The N-terminal amino acid sequence of this novel chemotactic protein shared significant homology with a number of proteases, such as human elastase (57.5%), cathepsin G (45%), bovine plasminogen (45%), and human complement factor D (42.5%). Rat mast cell proteases I and II and proteinases derived from cytotoxic T cells also share significant homology. The first 20 residues of CAP37 are also similar to the recently published sequence of an antibacterial substance named azurocidin, isolated from the membrane of azurophil granules of human PMN and reported to have a molecular weight of 29 kDa.

In inflammatory serine proteases, disulfide bonds are well conserved. A portion of this disulfide bond is also conserved in CAP37. However, a critical substitution of a serine for a histidine at position 41 appears to be responsible for its lack of serine protease activity. It is quite an important physiological observation that changing only one important residue for protease activity changed the function of the molecule. There is little doubt that CAP37 as well as CAP57 has an affinity to LPS, and preliminary evidence using photoaffinity cross-linking experiments is in support of their concept.[25a] Homology of CAP57 with BPI and LBP and of CAP37 with elastase and cathepsin G also strongly supports the strong affinity to LPS. However, modulation of the biological activity of LPS by these proteins has not been clearly demonstrated to date.

C. DEFENSINS

Defensins are a family of 3 to 4-kDa (29 to 34 amino acid) peptides found in the granules of mammalian phagocytes.[26-29] The members of this family are invariably arginine rich and all share six conserved cysteine residues that participate in intramolecular disulfide bonds. The peptides have been detected in a wide variety of phagocytic cells including rabbit alveolar macrophages, human neutrophils, and myeloid cell lines. Human neutrophils were found to contain four defensin peptides, three of which (HNP-1, 2, and 3) differed from each other only in a single amino-terminal amino acid. HNP-4 was approximately 100-fold less abundant in peripheral blood cells than HNP 1-3. Because the cysteine residue closest to the amino terminus was disulfide linked to the cysteine closest to the carboxy terminus of the peptide, the defensins manifest a cyclic configuration. Crystal and two-dimensional NMR spectral analyses revealed spatial segregation of charged and hydrophobic residues. Such an amphiphilic structure may equip these peptides for insertion into phospholipid membranes of biological organisms. Although defensin molecules are too small to be able to form membrane pores as monomers, defensins may form multimers and these may be the actual membrane-active species. By high-resolution X-ray crystallography, HNP-3 reveals a dimeric sheet that has an architecture very different from other lytic peptides, such as melittin, cecropin, magainin, alamethicin, and hemolysin.[30]

To date, the cDNA of two human and the three rabbit defensins have been cloned and sequenced. An abundant myeloid mRNA (myeloid-related sequence, MRS) found independently also codes for human defensins. The five defensins studied are synthesized as 94 to 95 amino acid preprodefensins with a characteristic hydrophobic signal sequence. In the precursor, the positive charge of the mature arginine-rich defensins is balanced by negative charges in the preprodefensins, an arrangement which may serve to neutralize the potential toxicity of defensins during protein processing and intracellular transport.

Defensins are among the principal constituents of granules and constitute about 5 to 7% of all cellular protein in human neutrophils and 18% in rabbit neutrophils. Defensin synthesis is regulated mainly by the defensin RNA content of phagocytes, and in the rabbit it is limited to immature neutrophils in the bone marrow and alveolar macrophages of adult rabbits. Immature neutrophils also appear to be the principal site of human defensin synthesis. Large amounts of mRNA for a defensin in the Paneth cells of the mouse small bowel also suggested the high concentration of this protein in these areas. Recently, partially homologous proteins (insect defensins) were purified from the hemolymph of infected or injured insects. Low molecular weight, amphiphilic antimicrobial peptides such as bactenecin were also identifed from bovine neutrophils.

Among the antimicrobial proteins and peptides in granulocytes, the defensin family has been rather precisely examined from the chemical, medical, biological, and pathophysiological point of view. Defensins show *in vitro* activity against Gram-negative and Gram-positive bacteria, fungi, mammalian cells, and enveloped viruses. In the case of Gram-negative bacteria, defensins permeabilize both the inner and outer membranes, and that permeabilization of inner membrane is found to strongly correlate with cell death. Because of the important contribution of the membrane permeabilization, a membrane potential is apparently required for the action and membrane-depolarizing agents such as carbonylcyanide M-chlorophenylhydrazone, and 2,4-dinitrophenol protect the cells. These membrane-acting properties strongly indicated the binding of defensins to LPS. Unfortunately, there is no direct evidence on the modulation of immunopharmacological and endotoxic activities of LPS. The available physicochemical properties of defensin may have strong similarity to those of other LPS-binding proteins. Extensive work in this field may reveal more interesting properties and biological roles of these peptides.

D. LYSOZYME

Lysozyme (LZM), first discovered by Fleming in 1922, is a low molecular weight enzyme. It is a cationic protein with bacteriolytic properties mediated by hydrolysis of the peptidoglycan of the bacterial cell wall. The enzyme is widespread in nature and has been isolated from phages, fungi, plants, and animals, and can be divided into several distinct families.[31-33] In man it is present in high concentrations in tears and in lesser amounts in saliva and plasma (5 to 9 mg/l) and only in trace amounts in urine (0 to 2 mg/l). The enzyme is normally present in several mammalian tissues, notably blood, lungs, and kidneys. LZM seems to be one of the major enzymes of granulocytes, monocytes, and macrophages. In addition to the physiological importance of this enzyme, LZM is one of the most widely investigated proteins used as a model compound in varous areas of research,[34-39] i.e., X-ray crystallography, protein folding, evolution of protein, mechanism of enzymic activity, chemical modification of protein, genetic engineering, immunochemistry, and antigen presentation. Recently, we have found that LZM binds to LPS and neutralizes endotoxicity and immunopharmacological activities of LPS.[40-42,44,45]

Complex formation of LPS with LZM inhibited various kinds of biological as well as immunopharmacological activities of LPS.[40] B-lymphocyte proliferation, B-lymphocyte differentiation, macrophage production of lymphocyte-activating factor activity, activation of the classical pathway of complement, polymyxin B-mediated lysis of LPS-sensitized erythrocytes, and gelation of the limulus amebocyte lyzates were significantly reduced. In addition to these *in vitro* analyses, the complex formation also inhibited the lethal toxicity to actinomycin D-sensitized mice. All of these results strongly suggest that LZM is a good candidate for neutralizing endotoxicity. Detailed studies of the interaction of LPS with LZM suggested that complex formation was significantly affected by various chemical, physical, and environmental factors, that is, while both Re LPS- and lipid A-dependent immunostimulatory activities were readily inhibited by LZM in a dose-dependent fashion, immunostimulatory activities of S LPS and Ra LPS were relatively unaffected by LZM. Inhibition of the immunopharmacological activities of S LPS was strongly dependent on the ionic strength of the reaction mixture. It is also quite interesting that the temperature used to prepare LPS-LZM complexes is quite important for inhibiting biological activity, i.e., preparation at 50°C is significantly more effective in inhibiting the activity than that at 37°C. Immunopharmacological activities of a synthetic monosaccharide analog of lipid A, GLA-60, were also inhibited by LZM, and the inhibition was dependent on the salt form in addition to the reaction temperature and pH.[46] These data suggest that the factors which dictate the initial interactions between LPS and LZM may not be identical for all LPS preparations and/or purified lipid A.

With respect to the inhibition of the biological activity of LPS, enzymic activity of LZM was also inhibited by LPS.[41,42] This inhibition was observed in almost all LPS tested including phenol-extracted wild-type LPS, Re LPS, and butanol-extracted LPS. Interestingly, Re LPS and butanol extracted-LPS required higher temperatures to complete the enzyme inhibition. Enzyme inhibition by polysaccharide-containing LPS (s LPS) and Ra LPS preparations was independent of temperatures between 37 and 50°C; in contrast, that of Rd LPS, Re LPS, and lipid A was temperature dependent. The inhibition of LZM by Rd LPS and Re LPS was increased by treatment with mild alkali which had little detectable effect on inhibition of LZM by s LPS and Ra-Rc LPS preparations, which may suggest a lower fluidity of the micelle structure in Re LPS and butanol-extracted LPS. It is also noteworthy that LZM activity in relevant biological fluid, which contained many other proteins of higher concentrations, such as saliva,milk, and blood, has been inhibited by the addition of LPS. This observation may support the physiological importance of this enzyme inhibition and may have relevance in Gram-negative infections. The other characteristic feature of the enzyme inhibition was that inhibition required preincubation with LPS. It was also correlated with the fact that the inhibition was noncompetitive. The dramatic change of intrinsic fluoescence of tryptophan residues in LZM by LPS, resistance to the proteolytic digestion of LZM in the presence of LPS, and also resistance to the thermal denaturation of LZM support the drastic changes on the LZM molecule in the presence of LPS. Addition of NP-40, disialoganglioside, and crude membrane fraction of 70Z/3 cell dissociate the LZM-LPS complex, resulting in the appearance of enzymic activity. These data suggest that hydrophobic interactions may be quite important for the complex formation. On the other hand, acetylated LZM prepared by chemical modification did not produce a complex with LPS, suggesting the critical importance of ionic interactions as well for complex formation. Considering these facts in relation to the binding of LZM to the substrate, ionic interaction may first allow surface interaction of both substances and then hydrophobic interaction may stabilize the resulting LPS-LZM complex.

Inhibition of the enzymatic activity of LZM by LPS required preincubation and the conditions were dependent on the chemotype of LPS. Similarly, the increase in Fl-intensity of dansylated LZM following binding to LPS was also significantly dependent on temperature. In the case of smooth LPS-LZM complex formation, the Fl-intensity reached maximum levels after 20 min or longer at incubation temperatures below 37°C, 10 min at 50°C, and only 1 min at 70°C. The maximum Fl-intensity was also dependent on the reaction temperature, and the higher reaction temperatures gave higher Fl-intensity. The maximum of Fl-intensity was also dependent not on the measurement temperature but on the reaction temperature, suggesting greater, stronger, and irreversible binding at higher temperatures. These results suggest that binding of LPS and LZM should not be characterized as only one binding site, but as multiple binding sites with different affinities. Thus, the binding would be time as well as temperature dependent. Fluorescence intensity of DNS-LZM was also significantly increased in the presence of rough (r) LPS. The intensity was also time and temperature dependent, but r LPS required higher temperatures to saturate the reaction. These differences were closely related to the characteristic features of the physicochemical properties of micelles of each LPS.

Collectively these results suggest that the binding interactions of LZM and LPS are complex and may involve irreversible processes; Scatchard analysis gave affinity constants at various reaction temperatures. To estimate affinities, it was hypothesized that Fl-intensity of the reaction mixture has a linear relationship to the quantity of LPS-bound DNS-LZM. The dissociation constants of the complex between smooth LPS and DNS-LZM were estimated to be 0.1 (25°C), 0.08 (37°C), and 0.02 (50°C) μM^{-1}. These results also supported the temperature dependency of the complex formation.

During studies to investigate the conditions of enzyme inhibition by LPS, it was found that inhibition of LZM activity was more significant at acidic pH. Because of the fact that LZM has a very basic isoelectric point and since LPS is well recognized as an anionic macromolecule, it is not surprising that Fl-intensity was significantly dependent on both pH and buffer, and a higher Fl-intensity was observed at high pH and low ionic strength (lower than 0.5 *M* NaCl). Since the enzymic activity of LZM was significantly inhibited at lower pH, it would be anticipated that the pH dependency of the Fl-intensity should be negatively correlated with enzyme inhibition. However, the molecular complex once generated under optimum condition was relatively refractory to subsequent dissociation following dilution to various concentrations of ionic strength or pH. Again, these data suggest stabilization of the complex after initial binding of LZM and LPS. Considering the ionic characteristics of both substances, initial interaction would be mediated by ionic interaction and then hydrophobic interaction would stabilize the complex.

Hydrophobic interactions involving LZM were further characterized using a variety of detergents instead of LPS. Interestingly, the maximum Fl-intensities of DNS-LZM obtained by cetavlon were significantly stronger than those with LPS. Increase in LZM Fl-intensity was also observed with CHAPSO, CHAPS, SDS, cetavlon, decansulfonic acid, and cardiolipin, but not observed with caplilic acid, deoxycholic acid, cetylpyridinium chloride, and DOD-glucoside. On the other hand, Fl-intensity of DNS-polymyxin B was also increased in the presence of LPS. However, Fl-intensities of none of the DNS amino acids, basic as well as hydrophobic, were increased by LPS. Considering these facts, the environment of DNS group should strongly influence the reactivity, and amino acids and peptides surrounding DNS group in DNS-LZM would determine the specificity of LZM-LPS complex.

LZM is known to show fusogenic properties and can be inserted into micelles. In these micelles, the LZM molecule was protected from proteolytic digestion. LZM is also known to bind to cardiolipin, SDS, and phosphatidic acid, all of which is consistent with the capacity of LZM to bind to LPS. In addition, studies of the reactivity of LZM-LPS complex with several anti-LZM monoclonal antibodies with specificities for various LZM domains (e.g., HyHEL-5, 8, 9, 10, and 11) have indicated that reactivity of all of these antibodies was significantly reduced.[45] Considering these facts, it is reasonable to speculate that the LZM molecule inserted relatively deeply into the macromolecular aggregate structure of LPS. The exact nature of the LPS molecule as well as the mechanism of immunomodulation following complex formation with LZM remains to be clarified by future studies.

E. LACTOFERRIN

Lactoferrin (LF) is an iron-binding protein present in granulocytes of neutrophils as well as in milk, tears, and serum. The complete amino acid sequence of LF has been determined. LF has been demonstrated to manifest significant antimicrobial activity which is believed to be mediated at least in part by chelation of iron from the environment.[47-52] During phagocytosis and digestion of *E. coli* by the neutrophil *in vivo,* iron released from the bacterium by the action of myeloperoxidase was trapped by LF, thus reducing the capacity of the bacterium to survive in the neutrophil phagolysosomal environment. In addition, a number of immunological roles for LF, especially as a myelopoietic regulator, have been reported.[50,51] LF negatively regulates myelopoiesis *in vivo* and *in vitro,* by suppressing monocytes/macrophage function including cytokine release mediated by specific cell surface receptors for LF. Binding of LPS to LF has recently been shown to abrogate the negative regulation of myelopoiesis by LF. This observation was explained, at least in part, by binding among LPS, LF, and these receptors on macrophages.[51] Using PMA-induced monocytes/macrophage-differentiated HL-60 cells, LPS enhanced LF binding on the HL-60 cells at 37°C, while LPS inhibited LF binding at 4°C. LPS-mediated enhancement of LF binding was abrogated by either polymyxin B, anti-CD11, or mAb5D3, the latter of which have been reported to be associated with, or directed against, candidate LPS receptors. These results indicate a dichotomous nature of LF binding on HL-60 cells mediated by LF receptor and LPS receptor to which LF-LPS complex bind. Considering these facts, LPS-mediated modulation would be the result of binding of LPS to LF, which abrogates LF binding to LF-receptor. LF was found to be directly bound to LPS and to modify the receptor-mediated function of differentiated HL-60 cells by PMA.

F. CATHEPSIN G

Cathepsin G is a chymotrypsin-like protease possessing antimicrobial activity against both Gram-positive and Gram-negative bacteria as well as fungi. The isoelectric point of cathepsin G is higher than 12.5, and this physicochemical property is thought to be quite important, at least a part, for the antimicrobial activity of this protein.[53] Cathepsin G also showed anti-gonococcal activity.[54] The ability of cationic proteins to kill susceptible bacteria requires binding of these proteins to the microbial cell surface. Cathepsin G was found to bind at least two major outer membrane proteins, PIII and the major iron-regulated protein (MIRP), and to digest a part of these proteins.[55] There is little question but that the cationic properties of cathepsin G are important for the bactericidal activity. Proteolytic activity of the enzyme would be also important for its antigonococcal activity. The susceptibility of the bacterium to cathepsin G strongly depends on the susceptibility of the outer membrane proteins to enzymatic digestion. Preliminary experiments[55a] also strongly indicated the binding of LPS to cathepsin G. Surface exposure of cathepsin G-sensitive proteins may be related to the phenotype of LPS, and, in their respect, the deep-rough LPS mutant appeared somewhat less resistant than the parent strain. Although it is, at present, uncertain whether cathepsin G binds directly to LPS on the bacterial surface, the above-cited evidence supports the concept that the structure of LPS may contribute to the susceptibility of bacteria to cathepsin G.

III. SUMMARY AND CONCLUSIONS

The host defense network consists of a variety of biochemical systems which variously contribute to protection against a broad spectrum of invading organisms and substances. Phagocytic cells, especially PMN, act nonspecifically at a very early stage of host defense. Both oxidative and nonoxidative mechanisms are equally important for the appropriate function of PMN. Various kinds of host defense proteins (several hundred) have been identified in lysosomes of phagocytic cells, including bactericidal protein, hydrolytic enzymes, chelating proteins, and binding proteins.[56,57] This multiplicity of proteins in lysosomes will require considerable time in order to fully understand their various functions and spectra of biological activity. The proteins discussed in this chapter have focused upon antimicrobial proteins, those which have LPS-binding properties. Interestingly, not only specifically designed protein but proteins with independent functions have the capacity to bind to LPS. These proteins, which include LF, elastase, and cathepsin G may have physiologically important functions in controlling biological activities of LPS.[47-55]

A common property of both LPS-binding and bactericidal proteins in granulocytes is the presence of strongly hydrophobic and strongly basic domains; however, there is in most instances strikingly little sequence homology between them (with the exception of the BPI and LBP). It is generally recognized that most of these proteins are likely to bind to the lipid A region of LPS. These general properties of neutrophil granule LPS binding proteins are quite similar to the antibiotic polymyxin B, which is a well-characterized LPS neutralizing antibiotic. In the recent extensive studies on the mechanism of binding between LZM and LPS, it was clearly shown that binding to LPS also required hydrophobic and ionic interactions similar to polymyxin B. However, LZM and polymyxin B do not directly compete with each other in binding to LPS, and qualitative differences have been postulated. Interaction of LZM and LPS is also highly dependent on the environment such as temperature, ionic strength, and pH, and multiple binding modes have been suggested. These observations strongly indicate that this complex is not a type of specific ligand-receptor interaction. However, the available evidence summarized in this chapter strongly supports the physiological importance of the complex formation.

The recent development of molecular biological techniques have allowed us to examine complete sequences of biologically important proteins and peptides. Homology searches of these proteins indicated the extensive similarity between LBP in the serum and BPI in granulocytes.[15,16] The functions of both proteins are similar; thus, a close evolutionary relationship can be postulated. A second interesting observation is the homology between CAP37 and serine proteases.[25] Substitution of only a single amino acid significantly altered the function of the protein. The sequencing of the complementary as well as genomic DNA of some of the proteins, such as defensin and bacte-necin, also clarified the processing of these molecules to the mature active products.[59-62] Understanding of the structure of proteins at the molecular level has become increasingly valuable in applying their biological function. Extensive studies will be required to fully appreciate and understand the non-oxidative antimicrobial systems and LPS-neutralizing systems. One of the most important goals is to reduce the endotoxin shock and related diseases. There must be no limitation of origin of LPS-binding and/or anti-LPS proteins. They will originate from biologic fluids, micororganisms, as well as syntheses.

In addition to the proteins summarized in this section, there would be many antimicrobial proteins in granulocyte lysosomes. As already discussed above, the bactericidal action would be closely related to the high affinity to LPS, thus, it would be probable that some of these proteins will demonstrate binding to LPS and that some other proteins will be discovered.

A final important question is that of which mechanism underlies protein-mediated neutralization of the endotoxicity of LPS. One possible mechanism is that the hydrophobic region of the protein inserts into the macromolecular LPS micellar structure via lipid A interactions and sufficiently stabilizes the micelle structure so that lipid A cannot be made accessible to host target molecules. LPS-mediated biological activities are now generally known to be mediated by cell surface-specific receptor proteins, which are closely associated with the machinery of cytoplasmic second messengers.[1] Thus, failure of LPS to bind to specific receptor molecules may result in reducing the biological activities. Another mechanism which also may reduce the signal transduction is failure of LPS to bind to LPS-binding protein such as LBP which has a specific receptor on the cell surface.[58] It is, in any case, clear that many questions remain to be answered to fully understand these mechanisms.

ACKNOWLEDGMENTS

I am grateful to Prof. David C. Morrison of University of Kansas Medical Center and Prof. Toshiro Yadomae of Tokyo College of Pharmacy for advice and useful discussions.

REFERENCES

1. **Morrison, D. C.,** The case for specific lipopolysaccharide receptors expressed on mammalian cells, *Microb. Pathogenesis,* 7, 389, 1989.
2. **Morrison, D. C.,** Nonspecific interaction of bacterial lipopolysaccharides with membranes and membrane components, in *Cellular Biology of Endotoxins,* Berry, L. J., Ed., Elsevier/North-Holland, Amsterdam, 1985, 25.

3. **Morrison, D. C. and Ryan, J. L.,** Endotoxins and disease mechanisms, *Annu. Rev. Med.*, 38, 417, 1987.

4. **Morrison, D. C. and Ulevitch, R. J.,** The effects of bacterial endotoxins on host mediation systems, *Am. J. Pathol.*, 93, 527, 1978.

5. **Burrell, R.,** Immunomodulation by bacterial endotoxin, *Crit. Rev. Microbiol.*, 17, 189, 1990.

6. **Weiss, J., Elsbach, P., Olsson, I., and Odegerg, H.,** Purification and characterization of a potent bactericidal and membrane active protein from the granules of human polymorphonuclear leykocytes, *J. Biol. Chem.*, 253, 2664, 1978.

7. **Weiss, J., Muello, K., Victor, M., and Elsbach, P.,** The role of lipopolysaccharides in the action of the bactericidal/permeability-increasing neutrophil protein on the bacterial envelope, *J. Immunol.*, 132, 3109, 1984.

8. **Shafer, W. M., Martin, L. E., and Spitznagel, J. K.,** Cationic antimicrobial proteins isolated from human neutrophil granulocytes in the presence of diisopropyl fluorophosphate, *Infect. Immun.*, 45, 29, 1984.

9. **Hovde, C. J. and Gray, B. H.,** Physiological effects of a bactericidal protein from human polymorphonuclear leukocytes on *Pseudomonas aeruginosa, Infect. Immun.*, 52, 90, 1986.

10. **Mannion, B. A., Weiss, J., and Elsbach, P.,** Separation of sublethal and lethal effects of the bactericidal/permeability increasing protein on *Escherichia coli, J. Clin. Invest.*, 85, 853, 1990.

11. **Weiss, J., Victor, M., and Elsbach, P.,** Role of charge and hydrophobic interactions in the action of the bacterial/permeability-increasing protein of neutrophils on Gram-negative bacteria, *J. Clin. Invest.*, 71, 540, 1983.

12. **Weiss, J., Hutzler, M., and Kao, L.,** Environmental modulation of lipopolysaccharide chain length alters the sensitivity of *Escherichia coli* to the neutrophil bactericidal/permeability-increasing protein, *Infect. Immun.*, 51, 594, 1986.

13. **Marra, M. N., Wilde, C. G., Griffith, J. E., Snable, J. L., and Scott, R. W.,** Bactericidal/permeability-increasing protein has endotoxin-neutralizing activity, *J. Immunol.*, 144, 662, 1990.

14. **Ooi, C. E., Weiss, J., Elsbach, P., Frangione, B., and Mannion, B.,** A 25-kDa NH$_2$-terminal fragment carries all the antibacterial activities of the human neutrophil 60-kDa bactericidal/permeability-increasing protein, *J. Biol. Chem.*, 262, 14891, 1987.

15. **Gray, P. W., Flaggs, G., Leong, S. R., Gumina, R. J., Weiss, J., Ooi, C. E., and Elsbach, P.,** Cloning of the cDNA of a human neutrophil bactericidal protein, *J. Biol. Chem.*, 264, 9505, 1989.

16. **Schumann, R. R., Leong, S. R., Flaggs, G. W., Gray, P. W., Wright, S. D., Mathison, J. C., Tobias, P. S., and Ulevitch, R. J.,** Structure and function of lipopolysaccharide binding protein, *Science*, 249, 1429, 1990.

17. **Tobias, P. E., Soldau, K., and Ulevitch, R. J.,** Isolation of a lipopolysaccharide binding acute phase reactant from rabbit serum, *J. Exp. Med.*, 164, 777, 1986.

18. **Stinavage, P., Martin, L. E., and Spitznagel, J. K.,** O antigen and lipid A phosphoryl groups in resistance of *Salmonella typhimurium* LT-2 to nonoxidative killing in human polymorphonuclear neutrophils, *Infect. Immun.*, 57, 3894, 1989.

19. **Farley, M. M., Shafer, W. M., and Spitznagel, J. K.,** Lipopolysaccharide structure determines ionic and hydrophobic binding of a cationic antimicrobial neutrophil granule protein, *Infect. Immun.*, 56, 1589, 1988.

20. **Shafer, W. M., Engle, S. A., Martin, L. E., and Spitznagel, J. K.,** Killing of *Proteus mirabilis* by polymorphonuclear leukocyte granule proteins: evidence for species specificity by antimicrobial proteins, *Infect. Immun.*, 56, 51, 1988.

21. **Farley, M. M., Shafer, W. M., and Spitznagel, J. K.,** Antimicrobial binding of a radiolabeled cationic neutrophil granule protein, *Infect. Immun.*, 55, 1536, 1987.

22. **Shafer, W. M., Martin, L. E., and Spitznagel, J. K.**, Late intraphagosomal hydrogen ion concentration favors the *in vitro* antimicrobial capacity of a 37-kilodalton cationic granule protein of human neutrophil granulocytes, *Infect. Immun.*, 53, 651, 1986.

23. **Casey, S. G., Shafer, W. M., and Spitznagel, J. K.**, Anaerobiosis increases resistance of *Neisseria gonorrhoeae* to O₂-independent antimicrobial proteins from human polymorphonuclear granulocytes, *Infect. Immun.*, 47, 401, 1985.

24. **Spitznagel, J. K., Pereira, H. A., Martin, L. E., Guzman, G. S., and Shafer, W. M.**, A monoclonal antibody that inhibits the antimicrobial action of a 57 kD cationic protein of human polymorphonuclear leukocytes, *J. Immunol.*, 139, 1291, 1987.

25. **Pereira, H. A., Spitznagel, J. K., Pohl, J., Wilson, D. E., Morgan, J., Palings, I., and Larrick, J. W.**, CAP 37, a 37kD human neutrophil granule cationic protein shares homology with inflammatory proteinases, *Life Sci.*, 46, 189, 1990.

25a. **Ohno, N., Morrison, D. C., and Spitznagel, J. K.**, unpublished.

26. **Ganz, T., Selsted, M. E., and Lehrer, R. I.**, Defensins, *Eur. J. Haematol.*, 44, 1, 1990.

27. **Ganz, T., Selsted, M. E., Szklarek, D., Harwig, S. S. L., Daher, K., Bainton, D. F., and Lehrer, R. I.**, Defensins, natural peptide antibiotics of human neutrophils, *J. Clin. Invest.*, 76, 1427, 1985.

28. **Selsted, M. E., Harwig, S. S. L., Ganz, T., Schilling, J. W., and Lehrer, R. I.**, Primary structures of three human neutrophil defensins, *J. Clin. Invest.*, 76, 1436, 1985.

29. **Wilde, C. G., Griffith, J. E., Marra, M. N., Snable, J. L., and Scott, R. W.**, Purification and characterization of human neutrophil peptide 4, a novel member of the defensin family, *J. Biol. Chem.*, 264, 11200, 1989.

30. **Hill, C. P., Yee, J., Selsted, M. E., and Eisenberg, D.**, Crystal structure of defensin HNP-3, and amphiphilic dimer: mechanisms of membrane permeabilization, *Science*, 251, 1481, 1991.

31. **Grutter, M. G., Weaver, L. H., and Mathews, B. W.**, Goose lysozyme structure: an evolutionary link between hen and bacteriophage lysozymes?, *Nature*, 303, 828, 1983.

32. **Stewart, C. -B., Schilling, J. W., and Wilson, A. C.**, Adaptive evolution in the stomach lysozymes of foregut fermenters, *Nature*, 330, 401, 1987.

33. **Smith-Gill, S. J., Mainhart, C., Lavoie, T. B., Feldmann, R. J., Drohan, W., and Brooks, B. R.**, A three-dimensional model of an anti-lysozyme antibody, *J. Mol. Biol.*, 194, 713, 1987.

34. **Hall, L. and Campbell, P. N.**, α-Lactalbumin and related proteins: a versatile gene family with an interesting parentage, *Essays Biochem.*, 22, 1, 1986.

35. **Kimelberg, H. K.**, Protein-liposome interactions and their relevance to the structure and function of cell membranes, *Mol. Cell. Biochem.*, 10, 171, 1976.

36. **Peeoeller, H., Yeomans, F. G., Kydon, D. W., and Sharp, A. R.**, Water molecule dynamics inhydrated lysozyme, a deuteron magnetic resonance study, *Biophys. J.*, 49, 943, 1986.

37. **Millar, M. R. and Inglis, T.**, Influence of lysozyme on aggregation of *Staphylococcus aureus*, *J. Clin. Microbiol.*, 25, 1587, 1987.

38. **Redfield, C. and Dobson, C. M.**, Sequential ¹H-NMR assignments and secondary structure of hen egg white lysozyme in solution, *Biochemistry*, 27, 122, 1988.

39. **Allen, P. M., Babbitt, B. P., and Unanue , E. R.**, T-cell recognition of lysozyme: the biochemical basis of presentation, *Immunol. Rev.*, 98, 171, 1987.

40. **Ohno, N. and Morrison, D. C.**, Lipopolysaccharide interactions with lysozyme differentially affect lipopolysaccharide immunostimulatory activity, *Eur. J. Biochem.*, 186, 629, 1989.

41. **Ohno, N. and Morrison, D. C.**, Lipopolysaccharide interaction with lysozyme, binding of lipopolysaccharide to lysozyme and inhibition of lysozyme enzymatic activity, *J. Biol. Chem.*, 264, 4434, 1989.

42. **Ohno, N. and Morrison, D. C.**, Effect of lipopolysaccharide chemotype structure on binding and inactivation of hen egg lysozyme, *Eur. J. Biochem.*, 186, 621, 1989.
43. **Wollenwever, H. -W. and Morrison, D. C.**, Synthesis and biochemical characterization of a photoactivatable iodinatable, cleavable bacterial lipopolysaccharide derivative, *J. Biol. Chem.*, 260, 15068, 1985.
44. **Ohno, N., Tanida, N., and Yadomae, T.**, Characterization of complex formation between lipopolysaccharide and lysozyme, *Carbohydr. Res.*, 214, 115, 1991.
45. **Ohno, N. and Morrison, D. C.**, Structural specificity of the binding of bacterial lipopolysaccharide (LPS) to hen egg lysozyme, in *The Immune Response to Structurally Defined Proteins: Lysozyme Model*, Smith-Gill, S. and Sercarz, E., Eds., Adenine Press, Guilderland, NY, 1989, 97.
46. **Tanida, N., Ohno, N., Yadomae, T., Matsuura, M., Kiso, M., and Hasegawa, A.**, Modification of immunopharmacological activities of synthetic monosaccharide lipid A analogue, GLA-60, by lysozyme, in Proceedings of Microbial Infections, Role of Biological Response Modifiers, Clearwater Beach, Florida, 1991.
47. **Molloy, A. L. and Winterbourn, C. C.**, Release of iron from phagocytosed *Escherichia coli* and uptake by neutrophil lactoferrin, *Blood*, 75, 984, 1990.
48. **Arnold, R. R., Cole, M. F., and McGhee, J. R.**, A bactericidal effect for human lactoferrin, *Science*, 197, 263, 1977.
49. **Arnold, R. R., Brewer, M., and Gauthier, J. J.**, Bactericidal activity of human lactoferrin: sensitivity of a variety of microorganisms, *Infect. Immun.*, 28, 893, 1980.
50. **Broxmeyer, H. E., Bicknell, D. C., Gillis, S., Harris, E. L., Pelus, L. M., and Seldge, G. W., Jr.**, Lactoferrin: affinity purification from human milk and polymorphonuclear neutrophils using monoclonal antibody (II2C) to human lactoferrin, development of an immunoradiometric assay using II2C, and myelopoietic regulation and receptor-binding characteristics, *Blood Cells*, 11, 429, 1986.
51. **Miyazawa, K., Mantel, C., Lu, L., Morrison, D. C., and Broxmeyer, H. E.**, Lactoferrin-lipopolysaccharide interactions. Effect on lactoferrin binding to monocyte/macrophage-differentiated HL-60 cells, *J. Immunol.*, 146, 723, 1991.
52. **Rochard, E., Legrand, D., Mazuerier, J., Montreuil, J., and Spik, G.**, The N-terminal domain I of human lactotransferrin binds specifically to phytohemagglutinin-stimulated peripheral blood human lymphocyte receptors, *FEBS Lett.*, 255, 201, 1989.
53. **Salvesen, G., Farley, D., Shuman, J., Przybyla, A., Reilly, C., and Travis, J.**, Molecular cloning of human cathepsin G: structural similarity to mast cell and cytotoxic T lymphocyte proteinases, *Biochemistry*, 26, 2289, 1987.
54. **Shafer, W. M., Onunka, V. C., and Martin, L. E.**, Antigonococcal activity of human neutrophil cathepsin G, *Infect. Immun.*, 54, 184, 1986.
55. **Shafer, W. M., and Morse, S. A.**, Cleavage of the protein III and major iron-regulated protein of *Neisseria gonorrhoeae* by lysosomal cathepsin G, *J. Gen. Microb.*, 133, 155, 1987.
55a. **Ohno, N.**, unpublished.
56. **Gabay, J. E., Scott, R. W., Campanelli, D., Griffith, J., Wilde, C., Marra, M. N., Seeger, M., and Nathan, C. F.**, Antibiotic proteins of human polymorphonuclear leukocytes, *Proc. Natl. Acad. Sci. U.S.A.*, 86, 5610, 1989.
57. **Gabay, J. E., Heiple, J. M., Cohn, Z. A., and Nathan, C. F.**, Subcellular location and properties of bactericidal factors from human neutrophils, *J. Exp. Med.*, 164, 1407, 1986.
58. **Wright, S. D., Ramos, R. A., Tobias, P. S., Ulevitch, R. J., and Mathison, J. C.**, CD14, a receptor for complexes of lipopolysaccharide (LPS) and LPS binding protein, *Science*, 249, 1431, 1990.
59. **Zanetti, M., Litteri, L., Gennaro, R., Horstmann, H., and Romeo, D.**, Bactenecins, defence polypeptides of bovine neutrophils, are generated from precursor molecules stored in the large granules, *J. Cell. Biol.*, 111, 1363, 1990.

60. **Frank, R. W., Gennaro, R., Schneider, K., Przybylski, M., and Romeo, D.,** Amino acid sequences of two proline-rich bactenecins, antimicrobial peptides of bovine neutrophils, *J. Biol. Chem.,* 265, 18871, 1990.
61. **Skerkavaj, B., Romeo, D., and Gennaro, R.,** Rapid membrane permeabilization and inhibition of vital functions of gram-negative bacteria by bactenecins, *Infect. Immun.,* 58, 3724, 1990.
62. **Zanetti, M., Litteri, L., Griffiths, G., Gennaro, R., and Romeo, D.,** Stimulus-induced maturation of probactenecins, precursors of neutrophil antimicrobial polypeptides, *J. Immunol.,* 146, 4295, 1991.

Chapter 18

PROCESSING OF LPS BY PHAGOCYTES

Alice L. Erwin and Robert S. Munford

TABLE OF CONTENTS

I. INTRODUCTION

The idea that animals might be able to modify the biological activities of lipopolysaccharides (LPS) has intrigued investigators for decades. Although host mechanisms for detoxifying LPS have received the most attention, the possibility that processing of LPS by animal cells might increase the potency of LPS has also been considered. In this chapter, we shall review the evidence that animal cells alter the chemical structure of LPS, with particular attention to the enzymatic degradation of LPS by phagocytes, and discuss various roles that such modifications of LPS structure might play *in vivo*. The noncovalent interactions of LPS with serum lipoproteins and with other LPS-binding molecules will not be dealt with here, nor will the general topic of LPS detoxification (reviewed by Skarnes[1]) except where there is evidence for structural alteration of the LPS.

As reviewed in previous chapters in this volume, virtually all of the biological responses of animals to LPS can be induced by lipid A. It seems reasonable to expect, therefore, that a catabolic reaction that altered the potency of LPS would change the structure of its lipid A moiety, rather than that of its polymeric carbohydrate. Deacylation, dephosphorylation, or hydrolysis of the glycosidic linkage between the two glucosamines of lipid A might be expected to reduce the biological activity of the LPS (Chapter 5). In contrast, removal or alteration of the O-antigen or the carbohydrate core would be expected to have little effect on the bioactivities of LPS, although this statement should be qualified by noting that the carbohydrate chains of LPS may modulate lipid A bioactivities and that there are reports that the KDO-heptose region of the carbohydrate core may also possess biological activity (Chapter 8). Further, the structures of the lipid A moiety of LPS and of the inner region of the carbohydrate core are highly conserved among Gram-negative bacteria, while the outer region of the core and the O-antigen (if present) are much more variable. Animal enzymes whose primary function is catabolism of LPS might be expected to attack the more highly conserved features of LPS structure.

II. STUDY OF LPS PROCESSING: METHODOLOGY

A. RADIOLABELING LPS

The primary reagent for studying the uptake and processing of LPS by tissues or cells is radiolabeled LPS; its use is often combined with antigenic detection of LPS (e.g., immunoperoxidase staining of tissue sections), with chemical analysis of LPS, or with assays of LPS bioactivity. Each of the methods used for labeling LPS has different advantages and limitations, and we will therefore discuss labeling methods in some detail.

1. Extrinsic Labeling

Much early work on the *in vivo* fate of LPS used LPS that was labeled after purification with ^{51}Cr or [125]I. There are two potential problems with this approach: first, the specificity of labeling depends on the purity of the LPS (if contaminants are present, they may be labeled also); second, removal of the label from the LPS *in vivo* might occur separately from processing of the LPS molecule itself. Because of the latter problem, extrinsically labeled LPS, if it must be used, is best suited to short-term studies. In another approach to labeling purified LPS, ^{3}H may be incorporated into various positions in the carbohydrate chain of many LPS following periodate treatment and reduction with [3H]sodium borohydride;[2] the precise location of the label is not known, and this method alters the structure of the carbohydrate. Relatively high specific activities have been obtained, however.[3]

2. Intrinsic Labeling

To study the catabolism of LPS, most workers have used intrinsically labeled LPS, usually produced by growing bacteria in radiolabeled precursor molecules; labeling distinct regions of LPS with different isotopes has been particularly useful. If glucose is the primary carbon source, growth of bacteria in [14C]glucose should label the LPS carbons uniformly. The *fatty acids* of Gram-negative bacteria may be labeled by adding radiolabeled acetate to the growth medium.[4,5] In the purified LPS, the specificity with which label is incorporated into acyl chains is usually very high, as shown by recovery of the radiolabel in fatty acids after chemical hydrolysis of the LPS. The use of radiolabeled acetate as an energy source may be minimized by increasing the concentration of (nonradioactive) glucose in the medium. The specificity of incorporation of radiolabeled *sugars* into LPS is increased by the presence of certain bacterial mutations. For example, *galE* mutations (resulting in deficiency of glucose-4-epimerase) facilitate labeling the O-antigen and oligosaccharide core of *Salmonella typhimurium* or certain *Escherichia coli* strains with labeled galactose;[6,7] similarly, *pmi* strains (deficient in phosphomannose-isomerase) incorporate mannose exclusively into O-antigens that contain this sugar;[6] *nag* strains lack the ability to deaminate glucosamine and will incorporate labeled glucosamine and *N*-acetylglucosamine primarily into LPS (labeling lipid A and in some cases the carbohydrate chain) and into peptidoglycan.[6] Since both the oligosaccharide core and the lipid A moiety of enteric bacteria contain *phosphate* substituents, radioactive phosphate would be expected to be incorporated into both these regions; use of deep rough mutants may result in more specific labeling of the lipid A phosphates.[8] ^{32}P can also be incorporated at the 4'-position of lipid IV$_A$ *in vitro*.[9]

During bacterial growth, incorporation of the radiolabel may not be confined to LPS. For example, growth of bacteria in radioactive phosphate or acetate will label both LPS and phospholipids; unless contaminating phospholipids are removed from the final LPS preparation, their presence will complicate the analysis of LPS catabolism. Another potential source of contamination is radiolabeled precursors of LPS, such as lipid-linked O-antigen that has not yet been ligated to the oligosaccharide core to form smooth LPS. As noted above, glucosamine and *N*-acetylglucosamine will be incorporated into peptidoglycan as well as into LPS. When *galE* or *pmi* mutants are used, radiolabeled galactose or mannose, respectively, is incorporated almost entirely into LPS; while these provide the best available methods for specifically labeling LPS in the bacterial cell, they do not label all molecules of LPS uniformly (see below).

For certain analyses, it may be desirable to quantify LPS uptake or catabolism on a molar rather than a weight basis, and the choice of radiolabel determines whether this is possible. For example, glucosamine is incorporated into smooth *S. typhimurium* LPS on an equimolar basis: regardless of the length of its O-polysaccharide, each molecule of LPS will contain three glucosamine residues (two in lipid A and one in the oligosaccharide core). In contrast, the galactose content of *S. typhimurium* LPS depends on the length of the saccharide chain (each O-antigenic repeat unit contains one galactose residue), and the degree of acylation of the lipid A moiety may also vary with saccharide chain length.[10] A finding that smooth *S. typhimurium* LPS recovered from animals cells was enriched in a galactose label and depleted in a fatty acid label might suggest deacylation, but it might also represent preferential recovery of naturally underacylated LPS or of molecules that have long saccharide chains. Since LPS preparations are heterogeneous, containing molecules that differ not only in carbohydrate length but also in degree of acylation and substitution with phosphates and other polar groups, it seems likely that cells process different molecules in an LPS preparation in different ways. For example, there is evidence that the length of the saccharide chain may affect the ability of LPS to bind to cells and its potency in inducing certain cellular responses.[11] Rarely does recovery of radiolabel approach 100%; it should not be assumed without experimental evidence that the fraction of LPS recovered from animal cells is representative of the total population of LPS molecules.

B. ANALYSIS OF LPS CATABOLISM

In general, control LPS should be subjected to the same experimental conditions as the processed LPS, except for the processing step. For example, if the impact of cellular processing on LPS is to be studied, one should also consider the effect of incubating control LPS (or control bacteria, if used) under identical conditions in cell culture medium that lacks cells. This is particularly important if intact bacteria, rather than purified LPS, are used: certain bacterial metabolic pathways continue to function after phagocytosis, though viability is lost rapidly.[12] It is therefore desirable to compare phago-cytosed bacteria not only with freshly cultured bacteria but with bacteria that have been maintained in conditions that do not support growth. If the impact of a degradative enzyme on LPS bioactivities is to be studied, the control LPS should be incubated for the same time period under identical conditions, without the enzyme or with inactivated (e.g., heat-denatured) enzyme. This approach takes into account alterations in LPS structure (such as deacylation during prolonged incubation at neutral pH) that may otherwise give misleading conclusions about the impact of the processing event under study.

For studies of LPS uptake and catabolism, the LPS should be radiolabeled to the highest possible specific activity so that experiments can be carried out using concentrations of LPS that are as low as possible — ideally approaching concentrations found *in vivo*. Macrophages and other cells can be stimulated *in vitro* with less than 1 ng of LPS per milliliter, but no reported method for labeling LPS results in a specific activity high enough to allow studies of uptake and degradation of such low concentrations.

LPS with high specific radioactivity is particularly useful when the *extent* of degradation is to be determined: unless degradation of LPS proceeds as rapidly as uptake, the fraction of LPS degraded during a given time period will be inversely related to the amount of LPS taken up by the tissues or cells (Figure 1). Most studies of LPS processing have used amounts of LPS much greater than those likely to be present *in vivo,* and may therefore have under-estimated the ability of cells to degrade LPS.

The rate of catabolism of LPS by cells *in vitro* can best be studied using an initial uptake (or "load") period, usually 60 min or less, after which the cells are washed and incubated further; thus analysis of the rate of degradation is not complicated by uptake of additional LPS from the medium. There may be release of LPS and reuptake during the chase period, however.[13,14]

1. Release of Radiolabel from LPS

The catabolism of LPS may be reflected by the appearance of radiolabel in a chemical fraction or physical state different from that of the unaltered LPS. For example, fatty acids released from LPS will be extracted into the chloroform phase of a Bligh-Dyer extraction, while LPS partitions into the aqueous phase. A second example is the release of ^{32}P from LPS with recovery of the radioactivity in a dialysate. When the appropriate controls are done to exclude spontaneous decomposition of the LPS, such observations provide evidence for catabolism. The evidence should be strengthened by further analysis of the released products (e.g., identification of released fatty acids

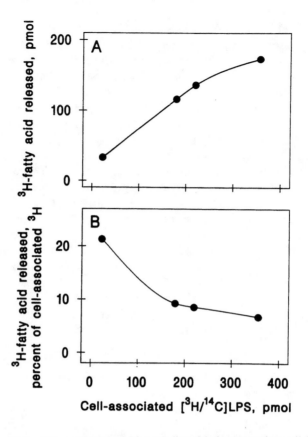

FIGURE 1. LPS uptake may outpace LPS catabolism. Murine macrophages (1.5 × 10⁶ per well) were incubated for 20 h with smooth *S. typhimurium* LPS (labeled with ¹⁴C in glucosamine residues and ³H in fatty acids, 0.8 μg per well), and the uptake of LPS and the release of ³H into a chloroform-extractable form were determined. The uptake of LPS was controlled by adding anti-*S. typhimurium* IgG in varying concentrations. While the *total* deacylation increased as LPS uptake increased (panel A), the *percent* deacylation (percent of cell-associated ³H that was chloroform extractable) decreased with increased uptake (panel B). The finding that only a small percentage of the LPS taken up by cells is degraded may thus result in part from choosing experimental conditions that increase the uptake of LPS. These data have been published previously in a slightly different form.[13]

by thin-layer or high-performance liquid chromatography) and by recovery and analysis of the partially degraded LPS.

Changes in isotope ratios of doubly labeled LPS (usually containing ³H and ¹⁴C in different structural components) are often interpreted as evidence of structural alteration. Such data are difficult to interpret, however, without careful attention to quench correction. Combustion of samples and separate recovery of ³H₂O and ¹⁴CO₂ for scintillation counting[15-17] is the preferred method for accurate quantitation of radioactivity. The use of a second LPS preparation in which isotope specificities are reversed provides an additional control.[16]

2. Alteration of Physical Properties of LPS

An alteration of physical properties, such as buoyant density, electrophoretic mobility, or behavior during gel filtration chromatography, is sometimes taken as evidence of LPS processing. However, such observations provide only indirect evidence of catabolism. Comparison of the physical properties of untreated LPS with those of LPS in cell culture supernatants, cell lysates, or tissue homogenates may be misleading, since the (noncovalent) association of LPS with animal-derived lipids and proteins may alter the property being examined. Even if the LPS recovered from cells is partially purified (usually by phenol extraction), it is likely that some cellular molecules remain with the LPS. Further, the aggregate structure of LPS is affected by the nature of the noncovalently bound cations that associate with it; these may well change during uptake of LPS by cells and/or the recovery of LPS from cells, resulting in an alteration of physical structure without any chemical change.

3. Alteration of LPS Biological Activites

There is a substantial literature on the detoxification of LPS by plasma and by cells;[1] while in some cases, loss of bioactivity may be associated with catabolism of LPS, the two processes need not be connected. For example, the association of LPS with lipoproteins and other plasma components decreases its potency substantially, but potency may be restored by reextraction of the LPS.[1,18]

The evaluation of the biological activity of LPS recovered from cells or from animals is complicated when the extracts may contain compounds of animal origin that have bioactivities similar to those elicited by the LPS. For example, the supernatants of macrophages incubated with LPS are likely to contain cytokines able to elicit cellular responses that may easily be mistaken for responses to the LPS in the supernatants; this problem is minimized by phenol extraction of the recovered LPS.[19] Comparison of the responses of target cells derived from LPS-resistant mice with those from LPS-responsive mice may also be useful.[19]

Comparison of the biological activity of processed LPS with that of the initial LPS preparation should ideally be done by constructing dose-response curves for each preparation. Any change in potency can then be quantified graphically by determining the horizontal distance between the linear regions of the curves (Figure 2). When it is not possible to assay more than one or two concentrations of the processed LPS, comparisons should be carried out using concentrations expected to give less than maximal responses. The choice of the LPS concentration for assay may determine whether a change in potency is detected, as illustrated in Figure 2.

The choice of bioassay is also important. The study of bioactivities of lipid A precursors and of chemically synthesized analogs of lipid A has established the structural requirements for several responses to lipid A (reviewed in Chapter 5). Only two activities clearly require the *complete* lipid

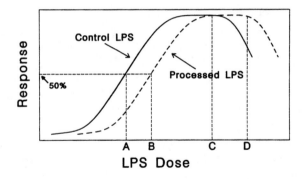

FIGURE 2. Role of the dose-response curve in evaluating LPS bioactivity. An idealized dose-response curve is presented schematically: for part of the range of doses tested, the measured response has an approximately linear relation to the log of the LPS concentration; the response plateaus at higher doses, and falls at extremely high doses. A hypothetical processing step results in a shift of the curve to the right. The reduction in potency can be expressed as the ratio of the concentrations required for half-maximal response, as illustrated by the horizontal dotted line: here potency is reduced by a factor equal to the ratio of concentrations B and A. If generation of complete dose-response curves is not feasible, the choice of concentrations to test must be made with care, since the reduced potency of the processed LPS is only apparent when the LPS preparations are tested at doses on the linear regions of both curves. If the processed LPS and the control LPS were assayed only at concentration C, the reduction in potency would not be detected; assaying only at concentration D would give the misleading result that the potency was actually increased by processing.

A structure: these are the chick embryo lethality test and the dermal Shwartzman reaction. Many responses to lipid A can be elicited by partial structures that lack one or both phosphate substituents or the secondary acyl chains. These responses include lethality for galactosamine-treated or actinomycin D-treated mice, pyrogenicity in rabbits, *Limulus* lysate clotting, and murine splenocyte mitogenicity. Certain responses can also be elicited by mono-glucosamine lipid A analogs. Moreover, in some cases, partial lipid A structures are able to inhibit responses to lipid A and to LPS (if both agonist and antagonist molecules were present in the same preparation of processed LPS, the net biological effect might be very difficult to interpret!). Thus the choice of bioactivity test may determine whether an effect of processing is seen. For example, LPS that had undergone a loss of secondary acyl chains would be predicted to have greatly reduced potency in the dermal Shwartzman reaction, yet be active when tested for *Limulus* activity or lethality for galactosamine-treated mice.

III. *IN VIVO* STUDIES OF LPS UPTAKE AND CATABOLISM

A. PERSISTENCE OF LPS IN ANIMALS; TISSUE AND CELLULAR LOCALIZATION

Numerous studies (reviewed by Vlevitch in Reference 20) have reported that a fraction of intravenously injected LPS is cleared from the circulation with a half-life of minutes, appearing primarily in phagocytic cells of the liver and spleen; a smaller part of the rapidly cleared LPS can be recovered from other organs, including lungs, kidneys, and adrenal glands. The remainder of the injected LPS is cleared from the circulation much more slowly (with a half-life of hours); it is bound to serum lipoproteins and is removed by cells that take up lipoproteins, primarily in the liver.[20] The reports described below are consistent with this general scheme, finding LPS primarily in hepatocytes and Kupffer cells of the liver, but also within phagocytes in other tissues (e.g., lungs) soon after injection. LPS is not degraded rapidly and can readily be found in tissues days or weeks after injection. Cellular LPS has been found in lysosomes or similar organelles in some cases, but there is also evidence that LPS can be found in the nucleus or in other parts of the cell.[21,22]

In a series of papers published in 1983 through 1985,[15-17,23,24] Freudenberg and colleagues described the tissue and cellular localization of LPS following intravenous injection into rats. Using a combination of radiologic and immunochemical techniques, they found that degradation of injected LPS occurs very slowly *in vivo*. (Their anti-LPS antisera were probably specific for the O-antigen, though this was not always described.) Kleine et al.[15] injected rats intravenously with *S. abortus equi* LPS labeled by growth of bacteria in [³H]acetate, [¹⁴C]glucose, and [¹⁴C]glucosamine. Both isotopes were excreted into feces and urine at a low, fairly constant rate over the 14 d of observation, with excretion of ¹⁴C greater than that of ³H. After 14 d, 27% of injected ¹⁴C and 18.5% of injected ³H were still present in the liver; much smaller amounts of label were recovered from other organs. Freudenberg et al.[23] used both immunoperoxidase staining and radioactivity to detect LPS in various tissues over 14 d following injection of *S. abortus equi* LPS (labeled in glucosamine residues) into rats. Liver, spleen, and adrenal glands took up approximately the same amounts of radiolabeled LPS (when compared per gram of tissue), but the fate of LPS in these tissues differed. In the liver, LPS-specific peroxidase staining was seen first in macrophages, followed after 3 d by an apparent redistribution of label to hepatocytes. The maximum radioactivity in liver tissue was seen at 3 d after injection; at 14 d, 50% of the maximum level was still present. In contrast, antigenic detection of LPS was greatest at 24 h and disappeared much more rapidly than radioactivity. The spleen contained as much radioactivity per gram of tissue as the liver, but immunoperoxidase staining was much weaker; antigen was detected primarily in macrophages of the splenic sinusoids. Again, the radioactivity of LPS persisted longer than its antigenicity. In contrast, LPS in the adrenals was detected antigenically in cortical epithelial cells; the antigenic reactivity did not wane over the 14-d period of observation.

The same laboratory described the localization of *S. abortus equi* LPS in the lungs following intravenous injection of a shock-inducing dose.[17] Within an hour after injection, 3% of alveolar macrophages contained LPS (determined by immunoperoxidase staining of tissue sections); by 3 d, 100% of alveolar macrophages were LPS positive, and even at 14 d 98% of alveolar macrophages contained LPS. LPS was found in a variety of cell types within lung sections, including mononuclear phagocytes, granulocytes, alveolar epithelial cells, and vascular endothelium. LPS-positive phagocytes were found in blood vessels, suggesting that some of these cells were already carrying endotoxin when they migrated into the lung. The LPS in lung phagocytes was found in phagocytic vacuoles. In endothelial cells and in alveolar epithelial cells, LPS was seen both in intracellular organelles and also diffusely (without a clear membrane demarcation) in the neighborhood of cell junctions. While the amount of LPS in the lungs was small (0.9% of injected radioactivity, after 3 d), its presence appeared to be correlated with histologic evidence of lung injury, such as interstitial edema.

Studies in other laboratories have examined the localization of LPS within the liver, spleen, and kidneys. Willerson et al. used autoradiography and cell fractionation to determine the sites of radioactivity after intravenous injection of uniformly labeled [^{14}C]LPS from *S. enteritidis* into mice and rats:[21] they found radiolabel primarily in nuclear and mitochondrial subfractions of both the liver and spleen. Autoradiographs of tissue sections prepared 3.5 h after injection showed label over the red pulp of the spleen, particularly in a small mononuclear cell; in the kidney, label was found over the glomeruli and the interlobular vessels; in the liver, label was found over the hepatocytes, the Kupffer cells, and vessels in the portal spaces. Hepatocytes had label over both the cytoplasm and nucleus. Zlydaszyk and Moon[25] reported also that, in livers of mice injected with LPS, radiolabel was found in both parenchymal cells and Kupffer cells (1 h after injection of ^{51}Cr-labeled *S. typhimurium* LPS). Of the radiolabel recovered in the liver, 40% was found in the nuclear fraction of a liver homogenate, 20% in a lysosomal-mitochondrial fraction, and 20% in a microsomal fraction. In contrast, Praaning-van Dalen et al.[26] found radiolabel associated only with Kupffer cells, and not with hepatocytes, 30 min after injection of ^{51}Cr-labeled LPS. The difference between these findings and those of other authors might be due to the short interval between injection and sacrifice, the type of the LPS (which was not specified), or the route of injection (into the mesenteric vein rather than into the tail vein).

The fate of injected LPS is influenced by its structure and physical state. Hopf et al.[27] reported that, 60 min after injection, smooth *S. abortus equi* LPS in the liver is confined to sinusoidal cells, while rough LPS or lipid A is detectable in both hepatocytes and Kupffer cells (by immunofluorescence). When smooth LPS was disaggregated by conversion to the triethylamine salt form, it appeared in the liver much more slowly than the more-aggregated sodium salt form. These data are consistent with other reports that rough LPS are removed from the circulation very rapidly after injection, while smooth LPS or LPS-lipoprotein complexes circulate much longer.[20] Although the explanation for these interesting results is not known, an intriguing recent report suggests that the uptake of LPS by hepatocytes may be mediated by KDO receptors on these cells;[28] this mechanism may favor uptake of rough

LPS more than smooth. The route of injection and the use of anesthesia may also affect the fate of LPS: in a paper cited above, Kleine et al.[15] reported that, when LPS (labeled with ^{3}H in fatty acids and ^{14}C in carbohydrate) was administered intraperitoneally without anesthesia, the appearance of both isotopes in urine and feces was much more rapid than when LPS was injected intravenously without anesthesia or intraperitoneally with anesthesia.

The rapid appearance of LPS in phagocytic cells following its injection into animals indicates that these cells are potential sites of LPS catabolism. The observations that the radioactivity of injected LPS, and in some studies its antigenic activity, can be detected for weeks after injection indicate that catabolism is incomplete; whether the lipid A moiety of the catabolized LPS retains biological activity after such long periods is not known, however. Evidence that LPS may retain immunogenicity for a long time *in vivo* was provided in an early study by Britton et al;[29] mice were immunized by intraperitoneal injection of either heat-inactivated *E. coli* O55:B5 or sheep red blood cells; at intervals after immunization, they were lethally irradiated and repopulated with nonimmune syngeneic lymphoid cells. Mice developed a primary immune response to sheep red blood cells if lymphoid cells were replaced within 14 d after immunization. The bacterial immunogen persisted much longer: antibodies reactive with the homologous LPS were produced even if irradiation and repopulation were delayed as long as 45 d after immunization. In an early study of *E. coli* O113 arthritis in rabbits, Braude et al. had reported a similar result:[30] immunogenic "somatic antigen" (O-antigen) persisted in synovial fluid exudates several days after the inoculated bacteria could no longer be cultured.

B. IMMUNOLOGIC IDENTIFICATION OF LPS IN CLINICAL SAMPLES

In addition to the studies of LPS in experimental animals described above, there is evidence that, in humans, LPS may sometimes be present in normally sterile sites, perhaps contributing to the pathogenesis of disease. Hopf et al. used monoclonal antibodies reactive with lipid A to analyze tissue from liver biopsies and reported that lipid A was present in the hepatocytes and Kupffer cells of patients with primary biliary cirrhosis, but not of patients with chronic viral hepatitis.[31] It is likely that small amounts of endotoxin enter the portal circulation from the intestine and are normally cleared by the liver; when liver clearance function is impaired, this portal endotoxemia may contribute to liver disease (reviewed by Nolan[32]). It is difficult to determine whether the presence of LPS (lipid A) in the livers of cirrhotic patients contributes to the pathogenesis of liver injury, as suggested by Hopf et al.[31] or, alternatively, may simply be a consequence of it.

Granfors et al.[33,34] used immunofluorescence to study inflammatory cells in synovial fluid collected from patients with reactive arthritis, reporting that they found O-antigenic material corresponding to that of the Gram-negative bacteria that were associated with the initial gastrointestinal illness (*Yersinia enterocolitica, Y. pseudotuberculosis, S. enteritidis,* or *S. typhimurium*). Bacterial cultures of the synovial fluids were negative. Synovial fluid cells from patients with rheumatoid arthritis and other inflammatory joint diseases were negative for these antigens; none of these control patients had experienced recent gastrointestinal infection with any of the implicated bacteria, however. These interesting observations are consistent with the *in vitro* observations (discussed below) that phagocytosed LPS can persist in cells for extended periods of time; they also are consistent with the hypothesis that (abnormally or incompletely degraded?) LPS may play a role in the induction of joint inflammation in reactive arthritis. It is not known whether the LPS detected immunologically in these studies were biologically active, although the clinical situation bears some resemblance to experimental *E. coli* arthritis in rabbits, in which sterile synovial exudates retained immunogenic endotoxin for many days.[30] In further studies, the same group found that monocytes and neutrophils from the peripheral blood of patients with *Salmonella*-associated reactive arthritis or with *Y. enterocolitica* infections also contained bacterial antigens,[33,35] raising the possibility that antigen-laden cells might home preferentially to synovium in genetically predisposed (HLA B27) patients, thereby eliciting the arthritic inflammatory response. Alternatively, the appearance of LPS-laden cells in the synovial fluids of reactive arthritis patients might be a secondary phenomenon that reflects merely the appearance of these phagocytic cells in a site of ongoing inflammation. Despite these uncertainties, reactive arthritis currently offers the best opportunity for investigating the possible contribution of LPS to a chronic human disease. Defining the role of LPS in the pathogenesis of this disorder should yield important information regarding the biological significance of LPS processing *in vivo*.

C. EVIDENCE FOR CATABOLISM OF LPS *IN VIVO*

A report suggesting dephosphorylation of LPS *in vivo* was published in 1956: Rowley et al.[36] injected [^{32}P]LPS from *E. coli* into mice and guinea pigs, and found that while 60 to 75% of the radiolabel was cleared into the liver, spleen, and other tissues within 6 to 8 min, the remainder disappeared from circulation slowly and appeared in the urine over a period of 48 h. Radioactivity in the urine was dialysable, suggesting that it had been removed from the LPS. Radioactivity in the initial LPS preparation and in plasma collected over the 48-h period after injection was not dialysable; however, incubation of LPS with sera from several species of laboratory animals or with human sera rapidly liberated dialyzable ^{32}P. This *in vitro* dephosphorylating activity was heat labile, was inhibited by EDTA, and had a pH optimum of about 8; it did not correlate with serum alkaline phosphatase activity. Unfortunately, the ^{32}P-LPS used for these studies was not shown to be free of labeled nucleic acid or phospholipid, nor was the location of the ^{32}P in the LPS molecule established. Results consistent with splenic dephosphorylation and detoxification of LPS were also noted by Wiznitzer et al.;[37] the location of the ^{32}P in their relatively crude preparation of LPS is also uncertain, however.

In another early study, Chedid et al.[38] injected ^{51}Cr-labeled *S. enteritidis* LPS into mice and followed the clearance of label from plasma and its appearance in the urine. Radioactivity in plasma samples was associated with toxicity, seen as lethality for adrenalectomized or pregnant mice; however, antigenic analysis of plasma LPS by agar immunodiffusion showed a gradual conversion from a slowly diffusing form that produced a radioactive precipitin line into rapidly diffusing antigenic forms that were not radioactive. This suggests that the extrinsic radiolabel was removed from the LPS during its circulation in the blood. Eighty percent of the radiolabel in the urine was dialysable (compared to 10% of plasma radiolabel); while LPS could be demonstrated in the urine antigenically, it was not toxic. Thus toxic and antigenic activities of LPS were separated. This might occur if the O-antigen were cleaved from the (toxic) lipid A region or if the LPS were detoxified by an alteration that did not affect its antigenicity.

Two papers by the Freiburg group[15,16] reported the fate of doubly labeled *S. abortus equi* LPS injected into rats. As noted above, the recovery of ^{3}H, the fatty acid label, in tissue and in excreta was less than the recovery of ^{14}C, the carbohydrate label, suggesting a partial removal of fatty acyl chains;[15] similar loss of the fatty acid label was seen when the isotopes were reversed.[16] Further study focused on LPS recovered from the liver.[16] Three days after injection of LPS, 44% of the carbohydrate label and 23% of the fatty acid label were located in the liver. LPS was recovered from liver homogenates following digestion with nucleases and proteases and a series of extraction and chromatographic steps; the final preparation was considered to be approximately 42% LPS, based on the concentrations of KDO and several other LPS-specific sugars per milligram, and contained 35% of the radioactivity present in the original liver homogenates, in the same ^{3}H/^{14}C ratio. LPS recovered from liver contained reduced amounts of abequose, when compared on a molar basis to other sugars, and reduced amounts of fatty acids. Consistent with the loss of abequose, reactivity with O-specific antisera was reduced approximately tenfold. Analysis of the starting LPS and the recovered LPS by sodium dodecyl sulfate-polyacrylamide gel elctrophoresis (SDS-PAGE) showed an alteration in the ladder pattern: the recovered LPS appeared to be enriched in medium-length molecules, while the starting LPS contained more long O-chain molecules and more molecules containing only a few O-antigen repeat units. In contrast, for several biological activities (pyrogenicity for rabbits, lethality for galactosamine-treated mice, *Limulus* activity, and ability to prepare the skin of rabbits for the dermal Shwartzman reaction), the LPS recovered from the liver was as potent as the starting LPS.

In summary, data obtained *in vivo* suggest that animals (rodents) are able to remove acyl and phosphate substituents from LPS[16,36,37] and to alter its O-antigen.[16,38] Importantly, the precise site(s) of LPS degradation *in vivo* have not been clearly defined. In particular, the relative roles of the liver (where catabolism appears to be limited, although nonspecific LPS-degrading hepatic enzymes have been reported[39,40]) and phagocytes in blood or other nonhepatic tissues have not been established.

IV. *IN VITRO* STUDIES OF LPS CATABOLISM

This section will review work which gives general evidence for processing of LPS by cells, then discuss work which describes specific structural alterations of LPS. Most studies followed the same general approach: radiolabeled LPS (or bacteria) were added to cells (usually freshly isolated phagocytes); after allowing time for initial uptake, the cells were washed and allowed to incubate further; cells and supernatants were sampled at intervals and analyzed for total radioactivity. In many cases, radioactive fractions were then studied using physical, chemical, and biological techniques.

A. PERSISTENCE OF LPS WITHIN CELLS, ANALYSIS OF PROCESSED LPS

Ultrastructural studies of the uptake of LPS by phagocytes *in vitro* are consistent with the *in vivo* reports cited above, showing that LPS is taken up rapidly and is found both within vacuoles and free in the cytoplasm and nucleus. Bona reported in 1973[41] that within 15 min of the addition of *S. typhimurium* LPS to guinea pig peritoneal exudate macrophages LPS was found adsorbed to the cell surface, and "small invaginations containing the labeled LPS were observed"; at later time points, LPS was in phagosomes or free in the cytoplasm. Bona also reported the transfer of pinocytosed LPS from macrophages to lymphocytes.[41] In a more recent electron microscopic study of the uptake of rough (J5) and smooth (O111:B4) *E. coli* LPS by human monocytes, Kang et al. reported that LPS appeared to be taken up both by nonspecific interactions with cell membranes and via coated pits, presumably by specific receptors; disruption of cell membranes by LPS was also noted.[22]

In general, *in vitro* study of the fate of LPS following uptake by phagocytes has shown that degradation of LPS is limited, yet granulocytes seem to destroy the antigenic properties of LPS more rapidly than do monocytes. Persistence of a poorly defined (but possibly LPS) *E. coli* agglutinogen in macrophages but not neutrophils was described by Cohn.[42] Mesrobeanu et al.[43] studied the antigenic and biological activities of smooth *S. typhimurium* LPS that had been taken up by either polymorphonuclear cells or mononuclear cells from guinea pig peritoneal exudates: following incubation of cells with LPS, they washed the cells, homogenized them, and treated them at 100°C for 10 min to inactivate cell-derived, heat-labile mediators of inflammation. The heat-treated homogenates were then used in intradermal (preparative) injections in the dermal Shwartzman reaction. They reported that LPS was inactivated within 3 h of uptake by polymorphonuclear leukocytes, but remained active as long as 48 h after addition to mononuclear cells. LPS was not inactivated during incubation *in vitro* with isolated lysosomes from polymorphonuclear cells. In contrast, incubation of LPS with lysosomes from either type of phagocyte was shown to alter the antigenic properties of LPS; this was studied by passive hemagglutination, using O-specific antisera, of sheep red blood cells that had been sensitized with the treated LPS. These observations suggested that the toxic and antigenic properties of LPS were distinct and could be altered selectively. In a subsequent paper,[44] these authors reported that *S. typhimurium* LPS was still immunogenic for rabbits 24 h after ingestion by guinea pig peritoneal granulocytes or macrophages. The LPS used in these studies was not radiolabeled, however, so the uptake of LPS by the cells could not be quantified, and no attempt was made to study structural alteration of the LPS.

Using a similar approach, Bona[41] immunized rabbits with lysosomes from guinea pig macrophages that had ingested *S. typhimurium* LPS 48 h previously and reported that the LPS was still immunogenic. Consistent with the report of Mesrobeanu et al.,[43] this macrophage-processed LPS was also still active in the preparative phase of the dermal Shwartzman reaction, at least at the single high dose tested.[41] In contrast, Filkins[45] found that rat macrophage sonicates could destroy the ability of LPS to kill lead-sensitized rats.

Duncan and Morrison[46] studied the events that followed phagocytosis by murine macrophages of *E. coli* labeled extrinsically with ^{125}I or intrinsically either with [^{3}H]galactose (incorporated specifically into LPS) or with [^{14}C]glucose. Following uptake of the bacteria, the ^{3}H was released into the culture supernatant much more slowly than the other isotopes: 100% of the iodine and 60% of the ^{14}C was released within 48 h, while even at 72 h 60% of the ^{3}H was still cell associated. They found similar results for macrophages from both C3HeB/FeJ (LPS-responsive) and C3H/HeJ (LPS-hyporesponsive) mice. This suggests that LPS was retained selectively; the extent to which the chemical structure of the LPS was altered within the cells is not clear. The isopycnic density profiles of LPS recovered from macrophage lysates or from culture supernatants differed from that of LPS extracted from unincubated bacteria; however, these differences appeared to be due primarily to noncovalent interactions of LPS with cellular components, since they were largely removed by phenol-water extraction of the processed LPS. SDS-PAGE analysis showed that the lysate LPS and the supernatant LPS both appeared to differ in subunit distribution from the control LPS, and chemical analysis showed a loss of colitose relative to galactose.[19] The biological activities of the processed LPS were also studied:[19,46] LPS recovered from cell supernatants and lysates was similar to control LPS (not incubated) in *Limulus* activity and in lethality for actinomycin D-treated mice; the processed LPS was actually increased in its ability to stimulate interleukin-1 release from murine macrophages and in murine splenocyte mitogenic activity. The increased potency was felt not to result from cellular components, since it was seen for LPS that had been phenol extracted, then further purified on cesium chloride density gradients. Moreover, the processed LPS prepartions had no mitogenic activity toward splenocytes from C3H/HeJ (LPS-hyporesponsive) mice. Lipid A derived from the processed LPS also had increased macrophage-activating activity, compared to lipid A from control LPS.

Fox et al.[47] studied the processing of *E. coli* LPS (labeled by growth of bacteria in [3H]galactose and [14C]acetate) by rat Kupffer cells and reported that LPS released from cells was enriched in ^3H, while LPS retained by the cells was enriched in ^{14}C. They interpreted these data to mean that the cells removed part of the O-polysaccharide (and presumably released it), without modification of the lipid. However, direct evidence for this interpretation was not presented; if [14C]fatty acids had been removed from the LPS and reutilized by the cells, the analysis described would not have detected it. The radio-labeled galactose or O-antigen found in the supernatants was not shown to be separated from lipid A. Analysis of cell lysates and of culture supernatants by density gradient centrifugation and by gel filtration showed differences between the initial LPS and the processed material; these differences may reflect alterations in physical structure of the processed LPS as a result of association with cellular components or with serum in the culture medium.

B. DEPHOSPHORYLATION

Rutenburg et al.[48] reported in 1960 that rabbit peritoneal macrophages were able to detoxify the LPS of *E. coli* O111:B4, as determined by lethality for chick embryos and for thorotrast-treated rabbits. Detoxification was par-alled by the release of 50% of ^{32}P from the LPS into a dialyzable form. While incubation of the LPS with plasma alone released approximately 20% of the ^{32}P (as seen also by Rowley et al.,[36] see above), plasma did not detoxify the LPS. Since smooth LPS from bacteria grown in ^{32}P is likely to contain radiolabel at several sites in the oligosaccharide core and in the lipid A moiety, it is not possible to determine whether the dephosphorylation by plasma was the same activity as that seen in macrophages.

Peterson and Munford reported the ability of murine macrophages to dephosphorylate the lipid A moiety of LPS.[8] The LPS used was from *E. coli* D21f2 (Re LPS chemotype) and contained, in addition to ^{32}P, a ^3H label in the glucosamine residues of lipid A. The ^{32}P was attached to lipid A, since the residues of the core that are phosphorylated in other strains are lacking in this deep-rough mutant. Dephosphorylation was detected by measuring ^{32}P in the supernatant following activated charcoal precipitation of LPS from cell lysates or culture supernatants. Approximately two thirds of the ^{32}P was removed from the LPS over 48 h of incubation; while most of the ^{32}P removed from LPS was released from the cells, part of it was incorporated into cellular lipids and other components. Dephosphorylation of LPS by cell lysates was also seen: this activity was heat labile, had a pH optimum of 4.5 to 5.5, and was inhibited by sodium fluoride, an inhibitor of acid phosphatases. The ability of RAW 264.7 cells, a murine macrophage-like cell line, to dephosphorylate lipid X, a monoglucosamine precursor of lipid A, was reported by Zoeller et al.[49]

The lipid A phosphates of many (but not all) LPS are substituted with polar groups such as ethanolamine and aminoarabinose. The influence of these polar groups on the ability of the molecules to undergo dephosphorylation is not known; they could be the basis for important differences in the processing of LPS from different bacterial sources *in vivo*.

C. DEACYLATION

In addition to the deacylation of LPS by mammalian cells, as described below, there are several reports of LPS deacylation by slime molds.[50-57] These primitive organisms feed on bacteria, which are digested within phagocyte vacuoles;[50] somewhat surprisingly, their ability to degrade LPS is limited. When amoebae of *Dictyostelium discoideum* were fed on *Salmonella london*, they secreted into the medium a degraded, water-soluble form of the LPS of this organism.[51] Chemical and antigenic analysis showed that the secreted material contained all of the constituents of the carbohydrate chain, including the O-antigen, the carbohydrate, core, and the diglucosamine lipid A backbone, but lacked the ester- and amide-linked fatty acyl chains of lipid A.[52] Some of the enzymes that catalyze the removal of these acyl chains have been characterized and found to be remarkably specific for removal of acyl chains from particular sites on the lipid A moiety.[54-57] The bulk excretion of partially deacylated LPS by *D. discoideum* is in keeping with the tendency of protozoa to expel undigested materials and contrasts with the tendency of animal macrophages to retain potentially dangerous materials intracellularly (in lysosomes) for long periods of time.[58]

Mammalian phagocytes also deacylate LPS. In 1969, Gimber and Rafter[59] reported that, following uptake by rabbit neutrophils of ^{14}C-labeled *E. coli* LPS, there was a transfer of radiolabel from LPS to a chloroform-soluble form. The radioactivity in the chloroform extracts was found to be in phospholipids, neutral lipids, and free fatty acids. Whether recovery of radiolabel in lipids could have been due in part to contamination of LPS by bacterial phospholipids was not addressed directly, but it was stated that the LPS contained no chloroform-extractable radioactivity. The appearance of ^{14}C in lipids was partially inhibited by EDTA, *N*-ethylmaleimide, and protamine sulfate. Although the fatty acid composition of neither the extracted lipid nor the remaining LPS was analyzed, these observations are most consistent with the enzymatic removal of fatty acyl chains from the lipid A moiety of LPS and the reutilization of some of these fatty acids for cellular lipid synthesis.

Hall and Munford reported the uptake and deacylation of opsonized, double labeled *S. typhimurium* LPS by human neutrophils;[4] as in the previously described paper, deacylation was detected as the release of radiolabeled fatty acids from LPS into a chloroform-extractable form. Most of the released radioactivity migrated on thin-layer chromatography as phospholipid or neu-

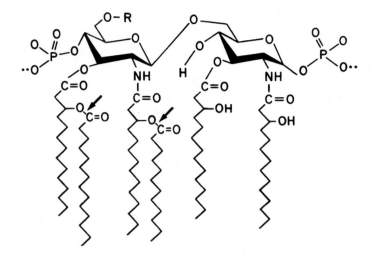

FIGURE 3. Structure of *S. typhimurium* lipid A. R indicates the site of attachment of the carbohydrate chain. The arrows show the site of cleavage by acyloxyacyl hydrolase. (From Munford, R. S. and Hall, C. L., *Science*, 234, 203, 1986. With permission.)

tral lipid and could be recovered as free fatty acid following chemical hydrolysis of these lipids. LPS-deacylating activity was then found in a granule fraction of neutrophils; activity was linear over time, was heat labile, and was strongly pH dependent, with optimum activity at approximately 4.8; an apparent Km of the granule preparation for the LPS substrate was approximately 0.6 μM. These observations suggested that deacylation was enzymatic. Analysis of the released fatty acids and of acyl chains remaining on the partly degraded LPS showed that deacylation by the granule fraction was remarkably specific: only the nonhydroxylated, or secondary, fatty acids were released, and the 3-hydroxy acyl chains that are substituted directly to the lipid A backbone were not removed (Figure 3). The ability of neutrophils to deacylate LPS thus appeared to be specific for hydrolysis of acyloxyacyl groups. The enzyme responsible for this reaction (acyloxyacyl hydrolase) has now been characterized in some detail;[5,60] the same activity has also been found in human monocytes and macrophages. Purified acyloxyacyl hydrolase is a heterodimer whose two disulfide-linked subunits are both glycosylated.[60] Cloning and expression of the cDNA for the enzyme[61] revealed that both subunits are produced from the same mRNA and that the recombinant enzyme, like purified LPS, can remove secondary acyl chains from more than one position on the lipid A backbone of LPS.

In another report,[13] Munford and Hall showed that LPS-deacylating activity was also present in thioglycollate-elicited peritoneal murine macrophages from both C3H/HeN and C3H/HeJ mice; both opsonized and unopsonized LPS were deacylated. In contrast to the specificity of the activity in the human neutrophils, elicited murine macrophages had a limited ability to remove the 3-hydroxy acyl chains as well as the secondary acyl chains from antibody-opsonized LPS. As seen for human neutrophils, fatty acids removed from LPS by the murine macrophages were reutilized for cellular lipid synthesis.

Deacylation of rough *E. coli* LPS by rat hepatocytes was reported by Fukuda et al.[62] Following uptake by cells of LPS labeled by growth in [14C]acetate and *N*-acetyl-[3H]glucosamine, 14C was found in a chloroform-extractable form — primarily in triglycerides and in 3-hydroxymyristate. Their observations differed from those described above in that (1) hydroxy fatty acids were released to a greater extent and (2) released lipophilic material was found in the culture supernatant; the cells contained no chloroform-soluble radioactivity.

D. PROCESSING OF LPS BY CELLS — SUMMARY

As previously seen in studies performed in living animals, the ability of animal cells to degrade and to inactivate LPS appears to be limited. This conclusion is probably due in part to the high concentrations of LPS used in these studies — in some cases, several milligrams per milliliter.[41,43] While phagocytes appear to be able to take up large amounts of LPS within a short period, processing these large amounts may require a long time; thus estimates of the ability of cells to degrade LPS, when expressed as a fraction of the cell-associated LPS, may be spuriously low. The finding that LPS is not readily degraded in cells is consistent with the finding of Lehrer and Nowotny that endotoxin is resistant to attack by a large number of commercially available hydrolytic enzymes.[63] Most of the studies cited here have found that catabolism of LPS by cells is selective, resulting in removal of specific sugar residues[19] or acyl chains,[4,62] for example. The relative rates of different catabolic steps (deacylation, dephosphorylation, alteration of carbohydrate structure) in the same cells have not been studied.

Several studies showed that LPS or LPS-derived material is released from cells after uptake.[8,13,19,46,47,62,64] Such release is consistent with the report of DeVoe, who found excretion of radioactivity following ingestion of 14C-labeled meningococci by mouse polymorphonuclear leukocytes; electron microscopy showed that the egested material consisted "mainly of membranous vesicles."[65] As noted above, Bona reported the transfer of uranyl acetate-labeled LPS from macrophages to lymphocytes in culture, observed by electron microscopy.[41] Exocytosis of LPS from phagocytes raises the possibility that partially degraded LPS released by these cells may possess altered biological activities, so that enzymatic processing may modulate certain interactions of LPS with the host.

V. EFFECTS OF PROCESSING ON LPS BIOACTIVITIES

As described above and as reviewed elsewhere,[1] numerous investigators have found evidence that plasma, tissue, and cell extracts can detoxify LPS.[38,43,48,49] Detoxification was detected using many different assays of LPS activity. On the other hand, there are many reports indicating that processing of LPS by cells does not destroy its bioactivity;[16,19,41,46] again, these studies used a number of different assays to measure LPS bioactivity. As noted above, some of the disparities between studies can be accounted for by the amounts of LPS used as starting material, which in some experiments may have exceeded the amount that could be degraded during the allotted time period (see Figure 1), and also by the doses of LPS that were tested in bioactivity assay: when complete dose-responses curves cannot be made, LPS preparations should be compared at doses expected to produce less than a maximal response (see Figure 2). Interpretation of the literature on LPS detoxification is complicated by the fact that, in many studies of the effects of LPS processing on bioactivity, the extent to which the chemical structure of LPS was altered is not known; in fact, much of the work in this area was carried out before the structure of LPS was well understood. LPS detoxification can now be considered in the light of the knowledge that has accumulated in the past decade about the structure/function relationships of LPS and lipid A. As noted above, only a few bioactivities require a complete lipid A structure, so a finding that LPS recovered from cells retains the ability to elicit a given response does not necessarily imply that the LPS structure has not been altered. In addition, certain chemical treatments of LPS remove toxicity (defined for these experiments as pyrogenicity and lethality for mice or for chick embryos) without affecting reactivity with O-specific antisera;[66,67] while the effects of these chemical treatments are very poorly defined, these observations suggest that a finding that the immunogenicity of LPS is retained for long periods following ingestion by phagocytes[29,30,41,44] does not imply that the LPS structure is unaltered or that the potency of the LPS in other bioassays is necessarily unaffected.

The remainder of this section will describe the effects on biological activities of clearly defined chemical alterations in LPS or related molecules.

A. DEPHOSPHORYLATION

Despite early reports suggesting that both plasma and cells are able to dephosphorylate LPS,[36,48] there has been little subsequent study of this phenomenon. A major technical problem has been the inability to determine the location of biosynthetically incorporated ^{32}P in the LPS molecule, which contains several phosphates; the recent dephosphorylation studies have used deep rough LPS molecules[8] or lipid A analogs[49] that have phosphates in defined positions, thus allowing interpretation of the site of dephosphorylation.

B. DEACYLATION

The ability of granule preparations from human neutrophils to catalyze the selective removal of secondary (acyloxyacyl-linked) fatty acyl chains from

LPS of *S. typhimurium*[4] was noted above; Munford and Hall[60] subsequently reported the purification of the enzyme responsible for this reaction, acyloxyacyl hydrolase (AOAH:Figure 3). The use of AOAH to deacylate LPS *in vitro* has allowed extensive characterization of the bioactivities of LPS that lack secondary acyl chains yet are intact with respect to carbohydrate structure and phosphate substitution. The lipid A moiety of enzymatically deacylated LPS resembles lipid A precursor IV_A; chemically synthesized compounds with the same structure are known as compound 406 or LA-14-PP.[68] The reports on studies of compound 406 allow certain predictions to be made about the bioactivities of enzymatically deacylated LPS. However, deacylated LPS and compound 406 differ in that the latter lacks a carbohydrate chain; there are no data that would predict the effect of the carbohydrate chain on the bioactivities of various lipid A analogs.

1. Bioactivities of Enzymatically Deacylated *S. typhimurium* Rc LPS

Most of the work on the effects of enzymatic deacylation of LPS has been carried out using Rc chemotype LPS from *S. typhimurium*. Munford and Hall reported that the ability of this LPS to prepare the skin of rabbits for the dermal Shwartzman reaction was reduced over 100-fold by enzymatic deacylation;[69] in contrast, the same enzyme-treated and mock-treated LPS preparations differed by less than 20-fold in murine splenocyte mitogenicity[69] and in chick embryo lethality.[69a] In later work it was found that enzymatic deacylation of *S. typhimurium* Rc LPS reduced its *Limulus* activity approximately 30-fold.[70] These observations can be compared with reports of the bioactivities of compound 406: deacylated LPS is similar to compound 406 in possessing *Limulus* and splenocyte-mitogenic activities while lacking activity in the dermal Shwartzman reaction; thus the presence of the Rc carbohydrate chain did not seem to alter the role of the secondary acyl chains in these activities of LPS. In contrast, the finding that deacylated LPS retain chick embryo lethality could not have been predicted from the study of lipid A analogs.

The effect of enzymatic deacylation on bioactivities of *S. typhimurium* Rc LPS was studied further, using several *in vitro* assays in which the target cells were human vascular endothelial cells, neutrophils, or monocytes. Pohlman et al. reported that enzymatic deacylation of LPS reduced over a hundredfold its ability to stimulate endothelial cells to promote neutrophil adherence; further, the deacylated LPS inhibited the activity of intact LPS.[71] Inhibition was specific for LPS; deacylated LPS did not inhibit neutrophil adherence induced in response to TNF or IL-1.[71] Further work on the interaction of deacylated LPS with endothelial cells was reported by Riedo et al.[72] As for neutrophil adherence, deacylated LPS was inactive and inhibitory for induction of plasminogen activator inhibitor-1, prostacyclin, and prostaglandin E_2 production in response to LPS; these responses to TNF were not inhibited. For plasminogen activator inhibitor-1, it was found that inhibition occurred at or before the accumulation of specific messenger RNA.[72]

Dal Nogare and Yarbrough studied the effect of enzymatic deacylation on the ability of LPS to stimulate several responses of human peripheral blood neutrophils.[73] Deacylated LPS was much less active than control (mock-treated) LPS in stimulating the adherence of neutrophils to plastic and the release of specific granule contents, and deacylated LPS was unable to prime neutrophils for superoxide release in response to a second stimulus. More recently, Kovach et al. reported that deacylated LPS is also greatly reduced in the ability to elicit TNF release from cells in human whole blood.[74]

The effects of enzymatic deacylation described above were for LPS that had undergone maximal deacylation; experiments using partially deacylated LPS have shown for several responses that the reduction in potency is proportional to the degree of deacylation.[69,71,74]

2. Bioactivities of Enzymatically Deacylated Nonenteric LPS

Most studies of the structure/function relationships of LPS have been carried out using LPS from *Salmonella* species or from *E. coli*. LPS from other Gram-negative bacteria differ from these enteric LPS in both lipid and carbohydrate structure; the effects of these differences on biological activities of LPS are poorly understood. It was noted in the previous section that for one activity tested (chick embryo lethality), the presence of the Rc oligosaccharide chain appeared to diminish the predicted effect of enzymatic deacylation of *S. typhimurium* Rc. It seemed possible, therefore, that the removal of secondary acyl chains might have different effects on LPS with diverse structures. AOAH carries out the same enzymatic reaction on LPS from several nonenteric bacteria as on *S. typhimurium* LPS: secondary acyl chains are removed specifically, without removal of 3-hydroxyl acyl chains.[5]

The effects of enzymatic deacylation of LPS from *Haemophilus influenzae* type b, *N. meningitidis,* and *E. coli* J5 were found to be similar to those described above for *S. typhimurium* Rc LPS in two assays: the *Limulus* reaction and the adherence of leukocytes to LPS-treated endothelial cells.[70] However, for spleen-cell mitogenicity, the effect of deacylation was dependent on the source of the LPS. For LPS from *E. coli* J5, *H. influenzae* type b, *Pseudomonas aeruginosa,* and smooth *S. typhimurium*, the reduction in potency was 15- to 20-fold, as seen previously for *S. typhimurium* Rc LPS; mitogenic activity of LPS from several isolates of *Neisseria meningitidis* and *N. gonorrhoeae*, in contrast, was reduced over 100-fold. Further, enzymatically deacylated *Neisseria* LPS were able to inhibit the mitogenic activity of mock-treated *Neisseria* and *S. typhimurium* LPS.[70] This appears to be the first example of a modified LPS or lipid A that can inhibit LPS-induced murine splenocyte mitogenesis.

The effect of enzymatic deacylation of LPS from *H. influenzae* type b was also studied in two animal models of meningitis.[75,76] Syrogiannopoulos et al. reported that intracisternal injection of LPS into rabbits resulted in meningitis similar to that seen following injection of live *H. influenzae* type b bacteria, assessed by cerebrospinal fluid leukocytosis and protein concentration and by pathologic examination of brain tissue.[75] Enzymatic deacylation of LPS reduced its ability to induce inflammation by a factor of approximately 100.[75] Wispelwey et al. evaluated the ability of intracisternally injected *H. influenzae* LPS to induce cerebrospinal fluid leukocytosis and to increase the permeability of the blood-brain barrier in rats. They reported that enzymatically deacylated LPS was significantly less potent than control LPS in inducing both of these indicators of meningitis.[76]

The study of enzymatically deacylated LPS indicates that catabolism of LPS (in this case, the removal of secondary acyl chains) may have a selective effect on biological activities, affecting some responses much more than others. The ability of LPS to elicit several responses from human cells (endothelial cells, monocytes, and neutrophils) is reduced substantially by deacylation, as are certain other activities; however, mitogenic activity for murine splenocytes, lethality for chick embryos, and *Limulus* activity are affected much less. (These data, like those of Golenbock et al.,[77] also are consistent with the conclusion that certain responses to lipid A analogs are species specific.) Further, the activities of LPS with different structures may be affected differently by catabolism. The effects on bioactivities of a structural alteration of the lipid A region of LPS cannot always be predicted from the activities of lipid A analogs.

VI. EVIDENCE FOR REGULATION OF LPS PROCESSING

A. *IN VIVO* STUDIES

Several studies have examined the extent to which the clearance of LPS is affected by altering the physiologic state of an animal. In several early reports (reviewed by Skarnes[1]), it was suggested that induction of tolerance to LPS increased LPS clearance from the circulation and that plasma from tolerant animals had increased LPS-detoxifying activity. As noted above, the clearance of injected LPS from the circulation depends in part on the extent to which it associates with serum lipoproteins;[20] this is altered by host factors, including specific antibody to LPS (which reduces the binding of LPS to HDL and increases its rapid clearance to phagocytic cells of the spleen and liver), levels of cholesterol and of lipoproteins, and other factors that affect uptake of HDL.[78] In addition, the association of LPS with HDL *in vitro* is much slower in acute-phase serum.[20] It is not known whether catabolism of LPS is affected by induction of tolerance to LPS, by induction of an acute-phase response, or by manipulation of lipoprotein metabolism.

We have recently found that intravenous injection of LPS into rabbits elicits a rapid increase in plasma acyloxyacyl hydrolase activity;[79] as noted above, this enzymatic activity was previously described in human peripheral blood neutrophils and is specific for removal of the secondary acyl chains from the lipid A moiety of LPS (Figure 3). The increase in plasma LPS-deacylating activity in response to LPS was prevented by pretreatment of rabbits with nitrogen mustard to induce leukopenia; it did not occur when the LPS injection was replaced by subcutaneous injection of silver nitrate to induce an acute-phase response.

B. *IN VITRO* STUDIES

Much of the work carried out on the catabolism of LPS by cells *in vitro* has used cells (usually macrophages) from peritoneal exudates; these cells are presumably at least partially activated, and may process LPS differently from unstimulated phagocytes. The effect of activation of cells on catabolism has not been studied very much. Hampton et al. reported that dephosphorylation of the lipid A precursor IV_A by RAW 264.7 cells was decreased when the cells were preincubated with LPS.[9] The incubation of rabbit peripheral blood neutrophils or mononuclear cells with LPS results in the release of acyloxyacyl hydrolase activity without stimulating intracellular AOAH activity.[79] One would expect that the processing of LPS by phagocytes might be affected by the presence of specific antibody to LPS, or by incorporation of LPS into liposomes or lipoproteins. However, while opsonization of LPS with antibody increased its uptake by murine macrophages substantially, the relation between uptake and rate of deacylation was the same for both opsonized and unopsonized LPS.[13]

VII. CONCLUSIONS

The experiments reviewed here demonstrate that animal cells are able to catabolize LPS, at least partially, both *in vivo* and *in vitro*, and that such catabolism can affect the biological activities of LPS. Until the past decade, much of the work in this area dealt with the inactivation of LPS, without much attention to the alterations in LPS structure that might be involved. Increasing knowledge of lipid A structure/function relationships has directed recent research toward specific structural alterations that are likely to affect LPS activity: deacylation and dephosphorylation of the lipid A moiety.

The extent to which LPS is catabolized either *in vitro* or *in vivo* has been difficult to determine; as a result of the low specific activities of intrinsically radiolabeled LPS, virtually all experiments have been carried out using amounts of LPS far greater than those likely to be present in animal tissues. Unfortunately, this limitation is likely to continue. In addition, nearly all studies have used purified LPS; little is known about the cellular processing of LPS that is taken up associated with bacteria or with bacterial membrane fragments.

What is the biological role of LPS processing by phagocytic cells? One potential result of the catabolism of LPS by phagocytes is the inactivation of LPS that is present in phagocytosed bacteria, released from bacteria during infection, or transported into the circulation from the intestine; the findings that phagocytes catabolize LPS and that enzymatic catabolism of LPS *in vitro* produces molecules with reduced biological activity are consistent with this hypothesis. On the other hand, the prolonged persistence of immunogenic LPS in phagocytes indicates clearly that LPS is not rapidly or completely degraded; the release of bioactive LPS from phagocytes that have ingested living bacteria *in vitro* is further evidence that LPS processing does not invariably lead to inactivation. How can these observations be reconciled? First, phagocytes seem to process LPS largely by modifying the lipid A moiety. With certain exceptions,[16,19] antigenic epitopes in the O-antigen seem to be retained. Second, as noted in Section V.B above, enzymatic deacylation of the lipid A moiety of LPS (using acyloxyacyl hydrolase) diminishes its potency in some bioassays much more than in others. It is likely, though not proven, that such deacylated LPS retain immunogenicity, just as they retain much of the ability of mock-treated LPS to stimulate murine splenocyte mitogenesis. Loss of toxicity, as measured by the dermal Shwartzman reaction (and possibly by most assays that test the responses of human cells to LPS), need not be accompanied by loss of immunogenicity, *Limulus* reactivity, or other bioactivity. The critical experimental issue seems to be the biological response assay chosen for measuring the impact of processing (see Section II.B.3). The ability of phagocyte-processed LPS to elicit certain biological responses is not inconsistent with the occurrence of structural modification(s) in the LPS that greatly alter its ability to elicit other responses: detoxification, with retention of immunogenicity (and/or adjuvanticity), would provide optimal benefit to the host.[69] Indeed, little is known about the mechanisms involved in the cellular presentation of LPS antigens to initiate the synthesis of anti-LPS antibodies by the host, and it is conceivable that enzymatic modification of the lipid A moiety is a step in this process.

It is also possible that LPS processing is an integral step in the sequence by which certain cells recognize LPS and respond to it. Although progress has been made recently in identifying cell-surface molecules that bind LPS,[80-83] almost nothing is known of the signal transduction that follows such binding. It is conceivable that, at least in some cases, a cellular response to LPS involves the transfer of fatty acyl or phosphate substituents of lipid A to acceptor molecules as yet unidentified. This concept is supported by the observation that for certain cellular responses, LPS or lipid A derivatives lacking secondary acyl chains or phosphate are not only inactive but are able to inhibit the activity of intact LPS: these inhibitors may be able to bind the appropriate cell surface molecule but be unable to supply the acyl chain or phosphate whose transfer might mediate signal transduction. Adequate tests of this hypothesis will require better methods for labeling LPS (to achieve much higher specific activity) and for synchronizing the uptake and processing of the LPS by target cells.

None of these potential roles for LPS processing is ruled out by the currently available evidence. Indeed, it is possible that deacylation or dephosphorylation that occurs at one site in the cell (e.g., the plasma membrane) could participate in signal transduction, while in other sites (e.g., phagosomes) the same reactions could be nonstimulatory and prepare LPS for presentation as an antigen.[84] Obviously, much more needs to be learned about the intracellular fate of LPS in phagocytes. Further studies in this area should clarify how the responses of animals to LPS are induced and how LPS signals are modified *in vivo*.

ACKNOWLEDGMENTS

This work was supported by U.S. Public Health Service grants AI18188 and HD22766. We thank Leon Eidels for several helpful suggestions.

REFERENCES

1. **Skarnes, R. C.**, *In vivo* distribution and detoxification of endotoxins, in *Handbook of Endotoxin*, Vol. 3, Berry, L .J., Ed., Elsevier Science, Amsterdam, 1985, 56.
2. **Watson, J. and Riblet, R.**, Genetic control of responses to bacterial lipopolysaccharides in mice. II. A gene that influences a membrane component involved in the activation of bone marrow-derived lymphocytes by lipopolysaccharides, *J. Immunol.*, 114, 1462, 1975.
3. **Peborde, J. P. and Samain, D.**, Preparation of tritiated lipopolysaccharides from *Escherichia coli* K12, *Biochim. Biophys. Acta*, 1033, 207, 1990.
4. **Hall, C. L. and Munford, R. S.**, Enzymatic deacylation of the lipid A moiety of *Salmonella typhimurium* lipopolysaccharides by human neutrophils, *Proc. Natl. Acad. Sci. U.S.A.*, 80, 6671, 1983.
5. **Erwin, A. L. and Munford, R. S.**, Deacylation of structurally diverse lipopolysaccharides by human acyloxyacyl hydrolase, *J. Biol. Chem.*, 265, 16444, 1990.
6. **Munford, R. S., Hall, C. L., and Rick, P. D.**, Size heterogeneity of *Salmonella typhimurium* lipopolysaccharides in outer membranes and culture supernatant membrane fragments, *J. Bacteriol.*, 144, 630, 1980.
7. **Schnaitman, C. A. and Austin, E. A.**, Efficient incorporation of galactose into lipopolysaccharide by *Escherichia coli* K-12 strains with polar *galE* mutations, *J. Bacteriol.*, 172, 5511, 1990.
8. **Peterson, A. A. and Munford, R. S.**, Dephosphorylation of the lipid A moiety of *Escherichia coli* lipopolysaccharide by mouse macrophages, *Infect. Immun.*, 55, 974, 1987.
9. **Hampton, R. Y., Golenbock, D. T., and Raetz, C. R. H.**, Lipopolysaccharide stimulation regulates lipid A metabolism in macrophage cell lines, *FASEB J.*, 4, A1908, 1990.
10. **Jiao, B., Freudenberg, M., and Galanos, C.**, Characterization of the lipid A component of genuine smooth-form lipopolysaccharide, *Eur. J. Biochem.*, 180, 515, 1989.
11. **Vukajlovich, S. W. and Morrison, D. C.**, Activation of murine spleen cells by lipid A: negative modulation of lipid A mitogenic activity of O-antigen polysaccharide, *J. Immunol.*, 135, 2546, 1985.
12. **Elsbach, P. and Weiss, J.**, Phagocytic cells: oxygen-independent antimicrobial systems, in *Inflammation: Basic Principles and Clinical Correlates*, Gallin, J. I., Goldstein, I. M., and Snyderman, R., Eds., Raven Press, New York, 1988, 445.

13. **Munford, R. S. and Hall, C. L.**, Uptake and deacylation of bacterial lipopolysaccharides by macrophages from normal and endotoxin-hyporesponsive mice, *Infect. Immun.*, 48, 464, 1985.

14. **Davies, M. and Stewart-Tull, D. E. S.**, The affinity of bacterial polysaccharide-containing fractions for mammalian cell membranes and its relationship to immunopotentiating activity, *Biochim. Biophys. Acta*, 643, 17, 1981.

15. **Kleine, B., Freudenberg, M. A., and Galanos, C.**, Excretion of radioactivity in faeces and urine of rats injected with ^{3}H, ^{14}C-lipopolysaccharide, *Br. J. Exp. Pathol.*, 66, 303, 1985.

16. **Freudenberg, M. A. and Galanos, C.**, Alterations in rats *in vivo* of the chemical structure of lipopolysaccharide from *Salmonella abortus equi*, *Eur. J. Biochem.*, 152, 353, 1985.

17. **Freudenberg, N., Freudenberg, M. A., Guzman, J., Mittermayer, Ch., Bandara, K., and Galanos, G.**, Identification of endotoxin-positive cells in the rat lung during shock, *Virchows Arch. Pathol. Anat.*, 404, 197, 1984.

18. **Rudbach, J. and Johnson, A. G.**, Alteration and restoration of endotoxin activity after complexing with plasma proteins, *J. Bacteriol.*, 92, 892, 1966.

19. **Duncan, R. L., Jr., Hoffman, J., Tesh, V. L., and Morrison, D. C.**, Immunologic activity of lipopolysaccharides released from macrophages after the uptake of intact *E. coli in vitro*, *J. Immunol.*, 136, 2924, 1986.

20. **Ulevitch, R. J.**, Interactions of bacterial lipopolysaccharides and plasma high density lipoproteins, in *Handbook of Endotoxin*, Vol. 3, Berry, L. J., Ed., Elsevier Science, Amsterdam, 1985, 372.

21. **Willerson, J. T., Trelstad, R. L., Pincus, T., Levy, S. B., and Wolff, S. M.**, Subcellular localization of *Salmonella enteritidis* endotoxin in liver and spleen of mice and rats, *Infect. Immun.*, 1, 440, 1970.

22. **Kang, Y. -H., Dwivedi, R. S., and Lee, C. -H.**, Ultrastructural and immunocytochemical study of the uptake and distribution of bacterial lipopolysaccharide in human monocytes, *J. Leukocyte Biol.*, 48, 316, 1990.

23. **Freudenberg, N., Freudenberg, M. A., Bandara, K., and Galanos, C.**, Distribution and localization of endotoxin in the reticuloendothelial system (RES) and in the main vessels of the rat during shock, *Path. Res. Pract.*, 179, 517, 1985.

24. **Freudenberg, M. A., Freudenberg, N., and Galanos, C.**, Time course of cellular distribution of endotoxin in liver, lungs and kidneys of rats, *Br. J. Exp. Pathol.*, 63, 56, 1982.

25. **Zlydaszyk, J. C. and Moon, R. J.**, Fate of ^{51}Cr-labeled lipopolysaccharide in tissue culture cells and livers of normal mice, *Infect. Immun.*, 14, 100, 1976.

26. **Praaning-van Dalen, D. P., Brouwer, A., and Knook, D. L.**, Clearance capacity of rat liver Kupffer, endothelial, and parenchymal cells, *Gastroenterology*, 81, 1036, 1981.

27. **Hopf, U., Ramadori, G., Möller, B., and Galanos, C.**, Hepatocellular clearance function of bacterial lipopolysaccharides and free lipid A in mice with endotoxin shock, *Am. J. Emerg. Med.*, 2, 13, 1983.

28. **Parent, J. B.**, Membrane receptors on rat hepatocytes for the inner core region of bacterial lipopolysaccharides, *J. Biol. Chem.*, 265, 3455, 1990.

29. **Britton, S., Wepsic, T., and Möller, G.**, Persistence of immunogenicity of two complex antigens retained *in vivo*, *Immunology*, 14, 491, 1968.

30. **Braude, A. I., Jones, J. L., and Douglas, H.**, The behavior of *Escherichia coli* endotoxin (somatic antigen) during infectious arthritis, *J. Immunol.*, 90, 297, 1963.

31. **Hopf, U., Möller, B., Stemerowicz, R., Lobeck, H., Rodloff, A., Freudenberg, M., Galanos, C., and Huhn, D.**, Relation between *Escherichia coli* R(rough)-forms in gut, lipid A in liver, and primary bilary cirrhosis, *Lancet*, 2, 1419, 1989.

32. **Nolan, J. P.**, Intestinal endotoxins as mediators of hepatic injury — an idea whose time has come again, *Hepatology*, 10, 887, 1989.

33. **Granfors, K., Jalkanen, S., Lindberg, A. A., Mäki-Ikola, O., Von Essen, R., Lahesmaa-Rantala, R., Isomäki, H., Saario, R., Arnold, W. J., and Toivanen, A.,** Salmonella lipopolysaccharide in synovial cells from patients with reactive arthritis, *Lancet*, 335, 685, 1990.

34. **Granfors, K., Jalkanen, S., Von Essen, R., Lahesmaa-Rantala, R., Isomäki, A., Pekkola-Heino, K., Merilahti-Palo, R., Saario, R., Isomäki, H., and Toivanen, A.,** Yersinia antigens in synovial-fluid cells from patients with reactive arthritis, *New Engl. J. Med.*, 320, 216, 1989.

35. **Granfors, K., Jalkanen, S., Lahesmaa-Rantala, R., Saario, R., Möttönen, T., and Toivanen, A.,** Bacterial antigens in peripheral blood cells in Yersinia-triggered reactive arthritis, *Scand. J. Rheumatol.*, 85 (Suppl.), 45, 1990.

36. **Rowley, D., Howard, J. G., and Jenkin, C. R.,** The fate of ^{32}P-labelled bacterial lipopolysaccharide in laboratory animals, *Lancet*, 1, 366, 1956.

37. **Wiznitzer, T., Better, N., Rachlin, W., Atkins, N., Frank, E. D., and Fine, J.,** *In vivo* detoxification of endotoxin by the reticuloendothelial system, *J. Exp. Med.*, 112, 1157, 1961.

38. **Chedid, L., Skarnes, R. C., and Parant, M.,** Characterization of a Cr^{51}-labeled endotoxin and its identification in plasma and urine after parenteral administration, *J. Exp. Med.*, 117, 561, 1963.

39. **Trapani, R. J., Waravdekar, V. S., Landy, M., and Shear, M. J.,** *In vitro* inactivation of endotoxin by an intracellular agent from rabbit liver, *J. Infect. Dis.*, 110, 135, 1962.

40. **Corwin, L. M. and Farrar, W. E.,** Nature of the endotoxin-inactivating principle in guinea-pig liver, *J. Bacteriol.*, 87, 832, 1964.

41. **Bona, C. A.,** Fate of endotoxin in macrophages: biological and ultrastructural aspects, *J. Infect. Dis.*, 128, S74, 1973.

42. **Cohn, Z. A.,** The fate of bacteria within phagocytic cells. III. Destruction of an *Escherichia coli* agglutinogen within polymorphonuclear leucocytes and macrophages, *J. Exp. Med.*, 120, 869, 1964.

43. **Mesrobeanu, L., Mesrobeanu, I., Bona, C., and Vranialici, D.,** Le destin des endotoxines thermostables pinocytees par les leucocytes, in *La Structure et les Effets Biologiques des Produits Bacteriens Provenant de Bacilles Gram-Negatifs*, Editions CNRS, Paris, 1969, 429.

44. **Mesrobeanu, I., Mesrobeanu, L., Bona, C., Vranialici, D., and Petrovici, A.,** Immunogenicity of lysosomal fractions prepared from leucocytes with pinocytized endotoxin, *Z. Immunitätsforsch. Allerg. Klin. Immunol.*, 139, 301, 1970.

45. **Filkins, J. P.,** Comparison of endotoxin detoxification by leukocytes and macrophages, *Proc. Soc. Exp. Biol. Med.*, 137, 1396, 1971.

46. **Duncan, R. L., Jr. and Morrison, D. C.,** The fate of *E. coli* lipopolysaccharide after the uptake of *E. coli* by murine macrophages *in vitro*, *J. Immunol.*, 132, 1416, 1984.

47. **Fox, E. S., Thomas, P., and Broitman, S. A.,** Clearance of gut-derived endotoxins by the liver. Release and modification of ^{3}H, ^{14}C-lipopolysaccharide by isolated rat Kupffer cells, *Gastroenterology*, 96, 456, 1989.

48. **Rutenburg, S. H., Schweinburg, F. B., and Fine, J.,** *In vitro* detoxification of bacterial endotoxin by macrophages, *J. Exp. Med.*, 112, 801, 1960.

49. **Zoeller, R. A., Wightman, P. D., Anderson, M. S., and Raetz, C. R. H.,** Accumulation of lysophosphatidylinositol in RAW 264.7 macrophage tumor cells stimulated by lipid A precursors, *J. Biol. Chem.*, 262, 17212, 1987.

50. **Rosen, O. M., Rosen, S. M., and Horecker, B. L.,** Fate of the cell wall of *Salmonella typhimurium* upon ingestion by the cellular slime mold: *Polysphondylium pallidum, Biochem. Biophys. Res. Commun.*, 18, 270, 1965.

51. **Malchow, D., Lüderitz, O., and Westphal, O.,** Polysaccharide in vegetativen und aggregationsreifen amöben von *Dictyostelium discoideum*. I. *In vivo* degradierung von bakterien-lipopolysaccharid, *Eur. J. Biochem.*, 2, 469, 1967.

52. **Malchow, D., Lüderitz, O., Kickhöfen, B., and Westphal, O.,** Polysaccharides in vegetative and aggregation-competent amoebae of *Dictyostelium discoideum.* II. Purification and characterization of amoeba-degraded bacterial polysaccharides, *Eur. J. Biochem.,* 7, 239, 1969.

53. **Saddler, J. N., Coote, J. G., and Wardlaw, A. C.,** Degradation of bacterial lipopolysaccharide by the slime mould *Physarum polycephalum, Can. J. Microbiol.,* 25, 124, 1979.

54. **Nigam, V. N., Malchow, D., Rietschel, E. Th., Lüderitz, O., and Westphal, O.,** Die enzymatische abspaltung langkettiger fettsäuren aus bakteriellen lipopolysacchariden mittels extrakten aus der amöbe von *Dictyostelium discoideum, Hoppe-Seyler's Z. Physiol. Chem.,* 351, 1123, 1970.

55. **Verret, C. R.,** Lipases specifically degrading lipopolysaccharide: fatty acyl amidases from *Dictyostelium discoideum, Rev. Infect. Dis.,* 6, 452, 1984.

56. **Rosner, M. R., Verret, R. C., and Khorana, H. G.,** The structure of lipopolysaccharide from an *Escherichia coli* heptoseless mutant. III. Two fatty acyl amidases from *Dictyostelium discoideum* and their action on lipopolysaccharide derivatives, *J. Biol. Chem.,* 254, 5926, 1979.

57. **Verret, C. R., Rosner, M. R., and Khorana, H. G.,** Fatty acyl amidases from *Dictyostelium discoideum* that act on lipopolysaccharide and derivatives. II. Aspects of substrate specificity, *J. Biol. Chem.,* 257, 10228, 1982.

58. **Holtzman, E.,** *Lysosomes,* Plenum Press, New York, 1989, 32.

59. **Gimber, P. E. and Rafter, G. W.,** The interaction of *Escherichia coli* endotoxin and leukocytes, *Arch. Biochem. Biophys.,* 135, 14, 1969.

60. **Munford, R. S. and Hall, C. L.,** Purification of acyloxyacyl hydrolase, a leukocyte enzyme that removes secondary acyl chains from bacterial lipopolysaccharides, *J. Biol. Chem.,* 264, 15613, 1989.

61. **Hagen, F. S., Grant, F. J., Kuijper, J. L., Slaughter, C. A., Moomaw, C. R., Orth, K., O'Hara, P. J., and Munford, R. S.,** Expression and characterization of recombinant human acyloxyacyl hydrolase, a leukocyte enzyme that deacylates bacterial lipopolysaccharides, *Biochemistry,* 30, 8415, 1992.

62. **Fukuda, I., Tanamoto, K., Kanegasaki, S., Yajima, Y., and Goto, Y.,** Deacylation of bacterial lipopolysaccharide in rat hepatocytes *in vitro, Br. J. Exp. Pathol.,* 70, 267, 1989.

63. **Lehrer, S. and Nowotny, A.,** Isolation and purification of endotoxin by hydrolytic enzymes, *Infect. Immun.,* 6, 928, 1972.

64. **van Bossuyt, H., Desmaretz, C., and Wisse, E.,** The fate of lipopolysaccharide in cultured rat Kupffer cells, *Virchows Arch. B,* 58, 89, 1989.

65. **Devoe, I. W.,** Egestion of degraded meningococci by polymorphonuclear leukocytes, *J. Bacteriol.,* 125, 258, 1976.

66. **Neter, E., Westphal, O., Lüdertiz, O., Gorzynski, E. A., and Eichenberger, E.,** Studies of enterobacterial lipopolysaccharides: effects of heat and chemicals on erythrocyte-modifying, antigenic, toxic and pyrogenic properties, *J. Immunol.,* 76, 377, 1956.

67. **Nowotny, A.,** Endotoxoid preparations, *Nature,* 197, 721, 1963.

68. **Takada, H. and Kotani, S.,** Structural requirements of lipid A for endotoxicity and other biological activities, *Crit. Rev. Microbiol.,* 16, 477, 1989.

69. **Munford, R. S. and Hall, C. L.,** Detoxification of bacterial lipopolysaccharides (endotoxins) by a human neutrophil enzyme, *Science,* 234, 203, 1986.

69a. **Erwin, A. L. and Munford, R. S.,** unpublished data.

70. **Erwin, A. L., Mandrell, R. E., and Munford, R. S.,** Enzymatically deacylated *Neisseria* LPS inhibits murine splenocyte mitogenesis induced by LPS, *Infect. Immun.,* 59, 1881, 1991.

71. **Pohlman, T. H., Munford, R. S., and Harlan, J. M.,** Deacylated lipopolysaccharide inhbits neutrophil adherence to endothelium induced by lipopolysaccharide *in vitro, J. Exp. Med.,* 165, 1393, 1987.
72. **Riedo, F. X., Munford, R. S., Campbell, W. B., Reisch, J. S., Chien, K. R., and Gerard, R. D.,** Deacylated lipopolysaccharide inhibits plasminogen activator inhibitor-1, prostacyclin, and prostaglandin E2 induction by lipopolysaccharide but not by tumor necrosis factor-alpha, *J. Immunol.,* 144, 3506, 1990.
73. **Dal Nogare, A. R. and Yarbrough, W. C., Jr.,** A comparison of the effects of intact and deacylated lipopolysaccharide on human polymorphonuclear leukocytes, *J. Immunol.,* 144, 1404, 1990.
74. **Kovach, N. L., Yee, E., Munford, R. S., Raetz, C. R. H., and Harlan, J. M.,** Lipid IVA inhibits synthesis and release of tumor necrosis factor induced by LPS in human whole blood *ex vivo, J. Exp. Med.,* 172, 77, 1990.
75. **Syrogiannopoulos, G. A., Hansen, E. J., Erwin, A. L., Munford, R. S., Rutledge, J., Reisch, J. S., and McCracken, F. H., Jr.,** *Haemophilus influenzae* type b lipo-oligosaccharide induces meningeal inflammation, *J. Infect. Dis.,* 157, 237, 1988.
76. **Wispelwey, B., Lesse, A. J., Hansen, E. J., and Scheld, W. M.,** *Haemophilus influenzae* lipopolysaccharide-induced blood brain barrier permeability during experimental meningitis in the rat, *J. Clin. Invest.,* 82, 1339, 1988.
77. **Golenbock, D. T., Hampton, R. Y., and Raetz, C. R. H.,** Lipopolysaccharide antagonism by lipid A precursor lipid IV$_A$ is species dependent, *FASEB J.,* 4, A2055, 1990.
78. **Munford, R. S. and Dietschy, J. M.,** Effects of specific antibodies, hormones, and lipoproteins on bacterial lipopolysaccharides injected into the rat, *J. Infect. Dis.,* 152, 177, 1985.
79. **Erwin, A. L. and Munford, R. S.,** Plasma lipopolysaccharide-deacylating activity (acyloxyacyl hydrolase) increases following lipopolysaccharide administration to rabbits, *Lab. Invest.,* 65, 138, 1991.
80. **Lei, M. and Morrison, D. C.,** Specific endotoxin lipopolysaccharide-binding proteins on murine splenocytes, *J. Immunol.,* 141, 996, 1988.
81. **Kirkland, T. N., Virca, G. D., Kuus-Reichel, T., Multer, F. K., Kim, S. Y., Ulevitch, R. J., and Tobias, P. S.,** Identification of lipopolysaccharide-binding proteins in 70Z/3 cells by photoaffinity cross-linking, *J. Biol. Chem.,* 265, 9520, 1990.
82. **Hampton, R. Y., Golenbock, D. T., and Raetz, C. R. H.,** Lipid A binding sites in membranes of macrophage tumor cells, *J. Biol. Chem.,* 263, 14802, 1988.
83. **Wright, S. D., Ramos, R. A., Tobias, P. S., Ulevitch, R. J., and Mathison, J. C.,** CD14, a receptor for complexes of lipopolysaccharide (LPS) and LPS binding protein, *Science,* 249, 1431, 1990.
84. **Munford, R. S., Erwin, A. L., Riedo, F. X., and Hall, C. L.,** Lipopolysaccharide signal modification by acyloxyacyl hydrolase, a leukocyte enzyme, in *Microbial Determinants of Virulence and Host Response,* Ayoub, E. M., Cassell, G. H., Branche, W. C., and Henry, T. J., Ed., American Society for Microbiology, Washington, DC, 1990, 271.

Index

INDEX

A

B